REACTION DYNAMICS IN CLUSTERS
AND CONDENSED PHASES

THE JERUSALEM SYMPOSIA ON
QUANTUM CHEMISTRY AND BIOCHEMISTRY

Published by the Israel Academy of Sciences and Humanities,
distributed by Academic Press (N.Y.)

1. *The Physicochemical Aspects of Carcinogenesis* (October 1968)
2. *Quantum Aspects of Heterocyclic Compounds in Chemistry and Biochemistry* (April 1969)
3. *Aromaticity, Pseudo-Aromaticity, Antiaromaticity* (April 1970)
4. *The Purines: Theory and Experiment* (April 1971)
5. *The Conformation of Biological Molecules and Polymers* (April 1972)

Published by the Israel Academy of Sciences and Humanities,
distributed by D. Reidel Publishing Company (Dordrecht, Boston, Lancaster, and Tokyo)

6. *Chemical and Biochemical Reactivity* (April 1973)

Published and distributed by D. Reidel Publishing Company
(Dordrecht, Boston, Lancaster, and Tokyo)

7. *Molecular and Quantum Pharmacology* (March/April 1974)
8. *Environmental Effects on Molecular Structure and Properties* (April 1975)
9. *Metal-Ligand Interactions in Organic Chemistry and Biochemistry* (April 1976)
10. *Excited States in Organic Chemistry and Biochemistry* (March 1977)
11. *Nuclear Magnetic Resonance Spectroscopy in Molecular Biology* (April 1978)
12. *Catalysis in Chemistry and Biochemistry Theory and Experiment* (April 1979)
13. *Carcinogenesis: Fundamental Mechanisms and Environmental Effects* (April/May 1980)
14. *Intermolecular Forces* (April 1981)
15. *Intermolecular Dynamics* (Maart/April 1982)
16. *Nucleic Acids: The Vectors of Life* (May 1983)
17. *Dynamics on Surfaces* (April/May 1984)
18. *Interrelationship Among Aging, Cancer and Differentiation* (April/May 1985)
19. *Tunneling* (May 1986)
20. *Large Finite Systems* (May 1987)

Published and distributed by Kluwer Academic Publishers
(Dordrecht, Boston, London)

21. *Transport through Membranes: Carriers, Channels and Pumps* (May 1988)
22. *Perspectives in Photosynthesis* (May 1989)
23. *Molecular Basis of Specificity in Nucleic Acid-Drug Interaction* (May 1990)
24. *Mode Selective Chemistry* (May 1991)
25. *Membrane Proteins: Structures, Interactions and Models* (May 1992)
26. *Reaction Dynamics in Clusters and Condensed Phases* (May 1993)

VOLUME 26

REACTION DYNAMICS IN CLUSTERS AND CONDENSED PHASES

PROCEEDINGS OF THE TWENTY-SIXTH JERUSALEM SYMPOSIUM ON
QUANTUM CHEMISTRY AND BIOCHEMISTRY HELD IN
JERUSALEM, ISRAEL, MAY 17–20, 1993.

Edited by

J. JORTNER

The Israel Academy of Sciences and Humanities,
Jerusalem, Israel

R. D. LEVINE

The Fritz Haber Center for Molecular Dynamics,
The Hebrew University of Jerusalem,
Jerusalem, Israel

and

B. PULLMAN

Institut de Biologie Physico-Chimique
(Fondation Edmond de Rothschild), Paris, France

KLUWER ACADEMIC PUBLISHERS

DORDRECHT / BOSTON / LONDON

06064772

CHEMISTRY

Library of Congress Cataloging-in-Publication Data

Jerusalem Symposium on Quantum Chemistry and Biochemistry (26th :
   1993)
   Reaction dynamics in clusters and condensed phases : proceedings
of the Twenty-Sixth Jerusalem Symposium on Quantum Chemistry and
Biochemistry held in Jerusalem, Israel, May 17-20, 1993 / edited by
J. Jortner, R.D. Levine, and B. Pullman.
      p.   cm. -- (The Jerusalem symposia on quantum chemistry and
biochemistry ; v. 26)
   ISBN 0-7923-2582-6 (alk. paper)
   1. Molecular dynamics--Congresses. 2. Chemical reaction,
Conditions and laws of--Congresses. 3. Solution (Chemistry)-
-Congresses.   I. Jortner, Joshua. II. Levine, Raphael D.
III. Pullman, Bernard, 1919-   . IV. Title. V. Series.
QD461.J47   1993
541.3'94--dc20   93-33794

ISBN 0-7923-2582-6

Published by Kluwer Academic Publishers,
P.O. Box 17, 3300 AA Dordrecht, The Netherlands.

Kluwer Academic Publishers incorporates
the publishing programmes of
D. Reidel, Martinus Nijhoff, Dr W. Junk and MTP Press.

Sold and distributed in the U.S.A. and Canada
by Kluwer Academic Publishers,
101 Philip Drive, Norwell, MA 02061, U.S.A.

In all other countries, sold and distributed
by Kluwer Academic Publishers Group,
P.O. Box 322, 3300 AH Dordrecht, The Netherlands.

Printed on acid-free paper

Printed in the Netherlands

TABLE OF CONTENTS

vi

PREFACE

The Twenty Sixth Jerusalem Symposium reflected the high standards of these distinguished scientific meetings, which convene once a year at the Israel Academy of Sciences and Humanities in Jerusalem to discuss a specific topic in the broad area of quantum chemistry and biochemistry. The topic at this year's Jerusalem Symposium was reaction dynamics in clusters and condensed phases, which constitutes a truly interdisciplinary subject of central interest in the areas of chemical dynamics, kinetics, photochemistry and condensed matter chemical physics.

The main theme of the Symposium was built around the exploration of the interrelationship between the dynamics in large finite clusters and in infinite bulk systems. The main issues addressed microscopic and macroscopic solvation phenomena, cluster and bulk spectroscopy, photodissociation and vibrational predissociation, cage effects, interphase dynamics, reaction dynamics and energy transfer in clusters, dense fluids, liquids, solids and biophysical systems. The interdisciplinary nature of this research area was deliberated by intensive and extensive interactions between modern theory and advanced experimental methods. This volume provides a record of the invited lectures at the Symposium.

Held under the auspices of the Israel Academy of Sciences and Humanities and the Hebrew University of Jerusalem, the Twenty Sixth Jerusalem Symposium was sponsored by the Institut de Biologie Physico-Chimique (Fondation Edmond de Rothschild) of Paris: We wish to express our deep thanks to Baron Edmond de Rothschild for his continuous and generous support, which makes him a true partner in this important endeavour. We would also like to express our gratitude to the Adminitrative Staff of the Israel Academy and, in particular, to Mrs. Avigail Hyam for the efficiency and excellency of the local arrangements.

Joshua Jortner
Raphael D. Levine
Bernard Pullman

ELUCIDATING THE INFLUENCE OF SOLVATION ON THE DYNAMICS OF CLUSTER REACTIONS

A. W. Castleman, Jr., S. Wei, J. Purnell, S. A. Buzza

Department of Chemistry
Pennsylvania State University
University Park, PA 16802 USA

1. Introduction

The ionization of neutral ammonia clusters mainly leads to the formation of protonated cluster ions,[1-9] although unprotonated species (intact $(NH_3)_n^+$) are sometimes observed in single photon experiments conducted under varying conditions.[4-6] By contrast, unprotonated species $(NH_3)_n^+$ are not observed in typical nanosecond multiphoton ionization experiments, and they have only been detected under very low fluence conditions. Some fraction of unprotonated clusters is always observed in femtosecond MPI, for which there is no dependence on fluence over a wide range of values.

In the formation of protonated cluster ions under multiphoton ionization conditions, two possible processes[1-3] have been proposed: an absorption-dissociation-ionization[10-12] (ADI) mechanism and an absorption-ionization-dissociation[12,13] (AID) mechanism. The ADI mechanism is proposed to involve formation of long-lived[14] radicals $(NH_3)_nNH_4$ in the intermediate state, followed by photoionization of these species. On the other hand, the AID mechanism assumes that the protonated clusters are formed through intracluster ion-molecule reactions which proceed upon direct ionization of ammonia clusters.

In order to elucidate these mechanisms we have investigated the ionization of ammonia clusters through the $\tilde{C}'$ and $\tilde{A}$ states using femtosecond pump-probe techniques.[15] The present paper summarizes our findings which resolve the mechanisms that have been the subject of some controversy.

2. Experimental

The apparatus used in these experiments is a reflectron time-of-flight (TOF) mass spectrometer[16] coupled with a femtosecond laser system. An overview of the laser system is shown in Fig. 1(a), and a schematic of the TOF mass spectrometer is presented in Fig.

1

J. Jortner et al. (eds.), Reaction Dynamics in Clusters and Condensed Phases, 1–12.

is shown in Fig. 1(a), and a schematic of the TOF mass spectrometer is presented in Fig. 1(b). Femtosecond laser pulses are generated by a colliding pulse mode-locked (CPM) ring dye laser. The cavity consists of a gain jet, a saturable absorber jet, and four

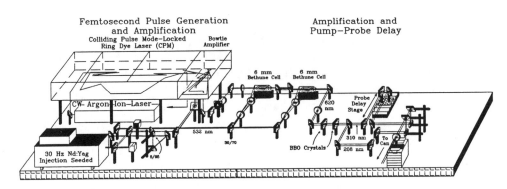

Figure 1(a). A Schematic of the Femtosecond Laser System

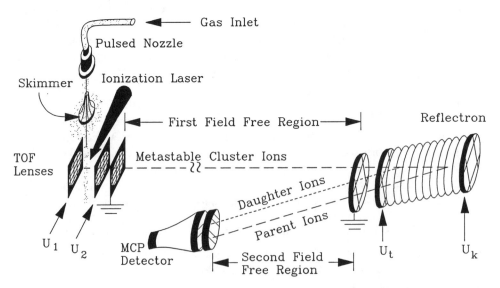

Figure 1(b). A Schematic of the Reflectron Time-of-Flight Apparatus

recompression prisms. The gain dye, rhodamine 590 tetrafluoroborate dissolved in ethylene glycol, is pumped with 5 W, all lines, from an Innova 300 argon ion laser. In order to generate short pulses, passive mode-locking is performed with DODCI, also dissolved in ethylene glycol, acting as the saturable absorber. Four recompression prisms are used to compensate for group velocity dispersion (GVD). The output wavelength, pulse width, and energy are ~624 nm, ~100 fs, and ~200 pJ, respectively.

In order to supply the large photon flux needed for multiphoton ionization (MPI), the laser pulses are amplified through three stages. Each of these is pumped with the second harmonic (532 nm) from an injection seeded GCR-5-30Hz Nd:YAG laser, which is synchronized with the femtosecond laser. The three sequential pump energies for the amplification stages are ~33 mJ, ~100 mJ, and ~250 mJ.

The first stage of amplification is a bowtie amplifier. The gain dye, sulforhodamine 640, is dissolved in a 50/50 mixture of methanol and water. The beam makes six passes through the dye cell giving a total amplification of ~10 µJ. The laser configuration to this point is the same for both the $\tilde{A}$ state and $\tilde{C}'$ state experiments, but the remaining arrangement varies depending on the state to be pumped. For the $\tilde{C}'$ state experiments, a wavelength of 624 nm is used while for the $\tilde{A}$ state experiments, the third harmonic of wavelengths-- 642 nm, 633 nm, and 624 nm--are used for accessing the v=0, v=1, and v=2 vibrational levels of ammonia molecules, respectively. The wavelength corresponding to the v=2 vibrational level is the third harmonic of the CPM fundamental wavelength; however, the wavelengths for the v=0 and v=1 vibrational levels are obtained by generating a white light continuum in a water cell. After continuum generation, the appropriate wavelength is selected with a 10 nm bandwidth interference filter.

The second and third stages of amplification are performed with 6mm bore prism dye cells, termed Bethune cells. For the amplification of 624 nm wavelength light, the gain dye is sulforhodamine 640 which is dissolved in a 50/50 mixture of methanol and water. For amplification of the 633 nm and 642 nm wavelengths, DCM, dissolved in methanol, is used. The output energy for the 624 nm wavelength light is ~200 µJ after the first dye cell and ~1.5 mJ after the second, with ~10% amplified spontaneous emission (ASE) and a pulse width of ~350 fs. The output energy for the wavelengths 633 nm and 642 nm is ~50 µJ after the first dye cell and ~1 mJ after the second, with ~10% ASE and ~350 fs pulse width. The second stage of amplification for the $\tilde{C}'$ state is the same as for the $\tilde{A}$ state (v=2); however, the third stage utilizes a 12 mm bore instead of another 6mm bore Bethune cell. The gain dye for the 12 mm bore cell is also sulforhodamine 640. The output energy and pulse width are the same as for the $\tilde{A}$ state; however, the effective pulse width involved in the multiphoton process is considerably shorter, depending on the number of photons absorbed in the excitation and ionization steps.

After amplification, the beam is split into pump and probe beams. For the $\tilde{A}$ state experiment, the pump beams are frequency-tripled and the probe beams are frequency-doubled. This gives pump wavelengths of 214 nm (v=0), 211 nm (v=1), and 208 nm (v=2) and probe wavelengths of 321 nm, 316.5 nm, and 312 nm. Using a 45° high beam and transmitting the probe beam. The probe beam is sent through a delay stage which can be varied from 0.1 μm to 1 nm. Thereafter, the beams are recombined using another 45° high reflector. For the $\tilde{C}'$ state experiment, the laser beam is split into identical pump and probe beams at a wavelength of 624 nm. A Michelson interferometric arrangement is used to set the time delay between the pump and probe beams.

After recombination, the laser beams are focused into the interaction region with a 50 cm lens, where they intersect the molecular beam containing the neutral ammonia clusters which are produced via supersonic expansion through a pulsed valve. The ions formed in the multiphoton ionization process are accelerated in a standard Wiley-McLaren double-electric field arrangement to an energy of 2000V. (See Figure 1(b).) The ions are directed through the first field free region, which is ~1.5 m long, toward a reflectron. Ions are then reflected, whereupon they travel through a second field free region which is ~0.5 m long. They are thereafter detected by a chevron microchannel plate detector. The signals received by the detector are directed into a digital oscilloscope coupled to a personal computer.

3. Elucidating the Mechanisms

Two mechanisms (Figure 2) have been proposed to account for the formation of rotonated ammonia clusters under multiphoton resonant ionization conditions. The ADI

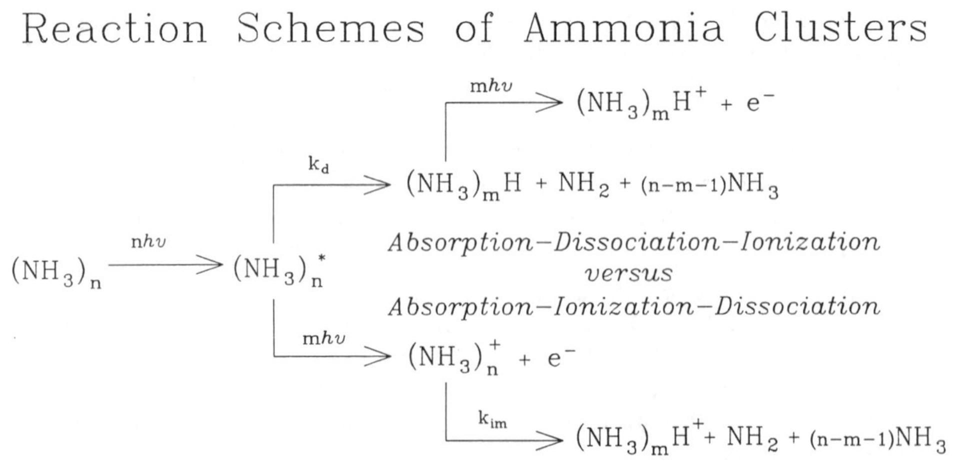

Reaction Schemes of Ammonia Clusters

$$
\begin{array}{c}
\xrightarrow{m h \upsilon} (NH_3)_m H^+ + e^- \\[4pt]
\xrightarrow{k_d} (NH_3)_m H + NH_2 + (n-m-1)NH_3 \\[4pt]
Absorption-Dissociation-Ionization \\
versus \\
Absorption-Ionization-Dissociation \\[4pt]
\xrightarrow{m h \upsilon} (NH_3)_n^+ + e^- \\[4pt]
\xrightarrow{k_{im}} (NH_3)_m H^+ + NH_2 + (n-m-1)NH_3
\end{array}
$$

$$(NH_3)_n \xrightarrow{n h \upsilon} (NH_3)_n^*$$

Figure 2. Two Possible Reaction Mechnisms of Ammonia Clusters

mechanism was initially proposed[10] based on theoretical calculations, and supported[14] by findings that hydrogenated ammonia clusters can have lifetimes of a few microseconds following neutralization of the cluster cations. Recent nanosecond pump-probe studies by

Mizaizu et. al.[11] also provided some evidence for the ADI mechanism for the case of large clusters ionized through the $\tilde{A}$ state. However, the fact that protonated ammonia clusters are formed under electron impact[7-9] and single photon ionization[4-6] conditions provides support that the AID mechanism must be operative at least under some situations. As discussed in what follows, the femtosecond pump-probe studies reported herein provide detailed insight into the mechanisms of formation of protonated ammonia cluster ions through the $\tilde{C}'$ state and $\tilde{A}$ state.

The possible profiles of intensity versus the delay between the pump and probe photons for the several potential processes operative in the mechanism of ionization are shown in Figures 3(a), (b) and (c). The ionization schemes employed in the present study are

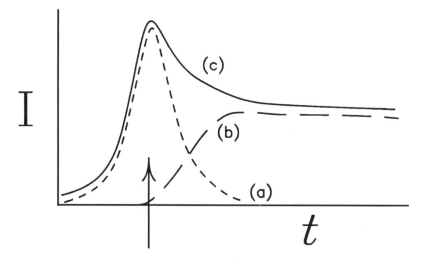

Figure 3. Possible Intensity Profiles for Two Mechanisms (a) Ionization Through AID Mechanism. (b) Ionization Through ADI Mechanism. The signal would persist for long times due to the lifetime of the NH_4 in the cluster, and its ensuing ionization. (c) Ionization Through both AID and ADI Mechanisms.

shown in Figure 4. The results of these studies presented in the next section reveal that different processes are operative in the $\tilde{C}'$ compared to the $\tilde{A}$ state.

4. Results and Discussion

In the first series of experiments, ammonia clusters are ionized by femtosecond laser pulses at 624 nm through intermediate states at energies corresponding to the $\tilde{C}'$ state of the monomer.[17] Pump-probe experiments are employed to distinguish these two

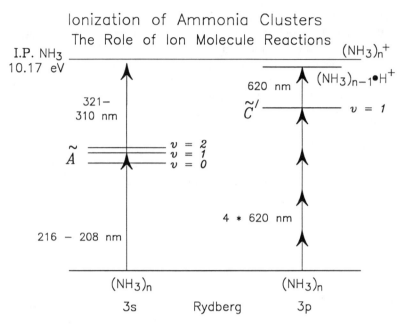

Figure 4. Ionization Schemes of the Present Experiments

mechanisms. Since the ADI mechanism involves a long-lived intermediate $(NH_3)_nH$ to the formation of the protonated species $(NH_3)_nH^+$, varying the delay time of the pump-probe laser (up to 1 picosecond) would have little influence on the protonated cluster signal, which should then persist for long times as shown in Figure 3(b). Note, since the protonated and unprotonated species have different intermediates, the lifetime of the unprotonated species would be expected to be much shorter. However, the measured lifetime for the intermediate to the formation of both $(NH_3)_m^+$ and $(NH_3)_nH^+$ would be expected to be identical if the AID mechanism is operative since there would be a common intermediate for the formation of both $(NH_3)_m^+$ and $(NH_3)_nH^+$, (n<m). See Figure 3(a).

To ascertain whether the ionization involves resonant or non-resonant multiphoton ionization, we carried out studies of the ionization signal versus the laser power dependence. Contrary to the case in typical nanosecond experiments,[3] it is found that the cluster distribution does not shift to small sizes with increasing fluence of the femtosecond laser pulses. A linear relationship between logarithm (laser power) and logarithm (ion intensity) is found[18] for all cluster ions studied and for laser powers ranging from the minimum power for observable ionization up to the maximum power obtainable in the current setup. The ionization potential[4] of ammonia is 10.17 eV, and depending on cluster size, a minimum of five to six photons are required for ionization. Hence, a slope of 4 which is observed for all five lines, indicates that ionization is achieved through a four-photon resonant intermediate state. This energy corresponds to the $\tilde{C}'$ (v=1) state of the ammonia monomer (8.04 eV).[19]

In order to find out whether the protonated cluster ions formed through the $\tilde{C}'$ state follow the absorption-ionization-dissociation mechanism or the absorption-dissociation-ionization mechanism, we performed femtosecond pump-probe experiments. One beam was used to excite the neutral clusters to the intermediate state, and the second laser beam to ionize the excited clusters. The laser power of each beam is carefully controlled so that the ion intensity resulting from either laser beam alone is less than 5%. Figure 5 shows a typical pump-probe spectrum. Since the pump and probe pulses are identical, the ion

Pump–Probe Spectrum of Ammonia Clusters

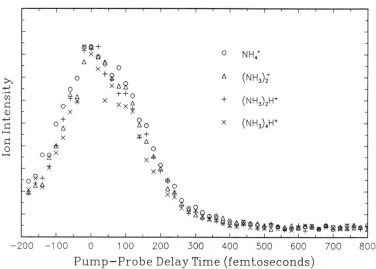

Figure 5. Pump-Probe Spectrum of Ammonia Clusters Through the $\tilde{C}'$ State

signal is symmetric at the zero time delay. It is seen that the response curves for all ions are identical within the experimental error. This finding suggests that the lifetime of the state leading to the formation of both the unprotonated and protonated clusters is the same. Since the lifetimes of the neutral species $(NH_3)_nH$ are measured to be on the order of microseconds,[14] the present results which show a rapid increase and decay in the signal versus pump-probe delay can only be explained on the basis of the absorption-ionization-dissociation mechanism. Otherwise, the signal arising from the ionization of $(NH_3)_nH$ would persist for long times corresponding to the microsecond lifetime of this intermediate. The rapid change in signal and the failure to observe ionization through NH_4 is attributed to the fact that the predissociation of ammonia through the $\tilde{C}'$ state has a dominant channel[20] via the formation of $NH + H_2$.

Studies to reveal the mechanisms operative through the $\tilde{A}$ state were conducted in a similar fashion, using the scheme given in Figure 4. Figure 6 shows typical pump-probe

Pump−Probe Spectrum of $(NH_3)_2H^+$

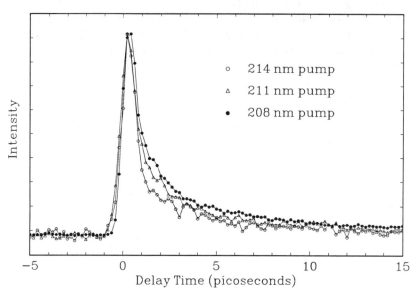

Figure 6. Pump-Probe Spectra of $(NH_3)_2H^+$ through the $\tilde{A}$ (v=0,1,2) States

spectra of protonated cluster ions $(NH_3)_nH^+$, n=2, through different vibrational levels of $\tilde{A}$ states (v=0,1,2) of ammonia molecules, at a time step-size of 200 fs/point. The zero-of-time is defined as the time when the pump and probe pulses are temporarally overlapped which results in the maximum protonated cluster ion signals. It is evident that when the probe is ahead of the pump (negative pump-probe delay), the signal retains a constant value. After that, the signal decreases in an exponential fashion and thereafter levels off to a finite, non-zero value. It is found that this pump-probe signal persists longer than 1 ns. To understand these common time response features which are generally in accord with Figure 3(c), we propose the following dynamical processes in the electronically excited $\tilde{A}$ state of ammonia clusters. The dominant predissociation channel in the $\tilde{A}$ state leads to NH_2 and H.

1. The neutral clusters are excited to the $\tilde{A}$ state through absorption of the first photon,

$$(NH_3)_n + h\nu_1 \rightarrow (NH_3)_n{}^*$$

The excited clusters undergo intracluster reactions as follows:

2. Predissociation of the excited ammonia moiety

$$(NH_3)_n{}^* \rightarrow (NH_3)_{n-2} \bullet H_3N \bullet (H \bullet\bullet\bullet NH_2)$$

3. The intermediate species can lead to the loss of H or NH_2, or reaction of the H to form NH_4.

$$(NH_3)_{n-2} \bullet H_3N \bullet (H \bullet \bullet \bullet NH_2) \rightarrow (NH_3)_{n-2} \bullet NH_4 + NH_2$$

4. Ionization of either $(NH_3)_n{}^*$ or the radicals $(NH_3)_{n-2} \bullet NH_4$ leads to formation of protonated cluster ions as follows:

    a. $(NH_3)_n{}^* + h\nu_2 \rightarrow (NH_3)_n + e^- \rightarrow (NH_3)_{n-2}NH_4{}^+ + NH_2 + e^-$

    b. $(NH_3)_{n-2} \bullet NH_4 + h\nu_2 \rightarrow (NH_3)_{n-2} \bullet NH_4{}^+ + e^-$

It should be noted that the rapid intensity drop observed for all protonated cluster ions when $n \geq 2$ is attributed to reaction 2, where the NH_2 or H containing species cannot be readily ionized. Reaction 3 leads to formation of long-lived radicals in accordance with the findings of non-zero ion intensity values at long pump-probe delays observed in the data for the $\tilde{A}$ state. The relative importance of the ionization of the NH_4 in the overall ionization of ammonia through the $\tilde{A}$ state at different vibrational levels is seen in Figure 6 and by comparing other data for the trimer and hexamer, detected as the protonated dimer and pentamer cluster ions.[22] The overall dependence of the decaying signal intensity on the vibrational level is indicative of the influence of the energetics on the predissociation and reaction forming NH_4, while the trend in the long-time tail reflects effects due to solvation and retainment of NH_4.

Proposed pseudo-potential wells, which are consistent with all of the experimental findings from our studies, are shown in Figure 7. Our findings that ionization through the $\tilde{A}$ state (v>0) displays a peak followed by rapid decay leveling off to a non-zero value of ion intensity, suggest that two processes are operating simultaneously. Since it is known[21] that ammonia clusters rapidly predissociate into NH_2 + H, the rapid decay that we observe would suggest that a similar predissociation mechanism is taking place in the clusters. It is also known[14] that the radicals $(NH_3)_nNH_4$ have long lifetimes (greater than 1 ns). Evidence suggests that formation of these radicals is taking place through intracluster reactions. For the case of $\tilde{A}$ (v>0), this is seen in the leveling off to a non-zero value of ion intensity which persists for longer than 1 ns. Unlike the $\tilde{C}'$ state which follows only the AID mechanism, it appears that the processes in the $\tilde{A}$ state compete between both the AID and ADI mechanisms. The AID is the dominant process when the pump and probe pulses are overlapped (t = 0), while the ADI is the major mechanism when the probe is at a longer time delay. The failure to see a definitive long-time tail for $\tilde{A}$ (v = 0) is attributable to the likelihood that the predissociation is probably endothermic for this situation.[11]

MPI of Ammonia Clusters

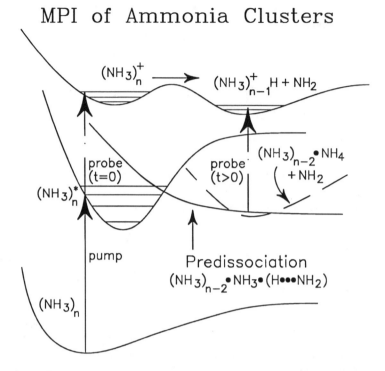

Figure 7. Proposed Potential Surfaces of Ionization of Ammonia
Clusters; the dashed potential curve represents a different
reaction coordinate for species formed by predissociation in
the Ã state (see text). The probe beam (dash-dot) is time
delayed by a selected value.

5. Conclusions

Femtosecond pump-probe techniques have revealed the mechanisms of ionization
operative for multiphoton ionization involving the $\tilde{C}'$ and $\tilde{A}$ states of ammonia clusters.
In the case of the $\tilde{C}'$ state, lifetimes of less than 100 fs for the species $(NH_3)_n^+$, and
$(NH_3)_nH^+$, (n = 1-4) establish the AID as the sole operative mechanism in the $\tilde{C}'$ state.
For the $\tilde{A}$ state (n=2-5), the pump-probe experiments show two distinct features with
respect to the pump-probe delays: a fast decay process, followed by a leveling off to a
non-zero value of ion intensity. These observations support a competing process between
the AID and ADI mechanisms. The first process, a maximum peak at t = 0 followed by a
rapid decay, supports the AID mechanism. These observations are consistent with those
observed for the $\tilde{C}'$ state. The second process, leveling off to a non-zero value persisting
for longer than 1 ns, supports the ADI mechanism. Work is in progress to model the
lifetimes of the individual steps in the reaction mechanisms.

Acknowledgments

Financial support by the U. S. Department of Energy, Grant No. DE-FGO2-88ER60648, is gratefully acknowledged.

References

1. (a) Echt, O., S. Morgan, P. D. Dao, R. J. Stanley, and A. W. Castleman, Jr.(1984) *Bunsen-Ges. Phys. Chem.* **88**, 217; (b) Echt, O., P. D. Dao, S. Morgan, and A. W. Castleman, Jr.(1985) *J. Chem. Phys.* **82**, 4076.
2. Shinohara, H., and N. Nishi (1987) *Chem. Phys. Lett.* **141**, 292.
3. Wei, S., W. B. Tzeng, and A. W. Castleman, Jr.(1990) *J. Chem. Phys.* **93**, 2506. (1990).
4. Ceyer, S. T., P. W. Tiedemann, B. H. Mahan, Y. T. Lee (1979) *J. Chem. Phys.* **70**, 4.
5. Kaiser, E., J. de Vries, H. Steger, C. Menzel, W. Kamke, and I. V. Hertel (1991) *Z. Phys. D* **20**, 193.
6. Shinohara, H., N. Nishi, and N. Washida (1985) *J. Chem. Phys.* **83**, 1939.
7. Stephan, K., J. H. Futrell, K. I. Peterson, A. W. Castleman, Jr., H. E. Wagner, N. Djuric, T. D. Märk (1982) *Int. J. Mass Spectrom. Ion Phys.* **30**, 345.
8. Buck, U., H. Meyer, D. Nelson, G. Fraser, W. Klemperer (1988) *J. Chem. Phys.* **88**, 3028
9. Peifer, W. R., M. Todd Coolbaugh, and J. F. Garvey (1989) *J. Chem. Phys.* **91**, 6684.
10. Cao, H., E. M. Evleth, E. Kassab (1984) *J. Chem. Phys.* **81**, 1512.
11. Misaizu, F., P. L. Houston, N. Nishi, H. Shinohara, T. Kondow, and M. Kinoshita (1993) *J. Chem. Phys.* **98**, 336.
12. Misaizu, F., P. L. Houston, N. Nishi, H. Shinohara, T. Kondow, and M. Kinoshita (1989) *J. Phys. Chem.* **93**, 7041.
13. Tomoda, S. (1986) *Chem. Phys.* **110**, 431.
14. Gellene, G. I., and R. F. Porter (1984) *J. Phys. Chem.* **88**, 6680.
15. Rosker, M. J., M. Dantus, and A. H. Zewail (1988) *J. Chem. Phys.* **89**, 6113; Khundkar, L. R., and A. H. Zewail (1990), *Annu. Rev. Phys. Chem.* **41**, 15; Dantus, M., M. H. M. Janssen, and A. H. Zewail (1991) *Chem. Phys. Lett.* **181**, 281.
16. Stanley, R. J., and A. W. Castleman, Jr (1991) *J. Chem. Phys.* **94**, 7874.
17. Glownia, J. H., S. J. Riley, S. D. Colson, and G. C. Nieman (1980) *J Chem. Phys.* **72**, 5998.
18. Wei, S., J. Purnell, S. A. Buzza, R. J. Stanley, and A. W. Castleman, Jr. (1992) *J. Chem. Phys.* **97**, 9480; Wei, S., J. Purnell, S. A. Buzza, and A. W. Castleman, Jr., "Ultrafast Reaction Dynamics of Electronically Excited Ã State of Ammonia Clusters, *J Chem. Phys.*, in press.

19.   Herzberg, G. (1960) Molecular Spectra and Molecular Structures, Vol. 3,
       Electronic Spectra and Electronic Structure of Polyatomic Molecules, Van
       Nostrand Reinhold, New York, pp. 463-466.
20.   Quinton, A. M., and J. P. Simons (1982) *Chem. Soc. Faraday Trans. 2* **78**, 1261.
21.   Ziegler, L. D. (1985) *J. Chem. Phys.*, **82**, 664.
22.   Purnell, J., S. Wei, S. A. Buzza, and A. W. Castleman, Jr., "Formation of
       Protonated Ammonia Clusters Probed by a Femtosecond Laser," *J. Phys. Chem.*,
       to be submitted.

THE SOLVATION OF HALOGEN ANIONS IN WATER CLUSTERS

Gil Markovich, Stuart Pollack, Rina Giniger, and Ori Cheshnovsky

School of Chemistry, Tel-Aviv University, 69978 Tel-Aviv University, Israel

Abstract: We have measured the photoelectron-spectra of I^-, Br^- and Cl^- solvated in water clusters-$(H_2O)_n$, where n is 1-60,1-16, and 1-7 respectively. The vertical binding energies of the solvated anions taken from the spectra were used to extract the electrostatic stabilization energies of the anion by the solvent. The photoelectron spectra of the solvated I^- indicate the formation of the first solvation layer with a coordination number of six. This conclusion is not born-out by the molecular dynamics calculation concerning the solvated anion clusters. Current calculations favor structures with a surface solvated anion (coordination number of 3-4) and claim to reproduce our vertical binding energies. The fitting of the binding energies of large $I^-(H_2O)_n$ to the classical electrostatic solvation energies with a centrally solvated anion is questionable. In the size range n=34-40 we have found evidence for the existence of special cluster structures with surface solvated anions.

1. Introduction

The energetics and structure of ion-water clusters are of great significance to the understanding of the solvation phenomena in bulk solutions as well as in confined systems. Alkali metal cations and halogen anions solvated in clusters have traditionally served as model systems for both experimental [1] and theoretical [2] studies of this problem. A convenient conceptual framework for modeling solvation in the bulk was the solvation-shells model [3]. According to this model, the first, highly perturbed shell of solvent molecules rearrange around the ion. Subsequent solvent layers are only slightly perturbed in structure by the central ion due to the weaker interactions with the ion. For most purposes these additional layers can be treated as a continuous dielectric medium. It is not obvious though, that by forming the solvated ion cluster, one solvent molecule at a time, the bulk solvation layer structure with a central ion is reproduced. Although most of the calculations support this picture for solvated cations, contrary structures of surface solvation are predicted from the calculations of water- halogen anion clusters.

Photoelectron-Spectroscopy (PES) is a most suitable technique for the investigation of anion-solvent interaction in clusters. This technique measures directly the binding energy of the electron in the solvated anion, and as such is very sensitive to the ion-polar solvent

13

J. Jortner et al. (eds.), Reaction Dynamics in Clusters and Condensed Phases, 13–19.

distance and orientation. Due to the vertical nature of the photodetachment, solvent-solvent interactions have hardly any effect on the energetics of the process. PES is a general technique and can be successfully applied to clusters and bulk alike. The photodetachment energetics can be easily simulated in calculations of ion solvation by suddenly switching off the interactions related to the electron on the anion, or alternatively by removing the HOMO electron in ab-initio calculations. This feature enables direct comparison of the PES binding energies with theory.

Recently, we have reported on PES solvation studies of iodine anion in up to 15 water molecules [4]. In this report we present the extension of these studies to larger clusters as well as to the chlorine and bromine anions.

2. Experimental

The anion-clusters are generated, using a pulsed electron-beam, in the early stages of a pulsed supersonic expansion. The expanding gas consists of 1-2 bar of Ar, 10 mbar of H_2O, and about 1 mbar of volatile compounds containing halogen. The newly formed charged clusters are cooled by further flow in the expansion, and mass-separated by a reflecting time-of-flight mass-spectrometer. This cluster source follows the design of Alexander et al [5]. The mass-selected ions are impulse decelerated to ~20 eV kinetic energy, to reduce Doppler-broadening.

All the PES are taken with 7.1 eV (the H_2 7th AS of the 3rd harmonic of a Nd:YAG laser). The kinetic energy of the electrons is analyzed with a 250 cm magnetic TOF photoelectron spectrometer [6] which follows the design principles of Kruit and Read [7]. The resolution is ~50 meV at 1.5 eV of electron kinetic energy. The spectrometer is calibrated with PES of the halogen anions (Cl-, Br-, I-) [8] taken with several laser photon-energies. A detailed description of the experimental apparatus is given elsewhere [4,9].

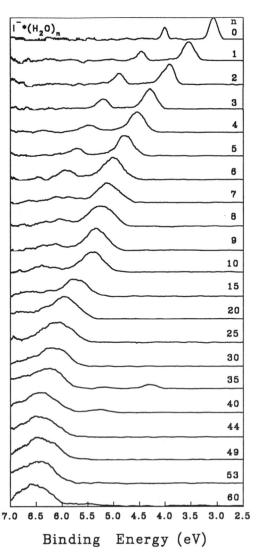

$I^- \cdot (H_2O)_n$

Binding   Energy (eV)

Fig. 1: The PES spectra of I- solvated in water Clusters Detachment energy is 7.1 eV.

3. Results and discussion

We have measured the PES of halogen anions solvated in n water molecules, where n = 1-60, 1-16 and 1-7 for I⁻ ,Br⁻ and Cl⁻ respectively. The cluster-size range of our measurements, is limited by the photon energy of our photodetachment laser (7.1 eV), due to increasing binding energies. Fig. 1 displays representative photoelectron spectra of the water solvated iodine anion. Notice that the typical spin-orbit split peaks of the bare iodine anion shift to higher binding energies with the increase of cluster size, reflecting the increasing electrostatic stabilization of the anion.

Our working hypothesis is that the peaks of these spectra represent the vertical binding energies of electrons in the clusters. The widths of the peaks originate from large geometry differences between the equilibrium states of the initial (ionic) and final (neutral) states, as well as from inhomogeneous broadening: namely, the contributions of different isomers with varying electron affinities. The difference between this vertical photodetachment energy and the electron affinity of the bare ion, $E_{stab}(n)$, is essentially the electrostatic stabilization of the solvated anion in its equilibrium configuration.

The extraction of genuine $E_{stab}(n)$ for large anion-water clusters may be obscured by "inelastic scattering" of the emerging photoelectron. This effect may cause reduction in the apparent electron-affinity. Using cross-sections of electron scattering by water [10] and model structure of centrally solvated anion, we have deconvoluted the "genuine" PES from our experimental spectra. This analysis shows that, even in our largest clusters the influence of electron scattering on E_{stab} is marginal (up to 0.14 eV for n~60). The values we are using throughout this discussion are the row experimental values of binding energies.

In a previous publication [4] we have advanced the idea that the PES of I⁻ $(H_2O)_{1-15}$ indicate the formation of a first solvation layer with six water molecules. This conclusion was based on an abrupt decrease in the differential change of E_{stab} at n=7. E_{stab} increases in steps, between 0.45 and 0.18 eV, averaging on 0.35 eV for the first six water molecules. In larger clusters the increase of E_{stab} with n becomes moderate (<0.07 eV). These changes are consistent with the picture that, solvent-molecules, in higher solvation layers, exert smaller interactions with the anion. As discussed later, the accumulated PES data on solvated Br⁻ and Cl⁻ as well as recent calculations do not necessarily support the notion of an anion that is centrally solvated in the cluster.

We have compared $E_{stab}(n)$ of the different ions (following the idea of Perera and Berkowitz [2b]) by using a

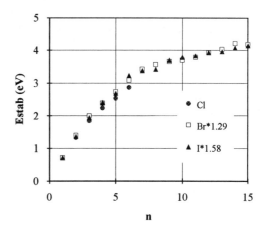

Fig. 2 Scaled stabilization energies of solvated anions

scaling transformation. We have chosen to scale the values of E_{stab} of I^- and Br^- by 1.58 and 1.29 respectively, so that the values of E_{stab} of all three anion complexes with **one** water molecule coincide [11]. Fig. 2 describes these scaled $E_{stab}(n)$ values for I^- ,Br^- and Cl^- solvated in water clusters with n=1-15, 1-15 and 1-6 respectively. Note that the distinct break in $E_{stab}(n)$ in I^- at n=6 is not discernible for Br^- and Cl^-, and that in this size I^- is better stabilized, probably due to its larger surface area. Also note, that I^- and Br^- show a similar solvation behavior up to n=15.

All molecular dynamics (MD) simulations indicate that, for small clusters the halide anion is only partially solvated, and is located close to the surface of the water cluster. Perera and Berkowitz simulated the system of $Cl^-(H_2O)_n$ with n ranging from 1 to 15[2b] and for n=20 [12]. Their calculations treat water-water interactions by including the self-consistent polarization energy. Their conclusion is that unlike cations, anions are surface solvated in clusters with a coordination number of 3-4. With these structures they reproduce well our experimental $E_{stab}(n)$. Although their calculations refer to Cl^- only, Perera and Berkowitz apply scaling arguments to compare with our measurements of Br^- and I^- as well. Dang and Garrett [13] obtained the vertical electron binding energies for $I^-(H_2O)_n$ (n=1 to 15) by using MD simulations. Their results compare well with our experimental results. They also find the anion to be on the surface of the water cluster.

However, note that the MD calculations suffer from considerable limitations: Because of the limited time-scale of the calculations the cluster temperature is chosen to be about 250 K in order to avoid trapping of the system in local minima. The temperature of our clusters is substantially lower. The MD calculations also underestimate the anion-water molecule distance by about 0.2 A as compared to the experimental bulk values [14]. Such an underestimate may artificially increase $E_{stab}(n)$ of solvated anions with small coordination numbers.

Recent Ab initio MO calculations[15] performed on $X^-(H_2O)_n$ (X=Cl, Br, I and n=1-6) are not conclusive about the geometry of the cluster. These calculation maintain reasonable agreement to the experimental $E_{stab}(n)$ data. However, the calculations do not contain finite temperature effects, they somewhat overestimate the anion-water bond length and may suffer from only partial geometry optimization.

In all the above calculations the sensitivity of $E_{stab}(n)$ to structure has been demonstrated. The absolute values of the calculations, however, are still doubtful as are their conclusions concerning the structure. It seems that higher accuracy quantum MD calculations are needed for conclusions on the structure of water solvated anions.

The PES of the solvated I^-, extending to sixty water molecules, provides an opportunity to analyze $E_{stab}(n)$ in terms of a classical continuous dielectric model. Such an analysis is simple in principle. For an anion solvated in the center of a cluster with a radius R, $E_{stab}(n)$ can be classically approximated [16] as:

$$E_{stab}(n) = E_{stab}(\infty) - \frac{e^2}{2R}(1 - \frac{2}{\varepsilon_s} + \frac{1}{\varepsilon_\infty}) \qquad (1)$$

where ε_s, ε_∞ are the static and optical dielectric constants of the solvent respectively. It is not clear however which physical constants should be applied to the cluster. Here we assume that the clusters consist of an anion (of finite volume) positioned in the center of a sphere, which is characterized by the liquid water density and dielectric constants. The radius of a cluster containing n water molecules is calculated using the following expression:

$$\frac{4\pi}{3}R^3 = \frac{4\pi}{3\eta}R_{ion}^3 + \frac{18}{\rho_w N_A}n \qquad (2)$$

where η is the volume fraction of the anion in its first solvation layer, and R_{ion} is the hard sphere ionic radius. The second expression on the right hand is the volume of n water molecules where ρ_w is the water density, and N_A is Avogadro number. Using this relation, the dielectric constants of liquid water at room temperature, the ionic radius of iodine (2.16 A), and 0.7 for η we obtain the following numerical expression for Eq. 1:

$$E_{stab}(n) = E_{stab}(\infty) - 5.76 \times (2+n)^{-\frac{1}{3}} \qquad (3)$$

Energy is given in eV. The value 2 in the equation represents the self-volume of the ion in terms of the water molecule volume. In Fig. 3 we display the experimental values of E_{stab} as a function of $(2+n)^{-0.33}$ together with Eq. 3 drawn with two values of $E_{stab}(\infty)$: 4.37 eV (taken from the measurements of Delahay and coworkers [17]) and 4.74 eV (taken from the measurements of Ballard and coworkers [18]). The dependence of the model on R^{-1} appears to agree well with the experimental results while the convergence to $E_{stab}(\infty)$ results in a higher value then that measured in both references 17 and 18.

Note, however, that conclusions about the validity of this simple model for a cluster containing a few tens of molecules should be taken with extreme care:

1. The location of the ion in the center of the cluster is doubtful.

2. The slope obtained from Eq. 3. depends on the chosen water density and ionic volume, and also weakly on the value of ε_s which is sensitive to the phase and temperature of water. In a

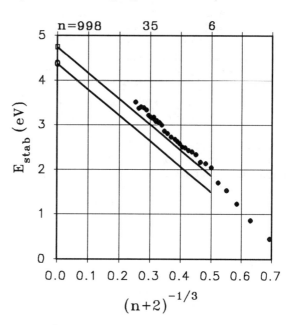

Fig. 3 R^{-1} dependence of the experimental results for $I^-(H_2O)_n$ vs. Eq. 3. with two $E_{stab}(\infty)$ values from Ref 17 and 18.

previous publication [19] we have applied the same model by assuming smaller water density in the first solvation layer as indicated by neutron diffraction studies. Such an assumption resulted in a slope which does not agree with the classical solvation model.

3. Although the slope of the plot fits well the classical model, it is questionable whether it extrapolates to the right bulk value-$E_{stab}(\infty)$. Our experimental results extrapolate better to the reported values of Ballard and coworkers[18] 4.74 eV. Note, however, that the reported values may be biased towards $E_{stab}(\infty)$ smaller than that of a real bulk state. The value of $E_{stab}(\infty)$ is substantially reduced in small distances (few hundreds A) from the surface, due to reduced electrostatic stabilization. In the experiment, the PES signals of these "close to the surface anions" are relatively amplified due to the finite depth of the light absorption (in ref. 18), and due to strong scattering of electrons emerging from deep anions (in ref. 17).

In the size range of $n=$ 34-40 small peaks, with typical spin orbit coupling of iodine, appear in the spectra at binding energies lower than the regular PES pattern (i.e. $n=35$ in Fig 1). The appearance of these peaks could be invoked only by special extreme, warm cluster generation conditions. We attribute these peaks to well stabilized water clusters with **surface** iodine ions attached to them. These surface anions are less stabilized by the water clusters, than the regularly solvated iodine anions. Since these surface solvated ions are difficult to generate, we believe that these clusters are the energetically unstable forms of water solvated anions.

4. Conclusions

We have studied the solvation of halogen anions in water clusters by using PES. We have identified the formation of the first solvation layer consisting of six water molecules in solvated iodide. Most calculations indicate that the lowest energy configurations of these clusters consist of surface solvated anions with small coordination numbers (3-4). In spite of the considerable calculational effort on the energetics of these solvation, final conclusions about their structure are still lacking. The solvation effect of large water clusters can be described in terms of classical polarizability of a dielectric water solvent. This conclusion, however is very sensitive to the applied model and should be taken with skepticism.

Acknowledgments: We wish to acknowledge useful discussions with J. Jortner, I. Rips, M. Levin, U. Even, A. Nizan, G. Makov, N. Kestner and M. Berkowitz. The research was supported by the Fund for Basic Research, administrated by the Israel Academy of Sciences and Humanities, by the James Franck German-Israeli Binational Program In Laser Matter interaction, and by the US-Israel Binational Science Foundation.

References:

1. P. Kebarle, Ann. Rev. Phys. 28, 445 (1977); T. D. Mark and A. W. Castleman, Advances in Atomic and Molecular Physics 20, 65 (1985); A. W. Castleman, and R. G. Keese, Acc. Chem. Res. 19, 413 (1986); A. W. Castleman, and R. G. Keese, Chem. Rev. 86, 589 (1986).

2a. S. Sung, and P. C. Jordan, J. Chem. Phys. **85**, 4045 (1986); S. Lin, and P. C. Jordan, J. Chem. Phys. **89**, 7492 (1988) and references mentioned there.

b. L. Perera, and M. L. Berkowitz, J. Chem. Phys. **95**, 1954 (1991); L. Perera, and M. L. Berkowitz, ibid, in press.

c. I. Rips, and J. Jortner, J. Chem. Phys. **97**, 536 (1992).

3. J. Bockris and A. K. N. Reddy., **Modern Electrochemistry**, Plenum, New York, (1970).

4. G. Markovich, R. Giniger, M. Levin, and O. Cheshnovsky, J. Chem. Phys. **95**, 9416 (1991).

5. M. L. Alexander, M. A. Johnson, N. E. Levinger, and W. C. Lineberger, Phys. Rev. Lett. **57**, 976 (1986).

6. O. Cheshnovsky, S. H. Yang, C. L. Pettiette, M. J. Craycraft, and R. E. Smalley, Rev. Sci. Instru. **58**, 2131 (1987).

7. P.Kruit, and F. H. Read, J. Phys. E. (Sci. Instrum.) **16**, 313 (1983).

8. H. Hotop and W. C. Lineberger , J. Phys. Chem. Ref. Data **4**, 539 (1975).

9 G. Markovich, S. Pollack, R. Giniger and O. Cheshnovsky (to be published).

10. M. Michaud and L. Sanche, Phys. Rev. A **36**, 4672 (1987); ibid 4684 (1987).

11. E_{stab} of $I^-(H_2O)$ $Br^-(H_2O)$ and $Cl^-(H_2O)$ are 0.45 , 0.55 and 0.71 eV respectivly.

12. L. Perera, and M. L. Berkowitz, J. Chem. Phys. **96**, 8288 (1992).

13. L. X. Dang, and B. C. Garrett, to be published.

14. See table 5.12 p. 117 in: Y. Marcus , **Ion Solvation**, Wiley, Chichester (1985).

15. J. E. Combariza, N. R. Kestner, and J. Jortner, Chem. Phys. Lett. **203**, 423 (1993); J. E. Combariza, N. R. Kestner, and J. Jortner, to be published.

16. R. N. Barnett, U. Landman, C. L. Cleveland, and J. Jortner, Chem. Phys. Lett. **145**, 382 (1988).

17. P. Delahay, Acc. Chem. Res. **15**, 40 (1982).

18. R. E. Ballard, J. Jones, D. Read, A. Inchley, and M. Cranmer, Chem. Phys. Lett. **134**, 177 (1987).

19. G. Markovich, S. Pollack, R. Giniger and O. Cheshnovsky, to appear in: Z. Phys. D, proceedings of the 6th International Symposium on Small Particles and Inorganic Clusters, held in Chicago, Sep. 1992.

EXCITATION AND IONIZATION OF CHLORIDE, IODIDE, BROMIDE AND SODIUM IN WATER CLUSTERS

NEIL R. KESTNER and JAIME COMBARIZA
Department of Chemistry
Louisiana State University
Baton Rouge, LA 70803

Abstract

In this paper we summarize recent ab initio calculations on halide ions in small water clusters. For most clusters the halide ion exists on the outside as a surface state but there is some evidence for the larger clusters that the interior state may be important. Studies of energy, vibrational spectra and vertical ionization energies are compared with experimental data; ionizational energies are the most information. Because the charge density at the halide changes with cluster size could limit molecular dynamics studies in these systems. Comparisons are also made with older studies on sodium water and negatively charged water clusters.

1. Introduction

Studying the chemical physics of clusters[1-6] allows us to explore the microscopic solvation phenomena of a molecule[7] or anion[8] embedded in a mass-selected cluster of solvent molecules. The experimental, computational and theoretical studies of such heteroclusters allow us to investigate specific cluster size effects, i.e, the buildup of the solvation layer(s) in small- and medium-sized clusters, and the exploration of the gradual "transition" from the finite system to the infinite bulk medium for large clusters[6]. In this context, the microscopic solvation of an ion in a water cluster is of considerable interest for the elucidation of ionic solvation in confined and bulk systems. Such studies are of central importance for the understanding of the structure, energetics, spectroscopy, dynamics and reaction kinetics in clusters and in condensed phases.

Solvation of ions in finite and infinite polar systems qualitatively modifies the characteristics of the ionic species. Four notable examples are a. the solvation and dissociation of an ionic diatomic molecule, e.g.,

a. NaCl, upon solvation of the bulk solvent or in a dielectric cluster[9],
b. the charge transfer to solvent (CTTS) absorption spectra of solvated anions, manifesting bound-bound electronic excitations[10], which are absent in the isolated anion,

J. Jortner et al. (eds.), Reaction Dynamics in Clusters and Condensed Phases, 21–36.

c. the chemical reaction of sodium with water, a reaction which does not occur in small clusters,

d. the solvation of the electron, a situation which is dramatically different in small clusters than in bulk water.

In this paper we will review some of the work underway in our research group and in conjunction with the University of Tel-Aviv experimental group on studies of anions and some atoms in water clusters. Some specifics will be discussed but we also want to discuss general issues of modeling of these systems. In particular we will discuss the limitations of classical molecular dynamic simulations as well as *ab initio* quantum chemistry applied to these systems. We will also briefly discuss the difficulty of using quantum simulations in many of these systems.

With the advent of supersonic and cluster beam techniques, microscopic solvation of ions[8,11-13] and of excess electron[8,14,15] in clusters of polar molecules became amenable to experimental studies. Those studies involve the energetics spectroscopy and reactivity. The experiments provided information on the microscopic solvation energies of cations and anions, which were determined by the sequential addition of solvent molecules to form a cluster[11-13]. Vertical ionization potentials of anions solvated in water clusters were determined from photoelectron spectroscopy by Cheshnovsky et al.[16]. Theoretical[17,18] and computational[19-23] studies of the microscopic aspects of the energetics, spectroscopy and dynamics of ion solvation in clusters allows us to understand the specificity vs the universality of ion-solvent interactions, i.e. the interplay between the specific short-range interactions and the universal long-range interactions.

Several computational studies addressed the quantum molecular dynamics (MD) of excess electrons in water[20] and ammonia[24] clusters and the classical MD of ion solvation in water cluster[19]. The energetic and structural information emerging from MD simulations for anion-water clusters[20-23] depends on the nature of the potentials. Berkowitz and Perera have shown[22] that the TIP4P potential[23], which represents the water-anion interaction in terms of an effective pair potential, results in an internal anion state for Cl^- $(H_2O)_n$ clusters,while the SPCE/POL potential[25] which incorporates two-center and three-center polarization interactions, predicts that a surface anion state for $Cl^-(H_2O)_n$ (n < 21) is preferred. The conclusions from finite temperature MD simulations are sensitive to the nature of the empirical potentials, whose accuracy cannot be readily assessed. Accordingly, *ab initio* quantum mechanical calculations are of considerable interest. Truhlar et al[26] carried out rather accurate ab initio calculations on the $Cl^-(H_2O)$ cluster and then proposed a new water ion potential fit not only to global *ab initio* data but also to some detailed properties of the ion-water complex such as the geometry, binding energy and frequency shifts. Unfortunately, these calculations were not extended to clusters with more than a few water molecules.

In this paper we will review some of our work on anions in water clusters and also refer to earlier work as well as ongoing work on sodium in water clusters and excess electrons on small water clusters. Recently we have undertaken an extensive quantum mechanical theoretical study of the equilibrium structure and the energetics of halide ion solvation in water clusters, $X^-(H_2O)_n$ (X=Cl, Br, I and n=1,6), obtaining information on the solvation energies and vertical ionization potentials of these clusters. A preliminary report of our work was already presented[27] and a complete paper has been submitted. In this paper we summarize some specific cluster size effects on anion solvation in water clusters, addressing the following issues:

(1) Structure-energetics relations.
(2) The existence of surface (S) and interior (I) states on and in water clusters.
(3) Physical properties of structural isomers.
(4) Solvation shell structure.
(5) Anion size effects.

We will then return to a comparison of other water cluster systems.

2. Computational Methods

Standard ab initio molecular orbital calculations were carried out at the SCF and MP2 level using extended basis sets. These calculations were carried out using the standard quantum chemistry packages Gaussian 90[28] and Gamess[29]. Structural optimization at the SCF level was performed for $X^-(H_2O)_n$, (X=Cl,Br,I and n=1,6) without any constraints on the geometry, using frequency checks to ensure a real minumum. Only SCF results are reported in this paper.

2.1. BASIS SETS

The water molecules were represented by the standard 6-31+G*[30] basis sets. In the bulk, the bound excited state of halide anions in water is characterized by a CTTS band and although we do not expect to observe this process in the small (n less than 6) clusters, the diffuse sp functions on oxygen were added to allow for the charge transfer to solvent process. Our choices of the electronic wavefunctions for the X^- anions were tested by comparison with experimental ionization potential data[31] for the bare ions to establish the quality of the basis set. All electron calculations were done for chlorine but effective core potentials were used for iodine and bromine. Details of the basis sets are in other publications.

3. Ground State Structures, Charge Distribution and Energies of Halide Water Clusters

3.1 STRUCTURES

The configurations of H_2O molecules relative to an ion in a cluster is determined by the intermolecular (hydrogen bonding) interactions between the strongly polar water molecules and by the nature of the ion. For hydrated cation clusters e.g. $Na^+(H_2O)_n$, the cation will interact most strongly with the negatively charged oxygen atom of H_2O, so that its hydrogen atoms point away from the cation. Accordingly, for hydrated cation clusters intrashell hydrogen bonding is not exhibited and only intershell hydrogen bonding prevails. On the other hand, for hydrated anion clusters, e.g. $I^-(H_2O)_n$ the interactions are much more complex. The anion will interact most strongly with one of the hydrogen atoms of H_2O and will result in an asymmetric configuration of H_2O relative to I^- in $I^-(H_2O)$. The second hydrogen atom can interact with other water molecules in the first solvation shell. This intershell hydrogen bonding in $I^-(H_2O)_n$ (n > 1) clusters gives rise to association of water molecules, which are held together by networks of hydrogen bonding (Fig. 1). Thus, the short-range interactions determine the structure, energetics and other electronic properties of these clusters and give rise to the existence of structural isomers. The simplest example pertains to the addition of (H_2O) to $I^-(H_2O)$, which gives rise to two types of distinct structures, with the second H_2O molecule interacting directly with the halide anion, or forming a hydrogen bonded structure with the first water molecule. These two isomers are portrayed on the top of figure 1. Further addition of water molecules to the $I^-(H_2O)_n$ (n > 3) cluster generates several size-specific structural isomers, which are shown in Figure 1.

3.2   SURFACE VS. INTERIOR ANION STATES

The asymmetric nature of these anion clusters produces two general types of isomers according to the conformation of the water molecules near the anion. An (S) state will correspond to the anion being located on top of the solvent molecules, which are strongly bound together by hydrogen bonds. In this type of isomer the hydrogen bonding between the water molecules is significant and the interaction of the halide anion with the water molecules is somewhat weaker. An (I) state is one in which the water molecules surround the anion producing a `cage' with the anion being located in the interior of that cage. The two hydrogen bonded and linear structural isomers of $I^-(H_2O)_2$ (Fig. 1) can be visualized as the precursors of the surface and interior states, respectively. Basically, for $X^-(H_2O)_n$ (n > 3) clusters, a transition from a surface state to an interior state can be visualized by breaking one of the hydrogen bonds in the network of water molecules and replacing it by a stronger water-halide interaction, as shown in figure 1 for n=4 and 5. For $X^-(H_2O)_n$ (n=4,5) structures the (S) isomer corresponds to a pyramidal configuration of $(H_2O)_n$ molecules, while the (I) structure corresponds to a [(n-1) + 1] configuration with a pyramid of (n-1) water molecules on one side of X^- and a single water molecule on its other side (Fig. 1). For $X^-(H_2O)_6$ clusters the (S) configuration corresponds to a hexagonal pyramid, while the (I) configuration is a distorted octahedron. (Fig. 1).

3.3   CHARGE DISTRIBUTION

In Table I we list the charge distribution for the iodine and water molecule components in several iodine water complexes. It is clear from this data that the electron density shifts from the iodine to the water molecule as the cluster grows in size. This means that the iodine electronic structure is being altered and we expect this to increase even more dramatically as the cluster size gets larger. This shift is consistent with the appearance of the CTTS spectra in the liquid state since it implies that the excited states resident on the iodine will decrease in energy as the cluster gets larger.

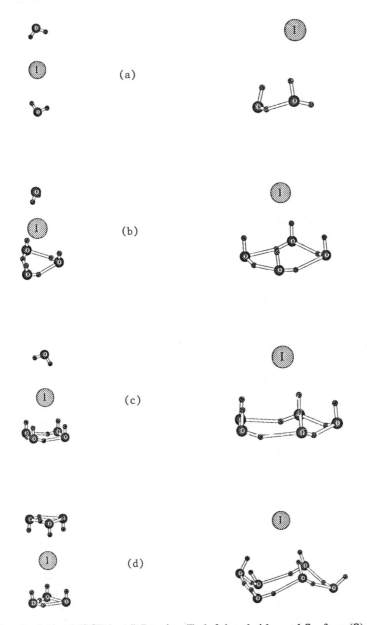

Figure 1.   Optimized (SCF level) Interior (I), left hand side, and Surface (S), right hand side, structures for I⁻H₂O)ₙ, n=2,4-6. Intrasolvation shell hydrogen bonding is represented by dotted lines.

(a) n=2   (b) n=4   (c) n=5   (d) n=6

Table I. Lowdin Charges (P_L)for the $I^-(H_2O)_n$, n=1,2 clusters.

Atom	$I^-(H_2O)^{(a)}$	$H_2O^{(b)}$	Atom	$I^-(H_2O)_2^{(c)}$	$(H_2O)_2^{(d)}$
	P_L	P_L		P_L	P_L
I	-0.958	-	I	-0.928	-
O1	-0.901	-0.835	O1	-0.908	-0.838
H1	0.444	0.418	O2	-0.910	-0.863
H2	0.415	0.417	H11	0.446	0.424
			H12	0.420	0.427
			H21	0.436	0.410
			H22	0.443	0.440

(a)   Charge distribution for the $I^-(H_2O)$ cluster.

(b)   Charge distribution for the H_2O molecule at the $I^-(H_2O)$ geometry.

(c)   Charge distribution for the $I^-(H_2O)_2$ cluster.

(d)   Charge distribution for the H_2O - H_2O fragment at the $I^-(H_2O)_2$ geometry.

3.4   GROUND STATE ENERGETICS

Of considerable interest are the relative stabilization energies of distinct size-selected isomers of fixed n (Figure 1). In Table II we present the characterization of the isomer structures, which correspond to (S) and (I) isomers, together with the relative stabilization energies of the (S) isomer relative to the (I) isomer for each cluster size.

$$\delta E(n) = E^S[X^-(H_2O)_n] - E^I[X^-(H_2O)_n]$$

where $E^S[X^-(H_2O)_n]$ and $E^I[X^-(H_2O)_n]$ correspond to the total energies of the surface and interior isomers, respectively, at their equilibrium nuclear configurations. For n=2 the H-bonded "(S)-type" structure is more stable than the linear "(I)-type" structure. For n=4 the (S) isomer is energetically more stable than the (I) isomer. For n=5 the (S) and (I) isomers are nearly isoenergetic for Cl⁻ and Br⁻ ions, while for the $I^-(H_2O)_5$ the (S) state seems to be energetically favored. The results for n=6 are qualitatively different, with the hexagonal pyramid (S) structure being less stable than the distorted octahedral (I) structure. For the two (S) and (I) n=6 isomers the transition from the (S) state to the (I) state does not involve the breaking of hydrogen bonds, unlike for the n=4,5 clusters (see Figure 1). There are six hydrogen bonds between the water molecules for both (S) and (I) isomers of n=6. The full optimization of the (S) structure, starting at a nearly symmetric hexagonal pyramid, ends up in a distorted hexagonal pyramid with three water molecules interacting more strongly with X⁻ than the other three water molecules (Fig. 1). On the other hand, the (I) structure is characterized by two separate subclusters, each consisting of three water molecules, with the $X^-(H_2O)$ being enhanced relative to the (S) state. The end result for n=6 is that the (S) state is marginally energetically stable relative to the (I) state.

The calculated values of ΔH for n=2,4 and 6 for all three anions and $I^-(H_2O)_5$ barely exceed the upper limit for the uncertainty of our computational results (2 Kcal mole⁻¹).

Thus, our conclusions regarding the relative stability of isomer structures must be considered as tentative. The relatively small values of $|\delta E(n)|$ prohibit the utilization of ground state energetic data for the definite assignment of the preferred isomer structures. Accordingly, theoretical and experimental total and sequential enthalpy of hydration data will be of very limited use for the assignment of structural isomers. In what follows we shall show that vertical ionization potential data provide a powerful diagnostic tool for the distinction between (S) and (I) isomers.

Table II. Characterization of Isomer Structures and Calculated Relative Stabilization Energies of (S) Structures Relative to (I) Structures[a].

n	Structure	Characterization	$\delta E(n)$ kcal mole$^-$		
			Cl^-	Br^-	I^-
2	H-bonded	(S)	1.6	1.6	2.0
	Linear	(I)			
4	Square Pyramid	(S)	3.1	3.2	4.3
	(3 - 1)	(I)			
5	Pentagonal pyramid	(S)	-0.3	0.2	1.8
	(4 - 1)	(I)			
6	Hexagonal pyramid	(S)	1.1	1.6	3.2
	Distorted Octahedral	(I)			

[a]   Data from total energies at the scf level.

3.5   VERTICAL IONIZATION POTENTIALS

To simplify the calculations we have utilized Koopmans' theorem[32],which represents the vertical ionization potential IP(n) of the $X^-(H_2O)_n$ cluster at its equilibrium nuclear configuration in terms of the one-electron energy of the highest occupied molecular orbital. However, the results obtained from Koopmans' theorem for the ionization potentials IP(0) of the bare anions and for the ionization potentials IP(1) of $X^-(H_2O)$ clusters are overestimated by ~ 0.4-0.5 eV and are useless for our purposes. We have therefore utilized Koopmans' theorem for the calculation of integral ionization potential shifts

$$dI(n) = IP(n) - IP(0)$$

and differential ionization potential shifts

$$DI(n) = IP(n) - IP(n-1)$$

i.e., the difference between the vertical ionization potentials of $X^-(H_2O)_n$ and $X^-(H_2O)_{n-1}$.

4. Size Dependence and Isomer Specificity of Vertical Ionization Potentials

The cluster size dependence and the isomer specificity of the IP(n) provides a diagnostic tool for the structural assignment of isomers, and, in particular, for the distinction between (S) and (I) isomers. In Table III we present the calculated data for the integral and differential ionization potential shifts together with the experimental data of Cheshnovsky, et. al[16]. The following conclusions emerge from the photoelectron spectra of $X^-(H_2O)_n$ clusters:

(1) For small n=1-3 (Cl^-, Br^- and I^-) clusters good agreement between the calculated and experimental dI(n) was achieved, with a maximum deviation of $\delta E=0.06$ eV. For the n=2 cluster we can not distinguish between the hydrogen bonded and the linear isomers as the differences between their vertical ionization potential is close to δE, although the differential shifts seem to favor the hydrogen bonded structure for n=2. For the n=3 cluster the pyramidal, (S) is favored on the basis of the ionization potential data.

(2) For n=4 and 5, the (S) states seem to be preferred on the basis of the comparisons with the experimental ionization potential data. For $Cl^-(H_2O)_n$ (n=4,5) and for $I^-(H_2O)_5$, the (S) pyramidal structures provide theoretical dI(n) data which are close to the experimental results while the 3-1 and 4-1 (I) structures provide higher dI(n) values, which considerably exceed the experimental dI(n) (n=4,5) data. This conclusion is supported by the differential shifts DI(n) (n=4,5) each calculated relative to the IP(n-1) for the (S) isomer, which indicates that for the n=4 and 5 (S) isomers are favored for Cl^-, Br^- and I^-.

(3) The overall agreement between the experimental vertical ionization potentials for n=4 and 5 and the calculated data for the (S) isomers is better than 0.1 eV.

(4) For the $X^-(H_2O)_6$ clusters, the calculated marked differences between the integral spectral shifts for the (S) and (I) isomers, together with the experimental data provide evidence for the occurrence of the (I) isomer. The calculated dI(6) data for the (I) isomer are closer to the experimental data for Cl^-, Br^-, and I^-. Even more striking is the observation that the predicted differential spectral shift DI(6) (calculated relative to the favored n=5 (S) isomer) is negative for the n=6 hexagonal pyramidal (S) structure. Experimental data have been interpreted as favoring the interior state. However, when our calculations are improved to the MP2 level (in work not yet completed) it is not longer so clear that the interior ionization potentials are in distinctly better agreement with experiment. Enthalpy values are also marginally in favor of the surface state.

(5) The experimental nonmonotonous increase of the differential DI(n) values between n=4-6 for $Cl^-(H_2O)_n$ and $I^-(H_2O)_n$ which reveal a decrease of DI(5) followed by an increase of DI(6) are consistent with the theoretical results for the prevalence of the (S) isomer for n=5 and the prevalence of the (I) isomer for n=6.

Table III. Structural Data and Vertical Ionization Potentials (in eV) for $[X^-(H_2O)_n]$, n=1-6, X=Cl, Br, I Clusters.

n	Structure[a]	Cl⁻ dIP(n)[b] Calc	Cl⁻ dIP(n)[b] Expt	Cl⁻ DIP(n)[c] Calc	Cl⁻ DIP(n)[c] Expt	Br⁻ dIP(n)[b] Calc	Br⁻ dIP(n)[b] Expt	Br⁻ DIP(n)[c] Calc	Br⁻ DIP(n)[c] Expt	I⁻ dIP(n)[b] Calc	I⁻ dIP(n)[b] Expt	I⁻ DIP(n)[c] Calc	I⁻ DIP(n)[c] Expt
1		0.642	0.71	0.642	0.71	0.527	0.55	0.527	0.55	0.422	0.45	0.422	0.45
2	Linear	1.246	1.32	0.604	0.61	1.069	1.08	0.540	0.53	0.830	0.86	0.410	0.41
	H - B	1.214		0.572		0.997		0.470		0.797		0.376	
3	Pyramidal	1.758	1.84	0.544	0.52	1.444	1.55	0.452	0.47	1.137	1.23	0.310	0.37
4	Sq. Pyramidal	2.166	2.24	0.409	0.40	1.774	1.85	0.325	0.30	1.398	1.53	0.261	0.30
	3 - 1	2.275		0.517		1.905		0.456		1.513		0.376	
5	Pent. Pyramidal	2.480	2.54	0.315	0.30	2.040	2.12	0.267	0.27	1.623	1.71	0.226	0.18
	4 - 1	2.637		0.470		2.20		0.430		1.768		0.370	
6	Hex. Pyramidal	2.258		-0.223		1.876		-0.164		1.480		-0.144	
	Dist. Octahedral	3.124	2.86	0.642	0.33	2.650	2.39	0.608	0.27	2.108	2.05	0.484	0.34

(a) Typical Structures presented in Fig 2,3,4. [(n-1) - 1] structures denote $(n-1)H_2O$ molecules in a planar configuration on one side and a single H_2O molecule on the other side of X⁻.

(b) $dI(n) = IP(n) - IP(0)$.

(c) $DI(n) = IP(n) - IP(n-1)$.

Our theoretical results in conjunction with experimental ionization potential data provide an estimate for the cluster size which could exhibit the (S) going to (I) transition, an issue of considerable importance for the elucidation of microscopic phenomena. From the foregoing analysis the following picture emerges for the dominating size dependent isomer structure of the $X^-(H_2O)_n$ (n=1-6) clusters. For n=2-5 the (S) isomers are dominating, while the "transition" from the (S) isomer to the (I) isomer is suggested for a cluster size n=6 for Cl^-, Br^- and I^-. However, work in progress suggests that higher level of calculations may temper these conclusions.

5. Vibrational Spectra

Both Truhlar and our work have evaluated the vibrational spectra in anion single water systems, but we have also been able to look at some larger halide water clusters. In the single water cluster, the vibrational energy corresponding to hydration is primarily observed at 150 cm^{-1}. In Figure 2, we show the vibrational spectra of both an interior and a surface state for the six water molecule iodide cluster. Notice that there is a distinct difference with the interior state having a set of higher frequencies characterisic of a more constrained state. it is hoped that experimentalists will soon be able to observe these frequencies and even pick up the signature of the surface to interior state transition.

6. Comparison with Other Cluster Systems

In previous work we have studied both sodium in water clusters and electrons on small water clusters. There are some useful parallels. We have also gone back to look at the trapped electron calculations in light of some of the anion hydration calculations.

In regard to electron attachment to water clusters quantum simulations[19] for $(H_2O)_n^-$ indicate the energetic stability of surface excess electron states for small clusters less than 32 and the "transition" from surface to an interior excess electron state for large clusters[19], while no definite experimental evidence for this transition could be yet inferred from photoelectron spectroscopy[8,14,15].

We have recently gone back to clusters of water and studied the electron affinity of small clusters. For many configurations of water an electron can be bound by up to a few hundredths to even a tenth of an electron volt. We have observed such states for water trimers and tetramers and in some hexamers. These are often not the lowest energy configurations of the water but states which are thermally accessible in most beams. Further studies are now underway to characterize these states. It is interesting that they were discovered by us only when we considered water configurations similar to those found in iodide water clusters.

In our previous work on sodium-water clusters we observed dramatic shifts in the ionization potentials as a function of cluster size. In Table IV we show some of these shifts in terms of Lowdin charge densities and HOMO orbital energies, the latter corresponding roughly to the ionization potential (using Koopman's theorem). Since the wavefunction has an extent proportional to the square root of the outer orbital energy (approximately), it is clear that the extent of the outer sodium orbital is roughly 1.4 times

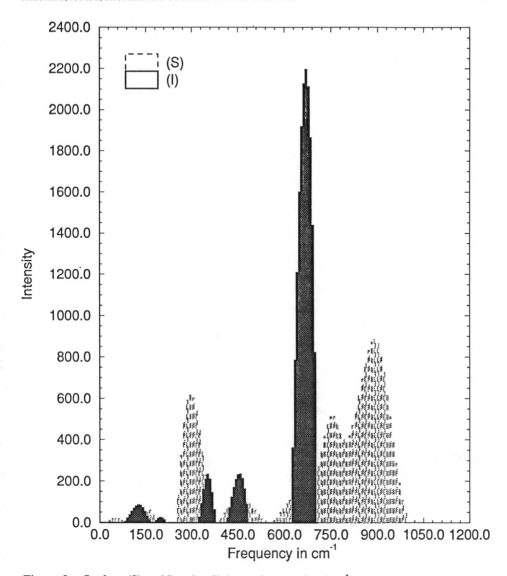

Figure 2.   Surface (S) and Interior (I) lower frequencies (c_m^{-1}) vs. Intensity, for the
Cl$^-$(H$_2$O)$_6$ cluster.

as large for n=4 as for n=1. Furthermore the effective charge on the sodium by this measure decreases greatly for that same range. Clearly we are dealing with major perturbations of the electron density due to the waters.

Table IV. Löwdin charges (P_L) and HOMO energies for the $Na(H_2O)_n$, n=1,2,4 clusters.

Atom	$Na(H_2O)$	Atom	$Na(H_2O)_2$	Atom	$Na(H_2O)_4$
	P_L		P_L		P_L
Na	-0.093	Na	-0.196	Na	-0.414
O1	-0.795	O1	-0.783	O1	-0.797
H1	0.444	O2	-0.802	O2	-0.762
H2	0.444	H11	0.439	O3	-0.791
		H12	0.454	O4	-0.799
		H21	0.444	H11	0.444
		H22	0.445	H12	0.442
HOMO	-0.074		-0.040		-0.033

7. Limitations Of Molecular Dynamics Studies

The results of the charge densities in the previous section and for the halide water clusters clearly demonstrate that the waters are having an effect on the electron density of the "impurity". One can even refer to these as "chemical" effects. Such a trend is required if the halides are to yield a CTTS spectrum in very large water clusters since the excited state energies must be significantly modified. These results indicate that molecular dynamics studies which use a fixed set of potentials to describe all of the interactions including the halide must ultimately fail in large clusters since they assume the halide electron density is unchanged. At best one can only empirically include this effect.

There are also problems in this system using quantum simulation techniques since there is no simple pseudopotential which can be used for the halide. The issue is again a change in electron density, this time between the anion and the neutral. This is in contrast to the sodium example where the outer electron is relatively independent of the core and thus the core for the atom and the cation are very similar. The hope here is that one can construct an empirical pseudopotential and use it in quantum similations. We have begun some work in this area and it appears promising.

8. Concluding Remarks

We have performed ab initio molecular orbital calculations on halide anions clusters using extended basis sets. The quality of the basis sets used in the quantum mechanical calculations is crucial for accurate predictions of the experimental observables. For the Cl^- anion an all electron calculation was performed and the basis sets was extended to include polarization functions on Cl^-. For the Br^- and I^- anions we have used pseudopotentials, sacrificing some accuracy in these calculations. It is well-known that the use of pseudopotentials results in equilibrium structures with bond lengths which differ by as much as 0.1 Å from the results of all electron calculations. We thus expect that the present calculations for $Br^-(H_2O)_n$ and $I^-(H_2O)_n$ may underestimate the halide-water molecule interactions.

Nevertheless, on the basis of comparison between the calculated and experimental results for the integral ionization potential shifts and the total enthalpies of cluster formation, it appears that the accuracy of the present calculations of these energetic observables for Cl^- as well as for Br^- and I^- is $\delta = 1$-2 Kcal mole^{-1}, approaching "chemical accuracy".

Our calculations provided information on the size dependence, on the enthalpies of hydration and vertical ionization potentials in good agreement with experimental data[16]. The isomer-specificity of the enthalpies of hydration is characterized by small energy differences between (S) and (I) isomers, precluding the utilization of total and sequential hydration enthalpies for the identification of structural (S) or (I) isomers for n=4-6. On the other hand, the calculation of the vertical ionization potentials, which manifest large isomer-specificity, in conjunction with the experimental photoelectron spectroscopy data[16], allow for the identification of the prevalent (S) isomers for n=4,5 clusters. In our analysis of vertical ionization potentials we considered the dominance of a single structural isomer for each n value. A more detailed comparison between theory and experimental photoelectron spectroscopy should consider the possible coexistence of several isomers of a fixed composition at a fixed value of n. A cursory examination of the experimental photoelectron spectra for $I^-(H_2O)_n$ (n=4-6) clusters indicates that these do not exhibit any marked fine structure and their spectral widths are similar to those for the n=2,3 clusters. Therefore, it appears that the coexistence of structural (S) and (I) isomers for n=4-6 is not excessive.

The utilization of the quantum mechanical calculations for the structure and energetic results in the relevant information at 0 K, which is subsequently confronted with the experimental finite-temperature results. Of course, entropic effects may be of importance in determining the finite-energy distribution of isomers. In this context, the temperature of the $X^-(H_2O)_n$ clusters is of considerable importance. Cheshnovsky et al.[33] have inferred from the binding of Ar atoms to $I^-(H_2O)_n$ clusters that the cluster temperatures are T < 100K. This cluster temperature is considerably lower than that utilized in MD calculations on $Cl^-(H_2O)_n$ clusters, which rested on rather approximate potentials. It is now clear whether the temperature variation will have a large effect on the MD simulations since in many cases quenched structures were used[22].

The central prediction emerging from the structure-energetics relations explored herein is that the "transition" from (S) to (I) isomer structures is suggested for $X^-(H_2O)_6$ with this cluster size being invariant with respect to the nature of the anion. The energetic stability of the $Cl^-(H_2O)_6$ anion state inferred from our calculations is in accord with the results of MD simulations which predicted that the (S) anion state is stable for $Cl^-(H_2O)_{20}$. Our predictions, based on quantum chemistry, are not subject tot he uncertainties of potentials as are MD calculations. An independent experimental support for their identification of the $X^-(H_2O)_6$ (I) isomers emerges from the resent work of Cheshnovsky et al. who have identified weak photoelectron spectra of (S) isomers of $I^-(H_2O)_n$ (n=34-40) clusters at energies lower than those corresponding to the vertical ionization potential for the

$I^-(H_2O)_6$ cluster. Accordingly, the photoelectron spectrum of the $I^-(H_2O)_6$ is dominated by the (I) isomer, according to the current interpretation of experimental work.

A further notable negative result of our computations is the absence of a bound electronically excited state for $X^-(H_2O)_n$ (n < 6) clusters. Thus the bound-bound electronic excitation in clusters, which constitutes the precursor of the CTTS transition in bulk water, sets in at n > 6. This conclusion is pertinent for a critical test of the empirical electronic pseudopotentials recently introduced for quantum simulations of CTTS optical spectra. Furthermore, a proper description of the halide-water interaction is crucial to obtain reliable results on the structure and electronic spectra of halide anions in the bulk. Currently, we are using the information from our *ab initio* calculations to construct and calibrate reliable interatomic potentials for classical and quantum MD simulations of energetics, structure and spectra in large clusters.

The chemical physics of the structure, energetics and optical spectroscopy of anions in solutions was addressed by James Franck and his colleages in the twenties[34,35]. This problems remains an exciting challenge for modern theorists.

9.   Acknowledgement

Very special thanks to Joshua Jortner who collaborated on much of this work and to Ori Chesnovsky for keeping us informed of recent experimental results.

10.   References

1. R. S. Berry Structure and Dynamics of Clusters, in: The Chemical Physics of Atomic and Molecular Clusters, pp 1 and pp 23. Edited by G. Scoles, Pub. Elsevier 1990, and references therein.
2. The Physics and Chemistry of Small Clusters, NATO ASI series, edited by P. Jena, B. K. Rao, and S. N. Khanna (Plenum, New York, N. Y. 1986)
3. Microclusters, edited by S. Sugano, Y. Nishima and S. Onishi, (Springer, Heidelberg, 1987).
4. Large Finite Systems, edited by J. Jortner, A. Pullman, and B. Pullman (D. Reidel, Utrecht, 1987).
5. Elemental and Molecular Clusters, edited by G. Benedek, T. P. Martin, and G. Pacchioni (Springer, Heidelberg, 1988).
6. Jortner, J. (1992) Z. Phys., D24, 247 .
7. A. Amirav, U. Even, and J. Jortner, J. Chem. Phys., 75, 2489 (1981); S. Leutwyler and J. Jortner, J. Phys. Chem., 91, 5558 (1987).
8. K. H. Bowen and J. G. Eaton in The Structure of Small Molecules and Ions edited by R. Naaman and Z. Vager (Plenum, New York, 1988); S. T. Arnold, J. G. Eaton, D. Patel-Misra, H. W. Sarkas, and K. H. Bowen in Ion and Cluster Ion Spectroscopy and Structure edited by J. P. Maier (Elsevier, Amsterdam, 1989).
9. G. Makov and A. Nitzan, J. Chem. Phys., wrong refernce!!!!
10. M. J. Blandamer and M. F. Fox, Chem. Rev. 70, 59 (1970); J. Jortner and A. Treinin, Trans. Faraday Soc., 58, 1503 (1962).

11. M. Arshadi, R. Yamdagni, and P. Kebarle, J. Phys. Chem. 74, 1475 (1970). R. Yamdagni, J. D. Payzant, and P. Kebarle, Can. J. Chem. 51, 2507 (1973). P. Kebarle, Ann. Rev. Phys. Chem. 28, 445 (1977). M. A. French, S. Ikuta, and P. Kebarle, Can. J. Chem. 60, 1907 (1982). T. F. Magnera, G. Caldwell, J. Sunner, S. Ikuta, and P. Kebarle, J. Am. Chem. Soc. 106, 6140 (1984).
12. A. W. Castleman Jr and I. N. Tang, J. Chem. Phys. 57, 3629 (1972). N. Lee, R. G. Keesee, and A. W. Castleman Jr., J. Colloid Interface Sci. 75, 555 (1980). R. G. Keesee, N. Lee, and A. W. Castleman Jr., J. Chem. Phys. 73, 2195 (1980). R. G. Keesee and A. W. Castleman Jr., Chem. Phys. Letters 74, 139 (1980). X. Yang and A. W. Castleman Jr., J. Phys. Chem. 94, 8500 (1990). Ibid, 95, 6182 (1991). X. Yang, X. Zhang, and A. W. Castleman Jr., J. Phys. Chem. 95, 8520 (1991).
13. K. Hiraoka and S. Mizuze, Chemical Physics 118, 457 (1987). K. Hiraoka, S. Misuze, and S. Yanabe, J. Phys. Chem. 92, 3943 (1988).
14. J. V. Coe, G. H. Lee, J. G. Eaton, S. T. Arnold, H. W. Sarkas, K. H. Bowen, C. Ludewigt, H. Haberland, D. R. Worsnop, J. Chem. Phys.92, 3980 (1990).
15. H. Haberland, H. G. Schindler, D. R. Worsnop, J. Phys. Chem.,88, 3903 (1984), H. Haberland, C. Ludewigt, H. G. Schindler, D. R. Worsnop, Phys. Rev. A}, 36, 967 (1987); H. Haberland in: Large Finite Systems, edited by J. Jortner, A. Pullman, B. Pullman (D. Reidel, Dordrecht, 1987).
16. G. Markovich, R. Giniger, M. Levin, and O. Cheshnovsky, J. Chem. Phys. 95, 9416 (1991). G. Markowitz, R. Giniger, M. Levin, and O. Cheshnovsky, Z. Phys. D. 20, 69 (1991). G. Markovich, R. Giniger, and O. Cheshnovsky. Proceedings of the VIIth Conference on Small Particles and Inorganic Clusters}, Chicago, September 1992 (in press). G. Markovich, R. Giniger, and O. Cheshnovsky, Z. Phys. D (in press. 1993).
17. J. Jortner, D. Scharf, N. Ben-Horin, U. Even, and U. Landman, Size Effects in Clusters, in: The Chemical Physics of Atomic and Molecular Clusters, Edited by G. Scoles, Pub. Elsevier, 1990, pp 43 and references therein.
18. B. K. Rao and N. R. Kestner, J. Chem. Phys. 80, 1587 (9184), N. R. Kestner and J. Jortner, J. Phys. Chem. 88, 3818 (1984).
19. D. F. Coker, D. Thirumalai, and B. J. Berne, J. Chem. Phys. 86, 5689,(1987), D. F. Coker and B. J. Berne, J. Chem. Phys. 89, 2128 (1988), B. Space, D. F. Coker, Z. H. Liu, B. J. Berne, and G. Martyna, J. Chem. Phys. 97, 2002 (1992).
20. R. N. Barnett, U. Landman, C. L. Cleveland, and J. Jortner, Phys. Rev. Lett. 59, 811 (1987); J. Chem. Phys.} 88, 4421 (1988); J. Chem. Phys. 88, 4429 (1988); Chem. Phys. Lett. 145, 382 (1988); R. N. Barnett, U. Landman, and A. Nitzan, Phys. Rev. A. 38, 2178 (1988); J. Chem. Phys. 89, 2242 (1988); ibib 90, 4413 (1989); Phys. Rev. Lett. 62, 106 (1989). R. Barnett, U. Landman, S. Dhar, N. R. Kestner, J. Jortner and A. Nitzan, J. Chem. Phys. 91, 7797 (1989).
21. S. Sung and P. C. Jordan, J. Chem. Phys. 85, 4045 (1986).
22. L. Perera and M. L. Berkowitz, J. Chem. Phys. 95, 1954 (1991).
23. J. Caldwell, L. X. Dang, and P. A. Kollman, J. Am. Chem. Soc. 112, 9144 (1990). L. X. Dang, J. E. Rice, J. Caldwell, and P. A. Kollman, J. Am. Chem. Soc. 113, 2481 (1991). J. Caldwell, L. X. Dang, and P. A. Kollman, J. Am. Chem. Soc. 112, 9145 (1990).
24. M. Sprik, M. L. Klein, and D. Chandler, Phys. Rev. B 31, 4234 (1985), J. Chem. Phys. 83, 3042 (1985).
25. W. L. Jorgensen, J. Chandrasekhar, J. D. Madura, R. W. Impey, and M. L. Klein, J. Chem. Phys. 79, 926 (1983).

26. X. G. Zhao, A. Gonzalez-Lafont, D. G. Truhlar, and R. Steckler, J. Chem. Phys. 94, 5544 (1991).
27. J. E. Combariza, N. R. Kestner, and J. Jortner, Chem. Phys. Letters 203, 423 (1993).
28. Gaussian 90 Revision J. M. J. Frisch, M. Head-Gordon, G. W. Trucks, J. B. Foresman, H. B. Schelegel, K. Raghavachari, M. Robb, J. S. Binkely, C. Gonzalez, D. J. Defrees, D. J. Fox, R. A. Whiteside, R. Seeger, C. F. Melius, J. Baker, R. L. Martin, L. R. Kahn, J. J. P. Steward, S. Topiol, and J. A. Pople. Gaussian Inc. Pittsburgh, PA 1990.
29. Gamess. Original program assembled by the staff of the NRCC; M. Dupuis, D. Spangler, and J. J. Wendoloski. national Resource for Computations in Chemistry Software Catalog, University of California, Berkely, CA 1980; Program QG01. This version of GAMESS is described in the Quantum Chemistry Program Exchange Newsletter: Schmidt M. W., Baldridge, K. K., Boatz, J. A., Jensen, J. H., Koseki, S., Gordon, M. S., Nyguyen, K. A., Windus. T. L., Elbert, S. T., QCPE Bull 10, 52 (1990).
30. R. Ditchfield, W. J. Hehre, and J. A. Pople, J. Chem. Phys. 54, 724 (1971). P. C. Hariharan and J. A. Pople, Theoret. Chim. Acta 28, 213 (1973).
31. Handbook of Chemistry and Physics, 71st ed., Lide, D. R., Ed. CRC Press: Boca Raton, Fl. 1990.
32. T. A. Koopmans, Physica, 1, 104 (1933).
33. O. Cheshnovsky. Private Communication, 1993.
34. J. Franck and G. Schiebe. Z. Phys. Chem., A139, 22 (1928).
35. J. Franck and F. Haber Sitzber. Preuss. Akad. Wiss. Phys. Math. Klasse 58/9, 342 (1927).

PHOTOELECTRON SPECTROSCOPY OF SOLVATED ANION CLUSTERS

S. T. Arnold, J. H. Hendricks, and K. H. Bowen
Department of Chemistry, Johns Hopkins University
3400 N. Charles St., Baltimore, MD 21218 USA

Abstract. The photoelectron spectra of $O^-(Ar)_{n=1-26}$ have been recorded, and total as well as sequential solvation energies have been extracted from the spectra. Plotting these values as a function of cluster size demonstrates that the first solvation shell for this system closes at n=12. Also, both the energetic data and the mass spectral data suggest structural information about the O^-Ar_n clusters. Specifically, these species appear to follow a pentagonal packing sequence, implying an icosahedral structure for n=12 and a capped icosahedral structure for n=18.

I. Introduction

Photodetachment studies of solvated anion clusters began in 1968, with the threshold photodetachment study of $OH^-(H_2O)$ by Golub and Steiner.[1] In 1985, we applied negative ion photoelectron (photodetachment) spectroscopy to the study of solvated anion clusters,[2] and this technique has proven to be very useful in examining the energetics of these cluster anions. Since then, we have applied negative ion photoelectron spectroscopy to the study of $H^-(NH_3)_{n=1-2}$, $NH_2^-(NH_3)_{n=1-2}$, $NO^-(N_2O)_{n=1-5}$, $NO^-(H_2O)_{n=1-2}$, and $NO^-(Rg)$, where Rg = Ar, Kr, and Xe;[3-6] Miller, Leopold, and Lineberger[7] have investigated $H^-(H_2O)$; Johnson et al.[8,9] have studied $O_2^-(O_2)$, $O_2^-(N_2)$, and $(NO)_2^-$; and recently Cheshnovsky and coworkers[10] have studied $I^-(H_2O)_{n=1-15}$, while Neumark and coworkers[11] have examined $I^-(CO_2)_{n=1-13}$. Also very relevant to anion solvation are the photodissociation studies of $Br_2^-(CO_2)_{n=1-24}$ and $I_2^-(CO_2)_{n=1-22}$ by Lineberger and coworkers.[12,13] Here, continuing our photoelectron studies of solvated anion clusters, we present the photoelectron spectra of $O^-(Ar)_{n=1-26}$. This particular solvated anion system was chosen for study because of its relative simplicity, an atomic anion interacting with rare gas solvent atoms, providing a system in which all the components are spherical and in which solvent-solvent interactions are minimized.

37

J. Jortner et al. (eds.), Reaction Dynamics in Clusters and Condensed Phases, 37–45.
© 1994 *Kluwer Academic Publishers. Printed in the Netherlands.*

II. Experimental

Negative ion photoelectron spectroscopy is conducted by crossing a mass selected beam of negative ions with a fixed-frequency photon beam and energy analyzing the resultant photodetached electrons. Our negative ion photoelectron spectrometer has been previously described in detail.[14] A supersonic expansion source coupled with an additional gas "pickup" line was used to generate the O^-Ar_n cluster anions. Typically, 8-10 atm of argon was expanded through a 12 μm nozzle into vacuum, while a small amount of N_2O was introduced into the plasma through a secondary "pick-up" line located just beyond the nozzle. A cooling jacket around the stagnation chamber allowed the source temperature to be maintained at -70 °C. A negatively biased filament (ThO_2/Ir) was used for ionization, forming O^- and NO^- anions, which then clustered with the argon in the expansion, forming O^-Ar_n and NO^-Ar_n as the primary cluster anions. A predominantly axial magnetic field confined the plasma and enhanced cluster anion production. Cluster anions were extracted into the spectrometer and transported through a Wien velocity selector, where they were mass selected. As shown in Figure 1, "magic numbers" were observed in the mass spectrum of O^-Ar_n, where n=10,12,15,18,22, and 25 were more intense than their neighboring clusters. The mass selected cluster ion beam was then crossed with an Ar^+ laser operated intracavity (488 nm or 457.9 nm), and the resulting photodetached electrons were energy analyzed using a hemispherical electron energy analyzer, with a typical resolution of 35 meV.

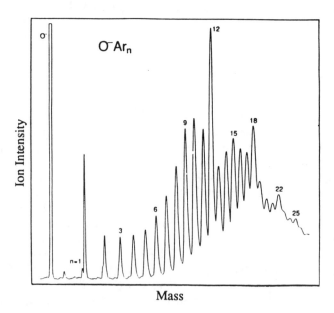

Figure 1. In the mass spectrum of O^-Ar_n, "magic numbers" were observed at n=10,12,15,18,22, and 25.

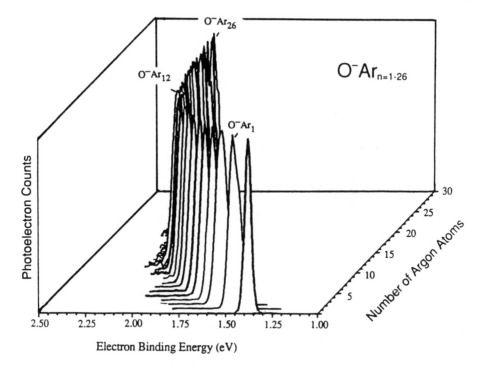

Figure 2. The negative ion photoelectron spectra of $O^-(Ar)_{n=1-26}$. All the spectra closely resemble that of O^-, each containing a single peak which shifts to higher electron binding energy with increasing cluster size. The electron binding energy at the peak maximum is the VDE of the anion.

III. Photoelectron Spectra

The negative ion photoelectron spectra of $O^-(Ar)_{n=1-26}$ are presented in Figure 2, along with the photoelectron spectrum of O^-, which was recorded before and after each cluster anion spectrum for calibration purposes. The photoelectron spectrum of O^- appears as a single, slightly broadened peak. Because O^- serves as the "chromophore" for photodetachment in O^-Ar_n, the cluster anion spectra each closely resemble that of O^-, except for being broadened and shifted toward higher electron binding energy with increasing cluster size. The electron binding energy at the peak maximum in each case is the vertical detachment energy (VDE) of the anion. This is the vertical photodetachment transition energy between the ground state of the anion and the ground state of the neutral at the equilibrium geometry of the negative ion.

IV. Cluster Anion Energetics

The energetics of the O^-Ar_n clusters are governed by the following relationship:

$$EA[O(Ar)_n] = EA[O] + \sum_{m=0}^{n-1} D_0[O^-(Ar)_m \cdots Ar] - \sum_{m=0}^{n-1} D_0[O(Ar)_m \cdots Ar] \qquad (1)$$

where EA denotes the adiabatic electron affinity, $D_0[O^-(Ar)_m \cdots Ar]$ is the ion-neutral dissociation energy for the loss of a single Ar atom from the cluster, and $D_0[O(Ar)_m \cdots Ar]$ is the analogous neutral cluster weak-bond dissociation energy for the loss of a single Ar atom. Since ion-solvent interaction energies generally exceed van der Waals bond strengths, it is evident from eq (1) that clustering can be expected to stabilize the excess electronic charge on a negative ion, ie. the electron affinities of these clusters should increase with cluster size. This is seen in the photoelectron spectra of O^-Ar_n, where the sub-ion is stabilized as the number of solvent atoms increases, shifting the spectra toward higher electron binding energies. It follows from eq (1), that the relationship between adjacent-sized clusters can be expressed as:

$$EA[O(Ar)_n] - EA[O(Ar)_{n-1}] = D_0[O^-(Ar)_{n-1} \cdots Ar] - D_0[O(Ar)_{n-1} \cdots Ar] \qquad (2)$$

Although these equations describe the cluster anion energetics rigorously, some approximations may be made to further simplify these expressions. The EA of each cluster can be approximated by the VDE of the O^-Ar_n anion. The VDE of O^- (1.465 eV) nearly equals the EA (1.462 eV), and because O^- serves as the photodetachment chromophore for the O^-Ar_n cluster anions, the EA of the cluster is taken to be nearly equal to the measured cluster anion VDE. In addition, ion-solvent interaction energies are often larger than van der Waals energies by an order of magnitude, so their relatively minor contributions in eqs (1) and (2) may be neglected. Thus, the total anion-solvent dissociation energy for a given cluster anion may be approximated as the difference between the VDE of that species and the VDE of the sub-ion:

$$\sum_{m=0}^{n-1} D_0[O^-(Ar)_m \cdots Ar] \approx VDE[O^-(Ar)_n] - VDE[O^-] \qquad (3a)$$

while cluster ion-single solvent dissociation energies for O^-Ar_n may be approximated by the difference between the VDEs of two adjacent-sized clusters:

$$D_0[O^-(Ar)_{n-1} \cdots Ar] \approx VDE[O^-(Ar)_n] - VDE[O^-(Ar)_{n-1}] \qquad (3b)$$

This is equivalent to saying that the spectral shift from that of O^- is an approximation to the total anion solvent dissociation energy, while the spectral shift between adjacent-sized cluster ions is an approximation to that particular anion-single solvent dissociation energy.

The approximation to neglect the neutral weak bond dissociation energy can be justified by examining the energetics of $O^-(Ar)_1$. Because there is information available about the neutral dissociation energy of O(Ar),[15] the dissociation energy of $O^-(Ar)$ can be

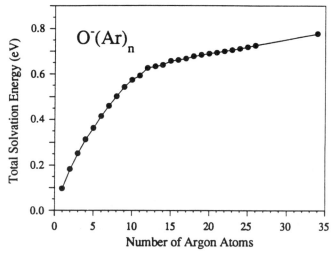

Figure 3. The total anion-solvent dissociation energy of $O^-Ar_{n=1-26}$ plotted as a function of cluster size. Preliminary data for n=34 is also included on the plot. The dramatic change in the slope at n=12 is consistent with the closing of the first solvation shell.

determined from eq(1). The vibrational frequency for the ground state of O(Ar) is ~20 cm⁻¹ and $D_e[O\cdots Ar]$ is 0.010 eV, so the neutral dissociation energy $D_0[O\cdots Ar]$ is 0.009 eV. In addition, the observed VDE of O^- is nearly equal to the EA of O, and this is expected for O^-Ar as well, so $EA[O(Ar)] \approx VDE[O^-(Ar)]$ and $EA[O] \approx VDE[O^-]$. This leads to an anion-solvent dissociation energy $D_0[O^-\cdots Ar]$ of 0.106 eV. Thus, the anion dissociation energy ($D_0[O^-\cdots Ar]$) is greater than that of the neutral ($D_0[O\cdots Ar]$) by an order of magnitude. The dissociation energy for $O^-(Ar)$, determined from eq(3) is 0.097 eV, which is slightly less than the value given from eq(1).

V. Interpretation

A. Solvation Shell Closing

The total anion solvent dissociation energy is the total amount of energy the oxygen sub-ion is stabilized by its association with a given number of argon atoms. This total solvation energy, shown as a function of cluster size for the O^-Ar_n system in Figure 3, increases smoothly up to n=12, where it changes slope. While the average stabilization energy per atom remains relatively large for n<12, it decreases significantly and remains relatively constant for n=13-26. The dramatic change in the slope of this plot at n=12 indicates a major change in the interaction of subsequent argon atoms with the cluster ion,

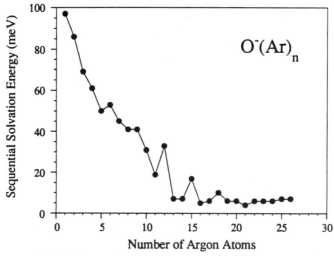

Figure 4. The sequential solvation energies of O^-Ar_n as a function of cluster size. Discontinuities in this plot at n=6,9,12,15, and 18 suggest structural information about these cluster anions. (An error bar of ± 5 meV is typical for the data.)

and it is consistent with the first solvation shell closing around the O^- sub-ion with 12 argon atoms. The decrease in the average stabilization energy for n>12 is a result of the additional argon atoms being shielded from the O^- sub-ion. As is further shown below, the first solvation shell closes at O^-Ar_{12}, as the O^- sub-ion is completely enclosed in a cage of argon solvents.

B. Cluster Structure

Additional structural implications for the O^-Ar_n system emerge upon examining cluster anion-single solvent dissociation energies as a function of cluster size. These dissociation energies are essentially stepwise (sequential) solvation energies because they quantify the effect each additional individual argon solvent has on the stability of the O^-Ar_n cluster. The interaction between O^- and the first few argon solvents is expected to be the strongest, while the stabilizing effect of each additional solvent should diminish as the cluster grows and the sub-ion's localized charge interacts with a greater number of solvent atoms. In the absence of structural effects, this should lead to a monotonically decreasing function of stepwise solvation energies with increasing cluster size.

The stepwise solvation energies for O^-Ar_n are shown in Figure 4 as a function of cluster size, and they follow the expected generally decreasing trend. However, discontinuities in this decreasing trend are observed at n=6,9,12,15, and 18, indicating that these particular clusters are being stabilized by some additional effect. The largest single discontinuity in the stepwise solvation energy trend occurs at n=12, and this is followed by an abrupt drop in sequential solvation energy at n=13. We interpret this as further evidence for the first solvation shell closing at n=12. The small sequential solvation energies for n=13 and n=14 are consistent with the sub-ion being shielded, resulting in weaker ion-neutral interactions.

The additional stability observed for the 12th argon atom is consistent with the cluster having an energetically favorable structure, which is likely to occur at a solvation shell closing. The mass spectrum presented earlier demonstrates that n=12 is the primary magic number observed, and while it may be misleading to suggest structural implications based on mass spectral magic numbers alone, it seems likely from the energetic data, as well as from the mass spectrum, that n=12 is forming an especially stable cluster anion. This suggests that the discontinuities in the sequential solvation energy plot at n=6,9,15, and 18 may also occur because of structural reasons. The mass spectrum at least partially supports this notion, with magic numbers occurring at n=10,12,15,18,22, and 25.

To better understand the structural implications of the sequential solvation energy plot (Figure 4), it is useful to examine the packing schemes that have been proposed in the past for the clustering of rare gas atoms. In particular, minimum energy structures for rare gas clusters of sizes n=6-60 were examined by Hoare and Pal using hard-sphere potentials, and three different packing sequences were found to compete: tetrahedral, pentagonal, and icosahedral.[16] For the size range of clusters we have examined here, ie. n<26, the pentagonal growth sequence consistently yielded the lowest energy structures. In this packing sequence, there are several particularly stable geometries which posses complete D_{5h} symmetry. These are the pentagonal bipyramid structure for the 7 atom cluster (which serves as the building block for larger clusters in this sequence), the icosahedral structure for the 13 atom cluster, and the capped (double) icosahedral structure for the 19 atom cluster. The tetrahedral sequence, on the other hand, predicts particularly stable structures for the 4, 8, 14, and 26 atom clusters, while the icosahedral sequence begins with the 13 atom icosahedron and predicts particularly stable structures for the 33 and 45 atom clusters (both of which are out of our size range). For larger clusters, n>26, these three sequences compete to give a variety of stable structures.

Sphere packing sequences such as these have been used to interpret the mass spectrum of Xe_m^+, which exhibits magic numbers at m=13,16,19,23,25,55,71,87, and 147.[17] A sequence of full shell icosahedra are expected for m=13,55, and 147. Adding one and two caps of 6 atoms each to m=13 yields m=19 and m=25, and adding one and two caps of 16 atoms each to m=55 yields m=71 and m=87, largely explaining the observed size distribution.

The "magic numbers" observed in the mass spectrum of O^-Ar_n are n=10,12,15,18,22, and 25, which correspond to clusters of 11,13,16,19,23, and 26 total atoms. This sequence of magic numbers is very similar to the observed pattern for Xe_m^+ clusters in the same size range, suggesting these two systems have similar structures. Moreover, the discontinuities observed in the stepwise solvation energy vs cluster size plot (Figure 4) for O^-Ar_n are n=6,9,12,15, and 18, which correspond to clusters with 7,10,13,16, and 19 total atoms. This observed sequence of 7,13, and 19 suggests that these cluster ions are following the pentagonal packing pattern described above. The intermediate magic numbers at 10 and 16 total atoms are probably due to the formation of partially capped structures (Adding a cap of 3 atoms to (n+1)=7 and (n+1)=13 yields (n+1)=10 and (n+1)=16, respectively).

For the case of O^-Ar_{12}, the evidence implies that this cluster has an icosahedral structure. Evidence for O^- being inside the cluster rather than on the surface comes from the dramatic drop in the stepwise solvation energy at n=13. If the ion were to reside on the surface of the cluster, the sequential solvation energy for n=12 and n=13 would be comparable. However, we observe a large drop in the sequential solvation energy at n=13, which is consistent with the 13th argon being shielded from the O^- core. Geometrical arguments also demonstrate that the O^- anion can fit into a cavity formed by 12 argon atoms given that the ionic radius of O^- is ~1.7 Å ,[18] the atomic radius of Ar is ~1.2 Å,[18] and the Ar-Ar bond is longer than the O-Ar bond.[15]

The evidence suggests that O^-Ar_n structures resemble those of pure rare gas clusters and that they follow the pentagonal packing sequence with O^-Ar_{12} as an icosahedron, O^-Ar_{15} as a partially capped icosahedron, and O^-Ar_{18} as a fully capped (double) icosahedron. The packing of atoms in accord with a packing sequence for spheres implies the domination of repulsive forces in these systems. However, since one of the components in the clusters in an ion, attractive forces must also play an important role. The competition between attractive and repulsive forces in O^-Ar_n, along with its structural implications, will be explored in a subsequent publication.

Acknowledgements. We gratefully acknowledge the support of the US National Science Foundation under Grant CHE-9007445.

References

1. Golub, S. and Steiner, B. (1968) "Photodetachment of [OH(H$_2$O)]$^-$", J. Chem. Phys. **49**, 5191-5193.

2. Coe, J.V., Snodgrass, J.T., Freidhoff, C.B., McHugh, K.M., and K.H. Bowen (1985) "Negative ion photoelectron spectroscopy of the negative cluster ion H$^-$(NH$_3$)", J. Chem. Phys. **83**, 3169-3170.

3. Coe, J.V., Snodgrass, J.T., Freidhoff, C.B., McHugh, K.M., and K.H. Bowen (1987) "Photoelectron spectroscopy of the negative cluster ions NO$^-$(N$_2$O)$_{n=1-2}$", J. Chem. Phys. **87**, 4302-4309.

4. Snodgrass, J.T., Coe, J.V., Freidhoff, C.B., McHugh, K.M., and K.H. Bowen (1988) "Photodetachment spectroscopy of cluster anions. Photoelectron spectroscopy of H$^-$(NH$_3$)$_1$, H$^-$(NH$_3$)$_2$, and the tetrahedral isomer of NH$_4^-$", Faraday Discuss. Chem. Soc. **86**, 241-256.

5. Bowen, K.H. and Eaton, J.G. (1988) "Photodetachment spectroscopy of negative cluster ions", in R. Naaman and Z. Vager (eds.), The Structure of Small Molecules and Ions, Plenum Publishing, pp. 147-169.

6. Eaton, J.G., Arnold, S.T., and Bowen, K.H. (1990) "The negative ion photoelectron (photodetachment) spectra of NO$^-$(H$_2$O)$_{n=1-2}$", Int. J. of Mass Spectrom. and Ion Proc. **102**, 303-312.

7. Miller, T.M., Leopold, D.G., Murray, K.K., and Lineberger, W.C. (1985) Bull. Am. Phys. Soc. **30**, 880.

8. Posey, L.A., Deluca, M.J., and Johnson, M.A. (1986) "Demonstration of a pulsed photoelectron spectrometer on mass-selected negative ions: O^-, O_2^-, and O_4^-", Chem. Phys. Lett. **131**, 170-174.

9. Posey, L.A., and Johnson, M.A. (1988) "Pulsed photoelectron spectroscopy of negative cluster ions: isolation of three distinguishable forms of $N_2O_2^-$", J. Chem. Phys. **88**, 5383-5395.

10. Markovich, G., Giniger, R., Levin, M., and Cheshnovsky, O. (1991) "Photoelectron spectroscopy of iodine anion solvated in water clusters", J. Chem. Phys. **95**, 9416-9419.

11. Arnold, D.W., Bradforth, S.E., Kim, E.H., and Neumark, D.M. (1992) "Anion photoelectron spectroscopy of iodine-carbon dioxide clusters", J. Chem. Phys. **97**, 9468-9471.

12. Alexander, M.L., Levinger, N.E., Johnson, M.A., Ray, D., Lineberger, W.C. (1988) "Recombination of Br_2^- photodissociated within mass selected ionic clusters", J. Chem. Phys. **88**, 6200-6210.

13. Papanikolas, J.M., Gord, J.R., Levinger, N.E., Ray, D., Vorsa, V., and Lineberger, W.C. (1991) "Photodissociation and geminate recombination dynamics of I_2^- in mass-selected $I_2^-(CO_2)$ cluster ions", J. Phys. Chem. **95**, 8028-8040.

14. Coe, J.V., Snodgrass, J.T., Freidhoff, C.B., McHugh, K.M., and Bowen, K.H. (1986) "Photoelectron spectroscopy of the negative ion SeO^-", J. Chem. Phys. **84**, 618-625.

15. Huber, K.P. and Herzberg, G. (1979) Constants of Diatomic Molecules, Van Nostrand Reinhold Company, New York, pp. 32-36.

16. Hoare, M.R. and Pal, P.(1971) "Physical cluster mechanics: statics and energy surfaces for monatomic systems", Adv. Phys. **20**, 161-196.

17. Recknagel, E (1984) "Production and properties of atomic and molecular microclusters", Ber. Bunsenges. Phys. Chem. **88**, 201-206.

18. Weast, R.C., Astle, M.J., and Bayer, W.H. (eds.) (1986) CRC Handbook of Chemistry and Physics, vol. 66, CRC Press, Boca Raton, Florida, p. E67, p. F164.

MAGIC NUMBERS AND GEMINATE RECOMBINATION DYNAMICS OF ANIONS IN WATER CLUSTERS

David J. Lavrich, Donna M. Cyr, Mark A. Buntine, Caroline E.Dessent
Lynmarie A. Posey, and Mark A. Johnson

Department of Chemistry
Yale University
225 Prospect St.
New Haven, CT 06511

1. Introduction

Several recent reports[1-3] involving the hydration of anions in water clusters have suggested that the solvation occurs in an asymmetric fashion where water molecules "wet" one side of the cluster, leaving the anion exposed at or near the surface of the cluster. This result appears to arise from the strength of the hydrogen bonded network compared to the ion-water interaction, raising the interesting question of how large the anionic clusters must be before the anion is indeed surrounded by the medium. At present, there are no good structural diagnostics capable of establishing the location of the solute anion in water clusters, and we therefore take this opportunity to report two properties of $A^- \cdot (H_2O)_n$ clusters which indirectly bear on the question of structure: the evolution of magic numbers in the parent ion distribution with the nature of the solute anion and the geminate recombination probability of a photodissociated anion within the water clusters. We typically find at least one magic number in the $A^- \cdot (H_2O)_n$ clusters in the vicinity of $n=15\text{-}20$. In order to explore whether these numbers are signalling encapsulated structures or "shell closings", we have taken a closer look at the $O_2^- \cdot (H_2O)_n$ system where we photodissociate the O_2^- superoxide anion within the cluster and monitor the geminate recombination probability to establish the ability of the water molecules to "stop" the ejected oxygen atom. We indeed find a break in the recombination propensity around $n=15$, somewhat above the magic number ($n=13$) in that system. This behavior is compared to that displayed by dissociation of dihalide anions in carbon dioxide clusters,[4-6] where the geminate recombination probability is strongly correlated to the location of magic numbers in the mass spectrum.

2. Results and Discussion

2.1 *Evaporative charge transfer reactions as a probe for magic numbers*

Several years ago, we demonstrated[7] a straightforward method of introducing anions into water clusters using the collisional charge transfer processes:

$$A + (H_2O)_n^- \rightarrow A^- \cdot (H_2O)_m + n\text{-}m \ H_2O \qquad [1]$$

47

J. Jortner et al. (eds.), Reaction Dynamics in Clusters and Condensed Phases, 47–55.
© 1994 *Kluwer Academic Publishers. Printed in the Netherlands.*

These reactions are the cluster analogue of electron scavenging reactions in aqueous radiation chemistry.[8] They are carried out after first creating $(H_2O)_n^-$ hydrated electron clusters by electron attachment onto neutral $(H_2O)_n$ clusters using electron impact ionization in the high density region near the throat of a supersonic expansion.[9] The collision partner (e.g. A=O_2) is then introduced into the source chamber outside the barrel shock boundary of the free jet, where it is entrained or "aspirated" into the expansion,[10,11] accelerated, and ultimately (hundreds of nozzle diameters downstream) interacts with the $(H_2O)_n^-$ clusters. The reaction exothermicity is then released as heat, eventually leading to the evaporation of water molecules.[7] The cascade down through the sequential evaporation events creates an evaporative ensemble[12] ideal for the observation of locally stable structures as intensity anomalies or magic numbers in the parent intensity distribution.[13] This synthetic methodology is quite general for any anionic species provided that it can be generated by low energy electron attachment onto a neutral molecule with significant vapor pressure ($> 10^{-5}$ Torr). Thus, the available synthetic reactions include dissociative attachment[14] reactions:

$$CH_3Br + (H_2O)_n^- \rightarrow Br^- \cdot (H_2O)_m + CH_3 + (n-m)H_2O \qquad [2]$$

for the preparation of closed shell anionic species (e.g. organic acids RCO_2^-, halide ions, etc.).

Raw data illustrating the formation of $CO_2^- \cdot (H_2O)_m$ by colliding CO_2 with $(H_2O)_n^-$ is shown in Fig. 1. The upper trace is the parent $(H_2O)_n^-$ cluster ion distribution before reaction while the lower trace is obtained after introducing 5×10^{-5} Torr of CO_2 into the source chamber. Note that almost all the $(H_2O)_n^-$ clusters are converted into $CO_2^- \cdot (H_2O)_n$ product ions, and the smooth $(H_2O)_n^-$ distribution evolves into a distribution with an anomalously low intensity of $CO_2^- \cdot (H_2O)_{17}$ relative to $n=16$. We therefore identify the $n=16$ peak as a species with special (thermodynamic) stability relative to $n=17$. In order to concentrate on the solute ion distributions, we report only the $A^- \cdot (H_2O)_n$ intensities in the stick spectra shown in Fig. 2. Interestingly, the magic number at $n=16$ (marked by *) in the upper trace increases to $n=17$ with the modification to the acetate anion, $CH_3CO_2^-$, indicating that the CH_3 does not simply replace one water molecule around the CO_2^- moiety. Another indication that the locations of the magic numbers is not simply related to size is indicated by the comparison between NO^- and O_2^-, which are discontinuous at $n=15$ and 13, respectively, despite their similar size. The SO_2^- anion presents the most spectacular case in which the large discontinuity at $n=15$ is preceded by an even/odd intensity alternation, while no such behavior was observed for the CO_2^- triatomic anion. On the other hand, the simplest anion, Br^-, displays only one discontinuity from $n=10$ to 40 at $n=21$, the largest magic number we have found in these species. All the other species exhibit magic numbers in the range from $n=13$ to 17.

It is important to acknowledge that the clusters in the evaporative ensemble are warm (probably about 150K) and undoubtedly contain several bond energies of internal excitation in excess of the energy required to eject one water monomer.[15] Furthermore, the magic numbers are sharp in the sense that often only one size is enhanced, usually at the expense of the next larger cluster. Note further that the reduced intensity of the $n+1$ cluster clearly indicates that the enhanced intensity at n is not due to accidental mass overlap of an impurity, but is a genuine property of the $A^- \cdot (H_2O)_n$ distribution.

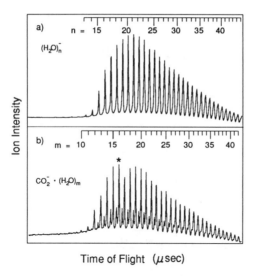

Figure 1

Castleman and co-workers[16,17] have also reported magic numbers in the hydrated O^- and OH^- anions (under thermal conditions) at $n=11$, 14, 17 and 20, in a similar range as found for the anions in Fig. 2. They suggest that the specific pattern of numbers in the O^- and OH^- system might be due to closed structures consisting of five and six membered rings around the anion. Interestingly, they do not find magic numbers for the O_2^- or O_3^- species under their conditions. It is possible that the intensity discontinuities we are discussing here (Fig. 2) are sufficiently subtle that they escaped notice relative to the stronger anomalies in the O^- and OH^- systems. In other related work, Cheshnovsky and co-workers[18] have used negative ion photoelectron spectroscopy (PES) to probe for especially stable $I^- \cdot (H_2O)_n$ clusters, and find a discontinuity in the vertical detachment energy at $n=6$, which they interpret to indicate an enclosing structure or solvation shell for the system. On the other hand, more recent calculations[2] indicate that the PES data are more consistent with asymmetric structures for the clusters. The propensity of these rather large clusters to exhibit magic numbers is therefore intriguing since they are not, at least superficially, easy to rationalize in the context of the "surface" anionic structures since the asymmetric structures do not appear ordered in any obvious way.

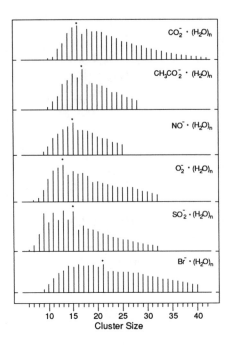

Figure 2

2.2   *Photodissociation and geminate recombination of O_2^- within water clusters*

Since the existence of magic numbers is often interpreted as indicative of stable geometrical structures,[19] we initiated a study of the photodissociation dynamics of the anions within the water clusters to establish the integrity or robustness of the cage they form around the anion. Lineberger and co-workers[4-6] have also recently reported magic numbers in the range of n=14-16 for the $X_2^- \cdot (CO_2)_n$ (X=Br, I) anionic clusters, and further demonstrated that 100% geminate recombination of the photodissociated X_2^- is achieved promptly at sizes corresponding to magic numbers. Simulations[20] of these experiments indicate that the CO_2 molecules completely surround the anionic chromophore at these sizes. Since magic numbers also appear in the $A^- \cdot (H_2O)_n$ systems in the same size range (n=13-21), we were lead to explore the structure of these clusters by monitoring the size dependence of the geminate recombination of photodissociated solute anions. Of the systems displayed in Fig. 2, only O_2^- possesses a simple dissociative excited state suitable for use in the photofragmentation experiments.

We have recently discovered[21] the region of the $A^2\Pi_u \leftarrow X^2\Pi_g$ transition in bare O_2^- corresponding to excitation of the repulsive portion of the bound $A^2\Pi_u$ state. Since this state lies about 3.5 eV above the electron binding energy of the bare O_2^- anion, it is autodetached in the gas phase species with only about 3% of the quantum yield resulting in photofragmentation. This minor channel is nonetheless sufficiently large to monitor the dissociation dynamics of the $O_2^- \cdot (H_2O)_n$ clusters. Interestingly, the matrix environment provides so much stabilization of the O_2^- anion that the $A^2\Pi_u$ becomes stable with respect to autodetachment and can be observed in fluorescence.[22] The disposition of the $A^2\Pi_u$ potential curve over $X^2\Pi_g$ is such that the photofragment recoil energy can be controlled from 0 to about 1 eV in the Franck Condon region of the absorption. In this brief report, we present the results for excitation at 4.66 eV, 0.57 eV above the dissociation energy (D_o=4.09 eV) of O_2^- into $O^-(^2P) + O(^3P)$. This excess energy is similar to that contained in the $I^- + I$ atoms (~0.5 eV) in the $I_2^- \cdot (CO_2)_n$ experiments of Lineberger.[6]

Typical photofragment distributions for parents n=6, 9 and 24 are shown in Fig.3.

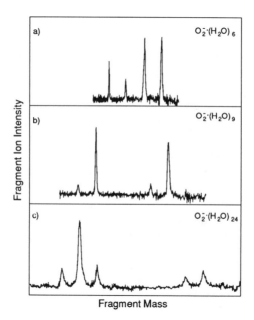

Figure 3

The complicated distribution of fragments from n=6 splits into two classes of fragments with increasing cluster size corresponding to the processes:

$$O_2^- \cdot (H_2O)_n + h\nu \quad \rightarrow \quad O_2^- \cdot (H_2O)_{n-10} + 10\ H_2O \qquad \text{[3a]}$$

$$\rightarrow \quad O^- \cdot (H_2O)_{n-2} + 2\ H_2O \qquad \text{[3b]}$$

While these experiments do not have unit mass resolution for the fragments in this size range, the evaporation of 10 water molecules in Eq. 3a indicates that the entire photon energy is accounted for by evaporation. This in turn requires the O_2^- to be reformed in the cluster. The appearance of two-photon dissociation fragments at high fluence confirms that the O_2^- chromophore is indeed intact in the lighter fragments. It is worthwhile to emphasize that the fragments only appear upon excitation of the O_2^- dissociation band. Excitation above the detachment threshold but below the onset of the $A^2\Pi_u \leftarrow X\ ^2\Pi_g$ band at 3.5 eV results only in detachment of the electron. This indicates that the $O_2^- \cdot (H_2O)_n$ fragment ions are *not* formed by charge transfer to solvent processes, for example, in which the electron is photodetached and recombined with the O_2 rather than by cleavage and reformation of the O-O bond. The heavier fragments have only lost two or three water molecules, indicating that most of the photon energy is stored as chemical potential energy. This requires that the O_2^- molecule has not reformed in the cluster, either due to ejection of the oxygen atom or chemical trapping of the system in a high-energy configuration (e.g. separated O_{aq} and O_{aq}^- moieties). While the mass spectra are again not sufficient to resolve the two mass units required to distinguish between water ejection and O atom ejection, in either case we can regard the heavy fragment as associated with the un-recombined O_2^- chromophore.

To quantify the fraction of photofragments in which the O_2^- core ion is regenerated after photofragmentation, we plot the relative yield of O_2^- containing fragments (compared to the sum of all fragments, not including electrons) versus cluster size in Fig. 4, along with the results from Lineberger and co-workers for the $I_2^- \cdot (CO_2)_n$ system.[6] At first glance, the similarity of the two curves ($I_2^- \cdot (CO_2)_n$, open squares, $O_2^- \cdot (H_2O)_n$, filled squares) is remarkable, with both curves rising approximately linearly from zero to an asymptotic value at about $n=16$. Note that $n=16$ is a strong magic number in $I_2^- \cdot (CO_2)_n$ but above the weaker magic number at $n=13$ in the $O_2^- \cdot (H_2O)_n$ case. Thus, there does not appear to be as strong a correlation between caging and magic in the superoxide system.

The similarity of the curves in Fig. 4 is surprising since these systems have such different mass (solute/solvent) ratios and solvent-solvent *vs* solute-solvent binding properties. In fact, the only parameter which is the same in both studies is the recoil energy (~0.5 eV in each case). Obviously, both studies should be extended to other wavelengths to see if the agreement is in some way fortuitous. In any event, the basic structure of the caging propensity is similar in both cases, and since this behavior has been associated with shell closing in the $I_2^- \cdot (CO_2)_n$ system, it is logical to suggest that a cage is responsible for the recombination above $n=16$ in the $O_2^- \cdot (H_2O)_n$ case as well. *Such a suggestion is, however, at odds with the current view of halide ion hydration[1-3] where the anion is thought to be near the surface.*

There is, however, a significant discrepancy in the high n asymptotes of the two data sets (see Fig. 4), which may provide an indication that something is fundamentally different in the two systems. Unlike the $I_2^- \cdot (CO_2)_n$ case,[6] the asymptotic "caging" fraction for the $O_2^- \cdot (H_2O)_n$ system does not approach unity, but rather remains constant at a value of 0.9 all the way out to $n=40$. This behavior is curious, indicating that the remaining water molecules are not incrementally effective in forcing recombination. It is tempting to associate this "uncaged" fraction with clusters whose anion resides near the surface where capture of the ejected O atom is inefficient. Apparently, such structures are not evident in the $I_2^- \cdot (CO_2)_n$ case where 100% caging is achieved after

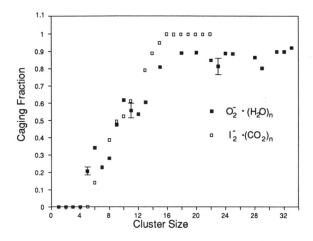

Figure 4

the magic number at $n=16$. It is also possible, however that this residual uncaged fraction of the photoproducts might result from the reaction of the nascent O^- and O radicals with the H_2O solvent molecules, providing a chemical trap for the intermediates. Indeed, the smaller clusters $O_2^- \cdot (H_2O)$ and $O_2^- \cdot (H_2O)_2$ are observed[23] to fragment into O_2H^- and OH^- species with about 5% of the fragmentation quantum yield. These photoproducts must arise from reaction with water molecules such as:

$$O^- + H_2O \quad \rightarrow \quad OH^- + OH \qquad [4]$$

$$(O_2^-)^* + H_2O \quad \rightarrow \quad O_2H^- + OH \qquad [5a]$$

$$\rightarrow \quad O_2H + OH^- \qquad [5b]$$

If these reactions are trapping the nascent photoproducts within the larger clusters, then these systems must be metastable relative to the back reaction. Since these back reactions are not likely to be activated in solution, it is likely that the "uncaged" fraction in Fig. 4 indeed corresponds to loss of either oxygen atoms, the hydroxyl radical or hydroperoxo radical so that the final anionic clusters are in fact stable.

3. Conclusions

Summarizing, magic numbers in the range $n=13-20$ commonly occur in the hydrated anion clusters $A^- \cdot (H_2O)_n$, with the clusters usually displaying only one major intensity discontinuity. Since the intensity discontinuities in the analogous $I_2^- \cdot (CO_2)_n$ clusters in this size range are thought to correspond to shell closings based upon geminate recombination data, we investigated the photofragmentation of the $O_2^- \cdot (H_2O)_n$ system to probe for similar structures. The size dependence of the recombination probability is similar to that found[6] for $I_2^- \cdot (CO_2)_n$ when excited with similar excess energy, indicating that the water molecules similarly surround the O_2^- core ion. There is a residual 10% fraction of the fragments in the $O_2^- \cdot (H_2O)_n$ case, however, in which the oxygen atoms do not recombine, unlike the situation in the $I_2^- \cdot (CO_2)_n$ system where the recombination probability is unity. This fraction most probably corresponds to escape of the neutral oxygen atoms, indicating that some of the O_2^- species may be near the surface. It is also possible that a small fraction of the photoproducts is chemically trapped after reaction with the water solvent molecules. This raises the possibility of studying the stepwise effects of solvent molecules on photoinduced ion-molecule reactions in the cluster medium.

References

1. Perera, L., and Berkowitz, M. L., (1991) "Many-Body Effects in Molecular Dynamics Simulations of $Na^+(H_2O)_n$ and $Cl^-(H_2O)_n$ Clusters", J. Chem. Phys., **95**, 1954.

2. Dang, L. X., and Garrett, B. C., (1993) "Photoelectron spectroscopy of hydrated iodine anion clusters from molecular dynamic simulations", accepted for publication, J. Chem. Phys.

3. Severence, D. and Jorgensen, W. L., (1993) "Limited effects of polarization for $Cl^-(H_2O)_n$ and $Na^+(H_2O)_n$ clusters", accepted for publication, J. Chem. Phys.

4. Alexander, M. L., Levinger, N. E., Johnson, M. A., Ray, D., and Lineberger, W. C., (1988) "Recombination of Br_2^- Photodissociated within Mass Selected Ionic Clusters", J. Chem. Phys., **88**, 6200.

5. Papanikolas, J. M., Gord, J. R., Levinger, N. E., Ray, D., Vorsa, V., and Lineberger, W. C., (1991) "Photodissociation and Geminate Recombination Dynamics of I_2^- in Mass-Selected $I_2^-(CO_2)_n$ Cluster Ions", J. Phys. Chem., **95**, 8028.

6. Papanikolas, J. M., Vorsa, V., Nadal, M. E., Campagnola, P. J., Gord, J. R., and Lineberger, W. C., (1991) "I_2^- Photofragmentation/ Recombination Dynamics in Size Selected $I_2^-(CO_2)_n$ Cluster Ions: Observation of Coherent $I \cdots I^-$ Vibrational Motion", J. Chem. Phys., **97**, 7002.

7. Posey, L. A., DeLuca, M. J., Campagnola, P. J., and Johnson, M. A., (1989) "Reactions of hydrated elctron clusters $(H_2O)_n^-$: Scavenging the excess electron", J. Phys. Chem., **93**, 1178.

8. Buxton, G. V., (1987) "Radiation Chemistry of the Liquid State: (1) Water and Homogeneous Aqueous Solutions", in Farhataziz and Micheal A. J. Rodgers (eds.), Radiation Chemistry, VCH Publishers Inc., New York, pp. 321-349.

9. Johnson, M. A., and Lineberger, W. C. (1988) "Pulsed Methods for Cluster Ion Spectroscopy", in J. M. Farrar and W. Saunders Jr. (eds.), Techniques of Chemistry, Wiley, New York, Vol. 20, p.591.

10.   Fenn, J. B., and Anderson, J. B. (1966) "Background and Sampling Effects in Free jet Studies by Molecular Beam Measurements", in J. H. de Leeuw (ed.), Rarefied Gas Dynamics, Academic Press, New York, Vol. 2, p. 311.

11.   Campargue, R., (1970) "Aerodynamic Separation Effect on Gas and Isotope Mixtures Induced by Invasion of the Free Jet Shock Wave Structure", J. Chem. Phys, **52**, 1795.

12.   Klots, C. E., (1985) "Evaporative Cooling", J. Chem. Phys., **83**, 5854.

13.   Klots, C. E., (1991) "Kinetic Methods for Quantifying Magic", Z. Phys. D. At. Mol. Clus., **21**, 3135.

14.   Massey, H., (1976) Negative Ions, Cambridge University Press, London.

15.   Campagnola, P. J., Posey, L. A., and Johnson, M. A., (1991) "Controlling the internal energy content of size-selected cluster ions: An experimental comparison of the metastable decay rate and photofragmentation methods of quantifying the internal excitation of $(H_2O)_n^-$", J. Chem. Phy.s, **95**, 7998.

16.   Yang, X., and Castleman, A. W. Jr., (1991) "Chemistry of Large Hydrated Anion Clusters $X^-(H_2O)_n$ $0 \leq n \approx 50$ and X = OH, O, O_2, and O_3. 1. Reaction of CO_2 and Possible Application in Understanding of Enzymatic Reaction Dynamics", J. Am. Chem. Soc., **113**, 6766.

17.   Yang, X., and Castleman, A. W. Jr., (1990) "Production and Magic Numbers of Large Hydrated Anion Clusters $X^-(H_2O)_{n=0-59}$ (X = OH, O, O_2, O_3) under Thermal Conditions", J. Phys. Chem, **94**, 8500.

18.   Markovich, G., Giniger, R., Levin, M., and Cheshnovsky, O., (1991) "Photoelectron spectroscopy of iodine anion solvated in water clusters", J. Chem. Phys., **95**, 9416.

19.   Harris, I. A., Kidwell, R. S., and Northby, J. A., (1984) "Structure of Charged Argon Clusters Formed in a Free Jet Expansion", Phys. Rev. Lett., **53**, 2390.

20.   Perera, L., and Amar, F. G., (1989) "Charge localization in negative ion dynamics: Effect on caging of Br_2^- in Ar_n and $(CO_2)_n$ clusters", J. Chem. Phys., **90**, 7354.

21.   Lavrich, D. J., Buntine, M. A., Serxner, D., and Johnson, M. A., (1993) "Observation of the A $^2\Pi_u \leftarrow X^2\Pi_g$ dissociative transition in isolated O_2^- using mass-selected photofragmentation spectroscopy", accepted for publication, J.Chem. Phys.

22.   Rolfe, J., (1979) "First excited state of the O_2^- ion", J. Chem. Phys., **70**, 2463.

23.   Lavrich, D. J., Buntine, M. A., and Johnson, M. A., (1993) "Photoinitiated reactions in the $O_2^- \cdot (H_2O)$ ion-dipole complex", manuscript in preparation.

THEORETICAL SPECTROSCOPY AND DYNAMICS OF TETRA-ATOMIC VAN DER WAALS CLUSTERS[1]

G. Delgado-Barrio, A. García-Vela, J. Rubayo-Soneira[2],
J. Campos-Martínez, S. Miret-Artés, O. Roncero and P. Villarreal
Instituto de Matemáticas y Física Fundamental, C.S.I.C.
Serrano 123, 28006 Madrid, SPAIN

Abstract

The dynamics of vibrational predissociation (VP) of the van der Waals (vdW) $Ne \cdots I_2 \cdots Ne$ cluster is studied by means of a Quasiclassical Trajectory approach. The initial conditions are selected according to quantum mechanical distributions associated to the ground state of the cluster for different vibrational excitations of I_2 . Predissociation lifetimes and final rotational distributions of the diatomic molecule are presented, and the dynamics of the process is discussed.

I   Introduction

A great deal of research effort has been addressed to the problem of understanding the structure and dynamics of van der Waals complexes [1-4]. The fragmentation of these clusters offers a unique opportunity for studying the half-collision dynamics in well-defined systems[5]. Several studies have been performed on vdW molecules composed of only rare-gas atoms bound by VdW forces using several teoretical approaches such as Monte Carlo and molecular-dynamics simulation[6-9]. Clusters of rare-gas atoms weakly bound to molecules with one or more chemical bonds have also been treated from both experimental[10-12] and theoretical[13-15] points of view.

In particular, considerable attention has been focused on triatomic vdW clusters[16-18] composed of a rare-gas atom and a conventional diatomic molecule. The reliability of various theoretical models has been tested on this type of complexes, providing lifetimes and linewidhts, spectral shifts, as well as final state

---

[1] Work supported by CICYT Grant No. PB87 - 0272, CAM Grant No. 064/92 and EEC Grant No. SC1.145.C

[2] Permanent address: I.S.P.de la Habana, Marianao, Ciudad Habana, CUBA

J. Jortner et al. (eds.), Reaction Dynamics in Clusters and Condensed Phases, 57–72.
© 1994 *Kluwer Academic Publishers. Printed in the Netherlands.*

vibrational and rotational distributions of the diatomic fragment[19-21].

Recently, some complexes with two weak bonds composed by I_2 and two rare gas atoms have been studied from the theoretical point of view[22-26] as well as experimentally[27-31]. The main goal of this work is to compare the results provided by theoretical methods with these very recent experimental data.

In this paper, for a total angular momentum $J = 0$, we solve the Schrödinger equation by means of a variational method. Afterthat the corresponding eigenfunctions are used to generate initial conditions in order to carry out classical trajectories. Then, we study the VP dynamics of $Ne \cdots I_2 \cdots Ne$, for different vibrational exitations of I_2 , within a Quasiclassical Trajectory formalism. It is worth to underline that this study, for $J = 0$, is performed without any restriction on the different motions.

The paper is organized as follows. Sec II presents the X..BC..Y Hamiltonian in the classical and the quantal frameworks. The quantum mechanical details, to obtain the tetra-atomic eingenfunctions, are shown in Sec III. In Sec IV the quasiclassical model used to select the initial conditions for the classical trajectories is discussed. Finally in Sec V the results and some concluding remarks are given.

II   The X···BC···Y Hamiltonian

We consider a four-particle system composed by two rare gas atoms, X and Y, weakly bound to an ordinary BC diatomic molecule. It seems quite natural to use, for this system, Jacobi coordinates for the diatomic partner and bond coordinates, which go from the BC center of mass to the respective X and Y atoms, for the latter ones.

II.1   Classical Hamiltonian Function

After separation of the center of mass motion of the whole system the Hamiltonian function for the X···BC···Y complex may be written[32]

$$H = \frac{\mathbf{p}_r^2}{2\mu_{BC}} + \frac{\mathbf{p}_1^2}{2\mu_{X,BC}} + \frac{\mathbf{p}_2^2}{2\mu_{Y,BC}} + \frac{\mathbf{p}_1 \cdot \mathbf{p}_2}{m_B + m_C} + V(\mathbf{r}, \mathbf{R}_1, \mathbf{R}_2) \qquad (1)$$

where $\mathbf{r}$ is the vector associated to the BC bond with conjugate momentum $\mathbf{p}_r$, meanwhile $\mathbf{R}_1$ and $\mathbf{R}_2$ are vectors going from the BC center of mass to the X and Y atoms, respectively, with conjugate momenta $\mathbf{p}_1$ and $\mathbf{p}_2$. The reduced masses appearing in Eq.(1) are $\mu_{BC} = (m_B m_C)/(m_B + m_C)$, $\mu_{X,BC} = m_X(m_B + m_C)/(m_X + m_B + m_C)$ and $\mu_{Y,BC} = m_Y(m_B + m_C)/(m_Y + m_B + m_C)$.

Regarding to the potential energy, we describe it like an addition of the diatomic interaction, $V_{BC}(r)$, plus two triatomic interactions, $V_{X,BC}(\mathbf{r}, \mathbf{R}_1)$ and

$V_{Y,BC}(\mathbf{r}, \mathbf{R}_2)$, plus the X-Y potential, $V_{X,Y}(\mathbf{R}_1, \mathbf{R}_2)$, that is

$$V(\mathbf{r}, \mathbf{R}_1, \mathbf{R}_2) = V_{BC}(r) + V_{X,BC}(\mathbf{r}, \mathbf{R}_1) + V_{Y,BC}(\mathbf{r}, \mathbf{R}_2) + V_{X,Y}(\mathbf{R}_1, \mathbf{R}_2) \tag{2}$$

In particular, for a total angular momentum $\mathbf{J} = 0$, choosing a body-fixed frame in which the Z-axis always points in the $\hat{\mathbf{r}}$ direction, and using polar coordinates $\mathbf{R} = (R, \theta, \varphi)$, we get the following Hamiltonian function[32],

$$
\begin{aligned}
H^{(J=0)} = &\frac{1}{2\mu_{BC}}\left[p_r^2 + \frac{\mathbf{j}^2}{r^2}\right] + \frac{1}{2\mu_{X,BC}}\left[p_{R_1}^2 + \frac{\mathbf{l_1}^2}{R_1^2}\right] + \frac{1}{2\mu_{Y,BC}}\left[p_{R_2}^2 + \frac{\mathbf{l_2}^2}{R_2^2}\right] + \\
&\frac{1}{m_B + m_C}\left[p_{R_1}p_{R_2}\cos\gamma - p_\varphi^2\frac{\cos\varphi}{R_1 R_2 \sin\theta_1 \sin\theta_2} + \right. \\
&p_{\theta_1}p_{\theta_2}\frac{\cos\theta_1 \cos\theta_2 \cos\varphi + \sin\theta_1 \sin\theta_2}{R_1 R_2} + \\
&p_{R_1}p_{\theta_2}\frac{\sin\theta_1 \cos\theta_2 \cos\varphi - \cos\theta_1 \sin\theta_2}{R_2} + \\
&p_{R_2}p_{\theta_1}\frac{\sin\theta_2 \cos\theta_1 \cos\varphi - \cos\theta_2 \sin\theta_1}{R_1} - \\
&- p_{R_1}p_\varphi\frac{\sin\theta_1 \sin\varphi}{R_2 \sin\theta_2} - p_{R_2}p_\varphi\frac{\sin\theta_2 \sin\varphi}{R_1 \sin\theta_1} - \\
&\left. - p_{\theta_1}p_\varphi\frac{\cos\theta_1 \sin\varphi}{R_1 R_2 \sin\theta_2} - p_{\theta_2}p_\varphi\frac{\cos\theta_2 \sin\varphi}{R_1 R_2 \sin\theta_1}\right] + \\
&V_{BC}(r) + V_{X,BC}(r, R_1, \cos\theta_1) + V_{Y,BC}(r, R_2, \cos\theta_2) + \\
&V_{X,Y}(R_1, R_2, \cos\gamma)
\end{aligned}
\tag{3}
$$

where we have included the angular momentum functions

$$
\begin{aligned}
\mathbf{l_1}^2 &= p_{\theta_1}^2 + \frac{p_\varphi^2}{\sin^2\theta_1} \\
\mathbf{l_2}^2 &= p_{\theta_2}^2 + \frac{p_\varphi^2}{\sin^2\theta_2} \\
\mathbf{j}^2 &= \mathbf{l_1}^2 + \mathbf{l_2}^2 + \mathbf{l_1}\cdot\mathbf{l_2} = \\
&= p_{\theta_1}^2 + p_{\theta_2}^2 + 2\cos\varphi\, p_{\theta_1}p_{\theta_2} - 2\sin\varphi(\frac{\cos\theta_2}{\sin\theta_2}p_{\theta_1} + \frac{\cos\theta_1}{\sin\theta_1}p_{\theta_2})p_\varphi + \\
&\quad (\frac{1}{\sin^2\theta_1} + \frac{1}{\sin^2\theta_2} - 2\cos\varphi\frac{\cos\theta_1 \cos\theta_2}{\sin\theta_1 \sin\theta_2} - 2)p_\varphi^2
\end{aligned}
\tag{4}
$$

In Eq. (4), $\varphi = \varphi_1 - \varphi_2$, meanwhile γ is the angle formed by the $\mathbf{R}_1$ and $\mathbf{R}_2$ vectors, so that

$$\cos\gamma = \sin\theta_1 \sin\theta_2 \cos\varphi + \cos\theta_1 \cos\theta_2$$

II.2 Quantal Hamiltonian

Resorting to the correspondence principle $\mathbf{p} \rightarrow -i\hbar\nabla$, we get from Eq.(1) the associated quantal Hamiltonian

$$H = -\frac{\hbar^2}{2\mu_{BC}}\nabla_r^2 - \frac{\hbar^2}{2\mu_{X,BC}}\nabla_1^2 - \frac{\hbar^2}{2\mu_{Y,BC}}\nabla_2^2 - \frac{\hbar^2}{m_B + m_C}\nabla_1 \cdot \nabla_2 + V(\mathbf{r}, \mathbf{R}_1, \mathbf{R}_2)$$

where the kinetic part is nothing but the kinetic energy operator for a three-particle system [33], in which the heavy particle has been substituted by a diatomic molecule, plus the relevant internal diatomic kinetic energy operator. Expressing now the ∇^2 operators in spherical coordinates , we finally get

$$\begin{aligned}
H = &-\frac{\hbar^2}{2\mu_{BC}}\left(\frac{\partial^2}{\partial r^2} + \frac{2}{r}\frac{\partial}{\partial r}\right) + \frac{\mathbf{j}^2}{2\mu_{BC}r^2} - \frac{\hbar^2}{2\mu_{X,BC}}\left(\frac{\partial^2}{\partial R_1^2} + \frac{2}{R_1}\frac{\partial}{\partial R_1}\right) + \frac{\mathbf{l}_1^2}{2\mu_{X,BC}R_1^2} \\
&-\frac{\hbar^2}{2\mu_{Y,BC}}\left(\frac{\partial^2}{\partial R_2^2} + \frac{2}{R_2}\frac{\partial}{\partial R_2}\right) + \frac{\mathbf{l}_2^2}{2\mu_{Y,BC}R_2^2} - \frac{\hbar^2}{m_B + m_C}\nabla_1 \cdot \nabla_2 \\
&+ V(\mathbf{r}, \mathbf{R}_1, \mathbf{R}_2)
\end{aligned} \qquad (5)$$

III   Variational procedure

As it was already mentioned, we consider a body-fixed (BF) system of reference in which the Z-axis always points in the $\hat{\mathbf{r}}$ direction. In this way, our basis functions are composed of products of radial by angular functions,

$$\Phi_{\ell_1\ell_2 L\Omega vmn}^{JM}(\mathbf{r}, \mathbf{R}_1, \mathbf{R}_2) = \phi_{vmn}(r, R_1, R_2)\mathcal{W}_{\ell_1\ell_2 L\Omega}^{JM}(\hat{\mathbf{r}}, \hat{\mathbf{R}}_1, \hat{\mathbf{R}}_2) \qquad (6)$$

The radial function ϕ is in turn

$$\phi_{vmn}(r, R_1, R_2) = \frac{\chi_v(r)\xi_m(R_1)\zeta_n(R_2)}{rR_1R_2}$$

where $\chi_v(r)$ is an eigenfunction of the non-rotating isolated diatomic molecule,

$$\left[-\frac{\hbar^2}{2\mu_{BC}}\frac{\partial^2}{\partial r^2} + V_{BC}(r)\right]\chi_v(r) = E_{BC}(v)\chi_v(r)$$

Likewise, the ξ, ζ functions are also vibrating basis functions for the R_1, R_2 coordinates.

On the other hand, the angular function $\mathcal{W}_{\ell_1\ell_2 L\Omega}^{JM}(\hat{\mathbf{r}},\hat{\mathbf{R}}_1,\hat{\mathbf{R}}_2)$, depending on the diatomic orientation $\hat{\mathbf{r}} \equiv (\theta_r, \varphi_r)$ with respect to a space-fixed (SF) reference system, and the orientations $\hat{\mathbf{R}}_i \equiv (\theta_i, \varphi_i)$, $i = 1, 2$ in the already mentioned BF reference system, is expressed as:

$$\mathcal{W}_{\ell_1\ell_2 L\Omega}^{JM}(\hat{\mathbf{r}},\hat{\mathbf{R}}_1,\hat{\mathbf{R}}_2) = \sqrt{\frac{2J+1}{4\pi}} D_{M\Omega}^{J^*}(\varphi_r, \theta_r, 0) \mathcal{Y}_{\ell_1\ell_2}^{L\Omega}(\hat{\mathbf{R}}_1,\hat{\mathbf{R}}_2)$$

It consists of an element of the Wigner rotation matrix, relating the SF and BF frames, and corresponds to a total angular momentum J with projections M and Ω on Z_{SF} and Z_{BF}, respectively, multiplied by an angular function in the coupled representation

$$\mathcal{Y}_{\ell_1\ell_2}^{L\Omega}(\hat{\mathbf{R}}_1,\hat{\mathbf{R}}_2) = (-1)^{L+\Omega}\sqrt{2L+1} \sum_{\omega} \begin{pmatrix} \ell_1 & \ell_2 & L \\ -\omega & \omega-\Omega & \Omega \end{pmatrix} Y_{\ell_1\omega}(\theta_1, \varphi_1)\, Y_{\ell_2\Omega-\omega}(\theta_2, \varphi_2)$$

where $\begin{pmatrix} \cdots \\ \cdots \end{pmatrix}$ denotes $3-j$ symbols and $Y_{\ell_i\omega}(\theta_i, \varphi_i)$ are spherical harmonics.

Thus, taking into account the conservation of the total angular momentum J and its $J_{Z_{SF}}$ component, the total wavefunction corresponding to a (J, M, k) level can be written as a linear combination

$$\Psi_k^{(JM)} = \sum_{\ell_1\ell_2 L\Omega vmn} A_k^{\ell_1\ell_2 L\Omega vmn} \Phi_{\ell_1\ell_2 L\Omega vmn}^{JM} \tag{7}$$

and the Schrödinger equation

$$H\Psi_k^{(JM)} = E_k \Psi_k^{(JM)}$$

is being solved by simple diagonalization of H represented on that basis.

III.1 Matrix elements

In the following, we shall implicitly use the identity

$$\left[\frac{\partial^2}{\partial x^2} + \frac{2}{x}\frac{\partial}{\partial x}\right]\left(\frac{f}{x}\right) = \frac{1}{x}\frac{\partial^2 f}{\partial x^2} \quad ; \quad x = r, R_1, R_2$$

for a generic function $f \equiv f(r, R_1, R_2)$.

Matrix elements of the $\mathbf{j}^2$ operator are readily obtained taking into account the relation

$$\mathbf{J} = \mathbf{j} + \mathbf{l}_1 + \mathbf{l}_2 = \mathbf{j} + \mathbf{L}$$

and introducing raising and lowering operators

$$\mathbf{J}_\pm = \mathbf{J}_x \pm i\mathbf{J}_y \quad ; \quad \mathbf{L}_\pm = \mathbf{L}_x \pm i\mathbf{L}_y$$

in such a way that

$$\mathbf{j}^2 = \mathbf{J}^2 + \mathbf{L}^2 - 2\mathbf{J}_z\mathbf{L}_z - (\mathbf{J}_+\mathbf{L}_- + \mathbf{J}_-\mathbf{L}_+)$$

As in the case of a triatomic system, $\mathbf{j}^2$ results diagonal in this representation with one exception; it also couples *tumbling* quantum numbers Ω diferring in one unit,

$$\langle \mathcal{W}^{JM}_{\ell_1\ell_2 L\Omega} |\mathbf{j}^2| \mathcal{W}^{JM}_{\ell_1'\ell_2' L'\Omega'} \rangle =$$

$$\hbar^2 \delta_{\ell_1\ell_1'} \delta_{\ell_2\ell_2'} \delta_{LL'} \{ \delta_{\Omega\Omega'} \left[J(J+1) + L(L+1) - 2\Omega^2 \right]$$
$$- \delta_{\Omega\Omega'\pm 1} \left[J(J+1) - \Omega\Omega' \right]^{1/2} \left[L(L+1) - \Omega\Omega' \right]^{1/2} \}$$

In this way, for the diatomic rotational energy we get

$$\langle \Phi^{JM}_{\ell_1\ell_2 L\Omega vmn} | \frac{\mathbf{j}^2}{2\mu_{BC}r^2} | \Phi^{JM}_{\ell_1'\ell_2' L'\Omega' v'm'n'} \rangle =$$

$$\frac{\delta_{mm'}\delta_{nn'}}{2\mu_{BC}} \langle \mathcal{W}^{JM}_{\ell_1\ell_2 L\Omega} |\mathbf{j}^2| \mathcal{W}^{JM}_{\ell_1'\ell_2' L'\Omega'} \rangle \int_0^{+\infty} dr \; \chi_v(r) r^{-2} \chi_{v'}(r)$$

On the other hand, the $\mathbf{l}_i^2$, $i = 1, 2$ operators are fully diagonal, and the matrix elements of the corresponding *end-over-end* angular energies are

$$\langle \Phi^{JM}_{\ell_1\ell_2 L\Omega vmn} | \frac{\mathbf{l}_1^2}{2\mu_{X,BC}R_1^2} | \Phi^{JM}_{\ell_1\ell_2 L\Omega v'm'n'} \rangle =$$

$$\frac{\hbar^2 \ell_1(\ell_1+1)}{2\mu_{X,BC}} \delta_{vv'} \delta_{nn'} \int_0^{+\infty} dR \; \xi_m(R) R^{-2} \xi_{m'}(R)$$

$$\langle \Phi^{JM}_{\ell_1\ell_2 L\Omega vmn} | \frac{\mathbf{l}_2^2}{2\mu_{Y,BC}R_2^2} | \Phi^{JM}_{\ell_1\ell_2 L\Omega v'm'n'} \rangle =$$

$$\frac{\hbar^2 \ell_2(\ell_2+1)}{2\mu_{Y,BC}} \delta_{vv'} \delta_{mm'} \int_0^{+\infty} dR \; \zeta_n(R) R^{-2} \zeta_{n'}(R)$$

The crossing kinetic term can be obtained by writing the gradient operators in terms of tensorial components

$$\nabla_1 \cdot \nabla_2 = \sum_{\nu=-1}^{1} (-1)^\nu \nabla^{(1)}_\nu \nabla^{(2)}_{-\nu}$$

in such a way that

$$\langle \Phi^{JM}_{\ell_1\ell_2 L\Omega vmn} | \nabla_1 \cdot \nabla_2 | \Phi^{JM}_{\ell_1'\ell_2' L'\Omega' v'm'n'} \rangle =$$

$$\delta_{vv'}\delta_{LL'}\delta_{\Omega\Omega'}(-1)^{L-\ell_1'+\ell_1} \begin{Bmatrix} \ell_2' & L & \ell_1' \\ \ell_1 & 1 & \ell_2 \end{Bmatrix} \times$$
$$G_1(m, m', \ell_1, \ell_1') G_2(n, n', \ell_2, \ell_2')$$

with $\left\{ \begin{array}{c} \cdots \\ \cdots \end{array} \right\}$ being $6-j$ symbols and

$$G_i(p,q,\ell,\jmath) = \delta_{\ell\jmath\pm1}(-1)^s\sqrt{s} \int_0^{+\infty} dR \ f_p^{(i)}(R)\left[\frac{df_q^{(i)}(R)}{dR} \mp s\frac{f_q^{(i)}(R)}{R}\right]$$

where

$$s = \max(\ell,\jmath) \qquad ; \qquad f^{(i)} = \begin{cases} \xi , & \text{if } i=1 \\ \zeta , & \text{if } i=2 \end{cases}$$

Note that this term is unable to change v, L or Ω and only couples states with $\ell_1 = \ell_1' \pm 1$ and $\ell_2 = \ell_2' \pm 1$. Since it increases with the angular excitations and, in addition, is divided by the diatomic total mass, it is expected to be almost negligible regarding the ground and first excited tetra-atomic levels.

In the BF reference system chosen, the triatomic interactions $V_{X,BC}$ and $V_{Y,BC}$ depend on $r, R_1, \cos\theta_1$ and $r, R_2, \cos\theta_2$, respectively. After an usual expansion in terms of Legendre polynomials,

$$V_{X,BC}(r, R_1, \cos\theta_1) = \sum_{\lambda_1} v_{\lambda_1}(r, R_1)P_{\lambda_1}(\cos\theta_1)$$

$$V_{Y,BC}(r, R_2, \cos\theta_2) = \sum_{\lambda_2} v_{\lambda_2}(r, R_2)P_{\lambda_2}(\cos\theta_2)$$

we have

$$\langle \Phi_{\ell_1\ell_2L\Omega vmn}^{JM} \mid V_{X,BC}\mid\Phi_{\ell_1'\ell_2'L'\Omega'v'm'n'}^{JM}\rangle =$$

$$=\delta_{\ell_2\ell_2'}\delta_{\Omega\Omega'}\delta_{nn'}(-1)^{\Omega-\ell_2}\sqrt{(2L+1)(2L'+1)(2\ell_1+1)(2\ell_1'+1)} \times$$

$$\sum_{\lambda_1} v_{\lambda_1}^{vm;v'm'} \left\{ \begin{array}{ccc} \ell_1 & L & \ell_2 \\ L' & \ell_1' & \lambda_1 \end{array} \right\} \left(\begin{array}{ccc} L & \lambda_1 & L' \\ \Omega & 0 & -\Omega \end{array} \right) \left(\begin{array}{ccc} \ell_1' & \lambda_1 & \ell_1 \\ 0 & 0 & 0 \end{array} \right)$$

$$\langle \Phi_{\ell_1\ell_2L\Omega vmn}^{JM} \mid V_{Y,BC}\mid\Phi_{\ell_1'\ell_2'L'\Omega'v'm'n'}^{JM}\rangle =$$

$$=\delta_{\ell_1\ell_1'}\delta_{\Omega\Omega'}\delta_{mm'}(-1)^{\Omega-\ell_1}\sqrt{(2L+1)(2L'+1)(2\ell_2+1)(2\ell_2'+1)} \times$$

$$\sum_{\lambda_2} v_{\lambda_2}^{vn;v'n'} \left\{ \begin{array}{ccc} \ell_2 & L & \ell_1 \\ L' & \ell_2' & \lambda_2 \end{array} \right\} \left(\begin{array}{ccc} L & \lambda_2 & L' \\ \Omega & 0 & -\Omega \end{array} \right) \left(\begin{array}{ccc} \ell_2' & \lambda_2 & \ell_2 \\ 0 & 0 & 0 \end{array} \right)$$

where the coefficients are two-dimensional quadratures

$$v_{\lambda_1}^{vm;v'm'} = \int_0^{+\infty}\int_0^{+\infty} dr \ dR_1 \ v_{\lambda_1}(r, R_1)\chi_v(r)\chi_{v'}(r)\xi_m(R_1)\xi_{m'}(R_1)$$

$$v_{\lambda_2}^{vn;v'n'} = \int_0^{+\infty}\int_0^{+\infty} dr \ dR_2 \ v_{\lambda_2}(r, R_2)\chi_v(r)\chi_{v'}(r)\zeta_n(R_2)\zeta_{n'}(R_2)$$

Note that the presence of $\begin{pmatrix} \ell_i' & \lambda_i & \ell_i \\ 0 & 0 & 0 \end{pmatrix}$ leads to these elements become zero unless the addition $\ell_i' + \lambda_i + \ell_i$ be *even*.

Finally, by expanding the $X - Y$ interaction

$$V_{X,Y}(R_1, R_2, \cos\gamma) = \sum_\Lambda v_\Lambda(R_1, R_2) P_\Lambda(\cos\gamma)$$

the corresponding matrix elements become

$$\langle \Phi^{JM}_{\ell_1\ell_2 L\Omega vmn} \mid V_{X,Y} \mid \Phi^{JM}_{\ell_1'\ell_2' L'\Omega'v'm'n'} \rangle =$$

$$= \delta_{vv'}\delta_{LL'}\delta_{\Omega\Omega'}(-1)^{\ell_1-\ell_1'+L}\sqrt{(2\ell_1+1)(2\ell_1'+1)(2\ell_2+1)(2\ell_2'+1)}\ \times$$

$$\sum_\Lambda v_\Lambda^{mn;m'n'} \begin{Bmatrix} \ell_1 & \Lambda & \ell_1' \\ \ell_2' & L & \ell_2 \end{Bmatrix} \begin{pmatrix} \ell_1 & \Lambda & \ell_1' \\ 0 & 0 & 0 \end{pmatrix} \begin{pmatrix} \ell_2 & \Lambda & \ell_2' \\ 0 & 0 & 0 \end{pmatrix}$$

where

$$v_\Lambda^{mn;m'n'} = \int_0^\infty \int_0^\infty dR_1\, dR_2\, v_\Lambda(R_1, R_2)\xi_m(R_1)\xi_{m'}(R_1)\zeta_n(R_2)\zeta_{n'}(R_2)$$

IV   Initial conditions

In order to generate the initial conditions necessary to start the classical trajectories, the ground state eingenfunction, Eq. (7), obtained assuming a diabatic separation of the diatomic vibration, is used to get the corresponding probability distributions. So that, we begin by choosing r and p_r according to an isolated diatomic molecule in a given vibrational level[23]. Regarding distributions of the different quantum numbers ℓ_i , L , etc., they can be readily obtained by using the $A_k^{\ell_1\ell_2 L\Omega vmn}$ coefficients.

Also, distributions of magnitudes on which the wavefunctions depend, as $R_i, i = 1, 2$, may be easily obtained by quadratures,

$$D(X) = \int d\tau \|\Psi_k^{(JM)}\|^2 \delta(R_i - X)$$

that reduce to first-order density matrices.

On the other hand, angular distributions are easier estimated by assuming an expansion in Legendre polynomials, in such a way that for $\alpha = \theta_1, \theta_2, \gamma$,

$$D(\cos\alpha) = \sum_n \frac{2n+1}{2}\langle \Psi_k^{(JM)} \mid P_n(\cos\alpha) \mid \Psi_k^{(JM)} \rangle P_n(\cos\alpha)$$

Fig.1: Distributions of radial variables

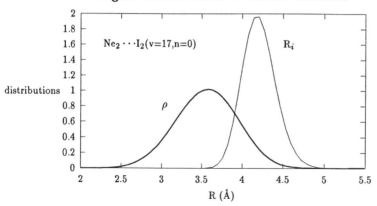

Fig. 2: Distributions of angular variables

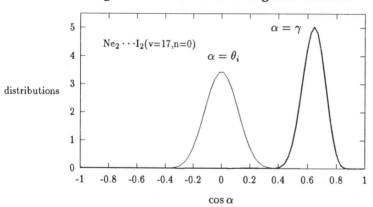

Fig.3: Distributions of rotational quantum numbers

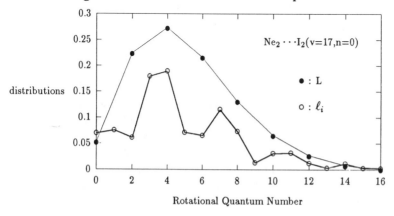

Finally, denoting by ρ the X-Y distance, we found useful to get the corresponding distribution performing the following two-dimensional quadratures

$$D(\rho) = \int \int dR_1 dR_2 \frac{\rho}{R_1 R_2} D(\cos \gamma(\rho, R_1, R_2))$$

where the $\rho/(R_1 R_2)$ factor comes from the Jacobian of the transformation $(R_1, R_2, \cos \gamma) \to (R_1, R_2, \rho)$ defined by the relation

$$\cos \gamma = \frac{R_1^2 + R_2^2 - \rho^2}{2 R_1 R_2}$$

These single mode probability funtions were used to select the initial positions according to a random procedure. For the system under study, Fig 1 shows the distributions corresponding to R_i, $i = 1, 2$ and ρ (the Ne-Ne distance) quantities. While the R distribution is relatively localized near the equilibrium position, that of the ρ distance is spread out over a larger range.

In Fig 2 the distributions associated with the angular variables θ_i, $i = 1, 2$ and γ are displayed. From the distribution is clear that the T-shape configuration for each vdW bond is the most probable as previously pointed out in several works[28].

Once the initial positions have been chosen with the above distributions, we still need to determine the corresponding initial momenta. The initial values of the angular conjugate momenta were obtained by solving the system of equations (4), ensuring a total angular momentum $J = 0$. To this end, the values of ℓ_1, ℓ_2 and $L = j$ were selected at random following the distributions of Fig 3, which were also built up from the ground state tetra-atomic wave function. Only those values subject to the restriction $|\ell_1 - \ell_2| < j < \ell_1 + \ell_2$ were accepted. Finally the P_{R_1} and P_{R_2} momenta were chosen by using a similar procedure to the diatomic case, the energy and the potencial being those correponding to the stretching of a triatomic $I_2 \cdots Ne$ cluster in a given configuration. The (small) difference between the total energy reached and the quantal one, $\approx -141. cm^{-1}$ with respect to the diatomic level, is then randomly shared by these two radial momenta.

V   Results

The potential energy surface used in the present calculations was modelled as a sum of pairwise atom-atom interactions, each one of them represented by a Morse function. The correponding parameters of the I-I[34] and Ne-Ne[35] interactions were taken from the literature. Regarding the I-Ne interaction potential, the associated Morse parameters were fitted by employing a close coupling model[36] in order to reproduce the experimental VP lifetimes for $I_2 \cdots Ne$ recently measured[27]. All these potential parameters are listed in table I.

	$D(cm^{-1})$	$\alpha(\text{Å}^{-1})$	$\gamma_{eq}(\text{Å})$
$I - I^a$	5245.7934	1.6832	3.0271
$I - Ne$	42.0	1.60	4.36
$Ne - Ne^b$	29.36	2.088	3.091

TABLE I. Morse potential parameters
[a] Parameteres from Ref. 34.
[b] Parameteres from Ref. 35.

v	N_T^{calc} (a)	τ (b)	τ_1 (c)	τ_2 (d)	$\% DC$ (e)	$\% Ne_2$ (f)
17	1291	291.64	158.95	138.31	1.63	0.46
23	1514	117.68	65.23	55.77	3.10	1.06

TABLE II. Different quantities depending on the I_2 vibrational quantum number v: (**a**) Total number of trajectories calculated. (**b**) Lifetimes (picoseconds) corresponding to complete dissociation.(**c**) Lifetime associated with the breaking of the first vdW bond. (**d**) Same as (c) for the last vdW bond. (**e**) Percent of trajectories leading to double continuum. (**f**) Percent of trajectories giving $I_2 + Ne_2$

Hamilton equations were numerically solved by means of an Adams-Moulton integrator initiated by a fourth-order Runge-Kutta-Gill integrator with a time step of $0.5 * 10^{-15}s$. The maximun time used in the integration was taken depending on the I_2 vibrational quantum number v, $i.e.$ $600ps$ for $v = 17$ and $250ps$ for $v = 23$. The integration of the trajectories was stopped when the two vdW bonds reached a maximun distance of $12\mathring{A}$.

Analysis of the trajectories provided the lifetime τ associated with total dissociation of both vdW bonds, as well as the lifetimes τ_1 and τ_2 , corresponding to partial dissociation of the first and the second vdW bonds, respectively. In table II these lifetimes are collected together with the number of trajectories calculated for each I_2 vibrational level. As is well stablished, the cluster lifetime decreases as the v quantum number increases. The ratio between the lifetimes associated with $v = 17$ and $v = 23$ is similar to that of the corresponding experimental lifetimes[29], albeit the calculated lifetimes are always longer than the experimental ones.

In these preliminary results we found that the first vdW bond breaking lifetime, τ_1 , is slightly longer than that associated to the breaking of the second vdW bond. In principle it could be expected a ratio $\tau_1/\tau_2 \approx 2$ since there is in principle double probability for the breaking of the first vdW bond compared to the second one. The deviation from the expected behavior can be explained by taking into account the importand role played by internal vibrational redistribution (IVR). In fact there is a quite good agreement between our calculated τ_2 and the experimental one for both $v = 17$ and $v = 23$. The disagreement stands basically for τ_1 , and then for τ . After the first diatomic quantum is lost, our calculations show that this energy is distributed among all the modes. The energy is stored in these modes for relatively long time before it finds its way towards the dissociative channel (the breaking of the first vdW bond). This is essentially the reason why the calculated τ_1 results larger than expected. The origin of this large IVR effect could be the Ne-Ne interaction potential. In this sense it would be very interesting to analyze the effect of the $Ne - Ne$ potential parameters on the calculated lifetimes.

Although in most of the trajectories dissociation of the two vdW bonds occurs through a sequential mechanism,

$$Ne - I_2(v) - Ne \rightarrow I_2(v - 1) - Ne + Ne \rightarrow I_2(v - 2) + Ne + Ne$$

an interesting dynamical feature of this process is the finding of non-sequential reactive pathways. So, it is possible the simultaneous breaking of both weak bonds, $i.e.$ a double continuum (DC) dissociation. Formation of I_2 and Ne_2 products constitutes an alternative pathway to the dominant sequential one. As previously discussed[26,37], the origin of these non-sequential mechanisms of dissociation is related to the statistical nature of the energy redistribution. Table II displays the percents of trajectories leading to DC dissociation and Ne_2 product formation. The ratio between the DC and Ne_2 percents is roughly 0.3 whatever be the iodine initial excitation. The same ratio was already found in a previous quasiclassical

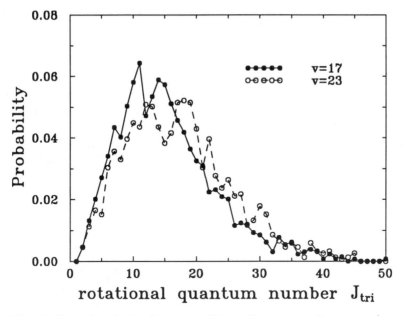

Fig. 4: Rotational distributions of $I_2 - Ne$ intermediate complex

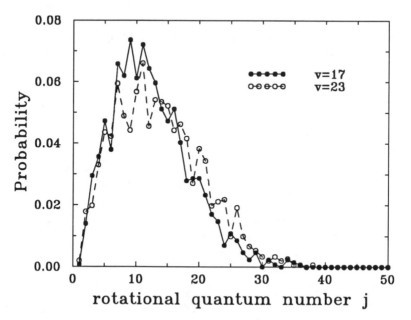

Fig. 5: Rotational distributions of final I_2 product

calculation[26] for $v = 28$ where both rare gas atoms were restricted to move along a perpendicular plane to the I_2 bond.

We calculated the distribution of the total angular momentum, J_{tri} , corresponding to the intermediate three-particle complex arising from the first dissociation. Also, the distribution of the rotational state, j of the final iodine product was obtained. In Fig.4 we show the J_{tri} distribution for the two diatomic excitations considered. It is peaked around $J_{tri} = 10$ and $J_{tri} = 17$ for $v = 17$ and $v = 23$, respectively. In addition, population of high rotational states, up to $J_{tri} = 45$, is observed. A similar behavior is found for the final j distributions shown in Fig. 5. They are peaked around $j = 10$ but do not reach so high rotational levels as compared with the previous distributions.

Concluding remarks

We have studied the vibrational predissociation dynamics of the $Ne \cdots I_2 \cdots Ne$ cluster. A Quasiclassical Trajectory method including all the degrees of freedom for zero total angular momentum was used. The preliminary results qualitatively agree with previous experimental and theoretical data. Non-sequential reactive pathways of statistical nature are predicted by the calculations: simultaneous breaking of both weak bonds and Ne_2 product formation with measurable rates. It would be both valuable and interesting the experimental investigation of these findings. Also of interest is to analyze the effect of the rare gas-rare gas interaction on the IVR process. Finally, the study of the variation of the predissociation rate constant versus the I_2 vibrational quantum number appears also very exciting, in particular on the region of the crossover level. Work in this direction is currently in progress.

References

[1] J.Jortner, Ber. Bunsenges. Phys. Chem.**88**, 188 (1984).

[2] Entire June issue of Chem. Rev.**86**, (1986).

[3] R.S. Berry, T.L. Beck. H.L. Davis, and J. Jellinek, in Advances in Chemical Physics, edited by I. Prigogine and S. A. Rice (Wiley, New York, 1987), Vol 70B.

[4] The Physics and Chemistry of Small Clusters, edited by P. Jena, B.K. Rao, and S.N. Khanna (Plenum, New York, 1978).

[5] J.J. Breen, D.M. Wilberg, M. Gutmann, and A.H. Zewail, J. Chem. Phys.**93**, 9180 (1990).

[6] T.L. Beck and R.S. Berry, J. Chem. Phys.**88**, 3910 (1988).

[7] M. Amarouche, G. Durand, and J.P. Malrieu, J. Chem. Phys. **88**, 1010 (1988).

[8] L.C. Balbás, A. Rubio, J.A. Alonso, and G. Borstel, Chem. Phys. **120**, 239

(1988).

[9] S. Stringari and J. Treiner, J. Chem. Phys. **87**, 5021 (1987).

[10] J.A. Blazy, B.M. Dekoven, T.D. Russell, and D.H. Levy, J. Chem. Phys. **72**, 2439 (1980); J.E. Kenny, K.E. Johnson, W. Sharfin, and D.H. Levy, *ibid.* **72**, 1109 (1980); D.H. Levy, Adv. Chem. Phys. **XLVII**, 323 (1981).

[11] K.C. Janda, Adv. Chem. Phys. **LX**, 201 (1985), and references therein; J.I. Cline, N. Sivakumar, D.D. Evard, and K.C. Janda, J. Chem. Phys. **86**, 1636 (1987); D.D. Evard, F. Thommen, J.I. Cline, and K.C. Janda, J. Phys. Chem. **91**, 2508 (1980); J.I. Cline, N. Sivakumar, D.D. Evard, and K.C. Janda, Phys. Rev. A **36**, 1944 (1987).

[12] See, e.g., Structure and Dynamics of Weakly Bound Molecular Complexes, NATO ASI Series C: Mathematical and Physical Sciences, Vol. 212, edited by A. Weber (Reidel, Dordrecht, 1987); J.M. Gardner and M.I. Lester, J. Chem. Phys. **85**, 2329 (1986); J.C. Drobits, J.M. Skene, and M.I. Lester, *ibid.* **84**, 2896 (1986).

[13] (a) J.A. Beswick and J. Jortner, J. Chem. Phys. **69**, 512 (1978); (b) J.A. Beswick, G. Delgado-Barrio, and J. Jortner, *ibid.* **70**, 3895 (1979); (c) J.A. Beswick and J. Jortner, Adv. Chem. Phys. **XLVII**, 363 (1981).

[14] (a) J.A. Beswick, and G. Delgado-Barrio, J. Chem. Phys. **78**, 3653 (1980); (b) M. Aguado, P. Villarreal, G. Delgado-Barrio, P. Mareca, and J.A. Beswick, Chem. Phys. Lett. **102**, 227 (1983); (c) O. Roncero, S. Miret-Artés, G. Delgado-Barrio, and P. Villarreal, J. Chem. Phys. **85**, 2084 (1986).

[15] P. Villarreal, S. Miret-Artés, O. Roncero, S. Serna, J. Campos-Martinez and G. Delgado-Barrio, J. Chem. Phys. **93**, 4016 (1990).

[16] P. Villarreal, G. Delgado-Barrio, O. Roncero, F.A. Gianturco, and A. Palma, Phys. Rev. A **36**, 617 (1987).

[17] G. Delgado-Barrio, P. Mareca, P. Villarreal, A.M. Cortina, and S. Miret-Artés, J. Chem. Phys. **84**, 4268 (1986).

[18] N. Halberstad, J.A. Beswick, and K.C. Janda, J. Chem. Phys. **87**, 3966 (1987); J.I. Cline, B.P. Reid, D. D. Evard, N. Sivakumar, N. Halberstad, and K.C. Janda, *ibid.* **89**, 3535 (1988).

[19] S.K. Gray, J. Chem. Phys. **87**, 2051 (1987).

[20] R.H. Bisseling, R. Kosloff, R.B. Gerber, M.A. Ratner, L. Gibson, and C. Cerjan, J. Chem. Phys. **87**, 2760 (1987).

[21] G.C. Schatz, R.B. Gerber, and M.A. Ratner, J. Chem. Phys. **88**, 3709 (1988).

[22] G.C. Schatz, V. Buch, M.A. Ratner, and R.B. Gerber, J. Chem. Phys. **79**, 1808 (1983).

[23] (a) G.Delgado-Barrio, P. Villarreal, P. Mareca, and G. Albelda, J. Chem. Phys. **78**, 280 (1983); (b) G. Delgado-Barrio, P. Villarreal, A. Varadé, N. Martín, and A. García-Vela, in Structure and Dynamics of Weakly Bound Molecular Complexes, NATO ASI Series C: Mathematical and Physical Sciences, Vol. 212, edited by A. Weber (Reidel, Dordrecht, 1987), p. 573.

[24] P. Villarreal, A. Varadé, and G. Delgado-Barrio, J. Chem. Phys.. **90**, 2684 (1989).

[25] (a) A. García-Vela, P. Villarreal, and G. Delgado-Barrio, J. Chem. Phys. **92**, 496 (1990); (b) Z. Bacic, M. Kennedy-Mandziuk, and J.W. Moskowitz, J. Chem. Phys. **97**, 6472 (1992).

[26] A. García-Vela, P. Villarreal, and G. Delgado-Barrio, J. Chem. Phys. **94**, 7868 (1991).

[27] D.M. Willberg, M. Gutmann, J.J. Breen, and A.H. Zewail, J. Chem. Phys. **96**, 198 (1992).

[28] M. Gutmann, D.M. Willberg, and A.H. Zewail, J. Chem. Phys. **97**, 8037 (1992).

[29] M. Gutmann, D.M. Willberg, and A.H. Zewail, J. Chem. Phys. **97**, 8048 (1992).

[30] E.D. Potter, Q. Liu and A.H. Zewail, Chem. Phys. Lett. **200**, 605 (1992).

[31] D.M. Willberg, M. Gutmann, E.E. Nikitin, and A.H. Zewail, Chem. Phys. Lett. **201**, 506 (1993).

[32] G. Delgado-Barrio et al, to be published.

[33] G.A. Natanson, G.S. Ezra, G. Delgado-Barrio, and R.S. Berry, J. Chem. Phys. **81**, 3400 (1984); *ibid.* **84**, 2035 (1986).

[34] P. Luc, J. Mol. Spectr. **80**, 41 (1980).

[35] R.A. Aziz and M.J. Slaman, Chem. Phys. **130**, 187 (1989).

[36] G. Delgado-Barrio in "Dynamical Processes in Molecular Physics", edited by G. Delgado-Barrio, Institute of Physics Publishing, Bristol, (1993).

[37] S.R. Hair, J.I. Cline, C.R. Bieler, and K.C. Janda, J. Chem. Phys. **90**, 2935 (1989).

CAGING AND NONADIABATIC ELECTRONIC TRANSITIONS IN I_2-M COMPLEXES

O. RONCERO
Instituto de Matemáticas y Física Fundamental, CSIC, Madrid, Spain

N. HALBERSTADT, and J.A. BESWICK
LURE and Laboratoire Photophysique Moléculaire, Centre Universitaire de Paris-Sud, 91405 Orsay, France

1. Introduction

One of the current goals in the field of chemical physics is to understand the role of weak intermolecular forces (solvation bonds) in chemical reactivity. Photodissociation is a primary process in photochemistry which can be studied in well defined conditions in condensed phases [Hynes (1985), Schroeder and Troe (1987), Harris et al. (1988), Lingle (1990), Schwentner and Chergui (1992)].

In particular, the photolytic cage effect where after light absorption a fraction of the fragments recombine within the cavity of the surrounding solvent molecules, is still the subject of intense experimental [Yan et al. (1992), Schwartz et al. (1993)] as well as theoretical [Alimi, Gerber, and Apkarian (1988), Adelman et al. (1993), Gersonde and Gabriel (1993)] research.

Iodine is a model system which has been studied under a wide range of experimental conditions in compressed gases and liquids [Dutoit, Zellweger, and van den Bergh (1983), Otto, Schroeder, and Troe (1984), Yan et al. (1992), Scherer et al. (1992)]. Photoexcitation studies of I_2 at wavelengths where the free I_2 molecule is known to dissociate directly show that the dissociation yield drops below unity at surprisingly low gas densities, namely, 1 or 2 orders of magnitude below liquid densities.

Otto, Schroeder, and Troe (1984), suggested that this effect could be due to I_2 van der Waals complexes with solvent particles. In their cluster model, it was assumed that photolysis is quenched in iodine molecules which were complexed in a I_2-M_n cluster with M being a solvent atom or molecule. That this process actually occurs in van der Waals complexes was already demonstrated in experiments conducted in supersonic beams [Saenger, McClelland, and Herschbach (1981), Valentini and Cros (1982)]. It was shown that recombination occurs even for complexes with only one solvent particle. Subsequently, Philippoz et al. (1986-90) conducted several other studies on this "single-solvent photolytic cage effect" on van der Waals complexes of I_2 with rare-gas atoms and simple molecules, including I_2 itself.

J. Jortner et al. (eds.), Reaction Dynamics in Clusters and Condensed Phases, 73–87.
© 1994 *Kluwer Academic Publishers. Printed in the Netherlands.*

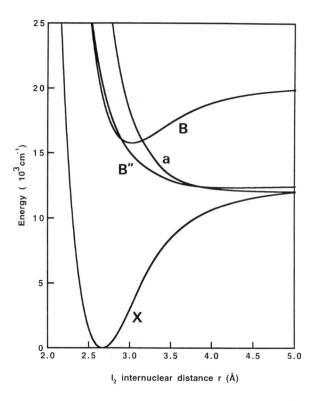

Fig. 1. Potential energy curves for I_2 used in the calculations.

The I_2-M complexes formed in supersonic beams have also been studied in the excitation region of the bound levels of the excited B state [Levy (1981), Breen et al. (1990), Willberg et al. (1992), Burke and Klemperer (1993a and b)]. These studies provided a wealth of crucial information not only on the structure of the complexes but also on dynamics of vibrational and electronic predissociation. Vibrational predissociation induced by van der Waals interactions has been extensively studied not only for I_2 complexes [Levy (1981), Beswick and Jortner (1981)], but also for other halogens [Janda (1985)], interhalogens and polyatomic molecules [Halberstadt and Janda (1990)]. In this paper we shall concentrate on the two other relaxation processes occurring in I_2-M complexes namely, caging and electronic predissociation induced by the presence of M.

2. One-atom cage effect

The "single-solvent particle cage effect" in the photodissociation of I_2-M van der Waals complexes was first studied by Saenger, McClelland, and Herschbach (1981). Complexes with Ar, N_2, benzene, and other polyatomic molecules, were excited in the region between 0-1400 cm^{-1} above the dissociation limit of the B state of I_2 (see

Fig. 1). Fluorescence was detected which was attributed to a blue-shifted absorption of the complex, followed by emission from free I$_2(B)$ released by dissociation of the complex. In subsequent experiments Valentini and Cross (1982), and Philippoz, Monot and van den Bergh (1986-90), found that the fluorescence emanates from vibrational levels of the B state of I$_2$ well below the dissociation limit.

It is well known that excitation of free I$_2$ molecules at these energies results in dissociation with a quantum yield of unity and no fluorescence. Thus, the presence of a single atom or molecule M bound to I$_2$ in a van der Waals complex induces some recombination into I$_2(B, v', j')$, according to

$$I_2\text{-}M + h\nu \longrightarrow I_2(B, v', j') + M; \qquad \left(h\nu > E_\infty^{(B)}\right) \qquad (1)$$

where $E_\infty^{(B)}$ is the dissociation energy limit of the B state. For excitation of I$_2$-Ar at $\lambda = 488$ nm for instance, it was found that the vibrational distribution peaks at around $v' = 42$ with a width of about $\Delta v' \sim 13$ [Philippoz et al. (1986-88)]. Furthermore, the rotational energy of the recombined I$_2$ was rather low so that in order to conserve energy, the fragments recoil energy in Eq. (1) should be of the order of 1000 cm^{-1}. Thus, as much as 1000 cm^{-1} is being transfered from the initially excited I-I coordinate to relative kinetic energy in the I$_2$-M degree of freedom.

The branching ratio between channel (1) and the dissociation of I$_2$ at the same total energy

$$I_2\text{-}M + h\nu \longrightarrow \begin{cases} I^*(^2P_{1/2}) + I(^2P_{3/2}) + M \\ I(^2P_{3/2}) + I(^2P_{3/2}) + M \end{cases} \qquad (2)$$

has not been measured. Even if the probability of channel (1) is small, it is rather surprising to observe it with such a large amount of energy going into relative translational energy of the fragments.

A purely ballistic mechanism was proposed by Valentini and Cross (1982), whereby the initial excitation is to the B state; impulsive energy transfer from I$_2$ to Ar, with dissociation of the complex, then dissipates enough energy to leave the I$_2$ fragment in the bound region of the B state. NoorBatcha, Raff, and Thompson (1984), have performed quasiclassical trajectory calculations on I$_2$-M, M = Ar, Kr, and Xe. They have found a much smaller extent of energy transfer from the I-I coordinate to the I$_2$-M coordinate than that observed experimentally, even assuming linear complexes which is the most favorable geometrical configuration for efficient kinetic energy transfer.

An alternative mechanism for the one-atom cage effect has been proposed by Beswick, Monot, Philippoz, and Van den Bergh (1987), where a nonadiabatic electronic transition was invoked [for a recent discussion on this model see Dardi and Dahler (1993)]. It is well known [Tellinghuisen (1982)] that in addition to the B state there is another state $B''(^1\Pi_{1u})$ that contributes significantly to the photon absorption cross section in the $\lambda = 500 - 450$ nm wavelength region (see Fig. 1). At the 488 nm excitation wavelength for instance, the strength of the $B'' \leftarrow X$ transition is about 0.6 that of the $B \leftarrow X$. The $B''(^1\Pi_{1u})$ potential energy curve is well known due to studies of the spontaneous and magnetic field induced predissociation of the B state [Vigué, Broyer, and Lehmann (1981)] and absorption measurements [Tellinghuisen (1982)]. The $B''(^1\Pi_{1u})$ and $B(^3\Pi_{0+u})$ states are weakly coupled by hyperfine interactions in the free I_2 molecule. The presence of a solvent molecule may induce a stronger coupling between these two states through spin-orbit interaction or the breakdown of the Born-Oppenheimer approximation. According to Beswick, Monot, Philippoz, and Van den Bergh (1987), these couplings can be responsible for nonadiabatic transitions between the initially excited dissociative B'' state and a final bound vibrational level v' of $I_2(B)$. The proposed mechanism for production of $I_2(B, v')$ was then

$$I_2\text{-M} + h\nu \longrightarrow I_2(B'', E)\text{-M} \longrightarrow I_2(B, v') + M \tag{3}$$

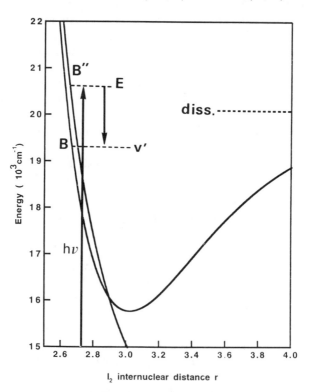

Fig. 2. Potential energy curves involved in the electronic nonadiabatic mechanism for $I_2(B)$ recombination.

A simple model calculation for this process can be performed assuming that electronic nonadiabatic coupling is a slowly varying function of the I_2 internuclear coordinate. The rates for I_2 recombination from the initial dissociative B'' state to the final vibrational levels v' of the B state will then be proportional to the

Franck-Condon factor

$$k_{(B'',E)\to(B,v')} \propto \left| \langle \chi_E^{(B'')} | \chi_{v'}^{(B)} \rangle \right|^2 \tag{4}$$

where $\chi_E^{(B'')}$ is the continuum wave function for the I$_2$ vibrational motion in the B''

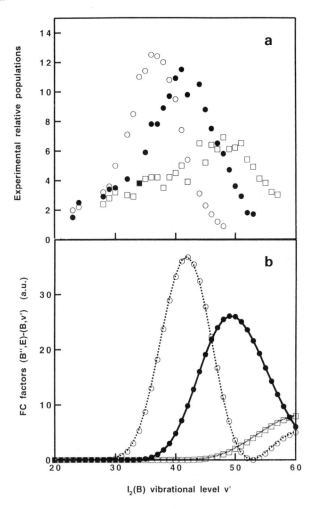

Fig. 3. a) Experimental final state vibrational distributions for recombined I$_2(B)$ fragments. Adapted from Philippoz (1988). Open circles: excitation wavelength $\lambda = 496.5$ nm. Full circles: $\lambda = 488$ nm. Open squares: $\lambda = 476.5$ nm. **b)** Franck-Condon factors between the vibrational bound wave functions of the B state and the three continuum wave functions of the B'' state corresponding to the excitation wavelengths of the experiments.

dissociative state at energy $E = E_0 + h\nu$, while $\chi_{v'}^{(B)}$ is a final bound vibrational wave function in the B state (see Fig. 2).

Using the B''-state potential of Tellinghuisen (1985) and the B-state data of Hutson et al. (1982), we have computed the Franck-Condon factors in Eq. (4) for the 3 different wavelengths used in the Lausanne experiments [Philippoz et al. (1986-88)]. The results are presented in Fig. 3b. The calculated distributions of $I_2(B, v')$ are very similar to the experimental results (see Fig. 3a). At 488 nm for instance, the experimental distribution peaks at $v' = 42$ and has a width of $\Delta v' = 13$, while the Franck-Condon model gives a maximum at $v' = 49$ and a width $\Delta v' = 14$. In addition, as the wavelength increases, the maximum and the width of the Franck-Condon distribution evolve very much like the experiments results.

It should be noted at this point, that this model assumes that the Ar interaction has no effect on the I_2 potentials nor on the dissociation dynamics. In particular, the final vibrational state distribution of I_2, and hence the final product recoil energy, does not depend on the nature of the solvent atom or molecule. This is because we have assume a diabatic separation of intermolecular and intramolecular vibrational degrees of freedom. In a more elaborated treatment the Franck-Condon factor in Eq. (4) should be multiplied by the square of the matrix element of the electronic coupling between the intermolecular vibrational (and rotational) wave functions. This will provide the dependence of the final vibrational state distribution of I_2, and hence the final product recoil energy, on the nature of the solvent atom or molecule.

3. Complex induced electronic predissociation

In addition to the caging effect discussed above, there is another process in which electronic nonadiabatic couplings induced by the formation of a complex are important. When the I_2-M complex is excited to the bound region of the B state of I_2, electronic predissociation

$$I_2(X)\text{-}M + h\nu \longrightarrow I_2(B, v')\text{-}M \longrightarrow I(^2P_{3/2}) + I(^2P_{3/2}) + M \qquad (5)$$

can compete with vibrational predissociation:

$$I_2(X)\text{-}M + h\nu \longrightarrow I_2(B, v')\text{-}M \longrightarrow I_2(B, v < v') + M \qquad (6)$$

Several repulsive states cross the B potential curve (see Fig. 1) in the region of the well and can thus be responsible for electronic predissociation.

Since channel (6) produces excited free I_2 molecules which can fluoresce, while channel (5) is a dark channel, measurements of the I_2-M/I_2 fluorescence quantum yield ratio, in conjunction with absorption spectra, can provide the relative importance of electronic predissociation as compared to vibrational predissociation [Burke and Klemperer (1993b)]. For I_2-He and I_2-Ne vibrational predissociation is much faster than electronic predissociation, while for I_2-Ar the two processes compete. For low vibrational levels ($v' < 12$) of I_2-Ar electronic predissociation prevails [Levy, Zewail], but for highly excited levels ($v' > 30$) vibrational predissociation dominates.

For the I_2-Kr complex, electronic predissociation is much more rapid than vibrational predissociation, since fluorescence following excitation of the discrete $B \leftarrow X$ bands seen in absorption, is completely quenched [Goldstein, Brack, and Atkinson (1986)].

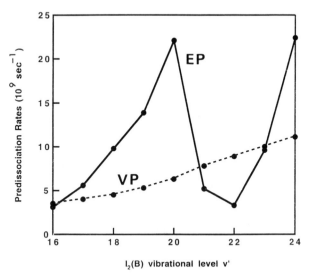

Fig. 4. Experimental rates for Vibrational Predissociation (dotted line) and Electronic Predissociation (full line) from an initially excited $I_2(B, v')$-Ar complex.

It was tempting to assume that the same repulsive electronic state was responsible for both caging above the B dissociation limit, and electronic predissociation below. Recently Burke and Klemperer (1993b) have studied in detail the rate for electronic predissociation of $I_2(B, v')$-Ar in the range $v' = 16\text{-}24$. They have found that the oscillations of the rate as a function of v' are very similar to the oscillations observed in electric field induced quenching of the B state of I_2 [Dalby, Levy, and Vanderlinde (1984)]. The state responsible for this quenching is believed to be the $a(^3\Pi_{1g})$ state (see Fig. 1). The B and the a states are directly coupled by electric dipole interactions in the T-shaped configuration. Assuming again that the electronic predissociation rates are proportional to the Franck-Condon factors, we have

$$k_{(B,v') \to (B'',E)} \propto \left| \langle \chi_{v'}^{(B)} | \chi_E^{(B'')} \rangle \right|^2 ; \qquad E = E_{v'}^{(B)} \tag{7}$$

The important difference with respect to Eq. (4) is that the continuum wave function $\chi_E^{(B'')}$ is calculated at the energy of the bound level of the B state and hence the Ar atom is not taking away any energy ("spectator model"). The results of this calculation are presented in Fig. 5. It is clear from comparison of Figs. 4 and 5, that the B'' state cannot be responsible for the pronounced oscillations observed in the electronic predissociation rates. As suggested by Tellinghuisen (1985) and by Burke and Klemperer (1993b), it is most probably the $a(^3\Pi_{1g})$ state which is producing electronic predissociation, at least in the T-shaped configuration.

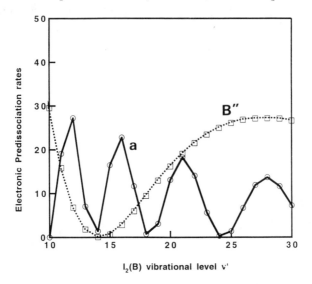

Fig. 5. Squares: calculated electronic predissociation rates Franck-Condon factors between a bound (B, v') level and the continuum wave function of the B'' state at total energy $E = E_{v'}^{(B)}$. Circles: same for the continuum wavefunction of the a state.

If the coupling is electrostatic, then it vanishes in the colinear configuration but it is maximum at T-shape. In addition, if the simple Franck-Condon model

$$k_{(B,v')\to(a,E)} \propto \left|\langle \chi_{v'}^{(B)} | \chi_E^{(a)} \rangle\right|^2; \qquad E = E_{v'}^{(B)} \tag{8}$$

is used again, the rate oscillates as a function of v' much more rapidly that the one calculated with the B'' state, as can be seen in Fig. 5. Actually, the frequency of the oscillations is very close to that of the experimental results (see Fig. 6). Only the position of the maxima and minima are somewhat shifted.

In order to study the influence of the intermolecular degrees of freedom (described by the vector $\boldsymbol{R}$ joining the Ar atom to the center of mass of I_2), we have recently performed quantum close coupling calculations on this system [Roncero, Halberstadt, and Beswick (1993)]. The final dissociative continuum of the a state was discretized in the intramolecular coordinate r of I_2. The results indicate that the right order of magnitude for the electronic predissociation rates (10^{10} sec^{-1}) is obtained for a coupling between the B and the a states of the order of 30 cm^{-1} at the equilibrium geometry of the B state. Another finding was that the Ar atom fragment has very little (essentially zero) kinetic energy. This is somewhat expected since in the T-shaped configuration the Ar atom induces the coupling between the

B and the a states but has very little direct energy exchange with the intramolecular vibrational motion of I_2.

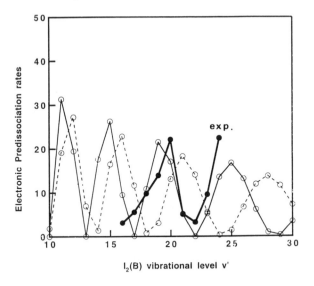

Fig. 6. Calculated electronic predissociation rates (open circles) and experimental results of Burke and Klemperer (full circles). The open circles joined by the dotted line corresponds to the simple Franck-Condon factors with undistorted I_2 potential energy curves for the B and a states. The ones joined by the full line are the calculations using the modified Franck-Condon factors described in the text.

Based on these quantum results it is possible to model electronic predissociation in this system by fixing the position of the Ar atom with respect to the center of mass of I_2 at the equilibrium position of the B state. This amounts to considering that the overall effect of the presence of the Ar atom is to induce a coupling between the B and the a states and to modify the potential energy curves of I_2. Applying again the Franck-Condon approximation,

$$k_{(\widetilde{B},v')\to(\widetilde{a},E)} \propto |\langle \chi_{v'}^{(\widetilde{B})} | \chi_E^{(\widetilde{a})} \rangle|^2 ; \qquad E = E_{v'}^{(\widetilde{B})} \tag{9}$$

where the notation $\widetilde{B}$ and $\widetilde{a}$ indicates that the calculation is performed with the B and the a states modified by the addition of the intermolecular potential at the perpendicular configuration with the distance between Ar and the center of mass of I_2 frozen at its equilibrium value in the B state.

In Fig. 6 we present the results of this modified Franck-Condon calculation. It is seen that the agreement between calculated and measured results is now very good.

4. Discussion and conclusions

From the calculations presented above it is clear that the $a(^3\Pi_{1g})$ state can satisfactorily describe electronic predissociation of $I_2(B)$-Ar complexes, as suggested by Burke and Klemperer (1993b). It is then very likely that the $a(^3\Pi_{1g})$ state is also involved in the collision induced predissociation of the B state [Brown and Klemperer (1964), Steinfeld and Klemperer (1965), Steinfeld (1966), Selwyn and Steinfeld (1969), Derouard and Sadeghi (1983), Nakagawa et al. (1986), Nicolai and Heaven (1986)]. On the other hand, the $a(^3\Pi_{1g})$ state cannot be involved directly in the nonadiabatic coupling model for caging above the dissociation limit of the B state, because it does not carry oscillator strength from the ground state. Apart from the B state, only the $B''(^1\Pi_{1u})$ state absorbs in that spectral region. Therefore, further analysis is necessary in order to ascertain the origin of the single-atom cage effect.

Recently, Burke and Klemperer (1993a) have conducted new experiments on I_2-Ar in an attempt to further characterize the caging effect. If the mechanism of nonadiabatic transitions between an initially $I_2(B'')$-Ar excited complex and final $I_2(B)$ + Ar fragments is operating above the B-state dissociation limit, it should also exist below that limit. However, in the bound region of the B state, the $B \leftarrow X$ transitions are very narrow in energy while the $B'' \leftarrow X$ transitions span a continuum. Hence, in contrast with the case of excitation above the dissociation limit of the B state, below that limit the two states can be independently excited. Burke and Klemperer (1993a) found that a significant continuum exists in the fluorescence excitation spectra taken in between the $B \leftarrow X$ discrete transitions. The intensity of the continuum was measured with respect to the I_2 $B \leftarrow X(v',0)$ transition in the near vicinity. For $v' = 9 - 13$ there was no evidence of fluorescence from continuum excitation. On the other hand, the fluorescence from continuum excitation became significant for $v' > 13$. Its intensity increased smoothly with v' in contrast with the intensity of the nearby I_2-Ar bands which oscillate as a function of v' due to electronic predissociation. These findings rule out the participation of the discrete I_2-Ar $B \leftarrow X$ transitions to the continuum excitation process. Furthermore, the introduction of Kr in the expansion led to a continuum intensity near $v' = 26$ which was 1.6 times greater than that found with Ar. For I_2-Kr, discrete $B \leftarrow X$ bands are seen in absorption, but the fluorescence is completely quenched due to very rapid electronic predissociation [Goldstein, Brack, and Atkinson (1986)]. It is thus not possible for the discrete I_2-Kr bands to contribute to the observed continuum excitation fluorescence.

The Ar-induced fluorescence measured by Burke and Klemperer (1993a) from continuum excitation in between the discrete I_2-Ar $B \leftarrow X$ bands, is emanating from the B state of I_2 and substantially red-shifted. Thus the continuum excitation process yields vibrational levels of $I_2(B)$ well below the level in the vicinity of the initial excitation.

All these findings are consistent with the nonadiabatic model which assumes that the I_2-Ar complex is initially excited to the dissociative B'' state of I_2, then undergoes an Ar-induced transition to the B state, from which it fluoresces. However, dividing the Ar-induced fluorescence by the total continuum absorption gives a minimum quantum efficiency of 0.07 for the generation of fluorescence following continuum excitation. This is a quite high lower bound for the probability of this process. Indeed, the measured intensity of fluorescence from continuum excitation is approximately a factor of 15 lower than the peak fluorescence intensity of the nearby discrete $B \leftarrow X(v' = 26, 0)$ band of I$_2$-Ar, but the integrated intensity of the continuum is relatively large. Taking the magnitude of the fluorescence and multiplying by the separation between vibronic bands, which is 78 cm^{-1} at this v', results in an integrated intensity for the continuum which is 2 times that of the discrete $B \leftarrow X$ band of I$_2$-Ar. However, at this excitation wavelength the transition intensity of the B'' state is only 0.13 of that to the B state. Even with a $B'' \rightarrow B$ crossing probability of unity, the $B'' \leftarrow X$ transition strength is a factor of 15 too low to account for the observed continuum intensity.

In order to account for the absolute intensity of the fluorescence from continuum excitation, the possible existence of a second, linear isomer was proposed [Burke and Klemperer (1993a)]. In contrast to the T-shaped isomer, which shows very minor changes in bond length and well depth upon excitation to the B state, it was proposed that there will be large differences for a linear isomer in the excited B state relative to the X state.

The changes in the bond energy and equilibrium center of mass separation for the linear complex upon $B \leftarrow X$ excitation would spread this transition over a continuum or a quasi-continuum. According to Burke and Klemperer (1993a), if the linear isomer is about 3 times more abundant than the T-shaped one, the observed fluorescence could be explained. If linear isomers of I$_2$-Ar are truly abundant in an Ar expansion, new quasiclassical trajectory calculations which incorporate the change in the I$_2$-Ar colinear potential upon excitation are necessary.

In conclusion, we can assert that single-solvent particle caging effects are interesting and intriguing problems in the field of reaction dynamics. These effects are not restricted to I$_2$ complexes. Recently for instance, one-atom effects in the photodissociation of Ar-HCl and Ar-HBr complexes has been studied theoretically and experimentally [Garcia-Vela et al. (1991-92), Segall et al. (1993)]. Also, the effect of a single-solvent atom on bimolecular reactions has been recently studied [Hurwitz et al. (1993)]. In the future, one may hope to bridge the gap between the one-solvent atom and the condensed phase dynamics by studying the same processes in medium and large clusters [Amar and Berne (1984), Alimi and Gerber (1990), Gerber et al. (1991), Fei et al. (1992), Potter, Liu, and Zewail (1992), Borrmann, Li, and Martens (1993)].

References

Adelman, S.A., Ravi, R., Muralidhar, R., Stote, R.H. (1993) "Molecular theory of liquid-phase vibrational energy relaxation", Adv. Chem. Phys. **84**, 73

Alimi, R., Gerber, R.B., and Apkarian, V.A. (1988) J. Chem. Phys. **89**, 174

Alimi, R., and Gerber, R.B. (1990) "Solvation effects on chemical reactions dynamics in clusters: Photodissociation of HI in $Xe_n HI$", Phys. Rev. Lett. **64**, 1453

Amar, F.G., and Berne, B.J. (1984) "Reaction dynamics and the cage effect in microclusters of $Br_2 Ar_n$", J. Phys. Chem. **88**, 6720

Beswick, J.A., and Jortner, J. (1981) "Vibrational predissociation of van der Waals molecules", Adv. Chem. Phys. **47**, 363

Beswick, J.A., Monot, R., Philippoz, J.-M., and van den Bergh, H. (1987) "On the possibility of nonadiabatic transitions in the photodissociation of $I_2 M$ clusters excited above the dissociation limit of the B state", J. Chem. Phys. **86**, 3965

Borrmann, A., Li, Z., and Martens, C.C. (1993) "Nonlinear resonance and correlated binary collisions in the vibrational predissociation of $I_2(B, v)$-Ar_{13}", to be published

Breen, J.J., Willberg, D.M., Gutmann, M., and Zewail, A.H. (1990) "Direct observation of the picosecond dynamics of $I_2 - Ar$ fragmentation", J. Chem. Phys. **93**, 9180

Brown, R.L., and Klemperer, W. (1964) "Energy transfer in the fluorescence of iodine excited by the sodium D lines", J. Chem. Phys. **41**, 3072

Burke, M.L., and Klemperer, W. (1993a) "The one-atom cage effect: Continnum prodesses in I_2-Ar below the B state dissociation limit", J. Chem. Phys. **98**, 1797

Burke, M.L., and Klemperer, W. (1993b) "Efficiency and mechanism of electronic predissociation of B state I_2-Ar", J. Chem. Phys. **98**, 6642

Dalby, F.W., Levy, C.D.P., and Vanderlinde, J. (1984) "Vibrational and rotational dependence of electric field induced predissociation of I_2", Chem. Phys. **85**, 23

Dardi, P.S., and Dahler, J.S. (1993) "A model for nonadiabatic coupling in the photodissociation of I_2-solvent complexes", J. Chem. Phys. **98**, 363

Derouard, J., and Sadeghi, N. (1983) "Collisional processes and corresponding rate constants in the $B^3\Pi(0_u^+)$, $v' = 15$ state of I_2", Chem. Phys. Lett. **102**, 324

Dutoit, J.-C., Zellweger, J.M., and van den Bergh, H. (1983) "The photolytic cage effect of iodine in gases and liquids", J. Chem. Phys. **78**, 1825

Fei, S., Zheng, X., and Heaven, M.C. (1992) "Spectroscopy and relaxation dynamics of $I_2 Ar_n$ clusters. Geminate recombination and cluster fragmentation", J. Chem. Phys. **97**, 6057

Garcia-Vela, A., Gerber, R.B., and Valentini, J.J. (1991) "Dynamics of molecular photodissociation in clusters: a study of ArHCl", Chem. Phys. Lett. **186**, 223

Garcia-Vela, A., Gerber, R.B., and Valentini J.J. (1992) "Effects of solvation by a single atom in photodissociation: Classical and quantum/classical studies of HCl photolysis in Ar-HCl", J. Chem. Phys. **97**, 3297

Gerber, R.B., Alimi, R., Garcia-Vela, A., Hurwitz, Y. (1991) "Dynamics of photoinduced reactions in clusters", in Mode Selective Chemistry, J. Jortner, R. Levine, and B. Pullmann (eds.), Kluwer Academic Publishers, p. 201

Gersonde, I.H., and Gabriel, H. (1993) "Molecular dynamics of photodissociation in matrices including nonadiabatic processes", J. Chem. Phys. **98**, 2094

Goldstein, N., Brack, T.L., and Atkinson, G.H. (1986) "Quantitative absorption spectroscopy and dissociation dynamics of I_2 van der Waals complexes with He, Ar, Kr, and Xe", J. Chem. Phys. **85**, 2684

Halberstadt, N., and Janda, K.C., Eds. (1990) "Dynamics of Polyatomic van der Waals complexes", Plenum, New York

Harris, A.L., Brown, J.K., and Harris, C.B. (1988) Annu. Rev. Phys. Chem. **39**, 341

Hutson, J.M., Gerstenkorn, S., Luc, P., and Sinzelle, J. (1982) "Use of calculated centrifugal distortion constants (D_v, H_v, L_v, and M_v) in the analysis of the $B \leftarrow X$ system of I_2", J. Mol. Spect. **96**, 266

Hurwitz, Y., Rudich, Y., Naaman, R., and Gerber, R.B. (1993) "Effect of a single solvent atom on bimolecular reactions: Collisions of $O(3^P)$ with hydrocarbons", J. Chem. Phys.

Hynes, J.T. (1985) Annu. Rev. Phys. Chem. **36**, 573

Janda, K.C. (1985) "Predissociation of polyatomic van der Waals molecules", Adv. Chem. Phys. **60**, 201

Levy, D.H. (1981) Spectroscopy of van der Waals molecules", Adv. Chem. Phys. **47**, 323

Lingle (1990) J. Chem. Phys. **93**, 5667

Nakagawa, K., Kitamura, M., Suzuki, K., Kondow, T., Munakata, T., Kasuya, T. (1986) Chem. Phys. **106**, 259

Nicolai, J.-P., and Heaven, M.C. (1986) J. Chem. Phys. **84**, 6694

NoorBatcha, I., Raff, L.M., and Thompson, D.L. (1984) "Cage effect in the dissociation of van der Waals complexes RgI_2 (Rg=Ar, Kr, Xe): a quasiclassical trajectory study", J. Chem. Phys. **81**, 5658

Otto, B., Schroeder, J., and Troe, J. (1984) "Photolytic cage effect and atom recombination of iodine in compressed gases and liquids: experiments and simple models", J. Chem. Phys. **81**, 202

Philippoz, J.M., Monot, R., and van den Bergh, H. (1986) "Product state distributions in the photodissociation of iodine-rare gas clusters", Helvetica Physica Acta **58**, 1089

Philippoz, J.M., van den Bergh, H., and Monot, R. (1987) "Product vibrational state distributions in the photodissociation of iodine-rare gas clusters", J. Phys. Chem. **91**, 2545

Philippoz, J.M., Melinon, P., Monot, R., and van den Bergh, H. (1987) "Solvent caging by diatomic molecules in the photodissociation of I_2M clusters", Chem. Phys. Lett. **138**, 579

Philippoz, J.M. (1988) "Une approche de l'effet-cage par la photodissociation de complexes de van der Waals", Ph. D. thesis EPFL No 727, Lausanne

Philippoz, J.M., Monot, R., and van den Bergh, H. (1990a) "Multiple oscillations observed in the rotational state population of $I_2(B)$ formed in the photodissociation of $(I_2)_2$", J. Chem. Phys. **92**, 288

Philippoz, J.M., Monot, R., and van den Bergh, H. (1990b) "Photodissociation of iodine complexes in relation to the cage effect", J. Chem. Phys. **93**, 8676

Potter, E.D., Liu, Q., and Zewail, A.H. (1992) "Femtosecond reaction dynamics in macroclusters. Effect of solvation on wave-packet motion", Chem. Phys. Lett. **200**, 605

Roncero, O., Halberstadt, N., and Beswick, J.A. (1993) "Quantum close coupling calculations for the electronic predissociation of $I_2(B)$-Ar complex", to be published

Saenger, K.L., McClelland, G.M., and Herschbach, D.R. (1981) "Blue shift of iodine in solvent complexes formed in supersonic beams", J. Phys. Chem. **85**, 3333

Scherer, N.F., Ziegler, L.D., and Fleming, G.R. (1992) "Heterodyne-detected time-domain measurement of I_2 predissociation and vibrational dynamics in solution", J. Chem. Phys. **96**, 5544

Schroeder, J., and Troe, J. (1987) Annu. Rev. Phys. Chem. **38**, 163

Schwartz, B.J., King, J.C., Zhang, J.Z., and Harris, C.B. (1993) "Direct femtosecond measurements of single collision dominated geminate recombination times of small molecules in liquids", Chem. Phys. Lett. **203**, 503

Schwentner, N., and Chergui, M. (1992) "Photochemistry, charge transfer states and laser applications of small molecules in rare gas crystals", in Optical Properties of Excited States in Solids, B. Di Bartolo Ed., Plenum Press, New York

Segall, J., Wen, Y., Singer, R., Wittig, C., Garcia-Vela, A., and Gerber, R.B. (1993) "Evidence for a cage effect in the UV photolysis of HBr in Ar-HBr. Theoretical and experimental results", Chem. Phys. Lett. **207**, 504

Selwyn, J.E., and Steinfeld, J.I. (1969) "Collision-induced predissociation by van der Waals interaction", Chem. Phys. Lett. **4**, 217

Steinfeld, J.I., and Klemperer, W. (1965) "Energy transfer process in monochromatically excited iodine molecules. I. Experimental results", J. Chem. Phys. **42**, 3475; **43**, 2926

Steinfeld, J.I. (1966) "Wavelength dependence of the quenching of iodine fluorescence", J. Chem. Phys. **44**, 2740

Tellinghuisen, J. (1982) "Transition strengths in the visible-infrared absorption spectrum of I_2", J. Chem. Phys. **76**, 4736

Tellinghuisen, J. (1985) "Potentials for weakly bound states in I_2 from diffuse spectra and predissociation data", J. Chem. Phys. **82**, 4012

Valentini, J.J., and Cross, J.B. (1982) "The photodissociation cage effect in van der Waals complexes", J. Chem. Phys. **77**, 572

Vigué, J., Broyer, M., and Lehmann, J.C. (1981) "Natural hyperfine and magnetic predissociation of the I_2 B state. I. Theory; II. Experiments on natural and hyperfine predissociation; III. Experiments on magnetic predissociation", J. Phys. (Paris) **42**, 937, 949, 961

Willberg, D.M., Gutmann, M., Breen, J.J., and Zewail, A.H. (1992) "Real-time dynamics of clusters. I. $I_2X_n (n = 1)$", J. Chem. Phys. **96**, 198

Yan, Y., Whitnell, R.M., Wilson, K.R., and Zewail, A.H. (1992) "Femtosecond chemical dynamics in solution. Wavepacket evolution and caging of I_2", Chem. Phys. Lett. **193**, 402

Phase Separation in Binary Clusters

A.S. Clarke*, R. Kapral[†], B. Moore*, G. Patey* and X.-G. Wu[†]

*Department of Chemistry, University of British Columbia,

Vancouver, BC, Canada V6T 1Z1

[†]Chemical Physics Theory Group, Department of Chemistry,

University of Toronto, Toronto, Canada M5S 1A1

September 17, 1993

Abstract

The dynamics and structural properties of clusters composed of two chemical species are investigated. The demixing process starting from the unstable fully-mixed state in the clusters possesses some features that are similar to domain formation and growth following a critical quench in a binary fluid. The structural properties of the phase separated clusters differ from those of the bulk fluid and are studied as a function of the temperature, concentration and interaction strength.

1 Introduction

Clusters are interesting materials with properties that differ from those of the bulk phase. Pure clusters and clusters containing "solute" particles have been studied often.[1, 2] The focus of the present work is on the structural properties and dynamics of clusters composed of two chemical species, which we denote by A and B.[3] The interaction potentials in such binary clusters often favor demixing and the cluster atoms will tend to segregate into two

J. Jortner et al. (eds.), Reaction Dynamics in Clusters and Condensed Phases, 89–100.
© 1994 *Kluwer Academic Publishers. Printed in the Netherlands.*

subclusters composed predominantly of A or B atoms. This segregation process is the cluster analog of binary mixture phase separation in the bulk fluid and its investigation along with the structural properties of the resulting phase-separated clusters are the topics of this paper.

We consider a simple model system consisting of two species interacting through Lennard-Jones (LJ) potentials with potential parameters $\epsilon_{\mu\nu}$ and $\sigma_{\mu\nu}$ with mass m_μ, for $\mu, \nu = A, B$. All of the calculations presented in this paper are restricted to the case of equal masses and LJ σ parameters with $m = m_A = m_B = 40$ amu and $\sigma = \sigma_{\mu\nu} = 3.4 \text{Å}$.

Even with these restrictions there is a wide variety of situations to consider. The concentrations of the two species in the cluster may vary as can the potential energy associated with the like-like and cross interactions. The temperature or energy can affect both the structure and dynamics, and the variations in these parameteres will also be considered.

In sec. 2 we present results on the dynamics of the demixing process in the cluster and contrast it with phase separation in the bulk. We discuss the time scales on which the demixing occurs and identify the physical processes that give rise to these time scales. Section 3 is devoted to an examination of the structural properties of the phase-separated clusters. We examine the structure as a function of the cluster energy, concentrations of the two components and relative interaction strengths as reflected in the values of the LJ ϵ parameters. The conclusions are presented in sec. 4.

2   Dynamics of Demixing

For simplicity we consider first the case of two mechanically identical species with equal concentrations and with a cross interaction less than that of the like-like interactions so that demixing is favored. We let $\epsilon = \epsilon_{AA} = \epsilon_{BB}$, $\alpha = \epsilon_{AB}/\epsilon$, with $N_A = N_B$ the numbers of each species and $N = N_A + N_B$. The parameter α which gauges the strength of the cross interaction will be used as the control parameter for these studies.

The demixing process in such clusters was studied by following the time evolution of the system from a fully mixed initial state prepared as follows: all N atoms in the cluster were taken to be identical and the system was evolved for a transient time until an "equilibrium" state was reached. Naturally, the cluster will eventually evaporate so all states considered are metastable; however, the time scale for such evaporation processes under our simulation conditions is long compared to those of interest here so we shall refer to such long-lived metastable states as equilibrium states. At time $t = 0$, N_A atoms chosen at random were labeled species A, the remainder were labeled B and the interaction potentials were assigned parameter values appropriate for the binary cluster. Demixing occurs as the system evolves in time and is shown in Fig. 1 for one realization of the dynamics for a cluster with $N_A = N_B = 55$. The potential parameters are given in the caption. The MD simulations were carried out at constant energy. Note that the demixing occurs by particle aggregation into small subclusters that, in turn, merge and form larger clusters (cf. panels a), b) and c)). Panels d), e) and f) give an indication of the magnitude of the fluctuations in the phase-separated state.

This is just the cluster analog of bulk-fluid domain formation and growth following a critical quench from the one-phase region to the two-phase region.[4] In the early stages of the phase separation, domains of the two phases begin to form but these domains are not defined by sharp interfaces. Fluctuations dominate or at least play an important role in the dynamics of this early stage growth. Once well-defined domains separated by sharp interfaces have formed, the evolution occurs on a slow time scale determined by the diffusive motion of the curved interfaces and their coupling to the bulk hydrodynamic modes of the fluid. In the cluster the microscopic character of the dynamics is evident in that the early and late stage growth processes are not as sharply distiguished. Fluctuations play an important role throughout the evolution, even in the phase-separated regime. Nevertheless, well-defined interfacial structure does exist, as discussed in the next section, and it is possible to observe incipient domain formation.

To monitor the phase separation we consider an order parameter defined in terms of the magnitude of the separation between the centers of mass of the two species. More specifically, we define $d(\mathbf{r}^N(t)) = |\mathbf{d}(\mathbf{r}^N(t))|$ with

$$\mathbf{d}(\mathbf{r}^N(t)) = N_A^{-1} \sum_{i=1}^{N_A} \mathbf{r}_{Ai}(t) - N_B^{-1} \sum_{i=1}^{N_B} \mathbf{r}_{Bi}(t) , \qquad (1)$$

where $\mathbf{r}^N = (\mathbf{r}_A^{N_A}, \mathbf{r}_B^{N_B})$ are the coordinates of the cluster atoms. At $t = 0$, $d(\mathbf{r}^N(0))$ has a small non-zero value $O(\sigma)$ and increases to a finite value in the phase separated cluster about which it oscillates. The average value of $d(\mathbf{r}^N(t))$ in this asymptotic regime will be denoted $\langle d \rangle$ and can be computed by an average over time or over realizations of the evolution process starting from different random initial conditions. The characteristics of the cluster in this "equilibrium" asymptotic regime will be discussed in the next section.

The average initial value of $d(\mathbf{r}^N(0))$ will be denoted d_0 and can be computed from an average over realizations.

From the above discussion it is clear that an order parameter, $\phi(\mathbf{r}^N(t))$, whose average, $\langle\phi\rangle(t)$, over realizations is zero in the fully-mixed initial state and non-zero in the phase-separated state can be constructed by defining

$$\phi(\mathbf{r}^N(t)) \equiv \phi(t) = d(\mathbf{r}^N(t)) - d_0 . \tag{2}$$

The time evolution of the order parameter is shown in Fig. 2a for several realizations of the demixing process for $\alpha = 0.6$. The plots of $\phi(t)$ show either direct evolution to a phase-separated state or evolution to the final state via a series of step-like jumps. Curves without steps typically correspond to realizations that form two subclusters of the two species directly from the mixed state, while those with steps correspond to realizations where microdomains of the phases first form and and diffuse on a longer time scale before aggregation into the final fully demixed state (cf. Fig. 1b and 1c).

The time evolution of the order parameter averaged over thirty realizations is shown in Fig. 2b. The figure shows that $\langle\phi\rangle(t)$ increases to its asymptotic value on a time scale of approximately $250\ \tau$ where $\tau = (m\sigma^2/\epsilon)^{1/2}$. The time scale is α and N dependent, increasing as α tends to unity and N increases.

3   Cluster Structure

We now turn to a characterization of the equilibrium properties of the binary cluster. The average value $\langle d \rangle$, the separation between the centers of mass of the two species in the equilibrium state, provides a crude characterization of the cohesive forces between the two subclusters in the demixed state. On physical grounds one would expect $\langle d \rangle$ to increase with decreasing α and this is the case as can be seen in Fig. 3 where $\langle d \rangle$ is plotted versus α. For α near unity the binary clusters are compact and nearly spherical while for smaller α the clusters are more elongated with the two subclusters being nearly spherical in shape. One may observe extremes which resemble two wetting phases for $\alpha \sim 1$ and non-wetting phases for very small α (cf. Fig. 1 of Ref. [3]).

The probability distribution of d values provides additional insight into the cluster structure. We may define the probability density $P(d)$ by

$$P(d) = \langle\delta(d(\mathbf{r}^N) - d)\rangle \equiv P_0 e^{-\beta W(d)} , \tag{3}$$

where $\delta(x)$ is the Dirac delta function, the angle brackets signify an equilibrium canonical ensemble average, d is a numerical value of $d(\mathbf{r}^N)$ and the last equality defines the potential of mean force along the many-body coordinate $d(\mathbf{r}^N)$ for a cluster with temperature $T = 1/k_B\beta$. The quantity P_0 is the probability density for a uniform distribution of d. While the calculation of $P(d)$ and $W(d)$ for all values of d requires special techniques, for d values lying near $\langle d \rangle$ it is possible to estimate these quantities directly by monitoring $d(\mathbf{r}^N(t))$ in a long MD run. The results of such simulations are shown in Figs. 4a and 4b for several values of α for clusters with $N_A = N_B = 25$. The simulations were carried out using constant temperature MD.[5] Both the probability density and the potential of mean force exhibit distinctive features. The distribution broadens as α decreases and tends to become more symmetric. For α values near unity the mean potential is "softer" for $d > \langle d \rangle$ than for $d < \langle d \rangle$. This implies that the reversible work required to decrease d from its minimum is more than that required to increase d and pull the two subclusters apart. Of course, if α is sufficiently small and the cluster temperature is such that $k_B T/\epsilon_{AB} > 1$ then thermal fluctuations can lead to dissociation of the phase-separated cluster into two subclusters. This has been observed in simulations. The full study of this dissociation process requires a knowledge of the free energy for d values outside the range of those studied here.

The local species density provides information on the inhomogeneous structure of the phase-separated cluster. We consider a cylindrical coordinate system (z, ρ, ϕ) with z-axis along the line joining the centers of mass of the two species and let $\rho_\mu(z) = \int_0^\infty d\rho\, \rho \int_0^{2\pi} d\phi \rho_\mu(\mathbf{r})$ be the local density of species $\mu = A$, B projected onto the z-axis. These densities are plotted in Fig. 5 for $N_A = N_B = 25$, $\epsilon/k_B = 121K$ and $\alpha = 0.5$, 0.65 and 0.8. Although the cluster is liquid-like there are strong srtuctural correlations for small α. For low α values there is little penetration of one species into the other and the correlations arise from the fact that one subcluster appears as a "wall" to the other subcluster, thus inducing spatial correlations. These correlations persist over long distances from the interface for larger clusters. We also observe that the correlations are much stronger in the cluster than the corresponding correlations observed in the structure of the interface in bulk binary fluids of this type.[6] As α increases towards unity these density correlations become less pronounced.

The structure of the cluster is a strong function of its temperature or energy. Solid-liquid phase transformations in pure clusters and clusters containing solute species have been studied in some detail.[1, 2] Above, we focused our attention on liquid-like clusters[7] where the cluster atoms diffuse freely throughout the cluster or subclusters. This temperature regime is rather narrow: for lower temperatures the cluster is solid-like where the atoms are largely confined to localized regions in the cluster, while for higher temperatures the cluster rapidly evaporates. Thus the regime in which phase separation of liquid-like clusters occurs with negligible evaporation is small. The change in the phase-separated cluster structure as the temperature varies can be seen in Fig. 6, which shows clusters with $N_A = 55$ and $N_B = 25$ at three different temperatures. Panels a) and b) are clusters with solid-like structures. The regular ordering of the atoms is visible in the figure, especially for the A (white) atoms. Panel c) is a liquid-like cluster. The differences between these solid-like and liquid-like phases is clearly seen in the structure of the species densities, $\rho_\mu(z)$. The densities in Figs. 7a and 7b correspond to the solid-like and liquid-like clusters in panels b) and c) of Fig. 6, respectively. Note that $\rho_\mu(z)$ for the liquid-like cluster has a structure like that of the $N_A = N_B$ liquid-like cluster described above while the new peaks appear at different locations for the solid-like phase.

The cluster geometry is also a function of the relative interaction strengths. Consider the case where $\epsilon_{AA} < \epsilon_{AB} < \epsilon_{BB}$. Here the strong $B - B$ interactions favor pure B clustering and the fact that $\epsilon_{AB} > \epsilon_{AA}$ means that it is energetically more favorable for phase A to interact with phase B than with itself. Figure 8 shows the results of a simulations for $\epsilon_{AA}/k_B = 80K$, $\epsilon_{BB}/k_B = 121K$ and $\epsilon_{AB}/k_B = 1.25\epsilon_{AA}/k_B$ for (a) $N_A = 55$ and $N_B = 25$ and (b) $N_A = 80$ and $N_B = 30$. In both cases the B atoms form a core coated by a sheath of A particles as expected from the physical arguments given above. For these clusters it is convenient to consider a spherical polar coordinate system $\mathbf{r} = (r, \theta, \phi)$ with origin at the center of mass of the cluster. The radial probability density, $R_\mu(r)$, where $R_\mu(r)dr = r^2 dr \int d\theta \sin\theta d\phi \rho_\mu(\mathbf{r})$ is presented in Fig. 9. The figure shows that in spite of the rather strong interpenetration of the two phases that exists for these parameter values distinct phases can be identified. There are also structural differences that arise as N changes. One may observe an additional peak near the origin for the the $N_B = 30$ subcluster.

4   Conclusion

The results presented in this paper show that binary clusters present a number of unique and interesting dynamical and structural features. Only a sketch of the phenomenology of the demixing or phase separation dynamics was presented. There are clear analogs of the early and late stage dynamics that are seen in bulk fluids. The extent to which hydrodynamic models, which have proved successful in the interpretation of bulk fluid phase-separation dyanamics, can be used to describe the more microscopic cluster demixing process remains to be investigated. Another interesting topic is the effects of transformations between solid-like and liquid-like structures on the demixing dynamics.

Both the liquid-like and solid-like binary clusters show strong structural correlations that are not present in the bulk fluid. Furthermore the cluster geometry is a sensitive function of the cluster concentration and relative interaction strengths among the different species.

The present calculations suggest that a wide range of binary mixture rate and a structural properties can be investigated in the cluster environment.

Acknowledgements

This work was supported in part by a grant funded by the Network of Centres of Excellence program in association with the Natural Sciences and Engineering Research Council of Canada and grants from the Natural Sciences and Engineering Research Council of Canada.

References

[1] For reviews see, for instance, *The Physics and Chemistry of Small Clusters*, P. Jena, B.K. Bao and S.H. Khanna, eds., (Plenum, NY, 1987); *Large Finite Systems*, J. Jortner, A. Pullman and B. Pullman, eds., (Reidel, Utrecht, 1987); A.W. Castleman and R.G. Keesee, Science, **241**, 36 (1988); R.S. Berry, T.L. Beck, H.I. Davis and J. Jellinek, Adv. Chem. Phys. **70**, 75 (1988).

[2] See, for example, C.L. Briant and J.J. Burton, J. Chem. Phys. **63**, 2045 (1975); H. Reiss, P. Mirabel and R.L. Whetten, J. Chem. Phys. **29**, 7214 (1988); H.-P. Cheng, X. Li, R.L. Whetten and R.S. Berry, Phys. Rev. A **46**, 791 (1992); M. Y. Hahn and R. L. Whetten, Phys. Rev. Let. **61**, 1190 (1988); J.E. Adams and R.M. Stratt, J. Chem. Phys. **93**, 1358 (1990); M. Bixon and J. Jortner, J. Chem. Phys. **91**, 1631 (1989); J.D. Honeycutt and H.C. Andersen, J. Phys. Chem. **91**, 4950 (1987); B. W. van de Waal, J. Chem. Phys. **90**, 3407 (1989).

[3] A.S. Clarke, R. Kapral, B. Moore, G. Patey and X.-G. Wu, *Structure and Dynamics of Binary Clusters*, Phys. Rev. Lett., May 26, (1993).

[4] J.D. Gunton, M. San Miguel and P.S. Sahni, in *Phase Transitions and Critical Phenomena*, eds. C. Domb and J.L. Lebowitz, (Academic Press, NY, 1983), Vol. 8.

[5] S. Nosé, Mol. Phys. **52**, 255 (1984).

[6] M. Hayoun, M. Meyer, M. Mareschal, G. Ciccotti and P. Turq, in *Chemical Reactivity in Liquids*, eds. M. Moreau and P. Turq (Plenum, NY, 1988), p.279; M. Meyer, M. Mareschal and M. Hayoun, J. Chem. Phys. **89**, 1067 (1988).

[7] D.J. McGinty, J. Chem. Phys. **58**, 4733 (1973); J.K. Lee, J.A. Barker and F.F. Abraham, ibid. 3166 (1973); C.L. Briant and J.J. Burton, Nature **243**, 100 (1973); J. Farges, M.F. de Feraudy, B. Raoult and G. Torchet, J. Chem. Phys. **78**, 5067 (1983); ibid. **84**,3491 (1986); J. Jellinek, T.L. Beck and R.S. Berry, J. Chem. Phys. **84**, 2783 (1986); T.L. Beck and T.L. Marchioro II, J. Chem. Phys. **93**, 1347 (1990).

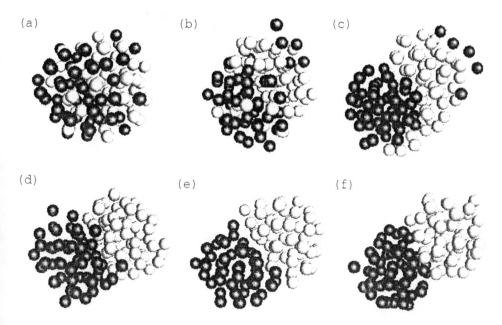

Fig. 1. Phase separation from the fully-mixed state in a constant energy MD simulation. The potential parameters are $\epsilon_{AA}/k_B = \epsilon_{BB}/k_B = 121K$ and $\alpha = 0.5$. The six frames are for times $t = 0$ (a), 360 (b), 520 (c), 800 (d), 1000 (e) and 1200 ps (f).

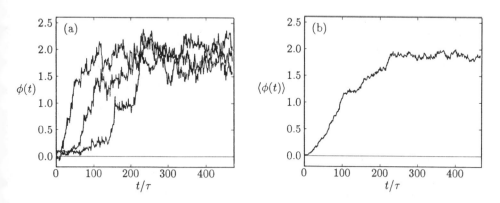

Fig. 2. Constant energy MD results for the order parameter $\phi(t)$ (measured in units of σ) of the phase separation process for $N_\mu = 25$ ($\mu = A, B$), $\epsilon_{\mu\mu}/k_B = 121K$ and $\alpha = 0.6$. (a) $\phi(t)$ versus t for three different realizations, (b) $\langle\phi\rangle(t)$ versus t computed from an average over thirty realizations.

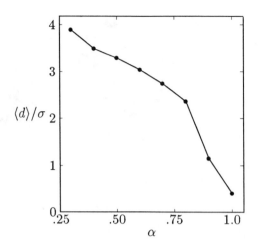

Fig. 3. Plot of $\langle d \rangle$ versus α for $N_\mu = 55$ ($\mu = A, B$) and $\epsilon_{\mu\mu}/k_B = 121K$. $T^* = Tk_B/\epsilon$ ranges from $0.32 - 0.34$ in the microcanonical simulations.

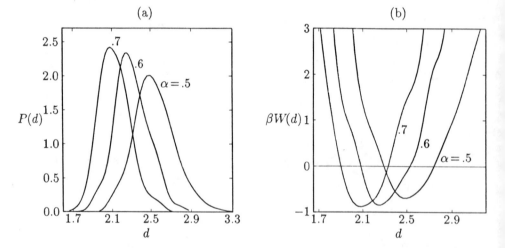

Fig. 4. (a) Probability density $P(d)$ for the binary cluster in the phase-separated state for $T = 36.3 \ K$ and several α values. (b) Free energy $W(d)$ computed from $P(d)$ in (a).

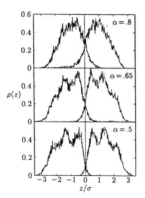

Fig. 5. $\rho_\mu(z)$ versus z for $N_A = N_B = 25$ and $\epsilon/k_B = 121\,K$ for three values of α.

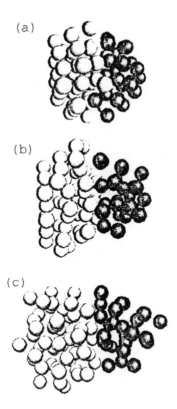

Fig. 6. Configurations of clusters with $N_A = 55$ and $N_B = 25$ for $T \approx 16\,K$ (a), $T \approx 25\,K$ (b), $T \approx 28\,K$ (c). Potential parameters are the same as in Fig. 1.

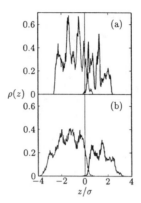

Fig. 7. Average densities $\rho_A(z)$ and $\rho_B(z)$ versus z direction for the clusters shown in Fig. 6b (a) and Fig. 6c (b). The $\rho_\mu(z)$ is normalized to have an area equal to N_μ/N_A.

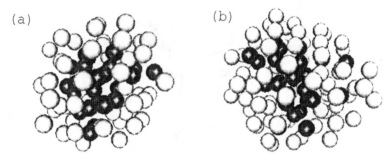

Fig. 8. Configurations of clusters with different $\epsilon_{\mu\nu}$ (see text). (a) $N_A = 55$, $N_B = 25$, $T \approx 30\ K$. (b) $N_A = 80$, $N_B = 30$, $T \approx 29.5\ K$.

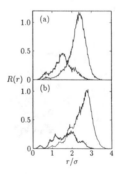

Fig. 9. Average radial densities $R_A(r)$ (thin solid line) and $R_B(r)$ (thick solid line) for the two clusters shown in Fig. 8. $R_\mu(r)$ is normalized to have an area equal to N_μ/N_A.

Molecular Collisions on Large Argon Clusters

J.M. Mestdagh, A.J. Bell[1], J. Berlande, X. Biquard, M.A. Gaveau,
A. Lallement, O. Sublemontier, and J.-P. Visticot

Service des Photons, Atomes et Molécules
C.E.A./C.E.N. Saclay
91191 Gif-sur-Yvette cedex, France

1 Introduction

For the experimental study of solution phase chemistry one would wish for a "perfect" system to be able to investigate. That is, a solution of known size, structure, and internal energy, with a knowledge of interaction energies of the solute and solvent. The need for such "ideal experiments" is made very acute by recent molecular dynamics calculations that open the route for modelling fundamental aspects of reaction dynamics in solution [1, 2].

Over the past two years we have been solvating atoms and molecules in large argon clusters with this goal in mind, investigating the effect of solvation on reaction dynamics [3, 4]. We have performed experiments designed to study mobility and reactions in argon clusters using one or two solvated species. Furthermore, these studies are quantitative – *i.e.* we know how many atoms or molecules we have in a solvent of known size [5, 6]. To a certain extent, we also know the amount of internal energy contained in the system [7, 8], and we know where the solutes are located, (either at the cluster surface or in the cluster interior [9, 10, 11, 12]).

The present paper reviews some of these investigations. It is divided into four areas. To begin with, we briefly describe the experimental apparatus, and the diagnostics that are necessary to characterize fully the clusters we are working with. Secondly, we give experimental evidence that neutral reactants attached to argon clusters have enough mobility to collide each other and react. This question is exemplified by studying chemiluminescence in the reaction

$$Ba(6s^2, {}^1S_0) + N_2O \rightarrow BaO + N_2 \quad \Delta H^0 = -4.11 eV \tag{1}$$

within argon clusters of average size $\bar{N} = 400$ to $\bar{N} = 4000$. We shall see also that the chemiluminescence observed in the cluster reaction differs markedly from that observed in the gas phase. Thirdly, we investigate the diffusion rate of the reactants within the cluster more quantitatively. Collisional quenching of excited barium $(Ba(6s6p\,{}^1P_1))$ by methane molecules is used as a probe for this mobility. After the question of intra-cluster mobility has been examined, we enter into the fourth area of the paper. This section deals with the simplest fundamental aspect of reaction dynamics in solution that can be investigated using the present experimental arrangement, *i.e.* the coupling between movements along a well identified reaction

---

[1]Department of Chemistry, University of Southampton, Southampton, SO9 5NH, UK

J. Jortner et al. (eds.), Reaction Dynamics in Clusters and Condensed Phases, 101–114.
© *1994 Kluwer Academic Publishers. Printed in the Netherlands.*

coordinate and disordered movements inside a size controlled thermal bath. This question is exemplified by showing that solvation of one solute ($Ba(6s6p\,^1P_1)$) enters into competition with desorption when excess internal energy is given to the *cluster-solute* system.

2   Experiment

2.1   APPARATUS

The apparatus used in our experiments is derived from the one described in references [13, 14]. It consists of a supersonic beam source[15] (operating with argon to generate argon clusters), an oven beam source (for generating a low energy, low density beam of barium), a collision chamber (where the two beams cross at right angles) and a time-of-flight arm containing a quadrupole mass spectrometer. The argon cluster beam passes through a pick-up chamber before entering the collision chamber. This allows us to attach molecules to the cluster using the pick-up technique first introduced by Scoles [10]. Finally, the time-of-flight chamber contains a quadrupole mass spectrometer and pressure gauges (ionization and viscosity gauges) that allow flux measurements of the beam by dispersing the molecular beam in an unpumped chamber.

Under this configuration the experimental apparatus acts as double pick-up machine where argon clusters are generated first; they then pick up a first reactant (CH_4 or N_2O molecules in our case); and finally they pick up the second reactant (one atom of barium).

To make quantitative experiments we have to characterize carefully both the cluster size and the number of molecules attached to the cluster. Due to the low density of the barium beam, those clusters containing barium contain, at the most, a single atom of barium.

2.2   CLUSTER SIZE

By varying the backing pressure of Ar between 2 bars and 35 bars, we can generate a supersonic beam comprising species in the range of pure monomers to large clusters of argon atoms ($\bar{N} \approx 4000$). Each expansion pressure results in a distribution of cluster sizes. Most thoughts initially turn to mass spectrometry for cluster size determination. However mass spectrometry cannot be used reliably. Difficulties arise due to mass discrimination in the spectrometer used and, more importantly, severe distortions of the cluster distribution due to fragmentation upon ionization [16]. For this reason we prefer measurement techniques that can be applied directly to neutral clusters and that is nondisruptive towards the neutral cluster distribution.

Two such methods have been developed in our laboratory that allowing us to determine the mean of the cluster distribution. In the first one, the clusters are passed through a region containing a low pressure of argon gas. Collisions between the clusters and gas occur resulting in the attachment of atoms of the gas to the clusters [11], which thus slows them down. The measurement of this retardation as a function of gas pressure allows us to determine an absolute mean size for the clusters [5]. Since this procedure is fairly cumbersome, we have also developed a simple Rayleigh scattering method for determining relative cluster sizes in the range $\bar{N} \approx 150\text{-}4000$ monomers [6]. Figure 1 shows the mean cluster sizes obtained at different stagnation

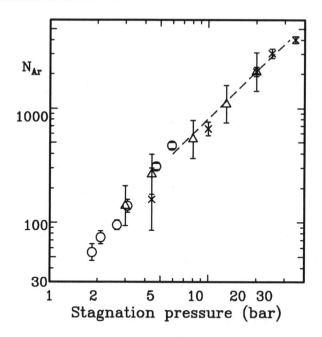

Figure 1: Average sizes of argon clusters determined from the Rayleigh scattering results ($\times$) and from the "slow-down" method ($\triangle$) for different stagnation pressures of argon. The size determination of Farges et al. [7] ($\bigcirc$) and the cluster size dependence measured in reference [17] (dashed curve) are shown for comparison. The relative sizes from the Rayleigh experiments have been scaled to the 20 bar result from the "slow-down" method.

pressures using our two measurement techniques. Our size determinations are in nice agreement with the absolute size determination of Farges et al. using the high energy electron diffraction technique [7] as well as with the size variation found in the work of Hagena and Obert [17].

Careful adjustment of the working conditions of a time of flight mass spectrometer (energy of the ionizing electrons, extraction, deflection voltages, and post acceleration of fragmented cluster ions) give insight into the width of the cluster distribution. For expansion conditions leading to an average cluster size $\bar{N}$, the full width at half maximum of the cluster distribution is estimated to be $\bar{N}$.

2.3 NUMBER OF MOLECULES ATTACHED TO THE CLUSTERS

As said above, CH_4 and N_2O molecules are deposited on the cluster by the pick-up technique. The cluster beam passes through a chamber containing methane up to a maximum pressure of 10^{-3} mbar. The flux measurements that are possible using the pressure gauges and the mass spectrometer installed in the time-of-flight chamber allow us to control and measure the relative flux of argon and molecules. With the knowledge of the average size of the cluster, the average number of molecules picked-up is determined. Up to 15 CH_4 molecules can be picked up by Ar_{4000} clusters without disturbing the cluster beam. This number is always very small compared to the average size of the clusters. Because the molecular pick-up is a random collision process, the number of molecules trapped in a cluster follows the Poisson distribution [18].

2.4 CLUSTER TEMPERATURE

The cluster temperature is not determined on our apparatus. However, Farges et al. [7], using a similar cluster beam, found a temperature of 32 ± 2K. Moreover, considering that our clusters are much more than a few microseconds old, a similar temperature can be found using the evaporative ensemble model of Klots [8]. As a result, we believe that the cluster temperature is about 30K in our experiment, even after the clusters have picked up a few molecules. Immediately after the pick-up, the cluster is certainly heated locally, but at a microsecond time scale, the cluster cools down by evaporating a few argon atoms. Considering the large size of the clusters this does not significantly affect their size.

2.5 LOCATION OF THE REACTANTS

The final point to be able to characterize the clusters with which we are working, is question of the location of the reactants (Ba, N_2O and CH_4). Experimental work of the group of Scoles [10] and molecular dynamics calculations of Perera and Amar [9] have addressed this question. The fate of our reactants was not investigated by Scoles, but the finding of this experimental work strongly supports the theoretical work of Perera and Amar. It is therefore possible to infer from the molecular dynamics calculations that both CH_4 and N_2O are solvated at the surface and in the interior of the cluster [19]. Methane may have a slight preference for the cluster interior.

The location of barium is investigated directly on our apparatus by laser spectroscopy. The idea is to observe the spectrum corresponding to fluorescence excitation of barium atoms trapped on argon cluster. A typical spectrum is shown in figure 2. The shape of this spectrum was interpreted in reference [12] as characteristics of surface location of the barium. The splitting of the barium resonance line into two broad components is indeed reminiscent of a Σ/Π splitting of the $Ba(^1P_1 \rightarrow ^1 S_0)$ resonance line in a *barium-cluster diatomic molecule*. Molecular dynamics calculations are under completion. They confirm this conclusion.

3   Experimental evidence of intra-cluster mobility of the reactants

Using the double pick-up technique we attached N_2O to large argon clusters and reacted it with barium atoms [3]. A low pressure of N_2O was used in the pick-up chamber to ensure that a single molecule is attached to the cluster. The chemiluminescence spectrum obtained from the collision of $(Ba + N_2O)Ar_{2000}$ is shown in figure 3. The spectrum has two superimposed components. One extending between 550 and 850 nm, is very structured. It is superimposed on top of a broad unstructured spectrum covering the whole visible range. These components do not scale together

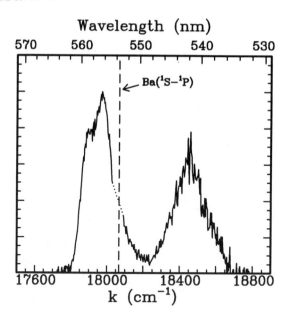

Figure 2: Spectroscopy of the $Ba(Ar)_n$ clusters. The curve is the excitation spectrum obtained by recording the full fluorescence emission when scanning the wavelength of the excitation laser. The position on the $Ba(^1P_1 \rightarrow{} ^1S_0)$ resonance line is marked in the figure. See reference [12] for more details.

when the Ar cluster size is varied, implying that they have independent origins. Moreover, the chemiluminescence intensity is very large. The maximum emission of the unstructured part of the spectrum at 500 nm is five times more intense than the maximum of chemiluminescence recorded when investigating the $(Ba + N_2O)$ reaction in the gas phase with the same flux of N_2O molecules. (This spectrum is shown in the inset of figure 3). Finally, the total chemiluminescence scales roughly as the cluster geometrical cross section as the size of the cluster is varied between $\bar{N} = 300$ and $\bar{N} = 2000$ [19].

A full discussion of these results is given in reference [3]. The broad unstructured part of the spectrum was attributed to fluorescence decay of vibrationally and rotationally hot $BaO(A^1\Sigma^+, a^3\Sigma^+)$ molecules that have left the cluster. This component of the spectrum is very similar to the spectrum recorded in the gas phase $(Ba + N_2O)$ reaction, except that it is much more intense. The large internal excitation of BaO arises due to the very large exoergicity of the reaction ($\approx$ 4eV), of which only 2eV are taken for electronic excitation of BaO. The structured component of the spectrum was attributed to vibrationally and rotationally cold $BaO(A^1\Sigma^+, a^3\Sigma^+)$ molecules that stay solvated within the cluster.

The fact that firstly, the chemiluminescent emission is observed with a large efficiency, and secondly that it scales with the geometrical cross section of the cluster implies that the two reactants Ba and N_2O picked up sequentially by the argon clus-

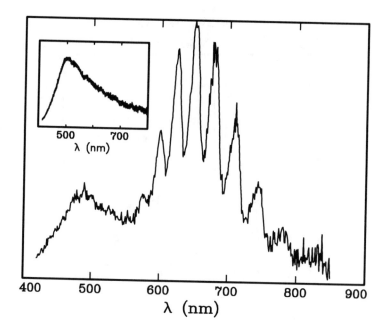

Figure 3: Chemiluminescence spectrum recorded in the $(Ba + N_2O)Ar_{2000}$ reaction. The spectrum is corrected for background signals, for the transmission of the optics and monochromator, and for the response of the photomultiplier. The inset shows the chemiluminescence of the normal gas phase $(Ba + N_2O)$ reaction.

ter are able to migrate within the cluster, to collide each other and to react in the time window of our experiment ($\approx 5\mu s$). The scaling of the chemiluminescence with the cluster geometrical cross section actually rules out the model where the reaction signal originates from barium atoms that hit the cluster directly in the vicinity of an N_2O molecule and therefore do not need migration to react.

   At this point, we are left with the conclusion that argon clusters act as a catalyst for the $Ba + N_2O$ reaction. The clusters play two roles to that end:

1. The argon cluster acts as a 'net' to catch the reactants with a large cross section (its geometrical cross section).

2. Then it acts as a solvent in which the reaction takes place, i.e. it allows the two reactants to migrate and to come close together and react. Considering the surface location of barium, the reaction is likely to occur at the surface of the cluster.

   The finite size of the cluster appears to be very important in the reaction, since it forces the reactants to explore a limited volume, and therefore allows them to find each other with a large probability. In addition it plays a second role since it allows the reaction product either to stay solvated within the cluster or to evaporate.

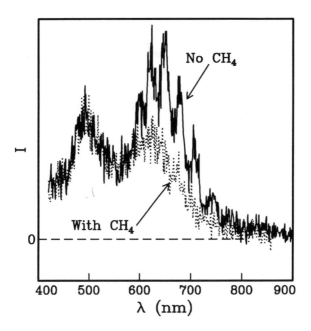

Figure 4: Chemiluminescence spectrum recorded in the $(Ba + N_2O)Ar_{2000}$ whether additional CH_4 molecules are present or not on the cluster.

The question whether reaction products leave the cluster or stay solvated is very interesting and deserves systematic study. We shall come back to this in section 5.

Let us return for a moment to the free migration of the reactants and on the formation of both a free and a solvated reaction product. We have illustrated these two points in very spectacular way by comparing the fluorescence signal obtained above in the $(Ba + N_2O)Ar_{2000}$ to that obtained when the argon cluster also contains CH_4 molecules. The results are shown in figure 4. In the experiment where CH_4 molecules are present, the unstructured signal corresponding to emission of free BaO is intact, whereas the signal originating from solvated BaO is much weaker. This result is well accounted by the above cluster model where reactants are free to migrate and to collide each other. Upon collision between Ba and N_2O, the reaction occurs as before when no CH_4 molecules were present. Part of the electronically excited BaO^* molecules that are formed leaves the clusters and never interact with the CH_4 molecules. The corresponding chemiluminescence signal is thus unaffected by the presence of methane as observed experimentally. In contrast, the BaO^* molecules that stay solvated have the opportunity to collide the CH_4 molecules and to be non-radiatively de-excited. This is consistent with the much weaker chemiluminescence signal that is observed in figure 4 between 550 and 850nm.

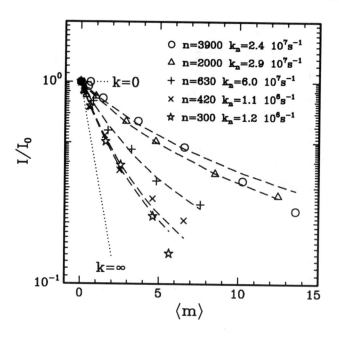

Figure 5: Relative fluorescence intensity as a function of the average number of methane molecules picked-up by the Ar_n cluster for various cluster sizes n as labelled in the figure. The dashed lines are fits to the data using expression 2. The parameter k_n of the fit is indicated in the figure for each cluster size. Note the logarithmic scale. The limiting behaviour for $k_n = 0$ and $k_n = \infty$ are also shown in the figure.

4   Quantitative study of reactants mobility

The former section was dealing with qualitative evidence of the relative mobility between the Ba atom and N_2O molecule in argon clusters. It is large enough for N_2O and Ba to meet each other and react in less than $5\mu s$ (time spent by the clusters in the observation zone). For this system, it is impossible to get better information on the time that is necessary for the Ba atom and the N_2O molecule to find each other. The mobility (observable because it leads to a chemiluminescent reaction) is not in competition with any other observable phenomenon (no signal without N_2O); thus to have an idea of the reaction rate, an absolute measurement of the signal is required. This is extremely difficult in our experiment. In contrast, if the diffusion is in competition with another observable phenomenon that has a known timescale (e.g. radiative decay) a relative measurement of the signal is sufficient.

The aim of the present section is thus to investigate collision quenching of $Ba(^1P_1)$ by CH_4 molecules so as to quantify the relative mobility of $Ba(^1P_1)$ and CH_4 within argon clusters. The timescale of this laser induced fluorescence experiment is set by the radiative lifetime of $Ba(^1P_1)$ ($\tau = 8$ ns [20]).

The double pick-up is used again to attach both CH_4 and Ba on clusters. The fluorescence $Ba(^1P_1 \rightarrow ^1 S_0)$ was monitored as a function of the number of CH_4

molecules for different Ar cluster sizes. The results are shown in figure 5. They are taken from references [4, 19]. For all cluster sizes the fluorescence signal is seen to decrease as we increase the number of CH_4 molecules. The curves in figure 5 can adequately be fitted by a model that assumes:

1. A first order kinetic law for collisional quenching. The quenching rate for cluster of size n is denoted k_n.

2. The number of CH_4 molecules attached to clusters is distributed according to a Poisson distribution.

3. The radiative lifetime τ of barium is unaffected by the vicinity of the argon cluster.

The fitting expression that follows these assumptions is derived in reference [4]. The fluorescence signal $I_{\bar{m}}$ observed when a cluster of size n contains an average number $\bar{m}$ of CH_4 molecules is given by

$$\frac{I_{\bar{m}}}{I_0} = e^{-\bar{m}} \sum_{i=0}^{\infty} \frac{\bar{m}^i}{(1 + k_n i \tau) i!} \qquad (2)$$

The parameter to fit using this expression is the quenching rate k_n. The quality of the fit and the value of the quenching rate obtained for each size of the cluster are shown in figure 5. The error bar on the absolute value of k_n is about 30%.

The quality of the fits shown in figure 5 suggests that the CH_4 molecules interact separately with excited barium. The important point that emerges from figure 5 is that the quenching rate depends on the cluster size. It is found that within 15%, the expression $k_n = 5.4 \times 10^9 n^{-2/3}$ s^{-1} fits the experimentally determined values of k_n. This corresponds to characteristic times for Ba and CH_4 to collide each other and quench electronic excitation ranging between 8 ± 3 ns ($n = 300$) and 47 ± 14 ns ($n = 4000$). These quenching times are comparable with the radiative life time of $Ba(^1P_1)$ (8 ns), and this is precisely the reason why these time measurements are successful.

The fact that the quenching rate depends on the cluster size suggests that for a large part the quenching is controlled by diffusion. It is interesting that the observed dependence $k_n \propto n^{-2/3}$ scales as the surface of the cluster, indicating movements of the quenchers at the surface of the cluster. This is not surprising considering the location of the reactants (discussed in section 2.5).

It is interesting at this point to relate the quenching rates we have measured to a constant of relative diffusion of Ba and CH_4. Two assumption must be made: i) the cluster radius R_n is a characteristic distance between two Ba/CH_4 collisions, ii) the quenching process is purely controlled by diffusion (*i.e.* quenching occurs at the first collision). With these assumptions, the diffusion constant is related to the quenching rate k_n by the expression:

$$D \approx k_n (R_n)^2 \qquad (3)$$

The radius of the cluster can be determined assuming it is spherical, and using the lattice parameter determined for argon clusters (a=0.534nm) [23].

$$R_n = a \sqrt[3]{\frac{3n}{16\pi}} \qquad (4)$$

These two expressions, plus the value found experimentally for k_n, lead to a diffusion constant of $2 \, 10^{-6} \, cm^2 s^{-1}$ that is independent of the size of the cluster. This value is precisely the one we expect for the diffusion of barium on an argon surface at 30K. This constant is indeed given by:

$$D \approx l \, V \, exp(-E/kT) \tag{5}$$

where l is the average hopping distance between trapping sites on the surface of solid argon. It can be estimated by the lattice parameter mentioned above. V is the thermal velocity of barium at 30K ($c.a.$ 70m/s). The exponential represents the probability to jump over the barrier E that separate two trapping sites. An approximate value of E is the energy needed to break one or two Ba-Ar bonds, $i.e.$ the energy to jump from one site with barium surrounded by four argon atoms to another site of the same type through a transition state where barium is bonded to only two or three argon atoms. The Ba-Ar bond strength is estimated to be 110K from references [21, 22]. The resulting value of D is ranging between $2 \, 10^{-7}$ and $8 \, 10^{-6} \, cm^2 s^{-1}$, in fair agreement with our experimental determination.

5   Competition between solvation and desorption

In the last section of this paper we report a very interesting phenomenon that was observed using an improved version of the apparatus described in section 2. We redesigned the barium pick-up to switch from a situation where only one cluster in 10^5 carries a barium atom, to a situation where 10% of the clusters contain one barium atom (1% of the cluster then contain 2 barium atoms). The idea is to pass the cluster beam through a second pick-up chamber that contains barium vapour.

The resulting strong increase of the signals allows us to detect what we call a half-collision between barium and the cluster at the surface on which it is solvated. Figure 6 reports the emission spectrum observed when exciting barium atoms in Ar_{125} clusters at various wavelengths chosen inside the excitation spectrum of figure 2.

The spectra mostly constitute a broad emission peaking 120cm^{-1} to the red of the barium resonance line. It arises due to the fluorescence decay of excited barium atoms that are at the surface of the cluster [12]. When the excitation laser is tuned to the red feature of the excitation spectrum ($\lambda > 548nm$), the fluorescence has no other component than this broad emission. In contrast, when the laser is tuned to a wavelength smaller than 546nm, the resonance line of free barium atoms appears besides the fluorescence decay of surface solvated atoms. This new line corresponds to barium atoms that have been excited at the surface of the cluster in the "Σ" component of the excitation spectrum, and that have left the cluster before emitting a photon. Remember that the Σ potential curve of a $Ba(^1P_1)$-$spherical \, perturber$ system is repulsive. We thus come to the conclusion that upon electronic excitation above a certain threshold, barium atoms start have the choice either to desorb out of the cluster or to stay solvated at the cluster surface and continue its migration. The bluer the laser excitation, the larger the barium resonance line, and therefore the greater the number of barium atoms desorbed from the cluster. Other results that are not shown in the figure indicate that large clusters (a few thousand argon atoms) lead to less desorption than observed in the figure for Ar_{125}.

Formation of both a solvated and free reaction product has been encountered above in section 3 when considering the full collision $Ba + N_2O \rightarrow BaO^*$. In this

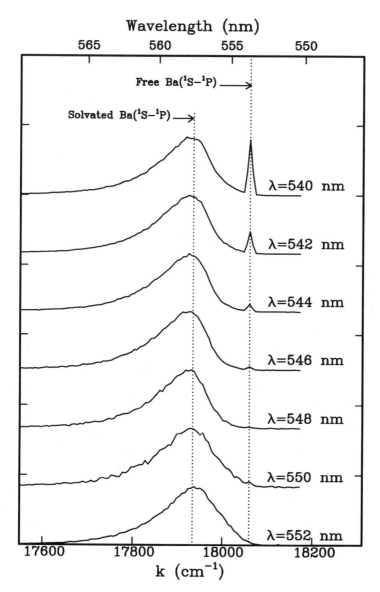

Figure 6: Spectroscopy of the $Ba(Ar)_{125}$ clusters. Emission spectrum as a function of the wavelength λ of the excitation laser as labelled in the figure.

case as well, the size of the argon cluster was playing a role in the branching between both exit channels.

In such experiments, and especially in the half-collision experiment reported in the present section, a well defined amount of energy is deposited into a reaction coordinate. For the half-collision experiment, the energy is deposited in the repulsive barium-cluster coordinate (remember, firstly that barium is at the surface of the cluster, and secondly that the laser is tuned to the Σ-like component of the excitation spectrum). Therefore, what figure 6 actually shows is the competition between the release of this energy as kinetic energy along the repulsive coordinate (the barium desorbs), and thermalization of this energy in the bath formed by the cluster (the barium stays solvated). The number of degrees of freedom of the bath can be varied by varying the size of the cluster. For us, such an experiment provide a *clean sample* in which to study the coupling between an isolated reaction coordinate and a thermal bath.

6   Concluding remarks

The techniques described in this article are a preview of the vast array of experiments that are becoming available for the study of elementary collision and half-collision processes at the contact of a thermal bath.

The processes that have been investigated so far include chemiluminescent reactions, collisional quenching and desorption/migration of electronically excited species. These investigations provide evidence for solvation and subsequent surface motion of barium, CH_4 and N_2O when picked-up by an argon cluster. Upon laser excitation, collision quenching or chemical reaction within the cluster, barium (or the reaction product BaO) has the choice either to stay solvated or to exit to the gas phase. The branching between these two events obviously says something about the coupling between movements along a normal gas phase reaction coordinate that should push the reaction product off the cluster (the processes we look at are very exoergic), and thermalization of the excess energy inside the cluster. From this point of view, our experiments offer a unique means to study how, in a very controlled way, movements along a reaction coordinate are modified by the coupling with a thermal bath. This aspect of our work seems especially promising in being able to stimulate the interplay between experiment and theory.

Another important output of the present study is the catalyst effect of argon clusters on the $Ba + N_2O$ reaction. The surface of the cluster plays an essential role in that respect since it cages the reactants until they react. The gas phase cross section of this reaction is thus increased to be as large as the geometrical cross section of the cluster that has sequentially trapped the two reactants, barium and N_2O.

References

[1] I. Benjamin, A. Liu, K.R. Wilson and R.D. Levine J. Phys. Chem. **94**, 3937-3944 (1990); D.M. Charutz and R.D. Levine Chem. Phys. **152**, 31-43 (1991); D.M. Charutz and R.D. Levine J. Chem. Phys. **98**, 1979-88 (1993).

[2] R. Alimi and R.B. Gerber Phys. Rev. Lett. **64**, 1453 (1990); R. Alimi, R.B. Gerber, A.D. Hammerich, R. Kosloff and M.A. Ratner J. Chem. Phys. **93**, 6484 (1990).

[3] A. Lallement, J. Cuvellier, J.M. Mestdagh, P. Meynadier, P. de Pujo, O. Sublemontier, J.P. Visticot, J. Berlande, and X. Biquard Chem. Phys. Lett. **189**, 182 (1992).

[4] A. Lallement, O. Sublemontier, J.-P. Visticot, A.J. Bell, J. Berlande, J. Cuvellier, J.-M. Mestdagh, and P. Meynadier Chem. Phys. Lett. **204**, 440 (1993).

[5] J. Cuvellier, P. Meynadier, P. de Pujo, O. Sublemontier, J.P. Visticot, J. Berlande, A. Lallement, et J.M. Mestdagh Z. Phys. D **21**, 265 (1991).

[6] A.J. Bell, J.M. Mestdagh, J. Berlande, X. Biquard, J. Cuvellier, A. Lallement, P. Meynadier, O. Sublemontier, and J.P. Visticot J. Phys. D In press.

[7] J. Farges, M.F. de Feraudi, B. Raoult, and G. Torchet J. Chem. Phys. **84**, 3491 (1986).

[8] C. Klots, Nature **327**, 222 (1987); C. Klots, Z. Phys. D **20**, 105 (1991).

[9] L. Perera and F. Amar, J. Chem. Phys. **93**, 4884 (1990).

[10] T. Gough, M. Mengel, P. Rowntree, and G. Scoles, J. Chem. Phys. **83**, 4958 (1985); D. Levandier, S. Goyal, J. McCombie, B. Pate, and G. Scoles, J. Chem. Soc. Faraday Trans. **86**, 2361 (1990); X. Gu, D. Levandier, B. Zhang, G. Scoles, and D. Zhuang, J. Chem. Phys. **93**, 4898 (1990); S. Goyal, G. Robinson, D. Schutt, and G. Scoles, J. Phys. Chem. **95**, 4186 (1991).

[11] P. de Pujo, J.-M. Mestdagh, J.-P. Visticot, J. Cuvellier, P. Meynadier, O. Sublemontier, A. Lallement and J. Berlande Z. Phys. D **25**, 357-362 (1993).

[12] J.P. Visticot, J. Berlande, J. Cuvellier, A. Lallement, J.M. Mestdagh, P. Meynadier, P. de Pujo, O. Sublemontier Chem. Phys. Lett. **191**, 107 (1992).

[13] J. Cuvellier, J.M. Mestdagh, J. Berlande, P. de Pujo and A. Binet, Rev. Phys. Appl. **16**, 679 (1981).

[14] J.M. Mestdagh, J. Berlande, J. Cuvellier, P. de Pujo and A. Binet, J. Phys. B **15**, 439 (1982).

[15] R. Campargue, J. Phys. Chem. **88**, 4466 (1984).

[16] M. Kappes and S. Leutwyler, in Atomic and Molecular Beam Methods Vol 1 p.380, G. Scoles ed, Oxford Univesity Press (1988).

[17] O.F. Hagena and W. Obert J. Chem. Phys. **56**, 1793 (1972).

[18] J.R. Green and D. Margerison, in Statistical treatment of experimental data, Elsevier, p.35 (1978).

[19] A. Lallement, Thesis, Université Paris Sud, Orsay (1993), unpublished; A. Lallement, et al. to be published.

[20] F.M. Kelly and M.S. Mathur, Can. J. Phys., **55**, 83 (1977).

[21] A. Kowalski, D. Funk, and W. Breckenridge, Chem. Phys. Lett. **132**, 263 (1986).

[22] R. Bennet, J. McCaffrey, and W. Breckenridge, J. Chem. Phys. **92**, 2740 (1990).

[23] J. Farges, M.F. de Féraudi, B. Raoult and G. Torchet Surf. Sci., **106**, 95 (1981).

SPECTRA OF CONFORMERS OF MASS SELECTED VAN DER WAALS CLUSTERS

H. L. SELZLE and E. W. SCHLAG
Institut für physikalische Chemie
Technische Universität München
Lichtenbergstr.4
D-8046 Garching, Germany

Abstract

Cluster formation in a supersonic jet leads always to a distribution of clusters of various sizes. Mass selection via resonant two-color soft photoionization yields spectra of a given size, but this is still insufficient as there are many conformational isomers produced simultaneously in a non thermodynamical distribution even for a given single cluster size. This polymorphism problem of clusters can be overcome by mass selected spectral hole-burning in the gas phase. In this paper we present spectra of conformational isomers of benzene...Ar_n (n = 1,5) complexes. From the experiment the contributions of the different conformations to the spectra could be separated and information about the structure obtained. Varying the time delay between resonant excitation and soft ionization in addition yields information concerning the stability of the different conformational isomers.

1. Introduction

A very important problem in the spectroscopy of well defined molecular clusters is the disentanglement of the various contributions of different complexes to the spectra. This arises from the fact, that beside the simultaneous formation of clusters of different sizes, clusters of a single size can additionally be found in different conformations. This is due to the fact that even at nearly constant energy several conformers are predicted and often found. These clusters cannot be distinguished even though one employs mass analyzed ionization detection; the spectra obtained thus represent a superposition of the spectra of conformational isomers. Due to the lack of rotationally resolved spectra for larger clusters, their structure must be concluded from vibrational spectra, if available, or from theoretical calculations. This problem will be dominant for clusters with many particles, but it is also important even for simple dimers of molecules which can be arranged in various positions of the molecular geometry. This is of fundamental interest in the case of organic molecules, even for the simple case of a benzene dimer. This dimer represents a prototype situation for the benzenoid interaction which is found in many structures even in proteins. For this dimer early experiments by Janda et al. [1]

115

J. Jortner et al. (eds.), Reaction Dynamics in Clusters and Condensed Phases, 115–136.
© 1994 *Kluwer Academic Publishers. Printed in the Netherlands.*

suggested a T-shape structure, whereas on general consideration a parallel-displaced structure was proposed . Recent experiments by Henson et al. [2] reported non equivalent molecules in the cluster and in addition theoretical calculations predicted a 'floppy' T-shape structure [3,4,5]. This simple complex already demonstrates the difficulty for the determination of the conformation of the cluster without further experimental improvements.

In a first step van der Waals (vdW) complexes can be mass analyzed using soft ionizing techniques such as multi photon ionization (MUPI) mass spectroscopy or resonance enhanced multiphoton ionization (REMPI) which permits one to identify spectra from a single mass [6]. But even though the mass may be identified, this does not uniquely identify the complex, since it may exist as several structural isomers. A clear example here is the benzene...Ar_2 cluster. The two Argon atoms can be on the same side of the ring, a (2|0) structure, or on different sides of the ring, a (1|1) structure. Both structures have been suggested since two ionization potentials (IP) were observed in the MUPI mass spectrum at that mass [7]. The structure of the (1|1) has also been observed by Neusser et al. [8] using very high resolution spectroscopy. We recently introduced a mass selective hole-burning spectroscopy in the gas phase to study vdW complexes and applied this method to the benzene dimer [9] and also in a first study to the benzene...Ar_2 complex [10]. The technique of hole-burning spectroscopy was initially developed for solid state spectroscopy to identify electronic and vibrational states below the inhomogeneous broadening of transitions due to different sites in the solid [11]. This method can also be applied to the gas phase in a pump-probe type experiment. These experiments have in common that exciting a specific transition in the molecular cluster with a first laser depopulates the corresponding ground state. If relaxation to different states other than the initial state occurs or if the molecule is photochemically destroyed or ionized, a scan with a second probe laser will find a reduced intensity for transitions which have the same ground state in common.

This method has been applied to study differential cross sections for rotational transitions in collisions which repopulate the depleted ground state [12] or to obtain Doppler free spectra from hole-burning in the ground state velocity distribution [13]. Photochemical dissociation with the first laser can also be used to separate spectra of clusters differing in mass and size. The probe laser will find a reduced intensity for all lines belonging to clusters of the same size as the one which was depleted with the first ionizing laser. This method was successfully applied for IR predissociation to study the ground state of clusters [14] and for MUPI depletion of the ground state to separate excited state spectra of different sized clusters with the probing laser by observing fluorescence dip spectra [15]. This fluorescence techniques here represents an alternative for mass selection of spectra, if MUPI mass spectroscopy is not available, and correlation to cluster size is obtained from indirect methods or theoretical considerations and requires confirmation.

A new important feature of hole-burning spectroscopy is to investigate different conformers of clusters of only a single size i.e. of identical molecular weight. Here the clusters when produced in the supersonic jet expansion can form different conformational isomers which cannot be separated from even the softest REMPI mass spectroscopy. In this case the persistent spectral hole-burning is a new tool to study such isomers. Colson et al. [16,17] applied this technique to clusters of phenol with water and phenol with ethanol, but could not find any conformational isomers, but could show that this technique distinguishes between cis- and trans-m-cresol. Topp et al. studied clusters of perylene [18,19]. They observed dips in the fluorescence excitation probing scan. From hole-burning where they used different bands in the absorption spectrum they separated spectra of different conformational isomers. Here the fluorescence experiment which also lacks mass identification and correlation to the cluster size was related to the size effect of the cluster shift in the optical spectra. When larger clusters of molecules with noble gases are prepared in a supersonic jet expansion, the temperature of the internal degrees of freedom can be still larger than the translational temperature in the jet. In large clusters the density of low frequency intermolecular modes increases and one can not assume that all clusters are in the vibrational ground state of the intermolecular vibrational manifold. This leads to a further complication as the cluster of different internal energy behave different and show rigid and fluxional structures depending on their internal energy. Hole-burning experiments with carbazole...Ar_n clusters (n = 4-6) showed that only the sharp features in the observed bands are effected by the hole-burning experiment [20] which were assigned to the rigid structure, but no further information about conformational isomers was obtained.

We have now applied this method for the first time to clusters of benzene with Argon atoms where we not only separate the spectra of different conformational isomers at a given cluster size, but in addition obtain information about the stability and the structure of the clusters from ionization delay-time dependent experiments.

1. Experimental

The preparation of the clusters is performed in a supersonic jet experiment (Fig. 1). The apparatus consists of a first vacuum chamber which contains the expansion nozzle and is separated from the second chamber via a skimmer with a 1 mm opening. The skimmer is placed 35 mm downstream from the nozzle and selects the center part of the molecular beam. The second chamber contains the ion optics and the interaction region with the lasers and here the pressure is maintained at about 10^{-6} Torr. This chamber is followed by a RETOF-MS at a pressure of about 10^{-7} Torr. To generate the clusters, benzene at $0°$ C corresponding to 32 mbar seeded in a mixture of He with 10 % Ar at a total pressure of 5.6 bar is expanded through a pulsed nozzle into the vacuum. Typical opening times of the nozzle are 50 to 100 μsec at a repetition rate of 10 Hz. The molecular beam enters the ionization region at a distance of 170 mm from the nozzle and is crossed with the hole-burning laser beam of a frequency doubled dye laser (DYE 1, Quanta Ray PDL-1) pumped by a Nd:YAG laser (Nd-YAG 1, Quanta Ray DCR-1A). With this laser the depletion of the ground state of a selected cluster is performed via REMPI by placing the frequency to the corresponding resonant 6_0^1 transition of benzene in the cluster. Probing occurs 400 nsec later somewhat downstream from the position of the first laser. The time delay is controlled via a digital pulse generator (DG, Stanford Research Model DG 535). During the supersonic expansion clusters of benzene...Ar_n are generated simultaneously with a reasonable amount of clusters up to n = 6. To avoid contamination of the spectra from decomposition of simultaneously excited larger clusters soft ionization has to be used for probing with a two-color REMPI technique [21]. For this the molecular beam is crossed with two frequency doubled dye lasers (DYE 2, DYE 3, Quanta Ray PDL-1) which are simultaneously pumped by a Nd:YAG laser (Nd-YAG 2, Quanta Ray DCR-1A), where the first laser is scanned in the region of the S_1 6_0^1 transition and the second laser is used at a fixed frequency slightly above the ionization threshold of the cluster. The timing sequence of the hole-burning experi-

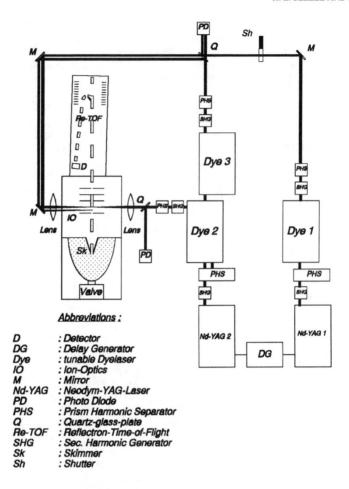

Figure 1:   Experimental setup for the hole-burning experiment.

ment is shown in fig. 2. At time 0 nsec the hole-burning laser is fired. To obtain a good hole-burning signal the laser intensity is high and produces a great amount of ions from REMPI of the selected isomer. To eliminate noise and saturation of the ion detector an electric field opposite to the direction of the jet is applied which removes the ions within 350 nsec out of the interaction region. At time
360 nsec the field is reversed, and at 400 nsec the probing laser is fired. At that time the jet has moved by about 0.6 mm due to the speed of about 1500 m/sec [22] which defines the position of the probing laser. The ionizing laser is optically delayed by 20 nsec relative to the resonant scanning laser. Thus the ions which are generated from the first scanning laser will start earlier towards the RETOF and this allows to separate one- and two-color contributions to the ion signal. The ions are extracted into RETOF-MS with

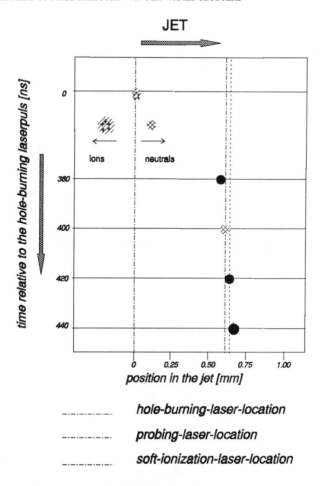

Figure 2: Timing diagram for the hole-burning experiment.

a resolution larger than 2000 M/ΔM. This high resolution is necessary for the separation of the ions produced by the different lasers. The ion signal is recorded with a transient recorder (Tektronix RTD 710) with a sampling rate of 200 Msamples/sec. The one-color and two-color signals of a selected mass are recorded separately from placing proper gates on the recorded time-of-flight mass spectrum and are averaged by a laboratory computer (FORCE MiniForce 3P37ZBE). This permits one to separate one- and two-color spectra to obtain pure two-color soft ionization spectra by detecting ions produced only by the second ionizing laser.

3. Results and Discussion

The benzene...Ar_n complexes have come to new importance as a model system for the study of intracluster dynamics [23]. Here at low temperatures well defined structures are proposed whereas for higher temperatures isomerization of different conformations should occur and no defined structure is maintained. As in systems containing larger aromatic molecules as phenanthrene...Ar_2 [24], perylene...Ar_2 [25] or substituted benzene as paraxylene...Ar_2 [26], where trans (1|1) and cis (2|0) structures are found, one expects that this should be also possible for benzene...Ar_2. Mons et al. [7,27] have proposed this from their experiments where they find different ionization thresholds for two bands of the S_1 6^1_0 transition in the complex and also from two different rotational contours of these bands. From Doppler-free spectra Neusser et al. [8] could show that the most red-shifted peak of this transitions corresponds to a trans structure of a well established geometry from a complete rotational analysis.

A much more direct approach to find different conformational isomers is given by the technique of mass selected hole-burning (MSHB) spectroscopy which uniquely distinguishes between complexes with the same stoichiometry but different structure. With a first laser which is positioned on a resonant transition of isomer A (Fig. 3) the ground state of this species will be depleted. When probing with two-color resonant

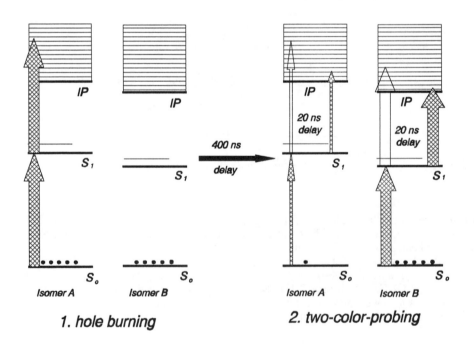

Figure 3: Hole-burning with two color probing.

ionization at a somewhat later time one will find a reduced number of isomers A in this state but for isomer B no change will have occurred. One then observes a reduced intensity in the REMPI spectrum for all lines belonging to species A which allows to identify the two different species A and B.

In a first experiment we have applied this new technique to the benzene...Ar_2 complex, where we found two different conformational isomers [10]. The corresponding spectra could be separated and the assignment to the one- and two-sided complexes are consistent with the high resolution measurements by Neusser et al. [8].

3.1. ONE- AND TWO-COLOR PROBING

The formation of clusters of benzene with noble gases in a supersonic jet expansion does not lead to a single cluster size: in contrast, many differently sized clusters are formed. The identification occurs via ionization and detection in a mass spectrometer. If this is done by multiphoton ionization or by one-color photoionization from an resonant intermediate state, the ions are always prepared with an excess energy in the ion. The ion distribution thus obtained does not reveal the distribution of the clusters in the jet and the spectra obtained by recording a given mass are contaminated from dissociation of larger

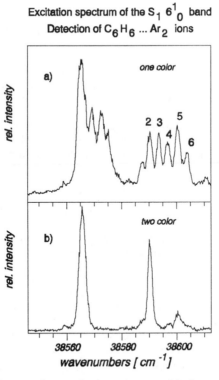

Figure 4: One- and two-color excitation spectra with detection at the mass of benzene...Ar_2.

sized clusters. In fig. 4a the one-color spectrum with detection at the mass of the benzene...Ar_2 cluster is shown. In this case the total energy from absorption of two photons is more than 77130 cm^{-1}, whereas the IP of the bare benzene molecule is only 74555 cm^{-1} [28] which leads to a lower limit of 2600 cm^{-1} for the possible excess energy even for the bare benzene molecule. The clusters have still a lower ionization threshold, which in the case of benzene...Ar is found to be 170 cm^{-1} lower than the benzene IP [29]. For the higher clusters the IP's were measured by Mons et al. [30] and they found a mean value of about 920 cm^{-1} for lowering of the IP for the possible isomers of the benzene...Ar_6 complex. This corresponds to an average stabilization energy of 160 cm^{-1} for every Argon atom added to the ion complex. This leads to a high excess energy in the cluster ions which thus will undergo fast vibrational predissociation after ionization. The spectrum obtained in fig. 4a therefore does not represent the spectrum of the benzene...Ar_2 cluster but is the superposition of the excitation spectra of benzene...Ar_n clusters. The spectrum shows comparable intense peaks arising from clusters with n up to 6. When one performs the experiment with two-color ionization, where the wavelength of the second laser is adjusted just above the ionization threshold of the cluster a much simplified spectrum is obtained (fig. 4b). Only two large peaks remain together with a weaker peak on the blue side at 38600 cm^{-1}. All contaminations of the spectrum from fragmentation of higher clusters ions have vanished and a clear pure spectrum of the benzene...Ar_2 cluster is obtained. In addition it revealed the weaker blue peak which was completely hidden under the fragmentation peak of a larger cluster.

3.2.   SPECTRAL SHIFTS OF BENZENE...AR_n CLUSTERS

In order to assign the peaks of the one-color spectrum of fig. 4a one has to perform soft ionization with detection at the selected mass of the benzene...Ar_n cluster with n larger than two. In fig. 5 two-color excitation spectra for the clusters of benzene with Argon are shown for n up to 5. The spectrum of the benzene...Ar complex is similar to the spectra obtained by Bernstein et al. [31], Neusser et al. [8] and Mons et al.[7] which also show spectra of the higher clusters. The spectra were obtained via resonant excitation of the 6^1_0 transition of the benzene molecule in the cluster. All the transition which are found show a single peak which indicates, that the degeneracy of the v_6 intramolecular vibration is not lifted by the complexation. The main Peak B for benzene...Ar is the transitions to the ground state of the vdW potential in the first excited singlet state and the other weak peaks belong to vdW modes which now have been completely analyzed and assigned from high resolution spectra of Neusser et.al. [32] and from theoretical analysis from van der Avoird [33]. The structure can be described as the conformation where the Argon atom is placed above the center of the benzene ring. From the spectrum a cluster shift of 19 cm^{-1} is obtained when one adds a single Argon atom to the benzene molecule. In the case of benzene...Ar_2 two main peaks A and B are found and also a weaker peak C to the blue side of peak B. The peak A shows about double the red shift of peak B of the benzene...Ar complex relative to the bare benzene molecule transition. From this it is predicted that this is the complex where the second Argon atom is added on the opposite side of the benzene molecule forming the (1|1) complex. Peak B was precluded to the case where both Argon atoms are on the same side of the benzene molecule[7] and represents the (2|0) conformation. This was based on the finding of different rotational conturs and IP's for these peaks serving as resonant intermediate states.

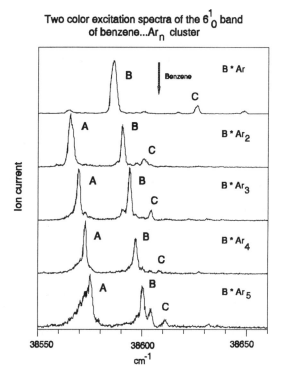

Figure 5: Excitation spectra of benzene...Ar_n clusters with soft ionization detection.

The pattern found in the spectrum of the benzene..Ar_2 complex pertains for all the larger clusters. The main difference is an incremental blue shift when adding a further Argon atom. This behavior is in contrast to the shifts found for clusters with larger aromatic molecules like anthracene, where one finds an incremental red shift if a further Argon atom is added [34].

The larger peak C of the vdW modes of the benzene...Ar complex which is found 40 cm^{-1} to the blue side of the main peak is assigned to the benzene Argon stretching mode. The peaks C found for the other higher clusters also have a blue shift of about 35 cm^{-1} relative to the peak A and a similar intensity, which indicates that this peak represents also a vdW mode transition. The peak A of the higher clusters first become narrower and starting from benzene...AR_3 show a shoulder on the red sid. This could be an indication that the higher clusters are not as efficiently cooled as the smaller ones during the supersonic expansion. Another possibility is that this shoulders represent absorptions bands of different conformational isomers formed in the jet which are not resolved. These conformational isomers cannot be separated from mass resolved two-color soft ionization and also a rotational contour analysis cannot be performed if the clusters get larger. The only indication so far for different isomers is the difference in the ionization threshold which is found for different resonant intermediate states [30]. Therefore a new experimental approach has to be performed which uniquely can identify a conformational isomer within a complex spectrum. This now can be achieved with the technique of persistent spectral hole-burning in the gas phase.

3.3   MEASUREMENTS OF CONFORMATIONAL ISOMERS WITH IONIZATION DELAY

3.3.1   Benzene...Ar$_2$

We have applied the mass resolved spectral hole-burning technique first to the benzene..Ar$_2$ complex. In this experiment the hole-burning laser was positioned to the major peak A of the two-color REMPI spectrum of the 6^1_0 band of the complex in fig. 5. After 400 nsec the remaining population of the clusters is probed with a first laser scanning the resonant intermediate state. After an additional delay of 20 nsec the resonant excited clusters are ionized and the ion signal is recorded. Peak A now shows a reduced intensity when the spectrum is scanned with the probing laser, whereas peak B is not affected. With our setup we could get a depletion down to 40 % of the original population. During the scan the hole-burning laser is applied for ten laser shots and then blocked with a shutter for the ten next laser shots at every wavelength of the scanning laser. The averaged signals with the hole-burning laser on and off are subtracted from each other and the difference is recorded as a function of the wavelength of the scanning laser. The resulting spectrum is shown in fig. 6a. The spectrum without hole-burning is drawn with a dotted line for comparison. When one performs the hole-burning at the position of peak A one finds the complete vanishing of peak B in the difference spec-

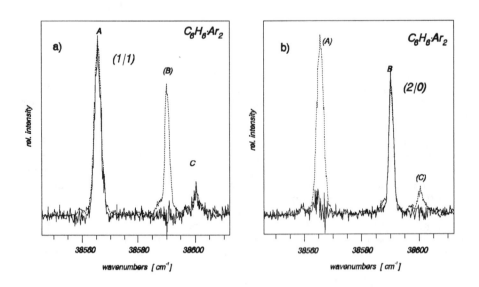

Figure 6:   Hole-burning for the benzene...Ar$_2$ complex with delayed ionization.

trum. The spectrum also shows that there are more than one conformational isomers for the benzene...Ar_2. Peak C belongs to the same isomer as peak A and arises from a vdW mode of this complex. When the hole-burning laser is positioned to peak B (fig. 6b) the difference spectrum gives the spectrum of the second conformer. The signal here at the wavelength position of peak A arises from incomplete subtraction due to laser fluctuations. With this experiment now all lines could be assigned and a complete separation into the separate spectra of the two conformers of the benzene...Ar_2 complex could be achieved.

3.3.2   Benzene...Ar_3

The complex of benzene with two Argon atoms was found to posses two different isomers, the (1I1) and the (2I2) structures. In the case with three Argon atoms there should be also two different isomers, the (2I1) , where two Argon atoms are on one side and one on the opposite side, and the (3I0) isomer, where all Argon atoms are on one side. To study the possible isomers we applied the hole-burning technique also to this cluster. The corresponding spectra are shown in fig. 7. Similar to the case with two Argon atoms we placed the hole-burning laser to the wavelength position of peak A (fig. 7a). The measured difference spectrum shows that the spectrum for this isomer is given by the two peaks A and C only as in the case of benzene...Ar_2. If one now places the hole-burning laser to the position of peak B the remaining difference spectrum clearly

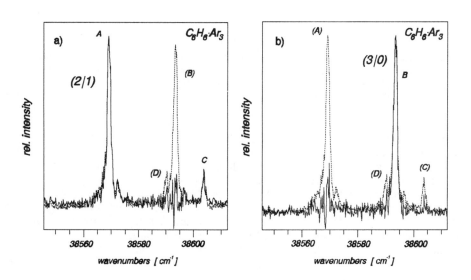

Figure 7:   Hole-burning spectra for the benzene...Ar_3 complex with delayed ionization.

shows that the spectrum of this isomer consists of only one sharp peak. It is interesting to note, that the weak feature D on the red side of Peak B does neither belong to one of the two isomers. To identify the corresponding structures of the two isomers found so far one cannot apply simple additivity laws with a constant shift per added atom. But from the spectra of benzene...Ar_2 one can conclude the following. If one adds one further atom on the same side to an existing atom then this will lead to a small blue shift and therefore peak B can be assigned to the (3|0) structure which is formed by adding one atom to the (2|0) structure. Peak A is red-shifted by about 20 cm^{-1} to the one-sided (2|0) cluster and therefore can be assigned to the cluster with two Argon atoms on one side and the third on the opposite side. The assignment can be compared with results of force field calculations [35]. For the calculation a high quality ab initio potential is used for the benzene-Argon interaction[36,37] and the Argon-Argon interaction is taken from Aziz et al. [38]. From this calculation two stable minima are found for the benzene...Ar_3 cluster. In the case of the (3|0) isomer, one Argon atom is placed above the benzene plane near to the C_6 axis of the benzene molecule, whereas the two other Argon atoms are placed on the outside above the ring between the hydrogen atoms. This explains the small interaction with the π-electrons and therefore the binding energy for the outer Argon atoms is not much different in the ground and the first excited electronic state of benzene molecule. From this the small blue shift for the (2|0) cluster relative to the (1|0) cluster and for the (3|0) cluster relative to the (2|0) cluster can be explained. It also explains the small blue shift of the (2|1) cluster relative to the (1|1) cluster. In addition to the structural information from the force field calculation for the (2|1) complex a total symmetric vdW stretch mode with a value of 35 cm^{-1} is also computed which nicely agrees with the 35 cm^{-1} shift of peak C and confirms the assignment of the spectrum consisting of peak A and C to the (2|1) structure.

3.3.3 benzene...Ar_4

In the case of the benzene...Ar_4 complex the number of possible conformers increases. From a topological view there exists the possibility of three different complexes, the (4|0) complex with all Argon atoms on one side, the (3|1) complex with three Argon atoms on one side and the fourth one on the opposite side, and the symmetric (2|2) complex. We also measured the hole-burning spectra for this cluster which are shown in fig. 8. The band at peak A consists of a sharp peak A with a half-width of less than 2 cm^{-1} and a shoulder to the red. If the hole-burning laser is now placed on the position of peak A the obtained difference spectrum shows only the sharp peak A together with the peak C (fig. 8a). Similarly the difference spectrum with the hole-burning laser positioned to the wavelength of peak B revealed a second isomer (fig. 8b). Hole-burning in the shoulder of peak A also leads to a spectrum of its own which is very weak. The assignment of the observed spectra is similar to the assignment for the clusters of benzene with two and three Argon atoms. From the observed shift the spectrum from hole-burning at B represents the (4|0) isomer, whereas the sharp peak at A could either be the (3|1) isomer or the (2|2) isomer. The peak C corresponding to a vdW mode of this isomer is also found to be 35 cm^{-1} to the blue of the peak A. A force field calculation for the clusters was also performed. Here only the (2|2) cluster shows a total symmetric vdW stretching vibration at 35 cm^{-1} and therefore peak A is tentatively assigned to this structure.

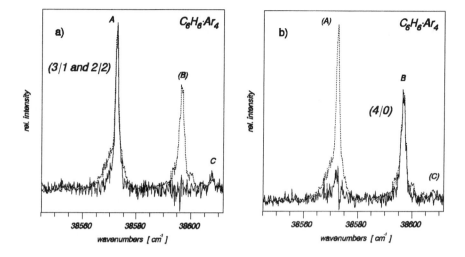

Figure 8: Hole-burning spectra for the benzene...Ar_4 complex with delayed ionization.

3.4 MEASUREMENTS OF CONFORMATIONAL ISOMERS WITHOUT IONIZATION DELAY

In all the cases discussed so far for the benzene-Argon complexes the assignments were performed for the topological isomers (n|m) which differ by the number of Argon atoms on either side of the benzene ring. The force field calculations then lead to the most stable structures for these isomers, where one always finds that the first Argon atom of one side is placed near the main C_6 rotation axis of the benzene molecule and all further Argon atoms which are attached at the same side are shifted out above the plane to places between the hydrogen atoms. Beside these most stable configurations there are more local minima for the clusters with less binding energy. The special formation of the clusters in a supersonic jet always leads to a strongly non thermodynamic distribution for this conformations. Due to the rapid cooling in the supersonic jet expansion the clusters can be trapped in local minima which will manifest themselves in different absorption spectra. Due to the different conformations vibronic coupling within the clusters will be different and can lead to variations of the lifetimes in the excited state of the clusters. In benzene the 0-0 transition is forbidden and this is also found for the clusters with Argon. Therefore, all REMPI experiments are performed via the 6^1_0 transition of the benzene molecule which will deposit a vibrational energy of 521 cm^{-1} in the cluster. From ab initio calculations one finds a stabilization energy of 419 cm^{-1} for the ground state of the benzene...Ar complex [36], which will be larger by 20 cm^{-1} in

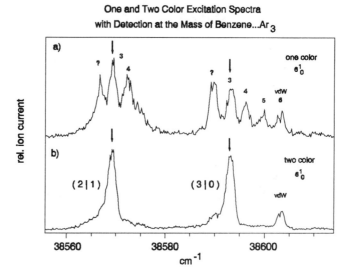

Figure 9:   One- and two-color excitation spectra of benzene...Ar$_3$. The peaks denoted with ? are also transitions of a benzene...Ar$_3$ isomer.

the first exited state according to the measured red-shift of the 6^1_0 transition. This energy is still smaller than the vibrational energy of the excited ν_6 intramolecular vibration and this could lead to vibrational predissociation. In the experiments so far the clusters were excited to the resonant intermediate ν_6 state and then ionized after a delay of 20 nsec. From the existence of ion signal after this delay it is found that the lifetime of the intermediate state has to be larger than 20 nsec and no fast decay due to vibrational predissociation had occurred. From this a very weak coupling of the intramolecular ν_6 vibration to the vdW modes must be concluded for the isomers which were found in the spectra. For structures which undergo a very fast predissociation the population after 20 nsec would have vanished and they could not be detected with the delayed two-color ionization probing. This is not true for the one-color probing. In this case the resonant excitation and ionization laser is the same and thus by principle a complete overlap in time is given leading to a zero time delay between probing and ionization. Despite the problem of contamination from ion fragmentation one can analyze the one-color REMPI spectrum for additional lines which are not seen in the two-color spectra of the clusters. This is demonstrated in fig. 9 for the case of the benzene...Ar$_3$, where the one- and two-color spectra are shown. The lower trace shows the two-color signal obtained after 20 nsec delay where the major peaks of the (3|0) and the (2|1) conformers are marked with an arrow. The upper trace shows the one-color signal which was recorded simultaneously in the same scan. The corresponding peaks of the two benzene...Ar$_3$ isomers are also marked with an arrow in this spectrum. The peaks to the right of the marked peaks arise from fragmentation of cluster ions of larger complexes which can be identified from their respective two-color spectrum of fig. 5. But there are also additional peaks to the left, where there is no transition of a higher cluster and the peaks therefore must arise from complexes of benzene with three Argon atoms. There is already an indication of these peaks in the two-color spectrum which shows up as a shoulder. This weak peak to the left of the main peak of the (3|0) isomer in the two-color signal was also be seen by Mons et al. [30] and was assigned to a (3|0) isomer with symmetric positions of the three Argon atoms above the benzene ring. In the one-color spectra these peaks are much more intense which indicates that the corresponding conformers must have a short lifetime which is less then 20 nsec, whereas the other conformers found in the two-color spectra live longer than 20 nsec.

3.4.1 *Two-color soft ionization spectra without ionization delay*

In order to prove that the additional transitions found in the one-color spectra are true cluster signals and no contaminations from other higher clusters we modified the experiment to achieve two-color excitation spectra without time delay between the resonant excitation and the ionization step. For this the intensity of the first resonant laser was reduced to such an amount, that the one-color signal was less then 10% of the two-color signal. Then the delay between the resonant laser and the ionization laser was adjusted in a way that the two lasers also overlap in time at the probing volume. To remove the rest of the one-color signal in the spectrum, the ionization laser was blocked for half of the laser shots during the scan and the difference spectrum was recorded, which now is a pure two-color spectrum recorded at zero ionization delay. With this setup we repeated the REMPI spectra for the benzene...Ar_3 cluster. The result for the two-color zero ionization delay is shown in the lower trace of fig. 10. The upper trace

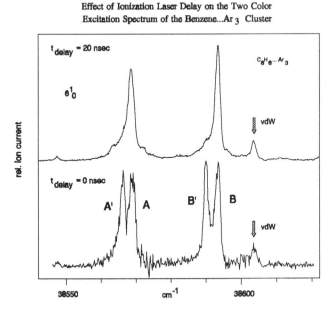

Figure 10: Comparison of two-color soft ionization spectra for the benzene...Ar_3 complex. Peak A' and peak B' refer to short-lived complexes.

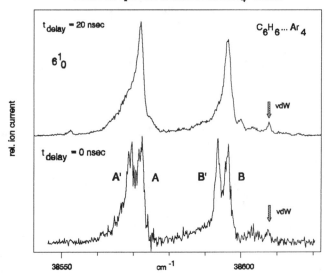

Figure 11: Comparison of two-color soft ionization spectra for the benzene...Ar_4 complex. Peak A' and peak B' refer to short-lived complexes.

shows the two-color spectrum with ionization delay for comparison. The spectrum now reveals two new peaks A' and B' with about equal intensity which are 3 cm^{-1} on the red side of the peaks found in the two-color scan with delayed ionization. It is also interesting to note that there is no additional vdW mode found in the spectrum which indicates that the structure of the clusters responsible for the new peaks must be different from the ones found earlier. We now performed the same experiment with ion detection at the mass of the benzene...Ar_4. The result is shown in fig. 11. There are also two new peaks of comparable intensity to the peaks found with ionization delay. The spectra still have a broad onset on the red side which can be a contribution from less cooled complexes in the jet. The small features found on the blue side at the onset of the main peaks in the upper spectrum arise from residual fragmentation signal of higher cluster ions and could be completely suppressed in the lower spectrum. From these spectra we now can prove that the additional peaks to the red in the one-color spectrum represent a true signal of the clusters for the selected mass.

3.4.2 *Hole-burning spectra without ionization delay*

In the case of the larger benzene...Ar_n clusters one expects to obtain more isomers with increasing number of Argon atoms. Beside the topological (n|m) classification of isomers there can be more isomers for the same set of m and n. This isomers are different in their configuration due to freezing in local minima other than the deepest one

Hole-burning Spectra of Benzene...Ar$_3$ with One-Color Probing

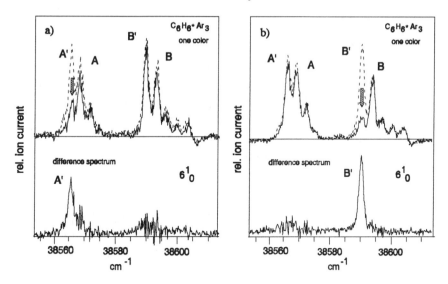

Figure 12:   Hole-burning spectra with one-color probing for the benzene...Ar$_3$ complex. Peaks A' and B' represent separate short lived isomers.

during the supersonic jet expansion. Mons et al. [30] differentiated between symmetric type and centered type clusters. In the latter case one Argon atom will be in a center position above the benzene ring.

We now performed the hole-burning experiment with one-color probing for the benzene...Ar$_3$ cluster, where we positioned the hole-burning laser on the A' peak (fig. 12). The dotted curves show the spectra without the hole-burning laser and the solid lines the spectra with the hole-burning laser switched on. Fig. 12a exhibits in the upper trace the reduced intensity of peak A' when the hole-burning laser is switched on. In the lower trace the difference spectrum is plotted which clearly shows that the peak A' belongs to a new isomer of the benzene...Ar$_3$ cluster. If one now performs the experiment with the hole-burning laser at the position of the peak B' (fig. 12b) one finds a strong reduction of the intensity of this peak and the difference spectrum reveals that this peak belongs to be another conformational isomer. From the red shift of the position of peak B' relative to the peak B one can conclude that the vdW interaction in the excited state is larger in this case which could arise from a symmetric (3|0) configuration as supposed by Mons et al. [30]. For peak A' one can also assume a symmetric configuration where both Argon atoms sitting on one side are shifted out from the center given by the benzene C$_6$ axis. This will give in total a stronger interaction of both Argon atoms with the π electron system and would explain the red shift compared to peak A. These structures are also less stable and due to their different structure they could better couple the intramo-

Hole-burning Spectra of Benzene...Ar $_4$ with One-Color Probing

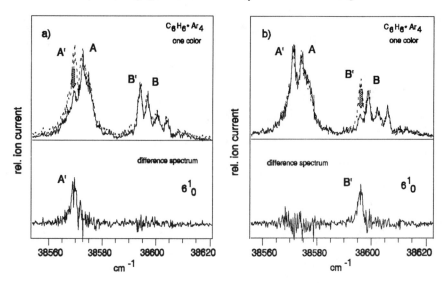

Figure 13: Hole-burning spectra with one-color probing for the benzene...Ar$_4$ complex. Peaks A' and B' represent separate isomers.

lecular ν_6 vibration to the vdW modes which would then explain the shorter lifetime by vibrational relaxation or vibrational predissociation. For a complete understanding of these structures more high precision ab initio data including the Argon-Argon interaction are required to describe the potential energy surface.

In the next experiment we performed the hole-burning with one-color probing for the benzene...Ar$_4$ cluster. Here the corresponding spectra are shown in fig. 13. If the hole-burning laser is positioned to peak A' (fig. 13a) one obtains a difference spectrum which also reveals a new isomer of this complex. The same is found for hole-burning at the position of peak B' (fig. 13b). Both isomers are red shifted corresponding to the peaks A and B which are found for the two-color probing with delayed ionization. This also indicates a stronger interaction with the π-electrons, which leads to a larger increment of the binding energy in the excited state than in the ground state compared to the A (2|2) and B (4|0) isomers. The isomer A' could either represent the (3|1) isomer with the three atoms in a symmetric position or an symmetric (2|2) isomer. The missing vdW modes in the 35 cm^{-1} region are also an indication for this assumption but here also a high precision potential surface is necessary to interpret the spectra.

4. Conclusion

A major information for the understanding of molecular clusters is derived from the spectroscopy. The weak binding energy prevents experiments at room temperature but vdW clusters can easily be formed in a supersonic jet expansion. Inherent to the process of the formation, clusters of different sizes are produced simultaneously and more important many different conformers at the same cluster size will be present in the jet. This problem is often ignored in the discussion of spectra of mass selected clusters. Mass selective techniques have been applied here to directly determine the spectra of single sized clusters where one-color REMPI spectra are often polluted from dissociation of larger cluster ions. Spectra obtained with two-color REMPI with soft ionization give spectra without this contamination but nevertheless the spectra then still are the sum of the spectra of conformational isomers which are also formed simultaneously in the jet. One can try to assign individual peaks from cluster shifts or from resolved fluorescence emission spectra, but this is only an indirect method and no mass resolution is included. Other method like rotational analysis of the spectra can be applied to resolve the different contributions, but this becomes more difficult for larger clusters and only highly symmetric structures can be determined for small clusters. The problem is here shown to be directly solved with the application of the mass selected hole-burning (MSHB) spectroscopy. Hole-burning was originally developed to distinguish between different sites of molecules in a solid or matrix. In this case a transition of molecules with a common ground state given by molecules positioned at equal sites in the solid is burnt out with a first laser leaving holes in the spectrum for all transitions which have this ground state in common. This then allowed for the identification of a special site in the solid or matrix. We have applied the hole-burning spectroscopy to mass selected clusters in the gas phase and show here results for the benzene...Ar_n clusters. The clusters are supposed to form different conformational isomers which for smaller clusters can be classified as one-sided (n|0) clusters where all Argon atoms sit on one side and two sided clusters (n|m) with variable numbers of Argon atoms on both sides of the benzene ring. The hole-burning experiment is now able to distinguish between these isomers and from the resulting spectra the structure can be assigned. In addition not only the topological cases (n|m) can be distinguished but even conformations with the same number of Argon atoms on the respective sides but being formed in different local minima of the potential energy surface are resolved. Our experiments also show different lifetimes for these clusters when they are excited to the first electronic S_1 state of the benzene molecule in the cluster. In this case two-color REMPI spectra are different if one applies the ionizing laser at the same time as the laser for the resonant excitation or if the ionizing laser is delayed. For benzene...Ar_3 and benzene...Ar_4 complexes two different types of isomers could be shown where one sort exhibits a very short lifetime which can be due to fast vibrational relaxation or vibrational predissociation. These short lived structures are presumed to be less stable configurations which are also formed in the jet due to the special feature that a non-thermodynamic distribution is generated in the expansion process. This polymorphism problem is inherently present for all cluster formations and the disentanglement of the spectra thus only can be performed with the hole-burning spectroscopy. This is a very prevalent problem - though typically ignored in spectroscopy of clusters. This problem is demonstrated here for the benzene...Ar_n clusters where the different isomers could be isolated and a complete disentanglement of the cluster spectra achieved. It could be shown that this new hole-burning spectroscopy together with mass resolution is a powerful technique to address the problem of the spectroscopy of conformational isomers.

Since part of this hole-burning experiment involves a time delay between two laser pulses, it is shown above that this technique in addition gives direct kinetic information for the clusters from time resolved experiments in the excited state.

Acknowledgement

This work was supported by the Binational German-Israeli James-Franck program for laser-matter interaction.

References

[1]   Janda, K. C., Hemminger, J. C., Winn, J. S., Novick, S. E.,[Harris, S. J., Klemperer, W. (1975) "Benzene Dimer: A Polar[Molecule", J. Chem. Phys. 63, 1419-1421

[2]   Henson, B. F., Hartland, G. V., Venturo, V. A., Hertz, R. A.,[Felker, P. M. (1991) "Stimulated Raman Spectroscopy in the v_1 Region of Isotopically Substituted Benzene Dimers: Evidence for[Symmetrically Inequivalent Benzene Moieties", Chem. Phys. Lett. 176, 91-98

[[3]   Pawliszyn, J., Szczésniak, M. M., Scheiner, S. (1984) "Interactions between Aromatic Systems: Dimers of Benzene and s- Tetrazine", J. Phys. Chem. 88, 1726-1730

[[4]   Carsky, P., Selzle, H. L., Schlag, E. W. (1988) "Ab Initio Calculations on the Structure of the Benzene Dimer", Chem. Phys. 125, 165-170

[5]   Hobza, P., Selzle, H. L., Schlag, E. W. (1990) "Floppy Structure of the Benzene Dimer: Ab Initio Calculation on the Structure and Dipole Moment", J. Chem. Phys. 98, 5893-5897

[6]   Fung, K. H., Selzle, H. L., Schlag, E. W. (1983) "Study of Isotope Effects in Benzene Dimers in a Seeded Supersonic Jet", J. Phys. Chem. 87, 5113-5116

[7]   Schmidt, M., Mons, M., Le Calvé, J. (1991) "Microsolvation of the Benzene Molecule by Argon Atoms: Spectroscopy and Isomers", Chem. Phys. Lett. 177, 371-379

[8]   Weber, Th., Neusser, H. J. (1991) "Structure of the Benzene-Ar_2 Cluster from Rotationally Resolved Ultraviolet Spectroscopy", J. Chem. Phys. 94, 7689-7699

[9]   Scherzer, W., Krätzschmar, O., Selzle, H. L., Schlag, E. W. (1992) "Structural Isomers of the Benzene Dimer from Mass Selective Hole-Burning Spectroscopy", Z. Naturforsch. A 47a, 1248-1252

[10]   Scherzer, W., Selzle, H. L., Schlag, E. W. (1992) "Identification of Spectra of Mixed Structural Isomers via Mass Selective Hole-burning in the Gas Phase", Chem. Phys. Lett. 195, 11-15

[11] Kharlamov, B. M., Personov, R. I., Bykovskaya, L. A. (1974) "Stable 'Gap' in Absorption Spectra of Solid Solutions of Organic Molecules by Laser Irradiation", Opt. Commun. **12**, 191-193

[12] Bergmann, K., Hefter, U., Witt, J. (1980) "State-to-State Differential Cross Sections for Rotationally Inelastic Scattering of Na_2 by He", J. Chem. Phys. **72**, 4777-4790

[13] Kiermeier, A., Dietrich, K., Riedle, E., Neusser, H. J. (1986) "Doppler-Free Saturation Spectroscopy of Polyatomic Molecules: Photochemical Hole Burning of Gas Phase s-Tetrazine", J. Chem. Phys. **85**, 6983-6990

[14] Heijmen, B., Bizarri, A., Stolte, S., Reuss, J. (1989) "IR-IR Double Resonance Experiments on SF_6 and SiF_4 Clusters", Chem. Phys. **132**, 331-349

[15] Dick, B., Zinghar, E., Haas, Y. (1991) "Spectral Hole-Burning of Tetracene and Tetracene-Argon Complexes in a Supersonic Jet", Chem. Phys. Lett. **187**, 571-578

[16] Lipert, R. J., Colson, S. D. (1989) "Persistent Spectral Hole- Burning of Molecular Clusters in a Supersonic Jet", J. Phys. Chem. **93**, 3894-3896

[17] Lipert, R. J., Colson, S. D. (1989) "Low Frequency Vibrations in Phenol-$(H_2O)_2$ Revealed By Hole-burning Spectroscopy in a Supersonic Jet", Chem. Phys. Lett. **161**, 303-307

[18] Wittmeyer, S. A., Topp, M. R. (1989) "Spectral Hole-Burning in Free Perylene and in Small Clusters with Methane and Alkyl Halides", Chem. Phys. Lett. **163**, 261-268

[19] Wittmeyer, S. A., Topp, M. R. (1991) "Vibronic Hole-Burning Spectroscopy of Small Clusters Involving Perylene", J. Phys. Chem. **95**, 4627-4635

[20] Knochenmuss, R., Leutwyler, S. (1990) "Selective Spectroscopy of Rigid and Fluxional Carbazole-Argon Clusters", J. Chem. Phys. **92**, 4686-4697

[21] Schlag, E. W., Selzle, H. L. (1990) "Weak Interaction in Benzene Clusters", J. Chem. Soc. Faraday Trans. **86**, 1-7

[22] Sengteller, S., Selzle, H. L., Schlag, E. W. (1990) "Study of the Metastable Decay of Single Size Benzene Clusters by an Ion Stopping Technique", Z. Naturforsch. **45a**, 169-172

[23] Ben-Horin, N., Even, U., Jortner, J. (1992) "Rigid and Nonrigid Benzene·Ar_2 van der Waals Heteroclusters", Chem. Phys. Lett. **188**, 73-79

[24] Troxler, T., Knochenmuss, R., Leutwyler, S. (1989) "Isomer- specific Spectra and Ionization Potentials of van der Waals Clusters", Chem. Phys. Lett. **156**, 554-558

[25]   Doxtader, M. M., Gulis, I. M., Schwartz, S. A., Topp, M. R. (1984) "Isomer
        Effects on Vibrational Energy Relaxation in Perylene-Argon Complexes", Chem.
        Phys. Lett. **112**, 483-490

[26]   Dao, P. D., Morgan, S., Castleman Jr., A. W. (1985) "Two-Color Resonance
        Enhanced Multiphoton Ionization of van der Waals Molecules: Studies of
        Spectroscopic Shifts and Ionization Thresholds of Paraxylene Clustered with
        Argon", Chem. Phys. Lett. **113**, 219-224

[27]   Schmidt, M., Mons, M., Le Calvé, J., Millié, P. (1991) "The Second Conformer
        of the Benzene-Argon$_2$ van der Waals complex", Chem. Phys. Lett. **183**, 69-76

[28]   Chewter, L. A., Sander, M., Müller-Dethlefs, K., Schlag, E. W. (1987) "High
        Resolution Zero Kinetic Energy Photoelectron Spectroscopy of Benzene and De-
        termination of the Ionization Potential", J. Chem. Phys. **86**, 4737-4744

[29]   Fung, K. H., Henke, W. E., Hays, T. R., Selzle, H. L., Schlag, E. W. (1981)
        "Ionization Potential of the Benzene-Argon Complex in a Jet", J. Phys. Chem.
        85, 3560-3563

[30]   Schmidt, M., Le Calve, J., Mons, M. (1993) "Structural Transitions in Benzene-
        Argon Clusters: Size and Temperature Effects", J. Chem. Phys. **98**, 6102-6120

[31]   Menapace, J. A., Bernstein, E. R. (1987) "Calculation of the Vibronic Structure
        of Solute/Solvent van der Waals Clusters", J. Phys. Chem. **91**, 2533-2544

[32]   Weber, Th., von Bergen, A., Riedle, E., Neusser, H. J. (1990) "Rotationally
        Resolved Ultraviolet Spectrum of the Benzene-Ar Complex by Mass-Selected
        Resonance-Enhanced Two-Photon Ionization", J. Chem. Phys. **92**, 90-96

[33]   Van der Avoird, A. (1993) "Van der Waals Rovibrational Levels and the High
        Resolution Spectrum of the Argon-Benzene Dimer", J. Chem. Phys. **98**, 5327-
        5336

[34]   Henke, W. E., Yu, W., Selzle, H. L., Schlag, E. W., Wutz, D., Lin, S. H. (1985)
        "Shifts in Fluorescence Excitation Spectra of Anthracene-Argon van der Waals
        complexes", Chem. Phys. **92**, 187-197

[35]   Warshel, A., Levitt, M. "Programm QCFF/PI by A. Warshel and M. Levitt",
        QCPE Program No. **247**

[36]   Hobza, P., Selzle, H. L., Schlag, E. W. (1991) "Ab Initio Calculations on the
        Structure, Stabilization, and Dipole Moment of Benzene-Ar Complex", J. Chem.
        Phys. **95**, 391-394

[37]   Bludsky, O., Spirko, V., Hrouda, V., Hobza, P. (1992) "Vibrational Dynamics of
        the Benzene...Argon Complex", Chem. Phys. Lett. **196**, 410-416

[38]   Aziz, R. A., Chen, H. H. (1977) "An Accurate Intermolecular Potential for
        Argon", J. Chem. Phys. **67**, 5719-5726

HIGH RESOLUTION IR STUDIES OF POLYMOLECULAR CLUSTERS: MICROMATRICES AND UNIMOLECULAR RING OPENING

David J. Nesbitt[*]
Joint Institute for Laboratory Astrophysics, National Institute
of Standards and Technology and University of Colorado,
Boulder, CO   80309-0440

1. Introduction

A long standing goal in the field of cluster research has been to elucidate the transition between the properties of isolated monomer species in the gas phase and the corresponding properties in the condensed phase.   One of the most exacting probes of small clusters in this transition region has recently been offered by high resolution spectroscopies, particularly in the microwave and IR region of the spectrum.[1-11]   Such high resolution studies at rotational resolution can provide detailed information on the distribution of the masses in the cluster, as well as clearly distinguish between various low lying isomeric structures of the same molecular composition.   High resolution spectroscopy in the IR also provides the opportunity to study dynamical phenomena such as predissociation,[1,12-14] isomerization, and IVR channels that can be accessed by quantum state selected vibrational excitation of the clusters.

The last decade has witnessed rapid advances in the experimental ability to detect and analyze dimer clusters via high resolution IR methods.   At the outset, we make the useful distinction between *far* IR direct absorption methods[4,5,11,17,18] which access primarily intermolecular "van der Waals" modes, with the *near* IR absorption techniques[2,12-16] which access both the high frequency intramolecular vibrations, as well as the low frequency intermolecular "van der Waals" modes built on these high frequency modes.   In particular, the high sensitivity of the slit jet expansion geometry[19] has made feasible studies of intermolecular vibrational modes in dimer clusters that extend up to and sometimes even above[20] the dissociation limit.

A key value in the IR based studies is that a wide range of vibrationally excited wave functions can be investigated, whose extreme amplitude motion of the nuclei extensively samples the full range of intermolecular geometries.   Inversion of large amplitude spectral data from such a series of levels, therefore, makes possible a direct determination of the *intermolecular* potential energy surface in several dimensions. Furthermore, by probing overtone vibrations and isotopically substituted clusters,[21,22]

---

[*]Staff Member, Quantum Physics Division, National Institute of Standards and Technology.   This paper is an official contribution of the National Institute of Standards and Technology, not subject to copyright in the U.S.

J. Jortner et al. (eds.), Reaction Dynamics in Clusters and Condensed Phases, 137–151.
© 1994 *Kluwer Academic Publishers. Printed in the Netherlands.*

these multidimensional potential surfaces[23] can be extended to include the high frequency, *intramolecular* coordinates as well. Specifically, this has been successfully demonstrated in a variety of triatomic rare gas-hydrogen halide dimers, in which the potential surface depends on three coordinates, i) van der Waals separation of the centers of mass (R), ii) the HX bond length (r), and iii) the angle between the HX vector (r) and the center-of-mass separation (R). There has also been a wealth of high resolution IR spectroscopic studies on dimers with more than three atoms. Due to the concomitant increase in intermolecular degrees of freedom, however, it has proven substantially more difficult to characterize the potential surfaces directly from spectroscopic observation. Nevertheless, studies have led in a few notable cases such as $(HF)_2$ to a detailed elucidation of the intermolecular potential surfaces,[24] particularly in combination with high level *ab initio* calculations.[25,26]

Any prospect for investigating the transition between gas and condensed phase properties clearly requires studies beyond the dimer level. Extending such high resolution studies to systems with increasing numbers of constituents (i.e. trimers, tetramers, etc.) requires considerable incremental effort, both experimentally and theoretically. In this paper, we address recent results from our laboratory which extend and apply these high resolution methods to larger cluster sizes. We will briefly discuss two parallel directions, i) rare gas oligomers of $(Ar)_n$ with HF/DF ($n = 1,2,3$ and 4) and ii) hydrogen bonded trimer clusters of DF. The organization of this paper is as follows. Section 2 provides a brief description of the experimental method. Section 3 presents experimental and theoretical results on Ar_n-HF and Ar_n-DF clusters, focusing on the n-dependent vibrational red shifts with respect to the results in an Ar matrix. Section 4 describes the spectroscopy and dynamics of vibrationally excited DF trimer, which provides direct evidence for a six-membered cyclic ring structure. Furthermore, IVR in the DF stretch excited trimer leads to a rapid ring opening on the 40 psec time scale.

2. Experimental

A more complete description of the experimental apparatus is available in the literature,[2,3,16,19] hence only details of specific relevance to these experiments are mentioned. The spectra are obtained by direct absorption of a single mode, cw tunable difference frequency laser in a 4 cm slit supersonic jet expansion with multiple pass optics. The expansion mix for studies of the Ar_n-HF and DF clusters is 0.8% HF/DF and 50-65% Ar in first run Ne diluent. The use of Ne diluent proves crucial to prevent runaway clustering in the slit expansion to form much larger oligomers that cannot be rotationally resolved. For the studies of DF trimer, a mix of 0.8% DF with 1:1 Ar and first run Ne is used, at a total backing pressure of 700 Torr. Under such conditions, the strongest absorptions of DF trimer correspond to approximately 0.5% in a 12 pass White cell.

The difference frequency laser is formed by optical subtraction of a cw ring dye and single mode Ar^+ laser in a temperature tuned $LiNbO_3$ crystal. The Ar^+ laser is frequency stabilized by fast locking onto a frequency transfer interferometer, which in turn is servo-loop locked to a polarization stabilized HeNe laser. This lock is quite robust, and can be maintained for periods up to a week without interruption.[27] The scanning of the dye laser, and hence the tunable difference frequency IR, is monitored by fringes on this stabilized cavity. The frequency stability of this cavity system has been demonstrated by repeated measurements on spectra of several weakly bound complexes, and which indicate an absolute consistency on the order of a few MHz over a one week time scale.[28] This capability proves absolutely crucial in the assignment of the rather complicated near IR spectra of multiple Ar clusters with HF and DF, by permitting the use of high precision combination differences from the previous microwave investigations on the ground vibrational states.

3. Near IR Spectra of (Ar)$_n$-HF/DF Clusters

The first high resolution spectroscopic detection of multiple Ar clusters attached to hydrogen halides was obtained by Gutowsky and coworkers[6-8] in FT microwave studies. This work provided the first glimpse of the vibrationally averaged structures of these species. The first low frequency van der Waals mode was observed in Ar$_2$-HCl complexes by Elrod et al.[4] in the far-IR, and more recently extended to include the three HCl bending vibrations[5,11] which correlate with the Σ and (doubly degenerate) Π bend in the corresponding Ar-HCl dimer species. The first detection of high frequency vibrational excitation of such clusters in the near-IR was obtained by McIlroy and Nesbitt,[2,3] using direct absorption in a slit jet apparatus. These studies focussed on (Ar)$_n$-HF species, with n=1,2,3 and 4. Rotational analysis of these spectra led to high precision rotational constants for both the v_{HF}=0 and v_{HF}=1 complexes. More recently, the corresponding spectra of (Ar)$_n$-DF complexes (for n=1,2 and 3) have also been observed.[29]

One interesting question to pose from these results is how the HF or DF frequencies are shifted from the gas phase values as a function of sequential "solvation" by the rare gas Ar atoms. The results are summarized in Fig. 1, and indicate a monotonic progression of red shifts for both HF and DF complexes. For comparison, the HF and DF absorption frequencies in an Ar rare gas matrix are also shown. Interestingly, the progression indicates the strongest red shift for the first Ar atom, and then successively smaller red shifts for each additional atom. Also, the accumulated red shift after only three Ar atoms is nearly 50% of the way toward the "bulk" value, even though the first "solvation shell" for an octahedral site in an Ar matrix would have 12 equivalent nearest neighbors. This would suggest a high orientational dependence on the HF/DF red shift, which is entirely consistent with pairwise potentials developed by Hutson[23] based on spectroscopic van der Waals data for Ar-HF (and Ar-DF) with v_{HF}=0,1,2 and 3. Furthermore, the incremental red shift for Ar$_4$-HF is far smaller than would be expected from the trend. This can be explained from the rotationally deduced structure of the complex, which has the fourth Ar atom nestled on the symmetry axis of the Ar$_3$ ring, and thus is in the second "solvation" shell with respect to the HF chromophore. This observation clearly supports a strong distance dependence on the HF red shift, and suggests that only a single fully occupied solvent shell might be necessary to achieve the red shift found in a bulk matrix environment.

Also of considerable interest to test in these clusters is the possible role of non-additive, "three body" effects in the intermolecular potential. As schematically shown in Fig. 2, the full potential in these rare gas clusters reflects the sum of *pairwise* additive potentials between the individual constituents in addition to the non-pairwise, or *three body* terms. For example, these three body terms can arise from the interaction *between* the dipoles in the Ar atoms induced by the permanent HF dipole, or even simply by the mutual "van der Waals" interaction between three polarizable species.[30]

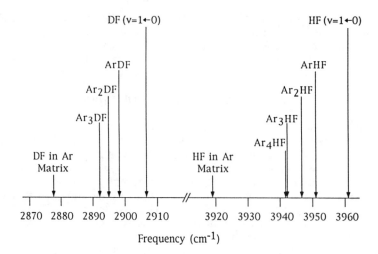

Fig. 1 Red shift of HF and DF $v=1\leftarrow0$ vibrational frequencies as a function of number of "solvent" Ar atoms. Note the monotonic progression of these shifts towards the matrix value, and that only three Ar atoms are sufficient to achieve nearly 50% of the matrix shift.

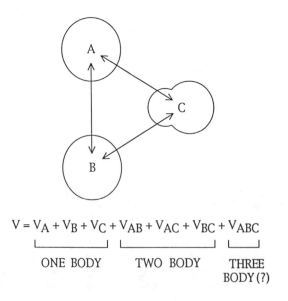

$$V = V_A + V_B + V_C + V_{AB} + V_{AC} + V_{BC} + V_{ABC}$$

ONE BODY   TWO BODY   THREE BODY (?)

Fig. 2 Schematic of the multibody contributions to the total potential in a trimer cluster.

The high resolution data on Ar_n-HF and Ar_n-DF permits both qualitative and quantitative investigation of these three body effects. We can address this issue qualitatively by predicting the equilibrium structures for the various complexes, based purely on pairwise additivity of the Ar-HF/DF[23,30] and Ar-Ar potentials.[32] Shown in Fig. 3 are the predicted lowest energy structures for n=1,2 3 and 4; the numbers in parentheses indicate the potential minimum (in cm^{-1}) with respect to total separation of the monomer subunits. In all cases, the lowest energy structure is indeed the one observed in the supersonic jet. It is worth noting that Quantum Monte Carlo (QMC) calculations have recently been performed on these pairwise additive potentials by Lewerenz,[33] and which therefore explicitly include significant zero point effects in both inter- and intramolecular coordinates. Even including these zero point effects, the lowest energy isomer in each case is in good agreement with the experimental observation. It is interesting to note that two nearly *isoenergetic* isomers are suggested for the pentamer, with only a low barrier separating them. For n>4, the predicted

PREDICTED Ar$_n$HF ISOMERS

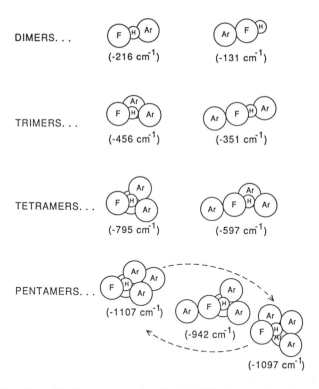

Fig. 3 Predicted equilibrium geometries for Ar$_n$HF clusters based on pairwise additive potentials for Ar-HF and Ar-Ar. In all cases, the lowest energy isomer is the one observed to be the ground state from the near-IR spectra.

number of nearly degenerate, multiple minima on the PES with low interconversion barriers grows rapidly.

A more quantitative test of the intermolecular potential can be made by probing the low frequency van der Waals modes in these species. Quantum calculation of these low frequency modes via close coupling methods on a known potential represents a major computational effort, and has only recently become feasible[34] for n≤2, i.e. up to the trimer Ar_2-HF species. As shown schematically in Fig. 4, we have been able to observe the two "Π" bends in Ar_2-HF, where the Π degeneracy in the Ar-HF dimer is strongly lifted by the anisotropic "ridge" of the two Ar atoms. The in-plane bend and out-of-plane bend are at 62.0 cm^{-1} and 88.9 cm^{-1}, respectively. These compare reasonably favorably, but are systematically *lower* than the predictions by Hutson and coworkers[34] from the pairwise additive potential, i.e. 67.4 cm^{-1} and 92.6 cm^{-1} respectively. This comparison indicates the influence of a slightly "repulsive" three body angular term, i.e. these pairwise additive potentials underestimate the degree of large amplitude vibrational motion. These data, along with the far-IR data on Ar_2-HCl complexes,[4,5,10,11] are beginning to provide a powerful benchmark for quantitative modelling of the three body forces in molecular systems with more than two species. It is interesting to note that inclusion of the "conventional" three body terms[30] in the potential, such as Axelrod-Teller and triple dipole terms,[35] shift the results in the correct direction, but only by less than one-third of the way toward agreement with experiment.

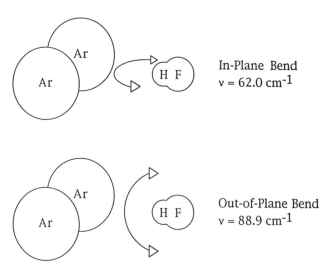

Fig. 4 Van der Waals intermolecular bending modes for Ar_2HF, corresponding to "in-plane" and "out-of-plane" motion. Spectra of these low frequency vibrations are obtained in the near IR by weak combination bands of the van der Waals modes built on top of the high frequency HF v=1←0 stretch.

4. DF v=1←0 excitation on (DF)$_3$: Structure, IVR-induced Ring Opening Dynamics

Hydrogen bonded networks are among the most interesting and important elements for determining short and long range structure in solutions and biomacromolecules.[36] By virtue of their relative simplicity, HF dimer and higher oligomers of (HF)$_n$ have long served as a prototype for understanding such phenomena, with a particular emphasis on the energetics and dynamics of hydrogen bond cleavage and formation. In another context, the sequence of (HF)$_n$ oligomers provides an appropriate coordinate along which to probe the transition between gas phase and condensed phase properties in systems with strong pairwise and non-pairwise additive interactions.

High resolution spectroscopic studies, in the microwave, far IR and near IR, have been a tremendously useful source of information on the well-studied HF dimer.[12-14] Unfortunately, there has been little to no success in the use of high resolution methods for HF systems for n > 2. The majority of the previous work at intermediate resolution has been based on mass spectroscopic, pulsed IR laser predissociation studies, combined with isotopic substitution investigations in order to distinguish between fragmentation from higher clusters under electron bombardment ionization.[37,38] High resolution efforts to detect (HF)$_3$ with cw IR laser sources using cooled cells and bolometric, optothermal methods have failed to reveal any rotationally resolved structure in the 3800-4000 cm^{-1} HF stretching region.[39] This has been attributed to fast predissociation of the trimer cluster upon vibrational excitation, which can energetically couple with the continuum and destroy any rotational structure by lifetime broadening.

Despite these experimental difficulties, theoretical studies on (HF)$_n$ have made considerable progress. Much of this progress has been triggered by the availability of an *analytical* potential surface for (HF)$_n$, pieced together from a high level ab initio pairwise potential for HF dimer, with non-pairwise additive corrections for the higher oligomers included at the Hartree-Fock level.[24-26] In such a large amplitude, multidimensional system, the influence of zero point energies and anharmonic effects are often of crucial importance, and thus QMC calculations on this potential energy surface prove necessary to make even qualitative predictions with which to guide further experimental investigations. In particular, QMC predicts that the fundamental DF (v=1) vibration in DF trimer at 2725 cm^{-1} should be energetically *bound*, whereas the corresponding HF stretch can energetically predissociate (HF)$_3$ into HF and (HF)$_3$ fragments. Hence, the high resolution, near IR spectrum of (DF)$_3$ could be rotationally structured, and thus yield new structural and dynamical information which had proven so elusive in the HF trimer system.

With this as our stimulus, we initiated a spectroscopic search in our laboratory for DF trimer and higher oligomers in the slit supersonic expansion.[40] The spectrum of DF trimer is shown in Fig. 5, and demonstrates the first evidence for clear rotational structure in any HF oligomer higher than the dimer. At the ≥ 0.1 cm^{-1} resolution level, the band consists of a typical planar oblate top perpendicular band envelope (A = B ≈ 2C), but with orders of magnitude more spectral congestion (on the 0.1 cm^{-1} - 0.001 cm^{-1} scale of resolution) than anticipated for a simple symmetric top spectrum under these 10 K jet temperatures.

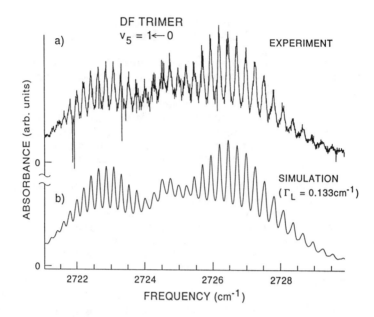

Fig. 5 (a) Survey spectrum of the asymmetric DF stretch (ν_5) in (DF)$_3$, displaying the clear perpendicular band characteristic of an planar, oblate symmetric top. Downward spikes are due to HDO traces in the chamber and are used for frequency calibration. (b) Simulation of the (DF)$_3$ band based on a Lorentzian envelope for each symmetric top transition with $\Gamma_L = 0.133$ cm^{-1}.

Our analysis strategy is two-fold; first, we focus on the coarse-grain rotational structure by assuming a Lorentzian broadening (Γ_L) of a spectrum which would be predicted for a symmetric top.   Such an analysis would be consistent, for example, with predissociation on each of the transitions with a J,K independent lifetime, or alternatively, with a coupling of the initial vibrational level into dark vibrational states with a J,K independent coupling width.   The fit is then adjusted in a least-squares fashion to obtain upper and lower rotational constants, rotational temperatures, etc. As shown in Fig. 5,   this fit provides an extremely good match, but requires a homogeneous broadening of $\Gamma_L = 0.133$ cm^{-1} to reproduce the observed spectral band contour.

The rotational constants from this fit indicate several interesting points. First, A, B, and C lead to a nearly zero inertial defect, characteristic of a nearly *planar* complex with C$_{3h}$ symmetry.   Analysis of the rotational constants indicates a nearly perfect hexagonal structure, with 0.93Å for the covalent F-H bond length, and 0.92Å for the hydrogen bond length.   Hence, although the nature of the bonding is qualitatively different than that of benzene, this trimer is structurally quite analogous. This explains the appearance of only one DF stretch band in the spectrum, since in C$_{3h}$ symmetry this asymmetric stretch would be doubly degenerate, while the third DF stretch vibration is totally symmetric and not IR allowed.

Comparison with theoretical predictions for the full 12-D surface looks quite good. For example, the sign of ΔC is consistent with the expected shortening of the hydrogen bond upon DF excitation. Second, the inertial defect is very small in the ground state, and increases $(B' > 2C')$ in the excited state. This is consistent with excitation of an in-plane DF vibration, which is the only IR-active DF stretch in the C_{3h} group. Furthermore, QMC calculations within the "clamped coordinate" approximation yields $C' = 0.121$ cm^{-1}. This is already in excellent agreement with the experimental value of 0.11992 cm^{-1} obtained from the spectral fits.

Next, we look at higher resolution and consider the origin of the fine-grained structure ($\lesssim 0.1$ cm^{-1}) in the experimental spectrum. As demonstrated by repeated scans, this structure is real, reproducible and not simply "noise." Specifically, the observed line widths are essentially dominated by reduced Doppler broadening (≈ 40 MHz) in the slit jet, which therefore represents a rigorous *upper* bound on the predissociation lifetime. This would be in support with theoretical predictions from the potential surface[24] of an upper DF trimer state which is energetically stable with respect to predissociation into DF + (DF)$_2$.

We thus are left with the need to account for the extensive spectral fine structure in the DF trimer spectrum. First of all, the requisite density of lines to achieve this degree of spectral congestion is orders of magnitude larger than can be generated by a simple symmetric top simulation. This is shown most clearly in a expanded region of the experimental spectrum in Fig. 6(a), along with the spectral simulation in Fig. 6(b) based on the best fitted rotational constants. In combination with the observation that the individual lines are only 40 MHz = 0.0013 cm^{-1} wide, the effective smearing out of rotational structure suggests *at least* a 100 fold excess density of states with sufficient oscillator strength to be visible in the spectrum. Since predissociation is energetically a closed channel, this additional structure must arise from unimolecular coupling of the "bright" DF stretch with the nearly isoenergetic "dark" states that reflect the true molecular eigenstates of the DF trimer. The linewidth Γ_L as determined in the spectral fit in Fig. 5, therefore, represents the reciprocal lifetime for intramolecular energy flow of the DF stretch into bath states that would occur for a coherently prepared initial state.

The contrast ratio, or "scratchiness," in this structured spectrum provides an important quantitative clue towards determining the density of dark states participating in the coupling. For example, as the density of dark states increases, the greater the loss of spectral structure by accidental overlap of lines. We can model this process as follows. A given bright state is allowed to couple with a density of bath states (ρ). For a given density of states, the magnitude of the average square coupling matrix elements ($<v^2>$) can be determined from the Fermi Golden rule,

$$\Gamma_L = 2\pi\rho<v^2> \tag{1}$$

where Γ_L is fixed at 0.133 cm^{-1} as determined in the spectral fit. The coupling of the dark states with the bright state, and thus the resulting spectral prediction, can then be obtained via simple matrix diagonalization methods.[41]

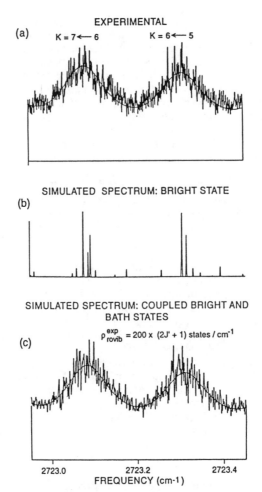

Fig. 6 (a) Expanded region of the spectra in Fig. 5 corresponding to K=7←6 and K=6←5. The "scratchiness" in the spectrum is completely real and reproducible, reflecting "bright" state mixing with background "dark" states due to strong intramolecular vibrational couplings. (b) Bright state spectrum predicted from the best fitted constants. (c) Simulation of the experimental spectrum in order to estimate the density of strongly coupled dark states (see text for details).

As shown in Fig. 7, however, the *statistics* of the dark state distribution and coupling matrix elements plays a major role in the resulting spectrum. For a *uniform* dark state spacing and fixed bright-dark state coupling matrix element, a classic result derived by Bixon and Jortner in early work on radiationless transitions,[42] the spectral structure is effectively lost for state densities in excess of the reciprocal experimental line

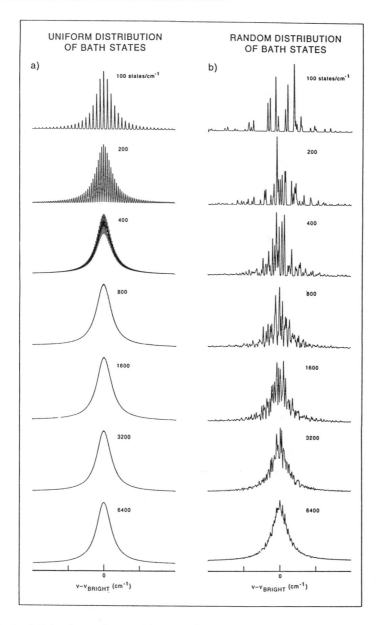

Fig. 7 A model for loss of spectral resolution due to strong IVR mixing of a single bright state with dark states with two different level spacings: a) *uniform* spacing and coupling (i.e. Bixon-Jortner) and b) *random* level spacings and couplings. The mean coupling strength in each case is determined from the average state density by the Golden Rule in order to maintain a constant FWHM of the resulting envelope.

width, in this case 100 state/cm^{-1}. However, for bath states with *random* spacings and coupling matrix elements, obtained from a Gaussian distribution around these mean values, the requisite state density to lead to a comparable level of "scratchiness" in the spectrum is more than 20 times larger. Stated simply, this is due to the diminished probability of coincidences filling in spectral holes for a random vs. uniform distribution of background level spacings. A more accurate description for strongly coupled dark states can be obtained from a Wigner distribution[43] of the level spacings, however, our conclusions are essentially identical to those obtained from a random distribution. A key point is that the necessary level density is 1-2 orders of magnitude higher than can be accounted for simply by anharmonic vibrational coupling, and thus indicating that some loss of K as a good quantum number (for example, due to Coriolis mixing of the 2J+1 different K states for a given J) must be taking place.

In order to investigate this situation more quantitatively, we have estimated the rovibrational density of states for DF trimer excited 2725 cm^{-1}. The method utilizes a direct state count based on the analytical 12-D potential energy surface and quantum Monte Carlo calculations. As anticipated for such large amplitude systems, the state density is somewhat dependent on the degree of sophistication of the model, ranging from ρ_{vib} = 70 states/cm^{-1} for a harmonic model up to ρ_{vib} = 500 state/cm^{-1} if anharmonic effects are included. If one allows for the "good" quantum numbers i) overall parity (1/2) and ii) nuclear spin (1/2) modification, then the effective state density for a single bright J' upper state could be as large as ρ_{rovib}^{tot} = 125·(2J'+1) states/cm^{-1}. In this expression, the factor of (2J'+1) assumes as an upper limit the total loss of K as a good quantum number.

Experimentally, we can estimate the density of coupled states from the ideas described above and shown in Fig. 6. First of all, the necessary line density is clearly greater than 125 states/cm^{-1}, and hence the spectra are simply inconsistent with any model that *only* considers anharmonic coupling between vibrational states, and neglects the maximum additional (2J'+1) factor in state density that can arise from rotationally mediated vibrational coupling such as Coriolis mixing. Hence, we can model the experimental spectrum in Fig. 6(a) by taking each bright upper state in the simulated spectrum (Fig. 6(b)), and coupling these states to a manifold of dark states for a given value of ρ_{rovib}^{exp} = ρ_{vib} · (2J'+1). As before, the mean square coupling matrix elements to these dark states is determined directly from Fermi's Golden Rule, such that Γ_L = 0.133 cm^{-1}. The value of ρ_{vib} is then adjusted till the correct contrast ratio in the observed spectrum is achieved. Best agreement between experiment and the simulation is obtained for ρ_{rovib}^{exp} ≈ 200 ·(2J'+1) states/cm^{-1}, as shown in Fig. 6(c).

This estimate is quite close to but already slightly in excess of the theoretical estimates of the total rovibrational state density from a direct anharmonic state count, i.e. ρ_{rovib}^{tot} ≈ 125·(2J'+1) states/cm^{-1}. The important conclusion is that the spectra indicate nearly *complete* IVR mixing of the DF stretching energy into the energetically available phase space through the manifold of low frequency hydrogen bond modes. Furthermore, the time scale for this IVR process can be directly estimated from the Γ_L = 0.133 cm^{-1} coupling widths to be roughly 40 psec. This is in stark contrast to the vibrational predissociation lifetimes in HF and DF dimer, where IVR is extremely incomplete even on the 1-100 nsec time scale of predissociation.

It is interesting to speculate on the nature of these coupled bath states into which the DF stretch state is so rapidly relaxing. At 2725 cm^{-1} these states are already well above the energy (1000-1500 cm^{-1}) for breaking a single hydrogen bond,[24] and therefore accessing the regime of floppy, chainlike conformations which dominate the rovibrational partition function due to their considerably higher entropy. By the direct state count, > 90% of the states at 2725 cm^{-1} can be conservatively classified as "open chains", i.e. with *one* rather than *two* hydrogen bonds broken.[40] Hence stated from a time dependent perspective, a coherent pulsed excitation of these eigenstates would therefore lead to a time dependent, unimolecular relaxation on a 40 psec time scale, corresponding physically to an IR induced unimolecular "opening" of the (DF)$_3$ ring, as represented schematically in Fig. 8.

We finish with two observations. First of all, it is interesting to compare the time scale for H-bond predissociation in DF dimer complexes, with the corresponding IVR induced H-bond breakage in DF trimer. In the former, the predissociation is energetically allowed, and occurs on the time scale of 3.2 nsec and 32 nsec for the "bound" and "free"-DF stretch vibrational modes, respectively. This is between 10 and 100 times *slower* than the corresponding IVR lifetimes and consequent hydrogen bond cleavage observed in DF trimer. This may reflect the role of strain in the trimer ring in enhancing the rate.

Secondly, the characteristic magnitude of the rovibrational coupling between the bright DF stretch state and dark states in DF trimer is quite small, with a rms matrix element only on the order of 0.002 cm^{-1}. However, from a Fermi Golden Rule perspective, the rapid 40 psec time scale of this IVR induced ring opening results from the large density of states into which the DF stretch can relax. This does not rule out alternate and more physically plausible descriptions such as strong coupling of the DF stretch into a small number of "doorway" states, which in turn can couple into the remaining low frequency

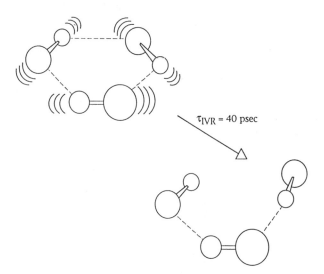

τ_{IVR} = 40 psec

Fig. 8 "Cartoon" of the IVR energy flow of DF stretch excitation into the full manifold of intermolecular vibrations of (DF)$_3$, leading statistically to hydrogen bond cleavage and hence a ring "opening" on the 40 psec time scale.

ring modes. Indeed, one might speculate that important doorway states in the ring opening of a hydrogen bond would sequentially be i) DF stretch-bend coupling to sever or weaken the hydrogen bond, followed by ii) slower opening of the F-F-F angle to prevent the bond from reforming. If so, the rate limiting dynamics of the bond breakage might be controlled by a sufficiently smaller number of degrees of freedom to permit a physically appealing framework for theoretical analysis. We intend to exploit such a reduced dimensionality description, and thereby probe the relevant dynamical pathways of such IVR induced ring opening processes.

Acknowledgment

This work has been supported by grants from the National Science Foundation (CHE90-00641 and PHY90-12244), and made possible by the diligent experimental efforts of John T. Farrell, Jr., Scott Davis, Dr. Steven H. Ashworth and Dr. Martin A. Suhm. We also gratefully acknowledge the help and stimulating interactions with Dr. Marius Lewerenz and Prof. Jeremy M. Hutson for their interest in performing Quantum Monte Carlo and close coupled quantum calculations on the Ar_n-HF and Ar_n-DF complexes. We would especially like to thank Dr. Martin A. Suhm for his QMC calculations on DF and HF trimer, and his encouragement and enthusiasm to search for the DF trimer IR spectra discussed in this work.

References

1. Nesbitt, D.J. (1988) Chem. Rev. **88**, 843.
2. McIlroy, A., Lascola, R., Lovejoy, C.M., and Nesbitt, D.J. (1991) J. Phys. Chem. **95**, 2636.
3. McIlroy, A. and Nesbitt, D.J. (1992) J. Chem. Phys. **97**, 6044.
4. Elrod, M.J., Steyert, D.W., and Saykally, R.J. (1991) J. Chem. Phys. **94**, 58.
5. Elrod, M.J., Steyert, D.W., and Saykally, R.J. (1991) J. Chem. Phys. **95**, 3182.
6. Klots, T.D., Chuang, C., Ruoff, R.S., Emilsson, T., and Gutowsky, H.S. (1987) J. Chem. Phys. **86**, 5315.
7. Gutowsky, H.S., Klots, T.D., Chuang, C., Schmuttenmaer, C.A., and Emilsson, T. (1987) J. Chem. Phys. **86**, 569.
8. Gutowsky, H.S., Klots, T.D., Chuang, C., Keen, J.D., Schmuttenmaer, C.A., and Emilsson, T. (1985) J. Am. Chem. Soc. **107**, 7174; ibid (1987) **109**, 5633.
9. Hutson, J.M., Beswick, J.A., and Halberstadt, N. (1989) J. Chem. Phys. **90**, 1337.
10. Cooper, A.R. and Hutson, J.M. (1993) J. Chem. Phys. **98**, 5337.
11. Elrod, M.J., Loeser, J.G., and Saykally, R.J. (1993) J. Chem. Phys. **98**, 5352.
12. Pine, A.S. and Lafferty, W.J. (1983) J. Chem. Phys. **78**, 2154.
13. Fraser, G.T. and Pine, A.S. (1989) J. Chem. Phys. **91**, 633.
14. Miller, R.E. (1990) Acc. Chem. Res. **23**, 10.
15. Schuder, M.D., Lovejoy, C.M., Lascola, R., and Nesbitt, D.J. (in press) J. Chem. Phys.
16. Lovejoy, C.M. and Nesbitt, D.J. (1987) J. Chem. Phys. **87**, 1450.
17. Puttkamer, K.v. and Quack, M. (1989) Chem. Phys. **139**, 31.

18. Quack, M. and Suhm, M.A. (1991) J. Chem. Phys. **95**, 28.
19. Lovejoy, C.M. and Nesbitt, D.J. (1987) Rev. Sci. Instr. **58**, 807.
20. Lovejoy, C.M., Hutson, J.M., and Nesbitt, D.J. (1992) J. Chem. Phys. **97**, 8009.
21. Suhm, M.A., Farrell, J.T. Jr., McIlroy, J.A., and Nesbitt, D.J. (1992) J. Chem. Phys. **97**, 5341.
22. Farrell, J.T. Jr., Sneh, O., McIlroy, A., Knight, A.E.W., and Nesbitt, D.J. (1992) J. Chem. Phys. **97**, 7967.
23. Hutson, J.M. (1992) J. Chem. Phys. **96**, 6752.
24. Quack, M., Stohner, J., and Suhm, M.A. (in press) J. Mol. Struct. (EUCMOS 21 Proceedings).
25. Karpfen, A., (1990) Int. J. Quantum. Chem. **24**, 129.
26. Chalasinski, G., Cybulski, S. M., Szczesniak, M. M., and Scheiner, S. (1989) J. Chem. Phys. **97**, 7048.
27. Riedle, E., Ashworth, S.H., Farrell, J.T. Jr., and Nesbitt, D.J. (submitted) Rev. Sci. Instr.
28. Farrell, J.T. Jr., and Ashworth, S.H. (private communication).
29. Farrell, J.T. Jr., Davis, S., and Nesbitt, D. J. (work in progress).
30. Meath, W. J. and Koulis, M. (1991) J. Mol. Struct. **226**, 1.
31. Nesbitt, D.J., Child, M.S., and Clary, D.C. (1989) J. Chem. Phys. **90**, 4855.
32. Aziz, R.A. (1989) Inert Gases (Springer-Verlag, Berlin).
33. Lewerenz, M. (work in progress).
34. Cooper, A.R. and Hutson, J.M. (work in progress).
35. Axilrod, B.M. and Teller, E. (1943) J. Chem. Phys. **11**, 299.
36. Pauling, L. (1940) The Nature of the Chemical Bond (University, Oxford).
37. Lisy, J.M., Tramer, A., Vernon, M.F., and Lee, Y.T. (1981) J. Chem. Phys. **75**, 4733.
38. Kolenbrander, K.D., Dykstra, C.E., and Lisy, J.M. (1988) J. Chem. Phys. **88**, 5995.
39. Pine, A.S. (private communication); Miller, R.E. (private communication).
40. Suhm, M.A., Farrell, J.T. Jr., Ashworth, S.H., and Nesbitt, D.J. (1993) J. Chem. Phys. **98**, 5985.
41. Lawrance, W.D. and Knight, A.E.W. (1985) J. Phys. Chem. **89**, 917.
42. Bixon, M. and Jortner, J. (1968) J. Chem. Phys. **48**, 715.
43. Wigner, E.P. (1967) Soc. Ind. Appl. Phys. **9**, 1.

ELECTRONIC SPECTRA OF A LITHIUM IMPURITY IN CLUSTERS,
THE BULK LIQUID, AND SOLID *para*- HYDROGEN

Daphna Scharf, Glenn G. Martyna and Michael L. Klein
Department of Chemistry, University of Pennsylvania
Philadelphia, PA 19102 - 6323, USA

1 Introduction

The similarities and differences between the site occupied by an impurity atom in a cluster and in the bulk is of fundamental importance to the understanding of evolution of physical properties as a function of system size. From the experimental point of view, it is possible to gain indirect information about the local environment and the trapping sites of an impurity in clusters and in the bulk by applying various spectroscopic techniques. Matrix isolated techniques have been developed to study guest chromophores in host solvents, [1, 2, 3, 4, 5, 6, 7] as well as a variety of inert guest atoms and molecular species [1, 8, 9, 10, 6, 11]. Similarly, chromophores have been used to probe the local environment in glasses [12, 13]. Complementary work on clusters, has probed the spectroscopic properties of small aggregates (homo and hetro dimers, trimers up to $n \leq 10$) [14, 15, 16, 17, 18] as well as larger size aggregates [19, 20, 21, 22, 23]. Theoretical work concentrated on the calculations of spectral shifts and line-shapes in classical clusters [24, 25, 26, 27, 28, 29, 30, 31]. In particular, the site occupied by an impurity, whether it is a surface or a bulk state, can in principle be resolved by understanding the spectroscopy of the guest chromophore.

In this study, we investigate the evolution of the inhomogeneously broadened dipole spectra and the ionization potential of a single lithium impurity atom in clusters of *para* -hydrogen with $n = 12, 33$ and 180. The inhomogeneously broadened dipole spectra are contrasted with the calculated spectra of a lithium impurity in liquid and in solid *para* -hydrogen. The quantum effects of the light atom and molecules play an important role in this evolution due to the large zero-point motion.

Structural and energetic properties of *para* -hydrogen (p -H_2) clusters, with and without a single lithium impurity, were calculated [32, 33, 34, 35] at finite temperature using the Feynman path-integral formulation of quantum statistical mechanics [36, 37] and the staging algorithm path-integral Monte Carlo sampling method [38]. To study the properties of bulk liquid and solid p -H_2, a constant pressure version of the path-integral formulation was developed [35]. The liquid p -H_2 was studied at $T = 14 - 20K$ and $P_{ext} = 0$. The solid p -H_2 was studies at $T = 4K$ and $P_{ext} = 0$. The inhomogeneously broadened dipole spectra of a single lithium atom in bulk liquid and solid p -H_2 were calculated by implementing the radial fast Fourier transform Lanczos method [39, 25]

J. Jortner et al. (eds.), Reaction Dynamics in Clusters and Condensed Phases, 153–168.

for configurations taken from the corresponding path-integral Monte Carlo runs. This procedure enabled us to directly compare our results with the recently measured absorption spectrum for Li/p-H_2 matrix [40].

Our results show that for a single lithium impurity in p-H_2 (a) the lithium always resides at or on the surface of the cluster; (b) the spectra in the clusters (dipole spectrum and the ionization potential) are very narrow (~ 0.1 eV and ~ 0.04 eV, respectively); (c) the widths of the calculated spectra broaden as a function of the size of the cluster; (d) the ionization potential shifts to lower energies as the size of the cluster increases; (e) the inhomogeneously broadened dipole spectrum becomes more asymmetric for larger size clusters, where a shoulder to the "blue" (high energy) of the atomic absorption becomes more prominent; (f) the "blue" shoulder of the atomic absorption line persists in the spectrum of the lithium in the smaller trapping sites in the solid (g) the overall widths of the bulk dipole spectra are much wider in the bulk than in the clusters; (h) in bulk liquid the inhomogeneously broadened dipole spectrum is more symmetric; (i) the corresponding spectra in the solid with variable number of vacancies attached to the lithium exhibit pronounced asymmetry and in some trapping sites a split first peak; (j) the experimental result for the spectrum of lithium laser ablated in p-H_2 matrix [40] corresponds best to our calculated dipole spectrum of a lithium impurity in a three-vacancy trapping site.

2   Numerical Methods

2.1   THE PATH INTEGRAL METHODS

Equilibrium thermodynamical properties of quantum systems at finite temperatures can readily be calculated by implementing the path-integral formulation of statistical mechanics [36, 37]. In the present work the staging Monte Carlo and standard Metropolis algorithms [38] were used combined to promote fast convergence of the calculated properties and more efficient sampling of the configurational phase space. Thus, in the cluster simulations two types of moves are performed, a sampling of a section of the path (length j) of a particle and a move of the particle centroid coordinate, $\mathbf{r}_{Ic} = 1/P \sum_{i=1}^{P} \mathbf{r}_{Ii}$. One pass consists of P/j moves of each particle path and one move of each particle centroid. In order to simulate the bulk, a sample of 180 molecules with periodic boundary conditions in a hexagonal-close-packed (HCP) lattice arrangement was used. Isothermal-isobaric calculations were performed for the bulk. Here, an additional move associated with sampling the volume is included. The logarithm of the volume, rather than the volume itself was sampled every two passes. This procedure is known to be more efficient [41]. All the step sizes in the simulations were chosen to give a 40 percent acceptance rate.

2.1.1   *The Canonical Ensemble*

In a system of N quantum particles, where the Hamiltonian is $\hat{H} = \hat{T} + \phi(\mathbf{R})$, the canonical partition function can be written

$$Q = Tr[\exp(-\beta H)] = \int d\mathbf{R}_1 \ldots d\mathbf{R}_P \prod_{i=1}^{P} \rho(\mathbf{R}_i, \mathbf{R}_{i+1}; \epsilon) \qquad (2.1)$$

where, $\rho(\mathbf{R}_i, \mathbf{R}_{i+1}; \epsilon)$ is the density matrix at inverse temperature $\epsilon = \beta/P$, $\beta = 1/k_B T$. Here, P is the number of discrete points ("beads") along the path such that $\mathbf{R}_{P+1} = \mathbf{R}_1$, and $\mathbf{R}_i$ is a coordinate representing the positions of the $i - th$ bead of all the quantum particles, $\mathbf{R}_i = \{\mathbf{r}_{1i} \ldots \mathbf{r}_{Ni}\}$. After introducing the Trotter product formula and the free particle density matrix, ρ_0, which is the product of the free particle density matrices for each of the N quantum particles,

$$\rho_0(\mathbf{R}_i, \mathbf{R}_{i+1}; \epsilon) = \prod_{I=1}^{N} \left(\frac{m_I}{2\pi\hbar^2\epsilon}\right)^{3/2} \exp\left[\frac{m_I}{2\hbar^2\epsilon}(\mathbf{r}_{Ii} - \mathbf{r}_{Ii+1})^2\right], \qquad (2.2)$$

the discrete path-integral partition function is given by

$$Q_P = \int d\mathbf{R}_1 \ldots d\mathbf{R}_P \qquad (2.3)$$

$$\exp\left[-\epsilon \sum_{i=1}^{P} \phi(\mathbf{R})\right] \prod_{i=1}^{P} \rho_0(\mathbf{R}_i, \mathbf{R}_{i+1}; \epsilon) + \mathcal{O}(\frac{\beta^3}{P^2}).$$

2.1.2 *The Isothermal − Isobaric Ensemble*

The properties of the cluster systems considered in this paper are calculated in the canonical ensemble using, Eq. (2.1). However, to examine the condensed phase systems, the isothermal-isobaric ensemble (constant N, P_{ext}, T) is used. In general, the isothermal-isobaric partition function can be written as [42]

$$\Delta = \int dV \exp\left(-\beta P_{ext} V\right) Q(V). \qquad (2.4)$$

In the discrete path-integral representation this becomes

$$\Delta = \int dV \int d\mathbf{S}_1 \ldots d\mathbf{S}_P V^{NP} \exp\left(-\beta P_{ext} V\right) \qquad (2.5)$$

$$\exp\left[-\epsilon \sum_{i=1}^{P} \phi(\mathbf{S}, V)\right] \prod_{i=1}^{P} \rho_0(\mathbf{S}_i, \mathbf{S}_{i+1}; \epsilon; V)$$

where $\mathbf{S} = \mathbf{R}/V^{1/3}$ and N is the total number of particles. In these calculations, $\phi(\mathbf{S}, V)$, is assumed to include the appropriate long range correction to the potential energy. This correction allows small system sizes (~ 200 particles) to achieve asymptotically correct results.

2.2   THE SPECTRA

Ionization and inhomogeneously broadened dipole spectra of an isolated lithium atom in *para*-hydrogen have been calculated. This was accomplished by taking the equilibrium configurations from the corresponding Monte Carlo runs, replacing the lithium atom by an electron and a lithium ion, and determining the electronic states using the radial fast Fourier transform Lanczos method. This approach accounts for the zero-point motion and thermal effects present in the system [35].

2.2.1   *The Fourier Projector*

A short imaginary time propagator is applied on a wave function by using the fast Fourier transform projection method [43]. The propagator is partitioned in the following way:

$$\exp\left[-\tau\left(\hat{H} - \hat{H}_0 + \hat{H}_0\right)\right]\psi = \exp\left[-\frac{\tau}{2}\left(\hat{H} - \hat{H}_0\right)\right] \quad (2.6)$$
$$\times \exp\left[-\tau\hat{H}_0\right]\exp\left[-\frac{\tau}{2}\left(\hat{H} - \hat{H}_0\right)\right]\psi + (0)\tau^3,$$

where $\hat{H}_0$ is chosen such that $(\hat{H} - \hat{H}_0)$ contains no differential operators. In this study, $\hat{H}_0$ is taken to be the kinetic energy operator in spherical polar coordinates. A grid in position space is used in conjunction with a finite number of basis functions in Eq. (2.6). This approach allows the use of non-local pseudopotential for an atom placed at the center of the radial grid, in our study the lithium atom [39, 25]. The pseudopotentials used to describe the $e^- - Li$ and the $e^- - H_2$ interactions are discussed in Section 4.

2.2.2   The Block Lanczos Method

The block Lanczos method can be combined with the fast Fourier projector method to obtain the n largest eigenvalues and eigenvectors of the short imaginary time propagator. [44]. The starting point is a vector or block of orthonormal wave functions. A new block of functions is created by applying the short time propagator to each of the n members of the old block. Each element in the resulting block is then Graham-Schmidt orthogonalized to the elements of the lower block and to each other. [45, 46]. This process is used to simultaneously determine the matrix elements of the short time propagator between blocks. Blocks are created until upon diagonalization of the resulting matrix the largest n eigenvalues are found to be converged.

Rather than performing one large block Lanczos procedure, one can perform successive evaluations of the eigenvalues, for smaller and smaller values of τ. The results for the larger τ value serve as an input for evaluating the eigenvalues with the smaller τ. The process is continued until τ is small enough, and the eigenvalues are converged. Note, the eigenvectors of the true propagator are the same as those of $\hat{H}$ and the eigenvalues are related by $E_\tau = \exp(-\tau E)$. Also, the largest eigenvalues of the propagator are the smallest eigenvalues of $\hat{H}$.

2.2.3   Dipole and Ionization Spectra

The inhomogeneously broadened spectrum of a lithium atom solvated by $para$-hydrogen molecules is

$$I(E) = \frac{1}{Q}\int d\mathbf{R}\rho(\mathbf{R},\mathbf{R},\beta)\sum_{i=1}^{\infty} |\langle\phi_0(\mathbf{R}) \mid \hat{\mu} \mid \phi_i(\mathbf{R})\rangle|^2 \delta(E_{i0}(\mathbf{R}) - E) \quad (2.7)$$

where $\hat{\mu}$ is the dipole moment operator. Here, the valence electron of the lithium atom is assumed to occupy the ground state at equilibrium. The quantity $\rho(\mathbf{R},\mathbf{R},\beta)$, is the quantum mechanical thermal density matrix for the coupled lithium-$para$-hydrogen system on the ground state electronic surface. The spectrum can, therefore, be obtained by calculating the electronic states of configurations taken from the path integral Monte Carlo simulations, which are distributed according to $\rho(\mathbf{R},\mathbf{R},\beta)/Q$. In practice, $N_c = 100$ - 200 configurations are used.

A path integral configuration contains P estimates of the distribution function (i.e. $< A >= 1/P \sum A_i$), one for each imaginary time slice. These contributions are not statistically independent. Therefore, the states of a number, p, of randomly chosen time slices are calculated for each configuration, where each time slice contains one lithium atom and N hydrogen molecules. A total of pN_c calculations are, thus, performed and used to construct the inhomogeneously broadened spectrum as in Eq. (2.7). However, a finite energy bin width is used instead of a delta function. This treatment includes all zero point and thermal effects of the nuclear degrees of freedom as can be seen from Eq. (2.7). A similar procedure can be used to determine the ionization spectrum

$$I(E) = \frac{1}{Q} \int d\mathbf{R} \rho(\mathbf{R}, \mathbf{R}, \beta) \delta(E_0(\mathbf{R}) - E). \tag{2.8}$$

3 Potential Functions

The implementation of the path-integral approach to the calculation of equilibrium statistical properties requires the knowledge of the potential functions representing the interparticle interactions. The interactions in this study are assumed to be additive pairwise and, moreover, only the spherical part of the interaction is explicitly incorporated. For the intermolecular p-H_2 interaction, these assumptions have been shown [47, 48] to be very good under normal conditions. The internal motions in p-H_2 can be largely ignored when one considers the properties of the bulk due to a pronounced mismatch between the intramolecular high-frequency strong interaction and the weak low-frequency intermolecular interactions. This is also reflected in the corresponding equilibrium and binding energies: $4.75 eV$ with $r_{eq} = 0.74$ Åfor the intramolecular interaction and $0.00278 eV$ with $r_{eq} = 3.44$ Åfor the intermolecular one. The large separations also enables us to consider the $J = 0$ as a good quantum number at low pressure. Thus, p-H_2 behaves with resemblance to rare-gas systems. The second type of interaction that has to be considered is between the Li and the p-H_2. In this case, only limited information is available. This interaction, taken from ab $initio$ calculations, is very weak and characterized by a large equilibrium distance ($r_{eq} = 5.21$ Åand $0.0015 eV$ binding energy) [49, 50, 51].

3.1 INTERMOLECULAR HYDROGEN POTENTIAL

At temperatures and pressures relevant to this study it is assumes that the $para$-hydrogen is in its ground rotational state, $J = 0$, and the molecule is spherical. The spherical part of the Silvera-Goldman potential [47] is used, as in the previous studies [32, 33, 34]. This potential accounts through an effective two-body term, C_9/r^9, for the three-body Axilrod-Teller-Muto triple-dipole dispersion interaction [47]. The potential energy function is written as

$$V_{H_2-H_2}(r) = \exp\left[\alpha - \beta r - \gamma r^2\right] - \left(\frac{C_6}{r^6} + \frac{C_8}{r^8} - \frac{C_9}{r^9} + \frac{C_{10}}{r^{10}}\right) f_c(r) \tag{3.9}$$

where

$$f_c(r) = \begin{cases} \exp\left\{-[1.28\,(r_m/r) - 1]^2\right\} & \text{if } r < 1.28\,r_m \\ 1 & \text{otherwise.} \end{cases} \tag{3.10}$$

The values of the parameters, all in atomic units, are

$$
\begin{array}{ll}
\alpha = 1.713 & C_6 = 12.14 \\
\beta = 1.5671 & C_8 = 215.2 \\
\gamma = 0.00993 & C_9 = 143.1 \\
r_m = 6.50 & C_{10} = 4813.9
\end{array}
\tag{3.11}
$$

3.2 LITHIUM - HYDROGEN POTENTIAL

Recently, state-of-the-art ISCF-CI calculation have been performed [51] to complement previous incomplete estimates and calculations [49, 50] on this interaction. The spherical part of the new results were fitted [35] to the same functional form as the intermolecular hydrogen potential, Eq. (3.1). The parameters, all in atomic units, are

$$
\begin{array}{ll}
\alpha = 2.511 & C_6 = 85.00 \\
\beta = 0.3040 & C_8 = 4049.98 \\
\gamma = 0.00420 & C_9 = 0.0 \\
r_m = 9.856 & C_{10} = 0.0
\end{array}
\tag{3.12}
$$

4   Pseudopotentials

The application of the fast Fourier transform Lanczos method requires the knowledge of two pseudopotentials, a e^- - Li pseudopotential and a e^- - H_2 pseudopotential.

4.1   THE ELECTRON - LITHIUM ION

The e^- - Li pseudopotential is taken from the compilation of Bachelet $et.al.$ [52]. It is a non-local pseudopotential having a term which explicitly depends on the angular momentum projection operator

$$
\hat{V}_{e^- - M^+}(\mathbf{r}) = \sum_{l=0}^{l_{max}-1} \mid l \rangle (V_l(r) - V_{l_{max}}(r)) \langle l \mid + V_{core}(r) + V_{l_{max}}(r).
\tag{4.1}
$$

In this potential the local piece, $V_{core}(r)$, is the screened long-range potential and $V_l(r)$ is the potential appropriate for angular momentum, l. The Radial Fast Fourier Lanczos method can accommodate this type of pseudopotential without loss of generality.

4.2 THE ELECTRON - MOLECULAR HYDROGEN

The e^- - H_2 pseudopotential is constructed for a spherical ($J = 0$) hydrogen molecule to comply with the lithium - hydrogen interaction through a radial pair potential. Therefore, a local pseudopotential of the form

$$
V_{e^- - H_2}(r) = A \exp(-ar) - \frac{\alpha e^2}{2(r^2 + d^2)^2}
\tag{4.2}
$$

has been fit to reproduce the aforementioned pair potential

$$V_{Li-H_2}(r) = E_0(r) - E_0(\infty) + V_{Li^+-H_2}(r), \tag{4.3}$$

where

$$V_{Li^+-H_2}(r) = -\frac{\alpha e^2}{2r^4} \quad r > 5au \tag{4.4}$$

and $\alpha = 5.5268927au$, $A = 4.52au$, $a = 1.6au$ and $d^2 = 1.0au$.

4.3 POLARIZATION ENERGY

For a system consisting of a single lithium ion and an electron in a cluster with N hydrogen molecules, the contribution to the electronic energy of the pair polarization energy can be directly evaluated [53, 54, 39, 25]

$$
\begin{aligned}
V_{pol}(\mathbf{r}_e) &= -\frac{\alpha}{2} \sum_{H_2} \mathbf{E}^0(\mathbf{r}_e - \mathbf{r}_{H_2}) \cdot \mathbf{E}^0(\mathbf{r}_e - \mathbf{r}_{H_2}) - 2\mathbf{E}^0(\mathbf{r}_e - r_{H_2}) \cdot \mathbf{E}^0(\mathbf{r}_{Li^+} - \mathbf{r}_{H_2}) \\
V_{pol}(\mathbf{r}_e) &= -\frac{\alpha}{2} \sum_{H_2} \frac{S(|\mathbf{r}_e - \mathbf{r}_{H_2}|)}{(|\mathbf{r}_e - \mathbf{r}_{H_2}|)^4} - 2\mathbf{E}^0(\mathbf{r}_e - r_{H_2}) \cdot \mathbf{E}^0(\mathbf{r}_{Li^+} - \mathbf{r}_{H_2}),
\end{aligned}
\tag{4.5}
$$

where α is the polarizability of the hydrogen molecules. Here, due care is taken to account for the finite core sizes of the ion and the molecules in the effective Coulomb field, $\mathbf{E}^0$, by using an adequate switching function, S, to screen the long-range polarization interaction [35].

In the bulk liquid and solid phases, the calculation is performed for a single lithium ion and a single electron at infinite dilution, namely, only the hydrogen molecules are replicated. The polarization contribution becomes [35]

$$
\begin{aligned}
V_{pol}(\mathbf{r}_e) &= -\frac{\alpha}{2} \sum_{H_2} \sum_{\mathbf{n}} 1\mathbf{E}^0(\mathbf{r}_e - \mathbf{r}_{H_2} - \mathbf{n}) \cdot \mathbf{E}^0(\mathbf{r}_e - \mathbf{r}_{H_2} - \mathbf{n}) \\
&\quad - 2\mathbf{E}^0(\mathbf{r}_e - \mathbf{r}_{H_2} - \mathbf{n}) \cdot \mathbf{E}^0(\mathbf{r}_{Li^+} - \mathbf{r}_{H_2} - \mathbf{n}).
\end{aligned}
\tag{4.6}
$$

The evaluation of the first term on the r.h.s. requires a lattice sum over $1/r^4$, which is performed following Ref. [55], and an additional approximation to evaluate the second term,

$$V_{pol}(\mathbf{r}_e) = I - \frac{\alpha}{2} \sum_{H_2} \sum_{\mathbf{n},\mathbf{n}'} -2\mathbf{E}^0(\mathbf{r}_e - \mathbf{r}_{H_2} - \mathbf{n}) \cdot \mathbf{E}^0(\mathbf{r}_{Li^+} - \mathbf{r}_{H_2} - \mathbf{n}'). \tag{4.7}$$

This last term, Eq. (4.7), can be evaluated using the Ewald method [41]. This approximation has the virtue that it preserves the symmetry of the lattice.

5 Results

5.1 CLUSTERS

5.1.1   *Structures and Energetics*

Neat $(p\text{-}H_2)_m$ and doped $Li(p\text{-}H_2)_n$ clusters ($n = m - 1$) were studied [33, 34, 35] in the temperature range $T = 2.5 - 6.0$, for cluster sizes $m = 12, 13, 14, 19, 32, 33, 34, 35, 55$ and 181. Only three representative cases are considered here explicitly, $n = 12, 33$ and 180. Very long runs were performed in the study of the clusters, up to 100000 passes at $T = 2.5K$, and about half as many at higher temperatures. The particular sizes were chosen because the configuration space of the 13− and 34− molecule neat clusters include high symmetry structures, as well as disordered ones [33, 34]. The largest cluster is used to demonstrate the size effects.

The configuration space of small and medium size neat $p\text{-}H_2$ clusters is dominated by disordered liquid like structures [33, 34]. However, high symmetry structures based on the icosahedral symmetry and pentagonal bi-pyramid could also be identified. The Li impurity, when attached to $p\text{-}H_2$ clusters, does not appear to serve as a nucleation center. Rather, the lithium atom is found at or near the surface of the clusters, see Fig. 1. This may be attributed to the combination of a large zero-point motion of the hydrogen molecules, the large equilibrium distance for the lithium interaction with $p\text{-}H_2$ and the small binding energy of lithium to $p\text{-}H_2$.

The energetics of the clusters suggest that the Li impurity reduces the overall stability of the clusters. The evaporation of a single $p\text{-}H_2$ molecule from the doped cluster may occur at a lower temperature, compared with a similar evaporation form a neat cluster of the same size.

5.1.2   *Dipole and Ionization Spectra*

The dipole and the ionization spectra of a lithium impurity in $p\text{-}H_2$ are compared with the corresponding atomic spectra of lithium, and with the lithium as guest atom in the bulk. The spectra of a lithium impurity in clusters are very narrow. In Fig. 2, we show the ionization spectra for a lithium impurity in various size clusters. The ionization spectra are shifted to lower energies compared with the atomic ionization potential, also shown in Fig. 2. The spectrum becomes broader for the larger size clusters. These "red" shifts are larger for the larger size clusters. However, per $p\text{-}H_2$ molecule, this shift is larger for the smaller size clusters. While the spectra can clearly distinguish between larger and smaller size clusters, unfortunately, only minor differences were found for clusters in the same size range.

The inhomogeneously broadened dipole spectra are shown in Fig. 3. For convenience, the spectra are presented on an energy scale relative to the isolated atomic absorption. The spectra consist of two bands. The central band is attributed to the two p-orbitals tangential to the surface of the cluster. The asymmetric band, at higher energies, is attributed to the third p-orbital directed radially to the surface. While only small variations between the spectra of closely spaced sizes could be found, the differences between large and small clusters are apparent. The spectrum of the larger clusters broaden and exhibit more pronounced asymmetry.

5.2   LITHIUM IN BULK *para* - HYDROGEN

The inhomogeneously broadened dipole spectra of a lithium impurity was calculated in bulk liquid $p\text{-}H_2$ and for a number of trapping sites in the HCP solid. In physical systems,

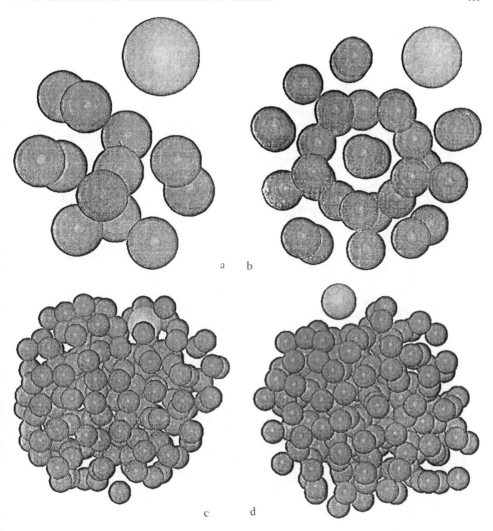

Figure 1: Snapshots from the simulations of a lithium impurity in p-H_2 clusters. The large and small circles represent the Li atom and a p-H_2 molecule, respectively. The ratio of the radii corresponds to the ratio of the inter-particle equilibrium distance. (a) $Li(p$-$H_2)_{12}$, $T = 2.5K$ (b) $Li(p$-$H_2)_{33}$, $T = 2.5K$ (c) and (d) $Li(p$-$H_2)_{180}$, $T = 4K$, showing various surface states for the Li near or in the surface.

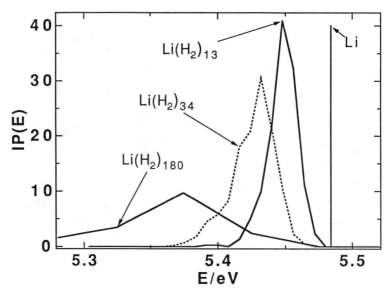

Figure 2: The ionization spectra in arbitrary units of Li in p-H_2 clusters compared with the atomic ionization potential of Li. The ionization spectra of a lithium impurity shifts to lower energies. Larger "red" shifts are exhibited for the larger size clusters.

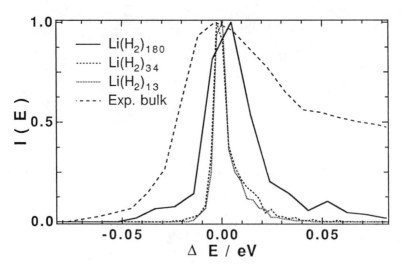

Figure 3: The inhomogeneously broadened dipole spectra of Li in $(p$-$H_2)_n$ clusters for $n = 12, 33$ and 180. The spectra are calculated as shifts from the atomic dipole spectrum. The spectra are peak normalized and show the development of a "shoulder" at higher energies as a function of the size of the cluster.

one does not expect to find atomic lithium in liquid p-H_2, because the high mobility of the lithium leads to fast dimerization and the formation of the energetically favored Li_2. However, the calculated spectrum of lithium in the liquid helps to demonstrate the effects of lower density and symmetry on the spectra in comparison to the highly ordered HCP solid. The contrast with the clusters is also instructive.

5.2.1 The Trapping Sites

In liquid p-H_2 the lithium impurity is found to be "solvated" by approximately twenty p-H_2 molecules [35]. This solvation shell is quite centro-symmetric. This si quite different from the cluster environment, where the lithium atom resides in a site near the surface that is mostly exposed.

In order to determine in a less biased, relatively simple, way the likely configuration of the vacancies bound to a lithium impurity in $para$-hydrogen HCP solid, the following method is used to select the likely trapping sites [56]. Starting with the result for a lithium impurity in liquid p-H_2, where the solvation of the lithium impurity can be sampled efficiently, several configurations of the lithium impurity and its immediate liquid solvation shell were stored. We refer to such configurations as "liquid core". A trapping site in a HCP solid with n_v vacancies attached to the lithium was prepared by replacing $20+n_v$ molecules from the sample of the solid by the liquid core, and starting the constant pressure path-integral Monte Carlo run. After equilibration (~ 5000 steps), the relaxed configuration was analyzed. Interestingly, relatively few relaxed configurations were found for a given n_v vacancies. In all cases the geometry of the trapping site was equilibrated by long runs, typically 40000-60000 steps, that started from an "idealized" trapping sites based on the sites from the runs with the liquid core. For example, when $n_v = 1$, the production runs started with the lithium in a substitutional lattice site. When $n_v = 2$ through 6, the corresponding number of molecules were removed, and the lithium, was positioned in an interstitial site.

The equilibrium trapping sites for lithium, following local relaxation, were found to be asymmetric lattice sites in all the cases studied [56]. The analysis of the energetics of the configurations showed that both the kinetic energy per particle and the average host lattice potential energy were roughly independent of n_v. The average energy per p-H_2 molecule, $\sim -87K$, was fairly close to its value in the neat solid calculated at the same level of approximation. Based on simple energetic considerations, vacancy sites with $n_v = 2, 3$ and 4 appeared to be favorable [56]. We note that the distortions of the lattice heal quickly as one moves away from the impurity. This is due to the large zero-point motion and the high compressibility of $para$-hydrogen. The lattice easily deforms to accommodate the presence of the large impurity. The lattice expansion can be calculated [56]. In units of the volume occupied by a p-H_2 molecule in the bulk solid p-H_2, the expansion is approximately seven for $n_v = 1$, decreasing monotonically to about two for the larger trapping sites, $n_v = 5$ and 6. These small perturbations of the lattice are found to be consistent with the small deviations in the radial distribution function for the p-H_2 molecules, compared with the radial distribution function in neat solid p-H_2 [56].

5.2.2 Dipole Spectra

The inhomogeneously broadened dipole spectra of a lithium impurity in bulk liquid and solid p-H_2 are much broader compared with the clusters spectra, Figs. 3 and 4. The liquid spectrum is both broader and more symmetric. The highly perturbative liquid contributes

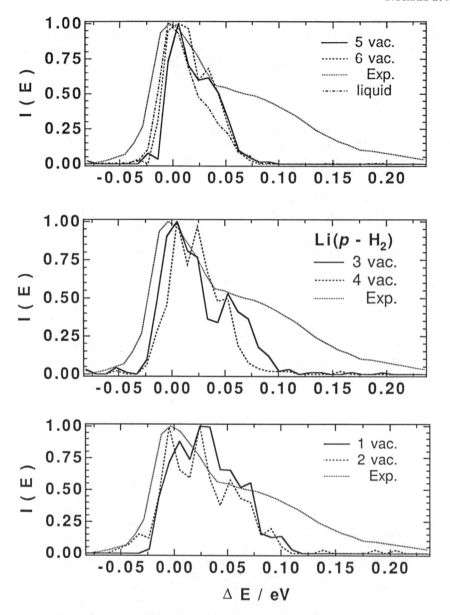

Figure 4: The inhomogeneously broadened dipole spectra of Li in solid p -H_2 at $T = 4K$ and $P_{ext} = 0$ compared to experiment (bold curve) from Ref. [41]: (a) $n_v = 5$ (dotted), $n_v = 6$ (dashed) and the spectrum at low temperature liquid $(T = 14K)$ (b) $n_v = 3$(dotted), $n_v = 4$ (dashed) (c) $n_v = 1$(dotted), $n_v = 2$ (dashed). All the spectra are calculated as shifts from the atomic dipole spectrum.

to the overall broadening of the spectra, as well as to the absence of structure, compared with the clusters. The calculated solid spectra are compared with the experimental results of Ref. [40] in Fig. 4. The calculated spectra resemble in their asymmetric line-shapes the cluster spectra and the main characteristics of the experiment. The trapping of the lithium in asymmetric lattice sites suggest that part of the asymmetry in the spectra may be attributed to the increased symmetry in the first "solvation" shell around the lithium atom in the solid, compared with the calculated result for the liquid. The calculated spectra for a lithium trapped in large vacancy sites, $n_v = 5$ and 6 resemble in widths the calculated spectra for the lithium in liquid $p\text{-}H_2$, see Fig. 4(a). The calculated dipole spectra for the sites with $n_v = 1, 2$ and 4 exhibit a split first peak, see Fig. 4(b) and 4(c). This calculated splitting is not seen in the experimental spectrum [40]. However, the calculated spectra in this size range are broader than the corresponding results for the sites with $n_v = 5$ and 6. Although altogether too narrow, based on the overall shape and width, the three vacancy trapping site seems to give the best agreement with experiment. The high energy "tail" of the experimental spectrum is absent in all the calculated spectra. The "average" energy of this tail region suggests that it may, in fact, be caused by a rotational Franck-Condon transition; such effects have been neglected in our simulation.

6 Conclusions

Path integral Monte Carlo studies of the energetics, structures and spectra have been performed for a single lithium impurity in $p\text{-}H_2$ solid, liquid and clusters. The large zero-point motion and favorable intermolecular $p\text{-}H_2$ interactions explain the tendency of the lithium to occupy surface states in doped clusters. In the bulk, the large compressibility of $para$-hydrogen explains the small structural deformations around the trapping sites.

The electronic spectrum of the lithium impurity atom was calculated for a variety of trapping sites in the solid, in bulk liquid and in clusters. The line-shapes of the inhomogeneously broadened dipole spectra are asymmetric in the solid and in the clusters, while more symmetric in the liquid. The surface states for the lithium in clusters result in a very narrow dipole spectra. The widths and the asymmetry in the cluster spectra become more pronounced for larger size clusters. Bulk perturbative interactions and higher density contribute to the broadening of the spectra. The calculated dipole spectrum for a three-vacancy trapping site in solid $p\text{-}H_2$ was found to give the best agreement with the available experimental results. The ionization spectrum was obtained for the lithium atom in clusters. These spectra have finite width and a band center that progressively red shifts from the atomic value and broadens as the number of $para$-hydrogen molecules increases. These predicted trends should be amenable to experimental testing using molecular beam experiments.

ACKNOWLEDGMENT

The research outlined herein was supported by the National Science Foundation under Grant No. CHE 92-23546, and the Air Force under Grant No. F04611-91-K-006. We thank Steve Rodgers for his interest and encouragement in pursuing this project. Some of the calculations were carried out at PSC under Grant NSF CHE 91-00027. G.M. would like to acknowledge a NSF Postdoctoral Research Associateship in Computer Science and Engineering (ASC-91-08812).

References

[1] N. Schwentner, E. E. Koch, and J. Jortner. *Electronic Excitations in Condensed Rare Gases.* Springer Verlag, Berlin, (1985).

[2] M. McCarty and G.W. Robinson. *Molecular Physics*, 2:415, (1959).

[3] B. Meyer. *Low Tempreture Spectroscopy.* Elsevier, New York, (1971).

[4] G. C. Pimental. *Ber. Bunsenges. Phys. Chem.*, 82:2, (1978).

[5] L. C. Balling, M. D. Harvey, and J. F. Dawson. *J. Chem. Phys.*, 69:1670, (1978).

[6] L. C. Balling and J. F. Dawson. *Phys. Rev. Lett.*, 43:435, (1979).

[7] H. H. von Grünberg and H. Gabriel. *Chem. Phys. Lett.*, 192:503, (1992).

[8] L. C. Balling and J. J. Wright. *J. Chem. Phys.*, 81:675, (1984).

[9] S. Ossicini and F. Forstman. *J. Chem. Phys.*, 75:2076, (1981).

[10] M. E. Fajardo and. G. Carrick and J. W. Kenney III. *J. Chem. Phys.*, 94:5812, (1991).

[11] J. F. Dawson and L. C. Balling. *J. Chem. Phys.*, 71:836, (1979).

[12] G. Fuxi. *Optical and Spectroscopic properties of Glass.* Springer-Verlag, Berlin, (1992).

[13] H. Scholtze. *Glass - Nature, Structure and Properties.* Springer-Verlag, New York, (1991). Translated by M. J. Lakin.

[14] F. G. Celli and K. C. Janda. *Chem. Rev.*, 86:507, (1986).

[15] T. E. Gough, M. Mengel, P. A. Rowntree, and G. Scoles. *J. Chem. Phys.*, 83:4958, (1985).

[16] X. J. Gu, D. J. Levandier, B. Zhang, G. Scoles, and D. Zhuang. *J. Phys. Chem.*, 93:4898, (1990).

[17] D. H. Levy. In R.G. Woolley, editor, *Quantum Dynamics of Molecules*, New York, (1980). Plenum.

[18] Ph. Bréchniac and B. Coutant. *Z. Phy. D*, 89:87, (1989).

[19] S. Goyal, G. N. Robinson, D. L. Schutt, and G. Scoles. *J. Phys. Chem.*, 95:4186, (1991).

[20] S. Goyal, D. L. Schutt, and G. Scoles. *Phys. Rev. Lett.*, 69:933, (1992).

[21] S. Goyal, D. L. Schutt, and G. Scoles. In P. Jena, editor, *Physics and Chemistry of Finite systems - From Clusters to Crystals.* Kluwer Academic Publishers, (1990).

[22] J.-P. Visticot, J. Berlande, J. Cuvellier, A. Lallement, J. M. Mestdagh, P. Meynadier, P. de Pujo, and O. Sublemontier. *Chem. Phys. Letts.*, 191:107, (1992).

[23] A. Lallement, O. Sublemontie, J.-P. Visticot, A. J. Bell, J. Berlande, J. Cuvellier, J. M. Mestdagh, and P. Meynadier. *Chem. Phys. Letts.*, 204:440, (1993).

[24] L. Perera and F. G. Amar. *J. Chem. Phys.*, 93:4884, (1990).

[25] G. J. Martyna, C. Cheng, and M. L. Klein. *J. Chem. Phys.*, 93:4386, (1990).

[26] K. Haug and H. Metiu. *J. Chem. Phys.*, 95:5670, (1991).

[27] C. Tsoo, D. A. Estrin, and S. J. Singer. *J. Chem. Phys.*, 93:7187, (1990).

[28] S. Leutwyler. *Chem. Phys. Lett*, 115:40, (1985).

[29] N. Liver, A. Nitzan, and J. Jortner. *J. Chem. Phys.*, 95:3516, (1988).

[30] E. Shalev, N. Ben-Horin, and J. Jortner. *Chem. Phys. Lett.*, 177:161, (1991).

[31] L. E. Fried and S. Mukamel. *Adv. Chem. Phys.*, (1993). to be published.

[32] P. Sindzingre, D. M. Ceperley, and M. L. Klein. *Phys. Rev. Lett.*, 67:1871, (1991).

[33] D. Scharf, G. J. Martyna, and M.L. Klein. *J. Chem. Phys.*, 97:3590, (1992).

[34] D. Scharf, G. J. Martyna, and M. L. Klein. *Chem. Phys. Lett.*, 197:231, (1992).

[35] D. Scharf, G. J. Martyna, and M. L. Klein. Path-integral monte carlo study of a lithium impurity in *para*-hydrogen clusters and bulk liquid. submitted, (1993).

[36] R. P. Feynman and A. R. Hibbs. *Quantum Mechanics and Path Integrals.* McGraw-Hill, New York, (1965).

[37] R. P. Feynman. *Statistical Mechanics.* Benjamin, Reading, (1972).

[38] E. L. Pollock and D. M. Ceperley. *Phys. Rev. B*, 30:2555, (1984).

[39] G. J. Martyna and B. J. Berne. *J. Chem. Phys.*, 90:4516, (1989).

[40] M. E. Fajardo. *J. Chem. Phys.*, 98:110, (1992).

[41] M. P. Allen and D. J. Tildesley. *Computer Simulations of Liquids.* Clarendon Press, Oxford, (1989).

[42] D. A. McQuarrie. *Statistical Mechanics.* Harpers & Row, New York, 1976.

[43] J. A. Fleck M. D. Feit and A. Steiger. *J. Comp. Phys.*, 47:412, (1982).

[44] F. Webster, P. Rossky, and R. Friesner. *Computer Physics Communications*, 25:780, (1990).

[45] J. K. Cullum and R. A. Willoughby. *Lancoz Algorithms for Large Symmetric Eigenvalue Problems.* Birkhauser, Boston, (1985).

[46] G. J. Martyna, C. Cheng, and M. L. Klein. *J. Chem. Phys.*, 95:1318, (1991).

[47] I. F. Silvera and V. V. Goldman. *J. Chem. Phys.*, 69:4209, (1978).

[48] M. J. Norman, R. O. Watts, and U. Buck. *J. Chem. Phys.*, 81:3500, (1984).

[49] J. C. Tully. *J. Chem. Phys.*, 59:5122, (1973).

[50] A. F. Wagner, A. C. Wahl, A. M. Karo, and R. Krejci. *J. Chem. Phys.*, 69:3756, (1978).

[51] J. Konowalow. ISCF calculation of the $Li - H_2$ pair potential. preprint, (1992).

[52] G. Bachelet, D. Hamann, and M. Schluter. *Phys. Rev. B*, 26:4199, (1982).

[53] J. D. Jackson. *Classical Electrodynamics*. John Wiley and Sons, New York, (1975).

[54] F. H. Stillinger. *J. Chem. Phys.*, 71:1674, (1979).

[55] D. E. Williams. *Acta Cryst.*, A27:452, (1971).

[56] D. Scharf, G. J. Martyna, D. Li, G. A. Voth, and M. L. Klein. Nature of lithium trapping sites in the quantum solids *para*-hydrogen and *ortho*-deuterium. submitted, (1993).

LOCALIZATION OF ELECTRONS AT INTERFACES

Robert Lingle, Jr., D. F. Padowitz, R. E. Jordan, J. D. McNeill,
and C. B. Harris
*Department of Chemistry, University of California, and
Chemical Sciences Division, Lawrence Berkeley Laboratory,
Berkeley, CA 94720*

Abstract. The states of image-potential electrons at the metal-dielectric interface were studied using angle-resolved two-photon photoemission in ultrahigh vacuum. The binding energy and dispersion of an electron in an image state were studied as a function of layer thickness for various metal-dielectric interfaces. Features representative of both delocalized and localized states were found. As best we can determine, this is the first observation of localized states at an interface with the type of nondispersive photoemission feature reported here. The propensity of layers of straight chain alkanes for localizing excess electrons at the metal-insulator interface correlates with the low electron mobilities observed for the corresponding molecular liquids. Excess electrons become localized as layers of aspherical hydrocarbons on Ag(111) are increased in thickness in a layer-by-layer fashion. However, layers of the nearly spherical neopentane molecule continue to primarily support a delocalized state at a coverage of three layers.

1. Introduction

The behavior of electrons at the interface of two materials is of fundamental importance to both the electronic and chemical properties of the system. Transport studies of conductance at varying temperatures and applied field strengths in liquids or amorphous solids directly measure the response of electrons to external fields. However, few techniques are available to determine the single particle states populated by an electron at an interface. In order to develop a microscopic understanding of the behavior of electrons in complex media, the energy levels associated with an image-potential electron at an interface have been studied.

The states available to excess electrons in alkane layers on Ag(111) have been measured by angle-resolved two-photon photoemission (TPPE). TPPE is well established as a technique for the study of electrons bound to a bare metal surface by the image force.[1] Work now focuses on applying TPPE to study metal/adsorbate systems.[2-5] It is observed that excess electrons placed at this metal-dielectric interface can occupy either localized or free electron-like states, depending on the thickness of the physisorbed layers. Adlayers of the straight chain alkanes support both localized and extended states at some thicknesses. Layers of the nearly spherical neopentane do not support a prominent localized state.

169

J. Jortner et al. (eds.), Reaction Dynamics in Clusters and Condensed Phases, 169–178.

Layer-Resolved TPPE for Alkane/Ag(111)

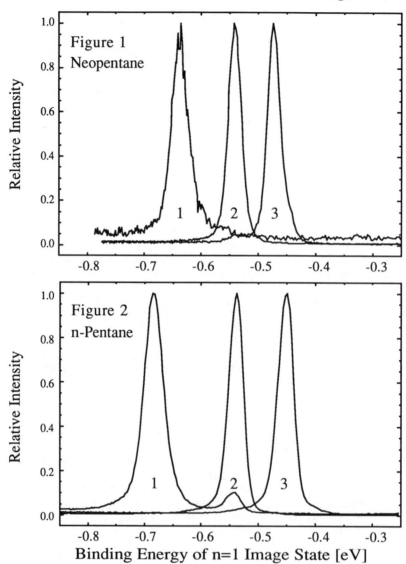

Figures 1 and 2 : Peaks in the TPPE spectrum shift to smaller binding energy as successive layers of dielectric material are adsorbed. The image state binding energy on bare Ag(111) is -0.77 eV. The monolayer peak in Figure 2 shows small patches of bilayer, illustrating the in situ monitoring of layer growth.

2. Experimental Technique

The details of the TPPE spectrometer have been given elsewhere [2-3] and are only briefly described here. The laser system is a Coherent 700 series cavity-dumped dye laser pumped by an Antares mode-locked YLF laser. The 6 ps visible pulses are doubled in BBO, and the temporally overlapped ultraviolet and visible pulses are focused collinearly into the UHV chamber and onto the Ag(111) sample. The uv pulse populates the image potential surface state with electrons promoted from the bulk bands of the metal. The visible photon ejects the electrons from the intermediate state, and the flight time is measured. Time-of-flight detection in conjunction with photon counting electronics yields a calculated energy resolution of 5 meV. Peaks are reproducible to within 7 meV. The *absolute* accuracy in assigning peak energies is limited by the accuracy of the work function measurements carried out in conjunction with the photoemission experiment. Good absolute energies can be obtained more easily by studying the convergence of the Rydberg series of higher quantum states (n=1,2,3,4) associated with the 1/z image potential, since these states rapidly converge to the vacuum. For calibrating the binding energies of n=1 image electrons in the presence of a monolayer, locating the n=2 peak determines the absolute energy within $\pm$ 20 meV. We have observed that no appreciable binding energy shift in the n=2 state occurs until three or more layers are deposited, probably due to the fact that the wave function for n=2 peaks much farther out in the vacuum than for n=1.

Molecular layers were grown by exposing the cold Ag(111) crystal to the sample gas. Layer by layer growth of the adsorbate can be identified and monitored in situ by TPPE.[2] Repeated dosing of the Ag(111) crystal at sub-Langmuir exposures results in the incremental growing in and dying out of well-separated peaks at binding energies consistent with increasing layer thickness. This sensitivity of TPPE to individual molecular layers is demonstrated in Figures 1 and 2. Physisorbed adlayers can also be grown in a well-controlled fashion by exposing the crystal to the sample gas under equilibrium pressure and temperature conditions, using the temperature to reversibly control the layer growth. Studies of this kind have in fact been correlated with low-energy electron diffraction (LEED) data, showing that TPPE is sensitive to surface phase transitions. [6]

Angle-resolved photoemission is accomplished by rotating the sample on a precision goniometer. The component of electron momentum in the plane of the surface, $k_\parallel$, is conserved in the photoemission process. Thus electrons with different values of $k_\parallel$ will be emitted at different angles. The relationship[7] between the kinetic energy E of the photoelectron, the in-plane wave vector $k_\parallel$, and the angle of photoemission θ is
$$k_\parallel = (2mE / h^2)^{1/2} \sin\theta.$$
By fixing θ and measuring E in the experiment, we arrive at the dispersion, $E(k_\parallel)$.

3. Results

Figure 1 shows the TPPE spectrum at $k_\parallel = 0$ (emission normal to the surface) for the image potential state that persists when the Ag(111) surface is covered with different adlayer thicknesses of neopentane. Figure 2 shows the same information for layers of n-pentane. These spectra illustrate the sensitivity of the image electron binding energy to an individual adlayer and demonstrate that well-defined alkane layers can be prepared

Electron Dispersion for Alkane/Ag(111) Monolayers

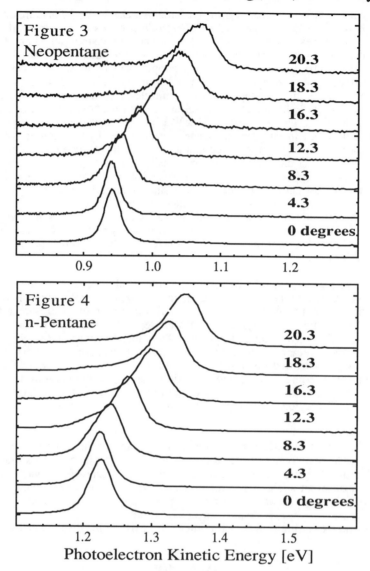

Figures 3 and 4: Angle-resolved TPPE measures the band structure of the image potential state which persists in the presence of a monolayer of n-pentane or neopentane on Ag(111). The image electron in either case has an effective mass ratio of one, indicating a perfect free electron. Curves heights have been normalized.

Electron Dispersion for Alkane/Ag(111) Bilayers

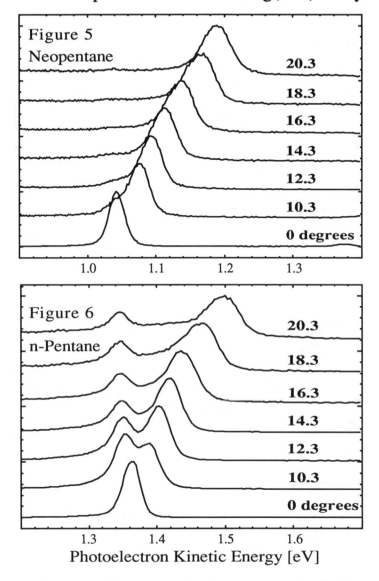

Figures 5 and 6 : A nondispersive feature (interpretted as a localized electron state) is prominent in Figure 6 for n-pentane bilayer, while the bilayer of neopentane is dominated by a parabolically dispersing band (extended electron state) of effective mass ratio 1.2 ± .1.

Electron Dispersion for Alkane/Ag(111) Trilayers

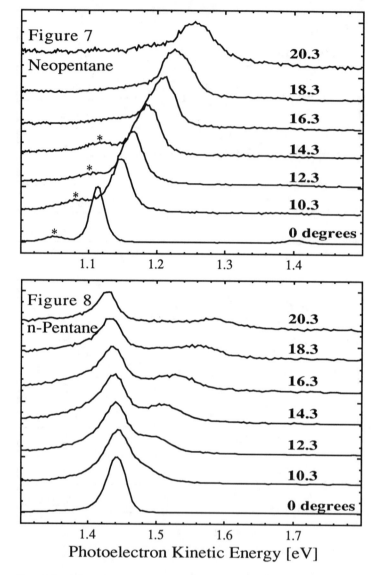

Figures 7 and 8 : The trilayer spectra continue the trend that the nondispersive feature (localized state) becomes more prominent in the n-pentane spectrum, while the delocalized state continues to dominate the neopentane spectrum. The features marked * in Figure 7 are due strictly to patches of bilayer interspersed with trilayer.

for study.

The dispersion data as a function of layer thickness are best illustrated by comparing and contrasting n-pentane/Ag(111) with neopentane/Ag(111). For a bare surface, the electron has parabolic dispersion with effective mass 1.3 times greater than that of a free electron, $m^*/m_e = 1.3 \pm 0.1$. For any physisorbed monolayer we have studied, the effective mass of the image electron is reduced to that of a free electron, $m^* \approx m_e$. This is evident in the almost identical monolayer spectra for neopentane and n-pentane in Figures 3 and 4. However, a bilayer of cyclohexane or any straight chain alkane shows a nondispersive feature (i.e. the energy of the photoelectron does not depend on angle) in addition to the parabolically dispersing peak, as shown in Figure 6 for the n-pentane bilayer. The TPPE spectrum of a trilayer of n-pentane is dominated by the nondispersive peak, showing only a remnant of the parabolic band, as seen in Figure 8. On the other hand, the bi- and trilayer spectra of neopentane show that the parabolic band remains the dominant feature in the spectrum. The parabolically dispersing bands in the bi- and trilayers show effective masses which are again greater than that of a free electron, unlike the monolayer results. These all show a ratio m^*/m_e in the range of 1.2 ± 0.1, with the notable exception of neopentane, which maintains free electron dispersion $m^* \approx m_e$ through the trilayer. n-Octane adlayers show weaker nondispersive features than n-pentane, n-hexane, or n-heptane. Its bilayer behavior seems to be intermediate to that of neopentane and the other straight chain alkanes.

We have measured dispersion curves of the n=2 image state in the presence of monolayers and bilayers of some of the straight chain alkanes (not shown). In none of these cases does the n=2 state show any nondispersive peak. Looking back at Figure 4, there is a hint of nondispersive feature even in the monolayer for n=1, especially in the 8.3, 12.3, and 16.3 degree scans. The n=2 dispersions of monolayers do not show this slight nondispersive peak.

4. Discussion

We attribute the nondispersive feature in these spectra to localized electron states, in contrast to parabolic bands which are indicative of electrons in extended states characterized by an effective mass close to the free electron value. (We speak here of localization and delocalization in the plane of the surface; the image potential obviously localizes the electron near the surface in the z-direction.) The wave functions are eigenstates of definite momentum, $p_\parallel = h k_\parallel$, with infinite spatial extent and energy $E(k_\parallel) = h^2 k_\parallel^2 / 2m^*$, where m^* is the electron effective mass. Since $k_\parallel$ is conserved in the photoemission process[7], the angle at which the ejected electron leaves the surface is fixed by the photon energy and $k_\parallel$. The extreme opposite case is that of a spatially localized electron. For a bound state, $k_\parallel$ is not a good quantum number. Instead, the localized state is a superposition of all $k_\parallel$ plane wave states, the coefficients of which are given by the spatial Fourier transform of the localized wave function. Photoelectrons emitted from a localized (bound) state can be ejected at various angles, with the relative intensity of emission at various angles dependent on the coefficients in the plane wave expansion of the localized state. However, $E \neq E(k_\parallel)$, so angle-resolved TPPE will yield a flat energy band as a function of angle. One can loosely consider a nondispersive feature in the parabolic band language by identifying it with a localized state having $m^*/m_e = \infty$.

Two experimental facts are relevant in determining the origin of the localized state. First, no amount of disorder that we could control had any significant effect on the ratio of the area of the dispersive feature to the that of the nondispersive feature for a bilayer. Annealing of bilayers to try to create better ordered bilayers did not affect the relative magnitude of the nondispersive peak. n-Pentane afforded the opportunity to study a rotationally disordered bilayer (consisting of randomly oriented patches) versus a more ordered bilayer, since each adlayer phase can be grown under specific equilibrium conditions.[8] These two bilayers gave approximately the same ratio of dispersive to nondispersive feature over several trials. Growing mixed layers of n-hexane and n-octane did not induce a nondispersive feature in the monolayer. Finally, TPPE of a Ag(111) surface roughened by sputtering did not reproduce a nondispersive feature.

Second, the only parameter that does seem to induce a nondispersive peak in the spectrum is layer thickness. This is illustrated by Figures 4, 6, and 8. The importance of wave function overlap with the layer is further demonstrated by the fact that n=2 dispersions do not show any nondispersive feature when the corresponding n=1 feature does. The hydrogenic model for an image potential electron on a bare metal predicts that the expectation value of the distance z from the surface is 3 Å for the n=1 wave function compared to 13 Å for the n=2 wave function[9]. *For a given molecular species, more or less electron overlap with the adlayer controlled the appearance of the nondispersive feature, regardless of our attempts to vary the order of that adlayer.*

A very interesting aspect of the data which is particularly obvious in the bilayer spectrum of Figure 6 is the fact that the localized and delocalized states coexist in the sample, being nearly degenerate at $k_{\parallel} = 0$. The gap between the bottom of the parabolic band and the localized state at $k_{\parallel} = 0$ is ≤ 10 meV. It is a key fact to be explained that the energy of the nondispersive feature follows the band edge of the dispersive feature. It appears that the energy of the localized state is pinned to the bottom of the extended state. For example, the n=1 binding energy shift when adding a third layer is ≈ 80 meV for n-pentane in Figure 2. Yet the localized state peak associated with the bilayer also moves ≈ 80 meV also, so that it is still ≤ 10 meV below the bottom of the band.

For the free electrons which exist in the case of monolayer coverages of these physisorbed hydrocarbons, $E(k_{\parallel}) = h^2 k_{\parallel}^2 / 2m^*$. The angles over which dispersion measurements were carried out in our experiments correspond to values of $k_{\parallel}$ ranging from 0 to 0.25 Å^{-1}. The kinetic energy associated with the momentum of the electron parallel to the surface thus falls in the range 0 to ≈ 0.25 eV. The Anderson criterion for strong localization of an electron by a fluctuating potential will be satisfied if the potential fluctuations exceed the average kinetic energy of the electrons.[10] Meeting this criterion would require that the potential fluctuations in an alkane layer exceed 250 meV in order to localize the full range of electrons detected in our experiment. Such potential fluctuations would most likely be caused in a dielectric material by the repulsive atomic cores of the molecular constituents and fluctuations in the electron-induced polarization potential of the medium. We do not know at present whether these could satisfy the strong localization criterion.

However, fluctuations in this potential might be of about the right size to produce shallow energy states yielding the localized states in the spectrum just beneath the delocalized states. This mechanism is consistent with the dominance of the localized state as layer thickness is increased, since formation of the bilayer and trilayer almost fills out the first two nearest-neighbor shells for the short range (r^{-4}) induced dipole term and the very short range repulsive term in the potential.

There exists a substantial body of literature from the field of excess electrons in nonpolar liquids with which to compare our results. Early theoretical work investigated the properties of electrons in rare gas fluids.[11] A summary of experimental and theoretical results for electrons in nonpolar liquids can be found in ref. 12, 13, and the articles cited therein. While many differences exist between studying the spectroscopy of the excess electron states of a frozen adlayer and measuring the electron mobility in a molecular liquid, the limited data so far do show that the molecule with the highest time-of-flight mobility in the liquid (neopentane) shows the least electron localization as an adlayer on Ag(111). The success of the liquid mobility data for postulating that neopentane would support primarily delocalized excess electron states indicates that further studies motivated by the liquid analogy should be pursued. In particular, the spherical methane has a very high excess electron mobility, while mobility drops precipitously for ethane. Further study will indicate whether the phenomena of these two experiments are being controlled by the same condensed phase physics.

5. Conclusions
We have studied the electronic structure of a metal-dielectric interface and found localized and delocalized states, which are near-degenerate at $k_\parallel = 0$. The appearance of the localized state as a nondispersive feature in angle-resolved photoemission is reported for the first time and encourages us to study other interfaces which may show effects of localization due to anion formation, strong disorder, local traps, or a Mott screening effect. Time-resolved data will be useful in addressing both the question of where the electron is spatially and as well as what its dynamics are at the interface. We are currently pursuing sub-picosecond dynamical measurements.

Acknowledgements
This work was supported by the U.S. Department of Energy, Office of Basic Energy Sciences, Chemical Sciences Division, under Contract No. DE-AC03-76SF0098.

References:
[1.] Steinmann, W. (1989) "Spectroscopy of Image-Potential States by Two-Photon Photoemission", Appl. Phys. A, **49**, 365-377.
[2.] Padowitz, D. F., Merry, W. R., Jordan, R. E., Harris, C. B. (1992) "Two-Photon Photoemission as a Probe of Electron Interactions with Atomically Thin Dielectric Films on Metal Surfaces", Phys. Rev. Lett., **69**, 3583-3586.
[3.] Merry, W. R., Padowitz, D. F., Jordan, R. E., and Harris, C. B. (1993) "Properties of Electrons at Metal-Insulator Interfaces I: The Effect of Xe Monolayers on the Image Potential States of Ag(111)", Surf. Sci., in press.
[4.] Fischer, R., Schuppler, S., Fischer, N., Fauster, Th., and Steinmann, W. (1993) "Image States and Local Work Function for Ag/Pd(111)", Phys. Rev. Lett., **70**, 654-657.
[5.] Wu, Z., Quiniou, B., Wang, J. and Osgood Jr., R. M. (1992) "Temperature and adsorbate dependence of the image-potential states on Cu(100)", Phys. Rev. B, **45**, 9406-9409.
[6.] Lingle Jr., R., Jordan, R. E., McNeill, J. D., Padowitz, D. F., Harris, C. B. "2D Phase Transitions Studied by Two-photon Photoemission", in preparation.
[7.] Smith, N. V. and Kevan, S. D. (1992) "Introduction to Angle-Resolved Photoemission", in S. D. Kevan (ed.), Angle-Resolved Photoemission: Theory and

Current Applications, Elsevier Science Publishers, Amsterdam, p. 5.

[8.] Firment, L.E. and Somorjai, G.A. (1978) "Low-energy electron diffraction study of the surface of thin crystals and monolayers of normal paraffins and cyclohexane on the Ag(111) crystal surface", J. Chem. Phys., 69, 3940-3952.

[9.] Merry, W. R. (1992) "Image Potential States at Metal-Dielectric Interfaces", Ph.D. dissertation, Department of Chemistry, University of California, Berkeley, pp. 12-14.

[10.] Anderson, P. W. (1958) "Absence of Diffusion in Certain Random Lattices", Phys. Rev., 109, 1492-1505.

[11.] Springett, B. E., Jortner, J. and Cohen, M. H. (1968) "Stability Criterion for the Localization of an Excess Electron in a Nonploar Fluid", J. Chem. Phys., 48, 2720-2731.

[12.] Allen, A. O. (1976) "Drift Mobilities and Conduction Band Energies of Excess Electrons in Dielectric Liquids", U.S. Dept. of Comm. / National Bureau of Standards, Report number NSRDS-NBS 58.

[13.] Ascarelli, G. (1985) "The Motion of Electrons Injected in Classical Nonpolar Insulating Liquids", Comments Solid State Phys., 11, 179-202.

SOLVATION AND CHARGE TRANSFER AT LIQUID INTERFACES

Ilan Benjamin
Department of Chemistry
University of California
Santa Cruz, CA 95064
U.S.A.

ABSTRACT: Molecular dynamics calculations are used to investigate the equilibrium solvation of small ions and solvent relaxation following a charge transfer at liquid interfaces. We consider the water-vacuum interface, a model liquid/liquid interface and the water/1,2 dichloroethane interface. The molecular dynamics results are compared with several theoretical models.

1. Introduction

One of the fundamental problems in condensed phase chemical dynamics, which is also of great practical importance, is to understand at the molecular level the effect of the medium on chemical reactions which occur at the interfacial region between different phases. In many systems, the interesting and relevant chemistry occurs at the interface between two immiscible liquids, at the liquid-solid interface and at the free liquid surface. Recent advances in picosecond and femtosecond time-resolved spectroscopic techniques are yielding intriguing new data on solvent effects on dynamical processes and chemical reactions in bulk solution. These techniques are also beginning to be applied to the study of chemical dynamics at interfaces. A molecular level understanding of the equilibrium and non-equilibrium behavior of small ions at liquid interfaces is of particular importance, because it forms the basis for understanding many of the fundamental chemical events which involve charge transfer at interfaces. As examples of recent experiments for which a molecular level understanding is particularly valuable, we mention the shift in the acid-base equilibria of nitrophenol at the air water interface,[1] the reaction of SO_2 with the water air/interface to form the HSO_3^- ion,[2] the dynamics of ion extraction through the liquid-liquid interface,[3] photochemical reactions at the liquid-liquid interface,[4] electron transfer at the liquid/liquid interface[5] and many other fast

J. Jortner et al. (eds.), Reaction Dynamics in Clusters and Condensed Phases, 179–194.

heterogeneous charge transfer reactions.[6]

The inhomogeneous boundary between coexistent phases has some unique features which can have a significant influence on reactive processes. Some important characteristics of the interface are the asymmetry in the intermolecular forces experienced by the molecules and the steep variations in density and dielectric properties. These directly affect the structure and orientation of molecules at the interface and the equilibrium solvation of ions.[7] This in turn can make the equilibrium constant and the activation energy for the reaction at the interface significantly different from their values in the bulk. The asymmetry in the forces, along with the change in the density and dielectric properties of the solvent in the interfacial region, can also alter dynamical properties of the system such as solvent friction and the rate of energy relaxation, all of which will affect the rate of chemical reactions.

In this work we focus on the utility of the continuum electrostatic model to account for the equilibrium free energy associated with ion adsorption and transfer at the interface as well as the non-equilibrium response of the solvent to a sudden change in the charge distribution. We do not focus on the details of a particular system, but rather consider the similarities and differences between several type of liquids interfaces. This will enable us to make some general conclusions which we believe are not very sensitive to the exact choice of potentials (which seldom are known exactly), as well as to help in identifying the unique characteristics of the interface which may affect the equilibrium and non-equilibrium properties of a chemically reactive system.

Our main theoretical tool for investigating the structure and dynamics of ions at interfaces is molecular dynamics simulation. Although significant advances have been made in the statistical mechanics of inhomogeneous fluids,[8,9] the progress has been mainly limited to the calculation of density and orientational profiles of the pure fluid and fluid mixtures. Although the application to a solute-solvent system has seen explosive growth in recent years for bulk homogeneous systems, it is still in its infancy as far as ionic solvation and chemical reactions at liquid interfaces are concerned.

The application of molecular simulation techniques to the interfacial fluid environment has mainly concentrated on the structure and thermodynamics of the *pure* liquid-liquid[10-14] and liquid-vapor[15-20] interfaces. In particular, there has been very little work done on the behavior of chemically reacting systems at liquid interfaces from a microscopic point of view. M. Mareschal studied the kinetics of adsorption of a diatomic molecule at the interface between two Lennard-Jones atomic solvents.[21] B. Smit used a similar system to investigate the change in the surface tension upon the introduction of a solute[22] and Wilson, Pohorille and Prat studied the interactions of small ions with the water surface.[23,24] More recently, computer simulation techniques which have been previously used to study solvation and chemical reactions in bulk liquids, have been extended to the study of relaxation and reactions at liquid interfaces.[25-31] The work presented below is the first attempt to systematically address the molecular aspects of equilibrium and non-equilibrium solvation in inhomogeneous fluid media.

2. Solvation and Relaxation at the Water Liquid-Vapor Interface

In order to understand solvent effects on reactions that involve charged species, an important first step is to understand the simpler problem of ionic solvation at the water liquid-vapor interface. With few exceptions,[23,24,26,27,32] studies of equilibrium solvation at interfaces have been limited to continuum models, and even fewer studies have been carried out on the dynamical aspect.[26,33] The dynamic of ion solvation at the interface is particularly relevant to fast charge transfer reactions at interfaces. In this section, we describe a comparison between the molecular dynamics and a continuum electrostatic model for the adsorption free energy of an ion and for the dynamics of solvent reorganization following a charge transfer at the water liquid-vapor interface. Our molecular model for the system includes 512 water molecules interacting via a flexible SPC model at $T = 300K$ and a single chloride ion interacting with the water molecules via Lennard-Jones and Coulomb potentials. More details about the potentials and the method of preparation of the liquid-vapor interface can be found elsewhere.[26] The density profile of the water in the system is shown in Fig. 1.

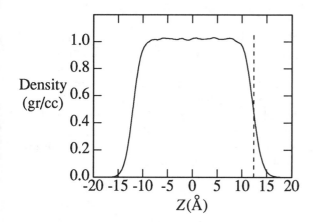

Fig. 1: Density profile of liquid water in equilibrium with vapor at 300 K. The dashed line represents the Gibbs dividing Surface.

In what follows, the concept of the Gibbs surface will be useful. It is defined as the planar surface $Z = Z_G$ such that

$$\int_{Z_B}^{Z_G} [\rho(Z) - \rho_B]dZ + \int_{Z_G}^{Z_V} [\rho(Z) - \rho_V]dZ = 0, \tag{1}$$

where $\rho_B = \rho(Z = Z_B)$ is the liquid bulk density and $\rho_V = \rho(Z = Z_V)$ is the vapor bulk density. For water, this is approximately at the point where the water density is 0.5 gr/cc.

2.1 FREE ENERGY PROFILE FOR IONIC ADSORPTION

The free energy profile is an equilibrium property which determines (together with the electrostatic double layer potential) the distribution of ions at the interface. Like any local thermodynamics property, care must be taken in properly defining it. Because of the cylindrical symmetry of the system, the work done to push an ion toward the surface is a function of Z only. We define:

$$A (Z_1) - A (Z_2) = - kT \ln P (Z_1)/P (Z_2) = - kT \ln \frac{\int e^{-\beta U (r_{ion} = Z_1)} d\mathbf{r}}{\int e^{-\beta U (r_{ion} = Z_2)} d\mathbf{r}}, \qquad (2)$$

where $\beta = 1/kT$, U is the total potential energy of the system and $\mathbf{r}$ represents all the solvent positions. Eq. 1 can be used to calculate $A(Z)$ by an umbrella sampling procedure: The simulation box is divided into a number of slabs perpendicular to the interface normal (Z direction). The thickness of each slab is 3Å, and their centers are located 2Å apart. Thus the distance along Z is divided into overlapping intervals. In each slab, the probability distribution is calculated by running a long trajectory while the ion is restricted to this slab (using a potential energy which is zero inside the slab and steeply rising outside its boundaries). The free energy $A(Z)$ is thus calculated by matching the overlapping regions of the individual $A_i(Z)$ calculated in each slab i. In order to insure a uniform sampling frequency of the Z_{ion} values, a biasing potential, which is a function of Z_{ion} only, is added to U. The effect of this on the left-hand side of Eq. 1 is simply to add a known function of Z. The free energy calculated with the biased distributions can be corrected by subtracting the biasing potential. The utility of this method has been demonstrated for the adsorption of surface active molecules, and good agreement with the experiment was obtained.[27, 32] Calculations similar to the one reported here have also been recently reported[24, 26] and used to develop a diffusion model for ion desorption from the liquid-vapor interface of water.[34]

In Fig. 2, the molecular dynamics free energy profile for a chloride ion is compared with a simple electrostatic model similar to models which are quite common in the literature.[7] In our particular implementation of this model, the chloride ion is represented by a charged sphere of a diameter which is equal to the distance at which the oxygen-ion radial distribution function has its first maximum. The liquid is described as a dielectric medium whose dielectric constant changes abruptly from ε(flexible SPC water) = 82.5 to $\varepsilon = 1$. The Poisson equation is solved numerically as the sphere is moved from the bulk across the interface. The result shown in Fig. 2 clearly demonstrates the inadequacy of this model.

One obvious reason for the failure of the continuum electrostatic model is that it does not take into account the change in liquid dielectric properties as one approaches the interface. Ellipsometric measurements on water can give some information about the dielectric profile.[35] However, we may also try to find a function $\varepsilon(Z)$ which will reproduce the molecular dynamics data for $A(Z)$. This can be done analytically if we assume that the inside of the sphere constitutes a dielectric whose dielectric constant is identical to the one of the liquid at this position. Otherwise, a numerical inversion must be done. This "success" in "fixing" the continuum model is, however, illusory, as will be discussed in the next section.

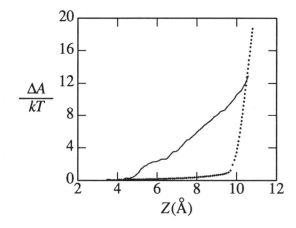

Fig. 2: Adsorption free energy of a chloride ion at the liquid-vapor interface of water. Solid line: molecular dynamics. Dotted line: a continuum electrostatic model.

2.2 SOLVATION DYNAMICS AT THE WATER LIQUID/VAPOR INTERFACE

The solvent response to a non-equilibrium charge distribution created at the interface is important for understanding fast, photoinduced chemical reactions at interfaces, as the process of producing stable products necessarily involves the relaxation of the solvent. The subject of solvent dynamics has received much attention recently for reactions in the bulk, where the relaxation can be followed experimentally through time-resolved fluorescence measurements.[36, 37] Here we use the measurement of solvent dynamics to furnish an additional check of the continuum electrostatic model of ion solvation at interfaces. More specifically, in the previous section we demonstrated that a continuum model with a sharp boundary between the liquid and the vapor underestimates the adsorption free energy at large distances from the interface and overestimates it near the surface, and that the inhomogeneous model with a distance-dependent dielectric constant is able to improve this equilibrium property. We now use this continuum model to calculate the dynamic of water reorganization following a sudden change in solute charge, and to compare it to the molecular dynamics calculations.

Consider a solvated chloride ion near the Gibbs surface of water. At $t = 0$, we set the charge on the ion to zero and follow the time development of the electrostatic potential $V(t)$ at the ion location. This somewhat artificial process has been used to investigate the rate of water reorganization following charge transfer in bulk water, and to test the applicability of continuum models to solvation dynamics in bulk water.[38] The dynamic is followed using the correlation function $S(t)$:

$$S(t) = \frac{V(t) - V(\infty)}{V(0) - V(\infty)}, \qquad (3)$$

which is directly related to dynamic spectral shifts[36] and can be easily calculated by molecular dynamics. One simple approach to computing the solvent dynamics is the Debye model,[39] which predicts that the relaxation is exponential and given by

$$S(t) = e^{-t/\tau_L}, \quad \tau_L = \varepsilon_\infty \tau_D / \varepsilon_0, \qquad (4)$$

where τ_D is the Debye relaxation time of the solvent which is related to the molecular dipoles reorientation dynamics, and ε_0 and ε_∞ are the static and infinite frequency dielectric constant of the liquid, respectively. For flexible SPC water in the bulk, these parameters are known: $\varepsilon_0 = 82.5$, $\varepsilon_\infty = 1$ and $\tau_D = 11$ ps.[40] At the water surface, we have estimated $\tau_D = 9$ ps from the water dipole reorientation time. Our inhomogeneous dielectric model gives for ε_0 near the Gibbs surface an average value of about 10. The molecular dynamics calculations were performed by averaging over 150 1ps trajectories, starting from independent initial conditions for the ion at the interface. As a comparison, the calculations have also been performed in the bulk.

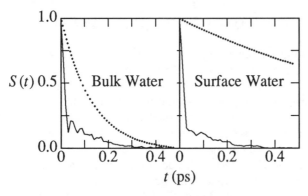

Fig. 3: Water relaxation following a charge neutralization reaction in the bulk and at the surface. Solid lines: molecular dynamics; dotted lines: continuum model.

Fig. 3 shows that the molecular dynamics data are dominated by the very fast initial relaxation due to the water librational modes. The appropriate comparison between the molecular dynamics and the continuum model should be the tail of the relaxation. In the bulk, the continuum model is in reasonable agreement with the tail of the molecular dynamics data. However, when the relaxation is monitored at the interface, the molecular dynamics give almost the same relaxation, but the continuum model predicts a much slower relaxation. Thus, the attempt to improve the agreement between the molecular dynamics and the continuum model for an equilibrium property (free energy of adsorption) results in marked disagreement for the *solvent dynamics*. In fact, the simpler, discontinuous dielectric model (sharp interface) would be much better as far as the dynamics are concerned because, expect for a small dielectric image effect, it would predict that the relaxation in the bulk is similar to the one at the interface. Then, however, the equilibrium adsorption free energy is miscalculated (as shown in Fig. 2).

The behavior depicted in Fig. 3 has a simple molecular explanation. As the ion is transferred from the bulk to the surface, it carries its solvation shell structurally (as judged by the pair correlation function) and dynamically (as judged by the reorientation dynamics of first shell water molecules) intact.[26] As a result, as the ion is transferred from the bulk to the surface, the electric potential at the location of the ion is decreased substantially, but the contribution from the first solvation shell is less affected (Fig. 4). Since it has been established that the dynamical response due to the first shell is dominating the total response,[38] the "robustness" of the ion's first shell is responsible for the small change in solvation dynamics upon transfer to the interface.

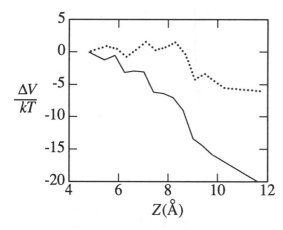

Fig. 4: The change in the electrostatic potential at the position of a chloride ion as a function of the ion position. Solid line: total change; dotted line: the change in the contribution of the first shell water molecules.

The ability of small ions to preserve the structure of the first solvation shell upon transfer to the interface has also been demonstrated by computing the ion-water radial distribution function,[26] and has been observed in calculations of ion adsorption at the water/metal interface.[33,41]

3. Solvation and Relaxation at the Liquid-Liquid Interface

Understanding the solvation structure, thermodynamics and dynamics of ions at the liquid-liquid interface is of fundamental importance to electrochemistry, analytical chemistry and biophysics, as reflected by much experimental and theoretical work. However, the interface between two immiscible liquids is a buried interface, and until recently, most of the experimental data about reactions in this system have been obtained via indirect means, such as surface tension measurements and electrochemical methods. The theoretical approaches have been almost exclusively limited to phenomenological continuum models. Both the experimental and theoretical treatments lack a molecular level understanding of this interesting system.

This situation is beginning to change with the application of new techniques for probing the structure of the interface and of adsorbed molecules at the interface,[3,42-44] and with the possibility of extending the fast reaction dynamics studies at the liquid-vapor[45,46] and liquid-solid[47] interfaces to the liquid-liquid interface. The new experimental data require a molecular level approach to the behavior of solute at the liquid-liquid interface. In this section, we essentially consider the analog of what we discussed above for the liquid-vapor interface: the free energy of ions at the liquid-liquid interface and solvent relaxation following a charge transfer at the interface. The comparison between the two environments will show that there are both similarities and important differences between the behavior of ions at the liquid-vapor and liquid-liquid interfaces.

3.1 FREE ENERGY PROFILE FOR IONS AT THE LIQUID-LIQUID INTERFACE

Consider an interface between two immiscible liquids which is planar, on average, with the Z-axis along the interface normal. The free energy profile for an ion as a function of Z is defined in Eq. 2 and can be calculated using an umbrella sampling procedure as described above. This free energy profile is important for calculating distributions of ions in the electric double layer and for the problem of ion transfer dynamics across the interface.

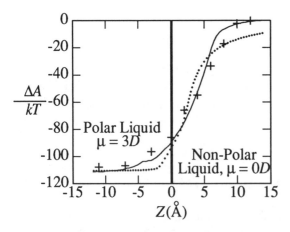

Fig. 5: Free energy profile for an ion at the liquid-liquid interface. Solid line: exact umbrella sampling (molecular dynamics); dotted line: continuum electrostatic model. The '+' are the result of an approximate Gaussian model. The solid vertical line is the approximate location of the interface.

We have previously calculated the free energy profile for an ion transfer across the interface between a model diatomic non-polar liquid and a model diatomic polar liquid[29] using the umbrella sampling technique and an approximate method which is based on assuming that the electric field fluctuations near the ion are Gaussian distributed.[48,49] The exact molecular dynamics results and the Gaussian approximation were found to be in close agreement. In Fig. 5, we compare these results with a solution of the Poisson equation for a charged sphere in a medium whose dielectric constant jumps from $\varepsilon = 1$ (in the non-polar liquid) to $\varepsilon = 35$ (calculated from the molecular dynamics of the polar liquid). The diameter of the sphere was selected to reproduces the molecular dynamics results for the net free energy of transfer. Fig. 5 demonstrates that as we move the ion from the polar liquid toward the non-polar liquid, the continuum model first underestimates the increase in the free energy, and then overestimates it with an opposite behavior when we consider the transfer from the non-polar to the polar liquid. The quantitative agreement is quite poor, with differences of many kT, even though the diameter of the ion is optimized to reproduce the total free energy change. The Gaussian fluctuation method is in much better agreement.

In the particular model for the liquid-liquid interface discussed above (and described in detail elsewhere[28]), the two liquids are identical except that one of the diatomic liquids has a dipole moment of $3D$. Thus, the ion interacts with each liquid via the same Lennard-Jones parameters. As a result, the overwhelming contribution to the free energy of transfer is electrostatic in nature. In reality, two immiscible liquids are usually made of different sized molecules. The water and 1,2-dichloroethane (DCE) are an example of two immiscible liquids with very different VdW interactions with small ions. The free energy of transfer will include, in addition to the electrostatic term, what is usually referred to as a cavity term. That such a contribution is necessary can be demonstrated by the fact that the miscibility of rare gases in water and in DCE are very different.[50]

In order to have an idea about the effect of the cavity term on the free energy profile, we use the scale particle theory[51] to calculate the "cavity profile", that is, the free energy required to create the cavity of a given diameter (D) as a function of the distance of the center of the cavity from the interface. For the purpose of calculating this term, we assume that that we have a mathematically sharp interface between two hard sphere liquids with different diameters d_1 and d_2, and we approximate the cavity work at the interface by a sum of two terms from the contribution from each liquid:

$$W_{cav} = W_{cav}(1) + W_{cav}(2) . \tag{5}$$

We use the expression for the cavity work in a bulk hard sphere liquid:

$$\beta W_{cav}(i) = -\ln(1-\eta_i) + 3\eta_i\chi_i(\chi_i + 1)/(1-\eta_i) \tag{6}$$

$$+ \tfrac{1}{2}9\eta_i^2\chi_i^2/(1-\eta_i)^2 + \eta_i(1+\eta_i+\eta_i^2-\eta_i^3)\chi_i^3/(1-\eta_i^3) ,$$

where $\chi_i = D_i/d_i$ and $\eta_i = (\pi/6)\rho_i d_i^3$. D_i is an effective diameter of the cavity in liquid i. The effective diameters $D_{1,2}$ depend on the actual diameter of the cavity D and the distance Z of the cavity center from the interface according to the fractional volume taken by the cavity in each liquid. A short calculation gives:

$$D_i = [\tfrac{1}{2}(D-Z)(D+2Z)^2]^{1/3} , \tag{7}$$

where, for liquid i, as Z goes from $Z = \tfrac{1}{2}D$ to $Z = -\tfrac{1}{2}D$, the cavity goes from being totally immersed in this liquid to being immersed in the other liquid.

Although the contribution of the cavity term is usually smaller than the electrostatic contribution to the free energy, it may be quite important. An example is presented in Fig. 6 for the transfer of a chloride ion across the water/DCE interface. The molecular model includes 343 water molecules and 108 DCE molecules at 300K in a box of cross-section $21.7 \times 21.7\text{Å}$. The interface is planar on average and perpendicular to the long axis of the box (the Z direction), such that liquid water occupies the region between -28Å and $+5\text{Å}$, and the DCE occupies the region between -5Å and 34Å. More details about the potentials and the structure of the neat interface can be found elsewhere.[52]

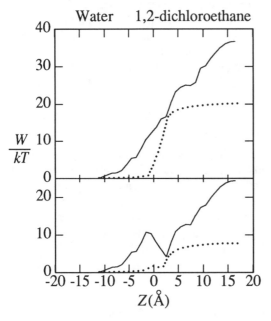

Fig. 6: Free energy profile for the transfer of a chloride ion across the water-dichloroethane interface. Top panel: electrostatic contribution; bottom panel: electrostatic + the work due to cavity formation. The electrostatic contribution is evaluated using the solution to the Poisson equation (dotted lines) and using the Gaussian model (solid lines).

This molecular model has been used to calculate the free energy profile for a transfer of a chloride ion across the interface using the Gaussian fluctuation model and the solution to Poisson equations with the dielectric constants calculated from independent molecular dynamics calculations of the two liquids. The results for the electrostatic component are shown using these two methods in the top panel of Fig. 6. The size of the hard sphere used in the Poisson equation was determined from the peak of the ion-oxygen radial distribution function. The bottom panel of Fig. 6 includes the cavity term calculated as explained above. The experimentally "accepted" value for the chloride ion net free energy of transfer is about 12.4 kcal/mol,[50] which is between the values predicted by these two methods, and somewhat closer to the value calculated using the molecular Gaussian method. More importantly, the molecular model predicts a barrier on the way (downhill) from the DCE phase to the water phase, which has been confirmed recently by molecular dynamics trajectory calculations for ion transfer across this interface.[53]

3.2 POLAR SOLVENT RELAXATION AT THE LIQUID-LIQUID INTERFACE

In the previous section, we have demonstrated that a molecular model for the interface between two immiscible liquids can be efficiently used for the study of the equilibrium solvation of ions at the interface, and that an electrostatic continuum model is in poor agreement with the exact results even if the continuum model parameters are tuned to produce the best agreement with the molecular dynamics. In this section, we consider the problem of solvent reorganization dynamics following a sudden jump in solute charge distribution. This is a preliminary report of results which will be published later.[54]

The molecular model for the liquid-liquid interface is the same as the one mentioned in the beginning of the last section and discussed in detail previously.[28] Briefly, it consists of 256 polar diatomic molecules ($\mu = 3$ Debye) and 256 non-polar diatomic molecules interacting via a combination of Lennard-Jones and coulomb potentials at $T^* = 1.5$. Two fixed atomic solutes, A and D, are located at positions R_+ and R_- at the interface, and they may participate in charge separation ($DA \rightarrow D^+ + A^-$) or charge recombination ($D^+ + A^- \rightarrow DA$) processes. We define a solvent coordinate[28,55-57] as the electrostatic potential energy of interaction between a given solvent configuration and two "test" charges of +1 and −1 located at the solute positions R_+ and R_-, respectively.

$$X(\mathbf{r}) = \sum_i \frac{q_i}{|\mathbf{r}_i - \mathbf{R}_+|} - \frac{q_i}{|\mathbf{r}_i - \mathbf{R}_-|} , \qquad (8)$$

where q_i is the charge of a liquid atom located at the position r_i. As the solvent dipoles fluctuate around the solute with a given fixed charge distribution, X can be approximately described as a Gaussian random variable, $P(x) = C \exp[-k(x - <x>)^2]$, and so the free energy associated with a fluctuation of size $x - <x>$ in X is approximately a parabola centered at $X = <x>$.

Initially, the solvent is equilibrated to one charge distribution. At $t = 0$, we switch to a new charge distribution and follow the dynamics by computing the function $S(t)$ defined in Eq. 3, except that we use $X(t)$ in that definition instead of $V(t)$. We examine both the charge recombination and charge separation processes at the interface and compare them to the same processes in the bulk.[58]

Fig. 7 summarizes the charge transfer process considered here.

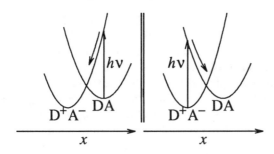

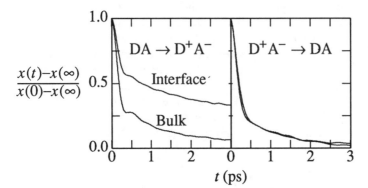

$$\frac{x(t)-x(\infty)}{x(0)-x(\infty)}$$

Fig. 7: Solvent relaxation following photo-chemically induced charge separation (left panel) and charge recombination (right panel) when the two charges are located in the bulk and at the liquid-liquid interface. Note the almost identical solvent response in the right panel.

The main difference between the bulk and the surface is that in the latter case, and only for the charge separation experiment, a smaller portion of the final solvation energy can be reached by the inertial motion of the solvent dipoles, and a larger portion must be supplied through more significant structural rearrangements. A simple rationale for this[54] may be the fact that the solvation structure around the ion-pair is less compatible with the neat liquid-liquid surface, thus requiring more structural adjustments that are less pronounced when a neutral pair is created. It is important to note that this effect is completely missed by the continuum electrostatic model. For example, the Debye model would predict that since the dielectric response of the polar liquid is almost the same at the interface and in the bulk (as suggested by molecular dynamics calculations on the neat interface[28]), and since the ion-pair is solvated by the polar solvent at the interface,[28] the relaxation rate should be the same at the surface and in the bulk (except for a small contribution due to image effects[59,60]).

4. Conclusions

The liquid-vapor and liquid-liquid interfaces are unique inhomogeneous environments characterized by the asymmetry of the forces experienced by the solute, the change in density and dielectric properties over distances that are comparable with molecular size, and the unique structure and dynamics of the solvent at the interface. All of these have important dynamical and thermodynamical consequences. In this paper we have demonstrated that continuum electrostatic models whose parameters have been adjusted to give a reasonable description of bulk solvation are inaccurate when applied at the interface when both the equilibrium and non-equilibrium behaviors of the system are considered. In general, this inadequacy can be traced to the robustness of the solvation structure as the solute is transferred to the interface. Thus, both the unique structure/dynamics of the interface and the surface deformation which must take place to accommodate the solute are important features that need to be quantitatively taken into account.

Acknowledgments

This work was supported by the National Science Foundation (CHE-9015106 and CHE-9221580) and by the Petroleum Research Fund, administered by the American Chemical Society (PRF-22862-G6).

References

[1] K. Bhattacharyya, E. V. Sitzmann, and K. B. Eisenthal, "Study of chemical reactions by surface second harmonic generation: p-Nitrophenol at the air-water interface," J. Chem. Phys. **87**, 1442(1987).

[2] J. T. Jayne, P. Davidovits, D. R. Worsnop, M. S. Zahniser, and C. E. Kolb, "Uptake of $SO_2(g)$ by aqueous surfaces as a function of pH: The effect of chemical reaction at the interface," J. Phys. Chem. **94**, 6041(1990).

[3] R. P. Sperline and H. Freiser, "Adsorption at the liquid-liquid interface analyzed by in situ infrared attenuated total reflection spectroscopy," Langmuir **6**, 344(1990).

[4] L. I. Boguslavsky and A. G. Volkov, "Redox and photochemical reactions at the interface between immiscible liquids," in *The Interface Structure and Electrochemical Processes at the Boundary Between Two Immiscible Liquids*, edited by V. E. Kazarinov (Springer, Berlin, 1987) p. 143.

[5] G. Geblewicz and D. J. Schiffrin, "Electron transfer between immiscible solutions. The hexacyanoferrate-lutetium biphthalocyanine system," J. Electroanal. Chem. **244**, 27(1988).

[6] M. Gratzel, *Heterogeneous Photochemical Electron Transfer* (CRC Press, Boca Raton, FL, 1989).

[7] B. E. Conway, "Electrical aspects of liquid/vapor, liquid/liquid, and liquid/metal interfaces," in *The Liquid State and its Electrical Properties*, edited by E. E. Kunhardt, L. G. Christophorou and L. H. Luessen (Nato ASI series B, Plenum, New York, 1988) Vol. 193.

[8] *Fluid Interfacial Phenomena*, edited by C. A. Croxton (Wiley, New York, 1986).

[9] *Fundamentals of Inhomogeneous Fluids*, edited by D. Henderson (Marcel Dekker, New York, 1992).

[10] M. Hayoun, M. Meyer, and M. Mareschal, "Molecular dynamics simulation of a liquid-liquid interface," in *Chemical Reactivity in Liquids*, edited by G. Ciccotti and P. Turq (Plenum, New York, 1987).

[11] M. Meyer, M. Mareschal, and M. Hayoun, "Computer modeling of a liquid-liquid interface," J. Chem. Phys. **89**, 1067(1988).

[12] P. Linse, "Monte Carlo simulation of liquid-liquid benzene-water interface," J. Chem. Phys. **86**, 4177(1987).

[13] J. Gao and W. L. Jorgensen, "Theoretical examination of hexanol-water interfaces," J. Phys. Chem. **92**, 5813(1988).

[14] I. L. Carpenter and W. J. Hehre, "A Molecular dynamics study of the hexane/water interface," J. Phys. Chem. **94**, 531(1990).

[15] J. S. Rowlinson and B. Widom, *Molecular Theory of Capillarity* (Clarendon, Oxford, 1982). Ch. 6.

[16] R. M. Townsend, J. Gryko, and S. A. Rice, "Structure of the liquid-vapor interface of water," J. Chem. Phys. **82**, 4391(1985).

[17] M. P. Allen and D. J. Tildesley, *Computer Simulation of Liquids* (Clarendon, Oxford, 1987).

[18] M. A. Wilson, A. Pohorille, and L. R. Pratt, "Surface potential of the water liquid-vapor interface," J. Chem. Phys. **88**, 3281 (1988).

[19] M. Matsumoto and Y. Kataoka, "Study on liquid-vapor interface of water. I. Simulational results of thermodynamic properties and orientational structure," J. Chem. Phys. **88**, 3233(1988).

[20] M. Matsumoto and Y. Kataoka, "Molecular orientation near liquid-vapor interface of methanol: Simulational study," J. Chem. Phys. **90**, 2398(1989).

[21] M. Mareschal, "Molecular dynamics simulations as a tool for the study of chemical physics and chemical kinetics: The kinetics of adsorption at a liquid- liquid interface," in *Chemical Reactivity in Liquids*, edited by M. Moreau and P. Turq (Plenum, New York, 1987) p. 265.

[22] B. Smit, "Molecular dynamics simulation of amphiphilic molecules at a liquid-liquid interface," Phys. Rev. A. **37**, 3431(1988).

[23] M. A. Wilson, A. Pohorille, and L. R. Pratt, "Interaction of a sodium ion with the water liquid-vapor interface," Chem. Phys. **129**, 209(1989).

[24] M. A. Wilson and A. Pohorille, "Interaction of monovalent ions with the water liquid-vapor: A molecular dynamics study," J. Chem. Phys. **95**, 6005(1991).

[25] I. Benjamin, "Molecular dynamics study of a model isomerization reaction at the liquid-vapor interface of a Lennard-Jones fluid," J. Chem. Phys. **94**, 662(1991).

[26] I. Benjamin, "Theoretical study of ion solvation at the water liquid-vapor interface," J. Chem. Phys. **95**, 3698(1991).

[27] A. Pohorille and I. Benjamin, "Molecular dynamics of phenol at the liquid-vapor interface of water," J. Chem. Phys. **94**, 5599(1991).

[28] I. Benjamin, "Molecular dynamics study of the free energy functions for electron

transfer reactions at the liquid-liquid Interface," J. Phys. Chem. **95**, 6675(1991).

[29] I. Benjamin, "Dynamics of ion transfer across a liquid-liquid interface: A comparison between molecular dynamics and a diffusion model.," J. Chem. Phys. **96**, 577(1992).

[30] D. A. Rose and I. Benjamin, "Femtosecond solvation dynamics following electron attachment at the Water-CH_3Cl interface. A molecular dynamics study," J. Phys. Chem. **96**, 9561(1992).

[31] I. Benjamin and A. Pohorille, "Isomerization reaction dynamics and equilibrium at the liquid-vapor interface of water: A molecular dynamics study.," J. Chem. Phys. **98**, 236(1993).

[32] A. Pohorille and I. Benjamin, "Structure and energetics of model amphiphilic molecules at the water liquid-vapor interface. A molecular dynamics study," J. Phys. Chem. **97**, 2664(1993).

[33] D. A. Rose and I. Benjamin, "Solvation of Na^+ and Cl^- at the water-platinum(100) interface," J. Chem. Phys. **95**, 6856(1991).

[34] K. J. Schweighofer and I. Benjamin, "Dynamics of ion desorption from the liquid-vapor interface of water," Chem. Phys. Lett. **202**, 379(1993).

[35] D. Beaglehole, "Experimental studies of liquid interfaces," in *Fluid Interfacial Phenomena*, edited by C. A. Croxton (Wiley, New York, 1986) p. 523.

[36] J. D. Simon, "Time-resolved studies of solvation in polar media," Acc. Chem. Res. **21**, 128(1988).

[37] P. F. Barbara and W. Jarzeba, "Ultrafast photochemical intramolecular charge and excited state solvation," Adv. Photochem. **15**, 1(1990).

[38] M. Maroncelli, J. MacInnis, and G. R. Fleming, "Polar solvent dynamics and electron-transfer reactions," Science **243**, 1674(1989).

[39] H. Frohlich, *Theory of dielectrics* (Clarendon, Oxford, 1958) p. 70.

[40] J. Anderson, J. J. Ullo, and S. Yip, "Molecular dynamics simulation of dielectric properties of water," J. Chem. Phys. **87**, 1726(1987).

[41] D. A. Rose and I. Benjamin, "Adsorption of Na^+ and Cl^- at the charged water-platinum interface," J. Chem. Phys. **98**, 2283(1993).

[42] S. G. Grubb, M. W. Kim, Th. Raising , and Y. R. Shen, "Orientation of molecular monolayers at the liquid-liquid interface as studied by optical second harmonic generation," Langmuir **4**, 452(1988).

[43] L. T. Lee, D. Langevin, and B. Farnoux, "Neutron reflectivity of an oil-water interface," Phys. Rev. Lett. **67**, 2678(1991).

[44] D. A. Higgins and R. M. Corn, "2nd harmonic generation studies of adsorption at a liquid-liquid electrochemical interface ," J. Phys. Chem. **97**, 489(1993).

[45] E. V. Sitzmann and K. B. Eisenthal, "Picosecond dynamics of a chemical reaction at the air-water interface studied by surface second harmonic generation," J. Phys. Chem. **92**, 4579(1988).

[46] E. V. Sitzmann and K. B. Eisenthal, "Dynamics of intermolecular electronic energy transfer at an air/liquid interface," J. Chem. Phys. **90**, 2831(1989).

[47] S. R. Meech and K. Yoshihara, "Picosecond dynamics at the solid liquid interface- a total internal reflection time-resolved surface 2nd-harmonic generation study," Chem.

Phys. Lett. **174**, 423(1990).

[48] T. Fonseca, B. M. Ladanyi, and J. T. Hynes, "Solvation free energies and solvent force constants," J. Phys. Chem. **96**, 4085(1992).

[49] R. M. Levy, M. Belhadj, and D. B. Kitchen, "Gaussian fluctuation formula for electrostatic free-energy changes in solution," J. Chem. Phys. **95**, 3627(1991).

[50] Y. Marcus, *Ion Solvation* (Wiley, New York, 1985). Ch. 6.

[51] H. Reiss, H. L. Frisch, and J. L. Lebowitz, "Statistical mechanics of rigid spheres," J. Chem. Phys. **31**, 369(1959).

[52] I. Benjamin, "Theoretical study of the water/1,2-dichloroethane interface: Structure, dynamics and conformational equilibria at the liquid-liquid interface," J. Chem. Phys. **97**, 1432(1992).

[53] I. Benjamin, "Mechanism and dynamics of ion transfer across a liquid-liquid interface ," submitted.

[54] I. Benjamin, "Solvent dynamics following charge transfer at the liquid-liquid interface," submitted.

[55] A. Warshel, "Dynamics of reactions in polar solvents. Semiclassical trajectory studies of electron-transfer and proton-transfer reactions," J. Phys. Chem. **86**, 2218(1982).

[56] R. A. Kuharski, J. S. Bader, D. Chandler, M. Sprik, M. L. Klein, and R. W. Impey, "Molecular model for aqueous ferrous-ferric electron transfer," J. Chem. Phys. **89**, 3248(1988).

[57] E. A. Carter and J. T. Hynes, "Solute-dependent solvent force constants for ion pairs and neutral pairs in polar solvent," J. Phys. Chem. **93**, 2184(1989).

[58] E. A. Carter and J. T. Hynes, "Solvation dynamics for an ion pair in a polar solvent: Time dependent fluorescence and photochemical charge transfer," J. Chem. Phys. **94**, 5961(1991).

[59] G. van der Zwan and R. M. Mazo, "Dielectric friction and ionic hydration near boundaries: Image charge effects," J. Chem. Phys. **82**, 3344(1985).

[60] M. Urbakh and J. Klafter, "Dipole relaxation near boundaries," J. Phys. Chem. **96**, 3480(1992).

SECOND HARMONIC AND SUM FREQUENCY STUDIES OF CHEMICAL EQUILIBRIA AND PHASE TRANSITIONS AT LIQUID INTERFACES

Kenneth B. Eisenthal and Jonathan H. Gutow

Chemistry Department
Columbia University
New York, NY 10027
USA

1. Introduction

An interface is the region between any two bulk media. All chemical systems contain interfaces and in many cases important chemistry, e.g. reactions at biomembranes, heterogeneous catalysis, etching of semiconductors and electrochemistry, occurs at these interfaces.[1] The primary characteristic differentiating an interface from the bulk is the asymmetry of the interface; the environment looks different as one moves to one side or the other of an interface. Asymmetric forces affect the chemical composition, the organization and orientation of species, equilibrium constants, pH and the motion of species at the interface. Thus interfaces exhibit unique properties that have spawned much interest and many technological applications.

Despite their importance and widespread interest, it remains difficult to probe the chemistry and physics of interfaces, and in particular liquid interfaces (vapor/liquid, liquid/liquid, liquid/solid), with the powerful techniques of spectroscopy. The reason is that the overwhelmingly large number of solute molecules in the bulk can dominate the signal originating from the interface. One way to circumvent this limiting characteristic of traditional methods is to use nonlinear spectroscopic methods that are now feasible with modern laser sources. In recent years it has been shown that the techniques of second harmonic generation (SHG) and sum frequency generation (SFG) are intrinsically sensitive to any interface that separates centrosymmetric media.[2-5] The key factor is that SHG and SFG are electric dipole forbidden in centrosymmetric media (liquids, gases, centrosymmetric solids) but not at the interface where inversion symmetry is broken. Optical nonlinearities occur in media exposed

J. Jortner et al. (eds.), Reaction Dynamics in Clusters and Condensed Phases, 195–206.

to very high light fluxes, such as those associated with lasers. The physical interpretation of SHG and SFG is that laser beams at frequencies ω_1 and ω_2 interact with a nonlinear medium (the interface) and generate a nonlinear polarization. This polarization acts as a collection of oscillating dipoles generating radiation at $\omega_1 + \omega_2$; if $\omega_1 = \omega_2$ then we have SHG.[6,7]

The sum frequency polarization P_i along axis i is proportional to the second-order nonlinear susceptibility of the surface $\chi^{s(2)}$,

$$P_i(\omega_1 + \omega_2) = \sum_{j,k=X,Y,Z} \chi^{(2)}_{ijk} E_j(\omega_1) E_k(\omega_2), \quad i=X,Y,Z. \tag{1}$$

$\chi^{(2)}$ is a frequency dependent tensor of rank three, E is the incoming electric field strength along axes j and k at frequency ω_1 or ω_2. If any of the frequencies (ω_1, ω_2 or $\omega_1 + \omega_2$) is in resonance with a molecular transition, $\chi^{(2)}$ increases and thus enhances the signal. In the experiments specific elements of the surface nonlinear susceptibility can be measured, which are directly related to the population of the molecules at the surface, N_s, and the nonlinear polarizability of the molecule, $\alpha^{(2)}$ averaging over the orientational distribution at the surface.

$$\chi^{(2)}_{ijk} = N_s < \alpha^{(2)}_{pqr} >. \tag{2}$$

Therefore, SFG and SHG can give information about molecular population and molecular orientation at the surface.[8-10]

In this paper we will discuss some of the results obtained with these nonlinear spectroscopic methods. Using SHG the free energies of adsorption, which determine the population of neutral and charged molecules at the air/water interface, were measured. In related acid-base equilibria studies we found that the neutral form of the acid-base pair is more stable in the interface. This result can be understood in terms of the increase in the free energy of a charged molecule in the interface relative to the bulk. Solvation favors the charged species being in the bulk and the image repulsion of the charged species in the interface raises its energy in the interface and thus favors the charged species being in the bulk. We found that we could drive charged species such as phenolate and anilinium to the air/water interface by attaching hydrophobic groups to the ions. A hydrocarbon chain with at least 5 methylenes was necessary to achieve a detectable population. We also discovered that charged species at the air/water and silica/water interface polarize the bulk water molecules and that this polarization effect can be used to obtain the electrostatic potential at these interfaces. Using the technique of infrared + visible SFG we have obtained vibrational spectra and the orientation of acetonitrile at the air/acetonitrile-water solution interface and seen evidence for an unexpected phase transition in the interface while the bulk solution remains homogeneous.

2. Experimental Methods

The SHG and SFG experiments are very similar. Ultrafast (femto- or picosecond) pulses of laser light are focused on the interface and the second harmonic or sum light is collected in reflection as depicted in figure 1. The ultrafast pulses are necessary to generate adequately intense electric fields in the interface to observe the nonlinear effects without depositing a large amount of energy in the interface. Although the examples discussed in this paper do not depend on time resolution the technique is inherently time resolved. In the case of SHG there is only one input laser beam, whereas in SFG the infrared and visible counterpropagate to the same point on the surface. The details of specific experiments are contained in the papers referenced as the work is discussed.

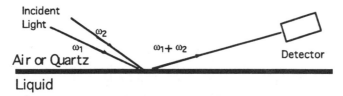

Figure 1: SHG and SFG experiment.

3. Chemical Equilibria at Interfaces

3.1 CHARGE INDUCED DESORPTION

For our first attempt at studying the effects of the interfacial environment upon acid-base equilibria we used SHG to interrogate para-nitrophenol (PNP) molecules at the air/H_2O interface.[11] At bulk pH's below the pK_a of 7.2 for this molecule the PNP is in its neutral form. The second harmonic signal is about a factor of 100 greater than that for pure water. However, when the pH is increased so that the bulk PNP is charged the second harmonic signal drops to that of pure water. Apparently, the anion (nitrophenolate) is not present at the air/water interface, but remains in the bulk water because the free energy of the bulk solvated anion is lower than in the interface. Although the absence of the anion, within our experimental sensitivity, precluded the study of the interfacial acid-base equilibrium, an interesting use of this can be made to study the balance between hydrophobicity and the energy of solvation of a charged species. If we substitute a long enough alkyl chain for the nitro group the charged species will stay at the interface. We studied both acidic para substituted alkyl phenols and basic para substituted alkyl anilines.[12, 13]

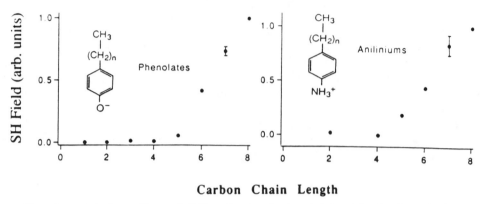

Figure 2: s-polarized second harmonic field as a function of carbon chain length for phenolate and anilinium ions.

By adjusting the bulk pH so that the bulk contained only the charged phenolate or anilinium forms we could monitor the effect of chain length on the surface activity of the charged species. It was found using SHG that not until the alkyl chain consisted of at least five carbons could the alkyl phenolate or anilinium at the interface be detected. The dependence of the second harmonic field on carbon chain length is shown in figure 2 for both alkyl phenolate and alkyl anilinium ions. The increment in the second harmonic field as the carbon chain length increases reflects the increase in the ion interface population. The interface population is determined, for a given bulk concentration, by the free energy of adsorption. If we view the adsorption process as the "reaction" of a bulk molecule with an empty surface "site" to give a filled "site", and desorption as the reverse process then we obtain the Langmuir adsorption equation,[1]

$$\frac{N_s}{N_{max}} = \frac{C}{C+a} \tag{3}$$

where N_s is the number/cm^2 of adsorbed solutes, N_{max} is the maximum number in a saturated monolayer, C is the bulk solute concentration and $a = 55.5 \exp(\Delta G^0_{ADS}/RT)$, where 55.5 is the molarity of bulk water and ΔG^0_{ADS} is the adsorption free energy. For the adsorption of a charged molecule the ΔG_{ADS}° consists of two parts. One is an electrostatic part ΔG_{EL}°, which is the reversible electrical work to bring a charge from the bulk where the electrostatic potential is zero to the interface where the electrostatic potential is $\Phi(0)$. The remaining part is not due to the electrostatic energy and is referred to as the chemical part ΔG_{CHEM}°,

$$\Delta G_{ADS}^\circ = \Delta G_{CHEM}^\circ + \Delta G_{EL}^\circ = \Delta G_{CHEM}^\circ + N_A z e \Phi(0) \tag{4}$$

where N_A is Avogadro's number, z is the valency of the adsorbed ion, and e is the charge on an electron.

The Gouy-Chapman model, which has been shown to be applicable under the conditions of our experiments[14], can be used to estimate $\Phi(0)$.[1] For all of the alkyl phenolates and alkyl aniliniums studied, (carbon chain lengths from 6 through 8), the adsorption isotherms yielded excellent fits to a Langmuir adsorption equation. From the fit the ΔG_{CHEM}° and the interface potential is obtained. An example fit is shown in figure 3. By taking the difference in ΔG_{CHEM}° between two consecutive members of the series of phenolates and aniliniums we estimate that the free energy of adsorption for a CH_2 group is -780 ± 20 cal/mol in good agreement with the results from surface tension measurements.[15-17]

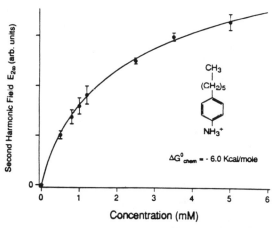

Figure 3: p-hexylanilinium adsorption isotherm. The line represents the best fit to the Langmuir equation including the electrostatic term.

3.2 ACID-BASE EQUILIBRIA

By increasing the chain length so that the charged species remain in the interface, we have examined the acid-base equilibria of insoluble amphiphiles, for example,

at the air/water interface. In addition to our fundamental interest in the effect of the forces at interfaces these systems are important as models for biological membranes where pK_a shifts may have profound effects on biological activity.[18,19] Using SHG we were able to measure the relative population of the acid and base form of the insoluble amphiphiles, para-hexadecylaniline[20] and para-octadecylphenol[21], at the air/water interface versus bulk pH (pH_B). This is almost enough information to determine the surface pK_a (pK_a^s). The missing piece of information is the surface hydronium ion concentration. This can be estimated using the Gouy-Chapman model[1,20] for the relation between the interface charge densities and the interface potential, $\Phi(0)$, of the charged interface. With the interface potential $\Phi(0)$ the surface population of hydronium ions can be calculated using the Boltzmann equation,

$$[H_3O^+]_s = [H_3O^+]_B \, \exp(-e\Phi/RT) \tag{5}$$

where the subscripts s and B refer to the surface and bulk regions and e is the charge on an electron. Given this surface pH (pH_s) we can easily calculate pK_a^s at the interface,

$$pK_a^s = pH_s + \log[HA]/[A^-]. \tag{6}$$

The SHG measurements yield the population ratio $[HA]/[A^-]$ together with a determination of the interface charge density (A^-). Using this approach the interface pK_a^s of hexadecylanilinium, $CH_3(CH_2)_{15}C_6H_4NH_3^+$, was found to be 3.6, which is a factor of 50 more acidic than in the bulk where the pK_a is 5.3. In contrast to the anilinium case it was found for the long chain phenol, $CH_3(CH_2)_{17}C_6H_4OH$, that the acidity decreased by a factor of 100. The pK_a at the interface is 12 versus 10 in the bulk. The observed decrease in acidity of the phenol derivative is attributed to the lower free energy of the neutral acidic form (HA) of the molecule versus its charged basic form (A^-) at the interface, whereas the acidic form of aniline (HA^+) is charged and has a higher free energy than its neutral basic form at the interface. Charges in the interface are destabilized by image charge repulsion, and the better solvation of charged species in the bulk improves the stability of charged molecules in the bulk versus those in the interface even more.

4. Polarization of Water by a Charged Interface

4.1 SILICA/WATER INTERFACE

In seeking to extend our studies of acid-base reactions and electrostatic potentials to other liquid interfaces we decided to investigate the silica water interface because of its widespread interest and importance.[22]

The acidic moieties are the silanol groups (-SiOH) that terminate the silica surface. Just as for the alkylphenols and alkylanilines at the air/water interface, we measured the second harmonic signal as a function of bulk pH and related this to the ratio of the charged to the uncharged populations. As can be seen from figure 4 there are two plateaus in the second harmonic electric field versus bulk pH, suggesting the presence of two silanol sites having different pK_a values. The very large increase (10^2) in the second harmonic signal as the pH increases indicates that the nonlinear polarizability of the $-SiO^-$ form is much greater than that of the neutral acidic form $-SiOH$; a result that was surprising. To check the validity of these various inferences we measured the dependence of the second harmonic signal on the bulk electrolyte concentration. At a given bulk pH we would expect that the number of SiO^- moieties would increase as electrolyte concentration increases due to the increased Debye screening. Since we inferred that $\alpha^{(2)}$ for –SiO$^-$ is greater than –SiOH from the higher second harmonic signal at high pH we would predict that the second harmonic signal should increase as electrolyte concentration increases. However, we found that the signal decreased as the electrolyte

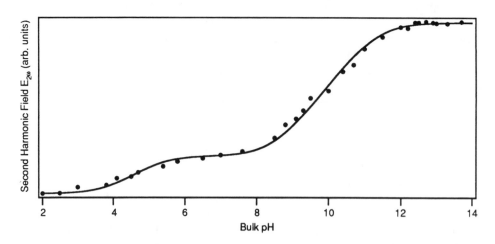

Figure 4: Second Harmonic electric field (in arbitrary units) from the SiO$_2$/H$_2$O interface as a function of pH in the bulk solution. The dots are experimental points, while the solid line is the theoretical fit using a simple electrostatic model.[22] The percentage of site 1 is 20% with a pK_a(1)=4.5; the percentage of site 2 is 80% with a pK_a(2)=8.5.

concentration increased. The temperature dependence of the second harmonic signal was also incompatible with our initial interpretation. At high pH (>12.5), where all (>99.9%) of the silanols are ionized we would predict that the signal should stay the same or decrease as the temperature was raised depending on the sign of the enthalpy change for the acid-base reaction. We find that the second harmonic signal increases with temperature at high pH, contrary to expectations. From the electrolyte and temperature results we concluded that the second harmonic signal could not be due simply to the populations of the -SiO$^-$ and SiOH forms. It turns out that the signal is generated not only by a second order process that depends on the populations and nonlinear polarizabilities $\alpha^{(2)}$ of all of the interfacial species, $\alpha^{(2)}_{H_2O}$, $\alpha^{(2)}_{SiO^-}$, $\alpha^{(2)}_{SiOH}$, but also by a third order process due to the very large electric field, E_0, arising from the charged silica interface (SiO$^-$),

$$E_{2\omega} \sim P_{2\omega}^{(3)} = \chi(3) \, E_0 \, E_\omega \, E_\omega \qquad (7)$$

The field E_0 extends into the bulk solution and polarizes the water molecules, both by a third order nonlinear polarizability $\alpha^{(3)}$ and by the alignment of water molecules that breaks the bulk inversion symmetry. Depending on the electrolyte concentration the field E_0 extends from tens to hundreds of Angstroms into the bulk region. The total second harmonic signal is given by

$$E_{2\omega} \propto P_{2\omega} = \chi^{(2)} \, E_\omega \, E_\omega + \chi^{(3)} \, \Phi(0) \, E_\omega E_\omega \qquad (8)$$

where $\Phi(0)$ is the electrostatic potential at the interface and results from the integration of the third order polarization $P_{2\omega}^{(3)}$ from the bulk to the surface. It was found that including the water polarization contribution, which dominates the second harmonic signal at alkaline pH values, yields quantitative agreement with the electrolyte and temperature dependent results. Briefly, the second harmonic signal decreases with increasing electrolyte concentration due to the more rapid decay of the electric field (increased Debye shielding) with distance from the interface. Thus fewer water molecules experience the polarizing electric field and the signal decreases. The increase in signal with higher temperature is due to the decrease in the dielectric constant, which increases the distance over which the electrostatic field extends. Fitting of the second harmonic signal versus bulk pH with this polarization model yielded two silanol sites, one occupying 20% of the sites with a pK_a of 4.5 and the other occupying 80% of the sites with a pK_a of 8.5. We believe that the more acidic site is hydrogen bonded to a neighboring silanol. One of the promising results of this study is that SHG appears to provide a new and perhaps powerful method for measuring the surface electrostatic potential at charged liquid interfaces.

4.2 CHARGED LANGMUIR MONOLAYERS

The silica/water results suggest that, if we had a charged monolayer at the air/water interface that had a smaller second order nonlinearity than the phenols and anilines, we should be able to observe the polarization of the bulk water manifested in the $\chi^{(3)}$ term and thus be able to obtain the interface electrostatic potential. Indeed, it was found that the third order contribution ($\chi^{(3)}$) is a major part of the second harmonic signal from the negatively charged monolayer, n-docosyl potassium sulfonate, $CH_3-(CH_2)_{21}SO_3^-...K^+$, and from the positively charged monolayer n-docosyltrimethylammonium bromide $CH_3-(CH_2)_{21}-N^+(CH_3)_3...Br^-$ at air/water interfaces.[23] The interface electric potential was determined to extend from -50mV to

-265mV for the negatively charged monolayer $CH_3-(CH_2)_{21}SO_3^-$ and from +50mV to +265mV for the positively charged monolayer $CH_3-(CH_2)_{21}-N^+(CH_3)_3$ depending on monolayer charge density and bulk electrolyte concentration.

5. Interfacial Phases

Infrared + visible sum frequency generation (SFG) adds the powerful analytical abilities of vibrational fingerprinting to the study of molecules at interfaces. It thus provides a valuable complement to second harmonic studies of interfacial molecular properties. Information on the orientation and environmental interactions of specific vibrational chromophores in a molecule can be obtained from measurements of the polarization of the sum frequency light and from the frequencies of the vibrational transitions.[2-5] We have used SFG to investigate the orientational structure and environment of CH_3CN and CD_3CN at the interface between air and a bulk acetonitrile-water solution.[24] At low acetonitrile bulk concentrations the $-C{\equiv}N$ vibrational frequency was shifted to higher energy with respect to neat bulk acetonitrile. This spectral shift is consistent with the effects of hydrogen bonding on the vibrational frequencies of nitriles in bulk aqueous or alcohol solutions. Upon increasing the acetonitrile bulk mole fraction to 0.07 the nitrile vibrational frequency changes abruptly to a lower energy, close to that observed for bulk neat acetonitrile, suggesting that the hydrogen

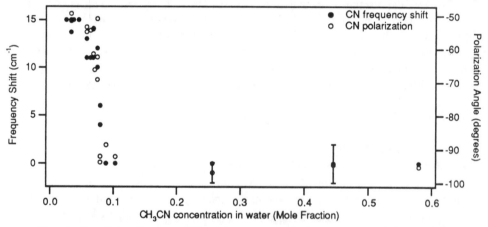

Figure 5: Compilation of frequency shifts of the resonance peak with respect to neat bulk acetonitrile absorption and polarization data for the CN transition of acetonitrile in the interface versus mole fraction of acetonitrile in the bulk water solution.

bonds to water have ruptured. At this same concentration of 0.07 mole fraction a sudden change in the polarization of the sum frequency signal occurred, which corresponds to the acetonitrile orientation changing from a more upright orientation (a tilt of about 40° from the normal) to laying nearly flat at the interface. These changes, which indicate a dramatic change in the environment at the interface, are shown in figure 5. This sharp change was attributed to a phase transition in the interface, which appears to be the first case of a phase transition involving small mutually soluble molecules at a liquid interface.

6. Summary

Utilizing SHG and SFG we have investigated equilibrium phenomena at liquid interfaces. We studied the balance between hydrophobic and solvation effects on the population of organic molecules at the air/water interface. The free energies of adsorption, which determine the interface population, were measured for para alkyl phenols, anilines, phenolates and aniliniums with various length alkyl chains using SHG. It was found that the alkyl chain had to be at least 5 carbons long for the charged forms of these molecules, i.e. phenolates and aniliniums, to be detectable at the interface. We were also able to measure the incremental free energy of adsorption at the air/water interface of a methylene group of the alkyl chain and found it to be about 800 cal/mole for both the neutral and charged species. Using these SHG methods we measured the surface pK_a of para-hexadecylaniline and para-octadecylphenol. We found that their acid-base equilibria in the interface are shifted strongly towards the neutral form of the acid-base pair, which we attribute to the image repulsion of the charged form at the interface and to the stronger solvation of the charged species in the bulk. Hexadecylanilinium, $CH_3(CH_2)_{15}C_6H_4NH_3^+$, is a factor of 50 more acidic in the interface than in the bulk and octadecylphenol, $CH_3(CH_2)_{17}C_6H_4OH$, is a factor of 100 more basic in the interface than in the bulk. We have also developed a method for measuring the interface potential at charged liquid interfaces by measuring the second harmonic signal from water molecules aligned by the electrostatic field near a charged interface. For insoluble monolayers of singly charged sulfonates and ammoniums the interface electric potential varied between an absolute value of 50 mV and 265 mV as the monolayer density was varied. Using this method we were also able to measure the pK_a of different acid-base sites at the silica aqueous electrolyte interface. One kind of silanol site comprises 20% of the surface and has a pK_a of 4.5. The other occupies 80% of the surface sites and has a pK_a of 8.5. We suggest that the more acidic sites are hydrogen bonded to a neighboring silanol. In our studies of the vibrational spectroscopies of molecules at liquid interfaces using SFG we observed an unexpected structural phase transition at the air/water-acetonitrile solution interface. Below a bulk acetonitrile mole fraction of 0.07 in water the acetonitrile at the interface appear to be hydrogen bonded to water and tilted about 40° from the surface normal. Above this concentration the sum frequency spectra indicate that the hydrogen bonds between the interface acetonitrile and water are broken and the acetonitrile assumes an almost flat orientation in the interface.

7. Acknowledgement

We thank the members of our group who contributed to this research. They are Kankan Bhattacharyya, Alonso Castro, Showei Ong, Eugene Sitzmann, Suchitra Subramanyan, Dina Zhang and Xiaolin Zhao. We also thank the Chemical Sciences Division of the U.S. Department of Energy, the U.S. National Science Foundation and the U.S. Air Force Office of Scientific Research for financial support.

8. References

1. Adamson, A. W. (1982) Physical Chemistry of Surfaces, 4th ed., Wiley, New York.
2. Shen, Y. R. (1989) Annu. Rev. Phys. Chem. **40**, 327-350.
3. Richmond, G. L., Robinson, J. M. and Shannon, V. L. (1988) Prog. Surf. Sci. **28**, 1-70.
4. Heinz, T. F. in (1991) Nonlinear Surface Electromagnetic Phenomena, ed. H. E. Ponath, G. I. Stegeman, North-Holland, Amsterdam, 353-416.
5. Eisenthal, K. B. (1992) Annu. Rev. Phys. Chem. **43**, 627.
6. Lee, C. H., Chang, R. K. and Bloembergen, N. (1967) Phys. Rev. Lett. **18**, 167.
7. Bloembergen, N. and Pershan, P. S. (1962) Phys. Rev. **128**, 606.
8. Guyot-Sionnest, P. and Tadjeddine, A. (1990) Chem. Phys. Lett. **172**, 341.
9. Superfine, R., Huang, J. Y. and Shen, Y. R. (1990) Chem. Phys. Lett. **172**, 303.
10. Zhu, X. D., Suhr, H. and Shen, Y. R. (1987) Phys. Rev. B **35**, 3047.
11. Battacharyya, K., Sitzmann, E. V. and Eisenthal, K. B. (1987) J. Chem. Phys. **87**, 1442.
12. Bhattacharyya, K., Castro, A., Sitzmann, E. V. and Eisenthal, K. B. (1988) J. Chem. Phys. **89**, 3376-3377.
13. Bhattacharyya, K., Castro, A. and Eisenthal, K. B. (1991) J. Chem. Phys. **95**, 1310.
14. Davis, J. T. (1951) Proc. Roy. Soc. A **208**, 224.
15. Ward, A. F. H. and Tordai, L. (1946) Nature **158**, 416.
16. Tokiwa, F. and Ohki, K. (1968) J. Coll. Inter. Sci. **26**, 457.
17. Lin, I. J. and Somasundaran, P. (1971) J. Coll. Inter. Sci. **37**, 731.
18. Sternberg, M. J. E., Hayes, F. R. F., Russell, A. J., Thomas, P. G. and Fersht, A. R. (1987) Nature **330**, 86.
19. Blesova, M., Cizmarik, J., Bachrata, M., Bezakova, Z. and Borovansky, A. (1985) Coll. Czech. Chem. Comm. **50**, 1133.
20. Zhao, X., Subrahmanyan, S. and Eisenthal, K. B. (1990) Chem. Phys. Lett. **171**, 558.
21. Subramanyan, S., Zhao, X. and Eisenthal, K. B. manuscript in preparation.
22. Ong, S., Zhao, X. and Eisenthal, K. B. (1992) Chem. Phys. Lett. **191**, 327.
23. Zhao, X., Ong, S. and Eisenthal, K. B. (1993) Chem. Phys. Lett. **202**, 513.
24. Zhang, D., Gutow, J. H. and Eisenthal, K. B. (1993) J. Chem. Phys. **98**, 5099.

ISOMERIZATION REACTIONS AT AQUEOUS INTERFACES

A. Pohorille and M. A. Wilson

Department of Pharmaceutical Chemistry, University of California,
San Francisco, CA 94143

Abstract. In this paper we discuss the transfer of small, flexible molecules across the water–hexane interface. The study is motivated by the biological and pharmacological importance of this process in water-membrane systems. We focus on three main issues: (a) what are the free energy profiles for transferring molecules across the interface, (b) how conformational equilibria at the interface differ from those in the bulk phases, and (c) how the rates of isomerization compare to the rates of transfer across the interface. We investigate these problems by molecular dynamics simulations of two systems — 1,2–dichloroethane and alanine dipeptide. Both molecules exhibit a free energy minimum at the interface. As a consequence, the molecules encounter an apparent "interfacial resistance", in violation of the solubility-diffusion model. For 1,2-dichloroethane the relaxation time of the isomerization reaction was calculated from the transition state theory and corrected for dynamic effects which included a contribution from quasi-periodic trajectories. This time was found to be much shorter than the lifetime of the solute at the interface, indicating that the conformational equilibrium in this region is readily reached during the transfer. For alanine dipeptide it was found that conformations present in water and in hexane are all populated at the interface, but energy barriers between them are markedly reduced. The description of the transfer across the interface by a simple diffusion model was tested. The model gives satisfactory results for 1,2-dichloroethane but is less accurate for alanine dipeptide.

I. INTRODUCTION

The interactions of small solute molecules with membrane–water interfaces and their transport across membranes are ubiquitous processes in cellular biology and pharmacology. They involve metabolites, neurotransmitters, agonists and antagonists to membrane–bound receptors, anesthetics, steroids, antibiotics and other drugs. Many of these molecules are conformationally flexible, and their conformational states at the interface may be relevant to their rates of transport and to their ability to bind to membrane receptors. To determine these conformational states, two types of conditions of particular biological interest should be considered. One corresponds to the equilibrium concentrations of the solute established across different cellular compartments. Such a situation is created, for example, when anesthetics are administered. Then, populations of different conformational states of a solute at the interface can be determined from the knowledge of its free

J. Jortner et al. (eds.), Reaction Dynamics in Clusters and Condensed Phases, 207–226.

energy surface in this region. In the second case, the solute occasionally arrives at the interface, crosses the membrane, possibly moves to another cellular compartment, and eventually becomes metabolized, hydrolyzed, destroyed in some other fashion, permanently bound by or built into a macromolecule or excreted from the organism. Under such conditions, at a low concentration limit, the equilibrium is never established. This is a common situation in drug delivery, where one of the main goals is to control the rate of release in order to maximize the efficacy of the active agent and minimize its potential side effects.

Despite its ubiquity, the transport of small, biologically relevant molecules across water–membrane or, in general, liquid–liquid interfaces has not been subjected to particularly extensive investigations at a molecular level. In the simplest picture, it is assumed that molecules arriving at the interface from a bulk phase, presumably in the equilibrium conformational distribution characteristic to this phase, encounter no specific interfacial resistance. It is further assumed that crossing the interface in the direction favored by the free energy is rapid and further motion of the molecules is diffusive on a flat free energy surface. This is essentially the conventional solubility–diffusion model. It implies that transfer across the interface is so fast that conformational equilibria at the interface are usually not reached by flexible molecules. In many instances, there is a growing body of evidence that the solubility–diffusion model is not adequate.[1,2] This is often interpreted as evidence for the existence of a free energy barrier at the interface. However, the same effect of apparent "interfacial resistance" would be also observed if solutes exhibited free energy minima at the interface. Then, of course, their conformational equilibria at the interface become relevant.

In this paper we explore the transfer of conformationally flexible molecules across a water–hexane interface with molecular dynamic calculations. There are several questions that we want to resolve. The first concerns the free energy surfaces experienced by solutes in the interfacial region. Then, given these surfaces, we investigate the rates of conformational equilibration at the interface *versus* the rates of transfer. Finally, we would like to know if a simple description of the transfer process can be developed to improve upon the solubility–diffusion model.

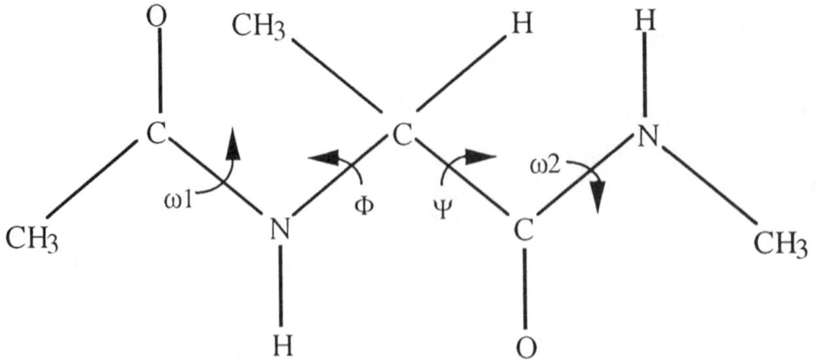

Figure 1. Chemical structure of alanine dipeptide.

For our study we have chosen two solute molecules— 1,2–dichloroethane (DCE) and alanine dipeptide (dialanine). DCE represents a simple model with only one conformational degree of freedom and two distinct conformational states. Dialanine is a markedly larger molecule but its main conformational features are described by only two torsional angles, ϕ and ψ (see Figure 1). One interesting feature of this molecule is that its stable conformational states depend on the solvent. Of course, there is also a much broader interest in this system related to conformational preferences of proteins.

Considering the complexities of the water–membrane system, we limited this study to a simple model system, the water–hexane interface, with an idea that the results will provide the baseline for future work on molecularly more complex interfaces.

II. METHODS

A. Description of the systems

The systems studied in this work consisted of a solute molecule located near the interface between lamellae of water and hexane. In our previous work[3], we have shown that the density profiles of the two liquids exhibit only slight overlap, indicating that these liquids are indeed immiscible. Furthermore, the penetration of molecules of one liquid into another is negligible. Thus, the interface is sharp at the molecular scale, but somewhat broadened by capillary waves.

The system was placed in a simulations box whose dimensions were 24.0 Å $\times$ 24.0 Å $\times$ 200.0 Å such that the interface was perpendicular to the z-axis. The water and hexane lamellae consisted of 480 and 83 molecules, respectively. This corresponds to the widths of 25 and 30 Å.

In addition to the simulations of interfacial system, some MD calculations on bulk solutions were also performed for comparisons. The aqueous solution consisted of one solute molecule and 480 water molecules in a cubic box whose edge length was 24.41 Å. This approximately yields water density equal to 1 g/cm^3. For hexane, the solute was placed in a box of size 26.41^3 Å^3, filled with 83 solvent molecules.

B. Potential energy functions

The water–water interactions were described by the TIP4P potential model.[4] This model has been successfully applied to several aqueous interfacial problems.[5–8] To represent the inter- and intra–molecular interactions involving hydrocarbon chains the OPLS potential functions of Jorgensen were used.[9] In this representation, all the CH$_n$ groups are treated as united atoms, i.e. hydrogen atoms bonded to a carbon atom are not considered explicitly and the carbon atom carries the total mass of the group.

Both solutes were described by an all atom, fully flexible potential energy function:

$$E_{pot} = \sum_{bonds} K_b(r - r_{eq})^2 + \sum_{angles} K_a(\phi - \phi_{eq})^2 + \sum_{tors.} V_n[1 + cos(n\theta - \theta_{eq})]$$

$$+ \sum_i \sum_{<j} \left[\frac{A_{ij}}{R_{ij}^{12}} - \frac{B_{ij}}{R_{ij}^6} + \frac{q_i q_j}{R_{ij}} \right] \tag{1}$$

Here r, ϕ and θ are bond lengths, planar angles and torsional angles, respectively, r_{eq}, ϕ_{eq} and θ_{eq} are their equilibrium values, R_{ij} is the distance between atoms i and j, and q_i is the partial charge on atom i. K_b, K_a, V_n, A_{ij} and B_{ij}, are empirical parameters depending on atom types. Parameters for water–hexane, solute–water and solute hexane van der Waals interactions were obtained from standard combination rules.[9]

The potential energy function for DCE was described in detail in our study of this molecule at the liquid–vapor interface of water.[8] Here we used exactly the same model. For dialanine, the partial atomic charges, van der Waals parameters and torsional parameters were derived to reproduce relative energies and ϕ, ψ angles obtained from *ab initio* quantum mechanical calculations on four conformational states, C_{7eq}, C_{7ax}, C_5 and α_R, with geometry optimization at the 6-31G** level.[10] The first three conformations correspond to minima on the ϕ, ψ energy surface and the fourth conformation represents a state markedly populated in aqueous solution and in proteins.

The derivation of the potential function parameters proceeded in two main steps. First, atomic partial charges, properly averaged over different conformations and equivalent atoms,[11] were fitted to reproduce electrostatic potential surface around the molecule. Then, given the charges, torsional parameters which yield conformational energies close to those calculated quantum-mechanically were determined. The results of the fitting procedure are summarized in Table I and a complete list of parameters is given in the Appendix.

Table I. Geometries and Energies of Four Conformational States of Alanine Dipeptide in the Gas Phase Obtained from *ab inito* Quantum Mechanical Calculations and from Molecular Dynamics Potential Energy Functions.

conformation	quantum mechanics[a]			MD potential functions		
	ϕ	ψ	energy[b]	ϕ	ψ	energy[b]
C_{7eq}	283.9	78.8	0.	285.6	61.0	0.
C_{7ax}	76.0	304.6	2.1	62.9	313.2	1.5
C_5	202.8	159.8	1.5	210.0	156.5	1.5
α_R[c]	300.0	320.0	4.0	300.0	320.0	4.1

[a] the MP2/TZVP//HF/6-31G** level; [b] energies in kcal/mol; [c] since this state does not correspond to a free energy minimum in the gas phase, the ϕ and ψ angles were constrained.

C. Molecular dynamics

The molecular dynamics (MD) equations of motion were integrated using the Verlet algorithm[12] with a time step of 2 fs for the DCE and 2.5 fs for the dialanine systems. The temperature of the system was 300 K. All intermolecular interactions were smoothly truncated with a cubic spline function between 8.0 and 8.5 Å. The bond and angular constraints in water and hexane molecules were handled using SHAKE.[13]

The free energy profile for transferring the solute across the interface was obtained in a series of calculations in which the center of mass of the solute was constrained, relative to the center of mass of the system, in windows along the z-coordinate.[14,15] For each window the probability distribution of finding the solute as a function of z, $P(z)$, was calculated and the free energy of the solute in this region, $A(z)$, was obtained as:

$$A(z) = -k_B T \ln P(z) \qquad (2)$$

where k_B is the Boltzmann constant and T is the temperature of the system.

Exploiting the fact that consecutive windows overlapped by at least 1.5 Å, the free energy profile for a full interfacial region was recovered by requiring that $A(z)$ be a continuous function of z. To improve statistical precision in regions where the free energy changes rapidly, even within a single window, an additional external potential, U_{ext}, was added to the potential energy, such that all values of z would be sampled approximately uniformly. Then $A(z)$ is obtained from the formula:

$$A(z) = -k_B T \ln P(z) - U_{ext}(z) \qquad (3)$$

To transfer both *trans* and *gauche* rotamers of DCE across the water-hexane interface, MD trajectories in 8 windows were generated, each 0.6 ns long (after equilibration). For dialanine, the same number of windows was used, but each trajectory was 1 ns long. The only exception was the case when the solute was located at the interface. In this instance, the total length of trajectories collected for analysis was 2 ns.

The same method was also applied to sample in detail conformational equilibria of dialanine at the interface and in bulk phases. In each case, 1 ns trajectories were obtained in three windows. The torsional potential of mean force for DCE at the interface and in hexane was calculated using U_{ext} determined before by Benjamin and Pohorille.[8] In that work the torsional potential of mean force in water was also given.

To test the diffusion model of transfer across the interface 300 trajectories for the *trans* and for the *gauche* rotamers were initiated in water, 5-6 Å from the interface, by sampling the velocities of all atoms from the Maxwell-Boltzmann distribution. Each trajectory was 20 ps long. No constraints were acting on the DCE during these trajectories. For dialanine, 100 trajectories initially located in hexane, 9 Å from the interface were generated. Since diffusion of dialanine is slower that DCE, the length of the trajectories was 40 ps.

III. 1,2-DICHLOROETHANE AT THE WATER-OIL INTERFACE

A. Free energy surface

To estimate the average time that a DCE molecule needs to cross the interface we first have to determine the free energy profile (potential of mean force), $A(z)$, of moving this molecule through the interfacial region. Since the dipole moments of the *trans* and *gauche* conformers differ significantly, the solvent effect on these two states in the inhomogeneous environment with changing polarity can be different. The resulting free energy profiles may then be markedly different and, therefore, we calculated them separately. These results do not allow us to determine the relative stabilities of the two conformers along z. To do so, we have to calculate the potential of mean force as a function of the torsional angle ϕ for at least one value of z. Such calculations were performed for DCE located in bulk hexane and at the interface. Combined with previous results in aqueous solution,[8] they allow for completing two independent thermodynamic cycles in which the molecule undergoes isomerization from *trans* to *gauche* at one position along the z-coordinate, is moved along the interface normal to the new value of z, isomerizes back to *trans* and then is transferred to the initial location along z. The total free energy change for this process should be zero and deviation from this value (closure error) provide an estimate of statistical uncertainties in the calculated free energy profiles. In both cases the closure error was small, indicating that the accuracy of the profiles is satisfactory.

The calculated free energy profiles as a function of z for both *trans* and *gauche* conformers are shown in Figure 2. As we can see, it is predicted that DCE preferentially partitions to the hexane phase. This is consistent with the fact that DCE is well soluble in organic solvents but rather poorly soluble in water. In hexane, the free energy difference between *trans* and *gauche* conformers is -1.3 kcal/mol and decreases at the interface to -0.4 kcal/mol. These values are very close to those obtained for DCE in the gase phase and at the liquid–vapor interface of water, respectively.[8] In bulk water, the *gauche* conformation is favored by 0.3 kcal/mol.[8] Thus, upon transfer from water to hexane the more polar *gauche* conformer is gradually destabilized. Also, the *gauche* to *trans* activation energy slowly decreases from 4.4 kcal/mol in water to 4.0 kcal/mol at the interface and to 3.5 kcal/mol in hexane.

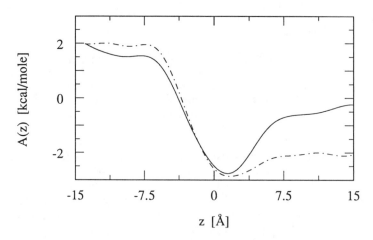

Figure 2.   Free energy of the *gauche* (solid line) and *trans* (dot-dash line) conformers of DCE, $A(z)$, as a function of the distance from the water–hexane interface. Water is located at $z < 0$ and hexane is located at $z > 0$.

Certainly the most interesting feature of the free energy profiles shown in Figure 2 is that they are non–monotonous. Both curves exhibit minima at the interface, but they differ in depth and width. For the *trans* conformer the minimum is wide and only 0.9 kcal/mol deep. In contrast, the *gauche* conformer exhibits a minimum which is narrower and 1.5 kcal/mol deeper than for *trans*. It appears that at least two effects contribute to the stabilization of DCE at the interface. One has to do with the reversible work needed to create a cavity in the solvent that can accommodate the solute. Recently, this quantity has been calculated for atomic–sized spherical solutes in the water–hexane system.[16] It was found that the reversible work exhibits a deep minimum at the interface. This means that the free energy cost of creating a cavity at the water–hexane interface is lower than in the bulk phases. The DCE molecule at the interface may be further stabilized by favorable electrostatic interactions with the net excess dipole moment of interfacial water.

The existence of the free energy minima undoubtly has an impact on the average time spent by DCE at the interface and may also influence the populations of conformers in the interfacial region. From the shapes of the profiles, we can anticipate that the lifetime of the *trans* conformer, τ_t, in the absence of the isomerization reaction will be shorter than the lifetime of the *gauche* conformer, τ_g. What actually happens depends on how τ_t and τ_g are related to the relaxation time of the isomerization reaction τ_i. Three limiting cases can be envisioned. If $\tau_i \ll \tau_t$ the solute will equilibrate at the interface and will move towards water along a free energy profile statistically averaged between *trans* and *gauche* states. If $\tau_t \ll \tau_i \ll \tau_g$ the molecules leaving the interface will be almost exclusively in the *trans* conformation. Finally, if $\tau_t < \tau_g \ll \tau_i$ the *trans* and *gauche* conformers will leave the interface in succession and their interfacial populations will closely correspond to those in the hexane phase. To decide which possibility approximates the real situation best we have to estimate the three lifetimes involved.

B. The isomerization reaction

In the transition state theory (TST) approximation the relaxation time of the isomerization reaction, τ_{ts}, is given as[17,18]

$$\tau_{ts}^{-1} = (2x_g x_t)^{-1} \langle | \dot{\phi}_{ts} | \rangle \frac{e^{-\beta A(\phi_{ts})}}{\int_0^{2\pi} d\phi\, e^{-\beta A(\phi)}} \tag{4}$$

where x_g and x_t are mole fractions of the *gauche* and *trans* conformers, $\beta = k_B T$ and $\langle | \dot{\phi}_{ts} | \rangle$ is the statistical average of the absolute value of the velocity along the reaction coordinate ϕ at the transition state ϕ_{ts}. The last term in Eq. (4) represents the probability of finding the molecule in the transition state. The formula also includes the fact that the system can isomerize from two *gauche* states. Using the potential of mean force as a function of ϕ calculated for DCE at the interface and $\langle | \dot{\phi}_{ts} | \rangle = 4.5$ ps^{-1} derived from MD trajectories initiated at the transition state, τ_{ts} was estimated at 30 ps. For comparison, the corresponding value in bulk hexane is 60 ps.

The actual relaxation time, τ_i, can be obtained by dividing τ_{ts} by the transmission coefficient, κ, to correct for the fact that some trajectories which cross the transition state may return to the reactant well and, therefore, should not be considered as reactive. The value of κ for the *gauche–trans* isomerization of DCE was estimated by the reactive flux method

$$k(t) = \langle \delta(\phi - \phi_{ts}) \dot{\phi} H_g[\phi(t)] \rangle \tag{5}$$

where $k(t)$ is the reactive flux correlation function, H_g is the characteristic function for the *gauche* state equal to 1 whenever ϕ is in the reactant well and zero otherwise, and the angle brackets indicate the equilibrium ensemble average over the initial conditions. All quantities without the time variable are calculated at $t = 0$. The transition state was located at $\phi_{ts} = 120°$.

To obtain κ, 400 uncorrelated MD trajectories were initiated from the transition state and followed for 1 ps. By then $k(t)$ reached a plateau value $k(t_p)$ which is related to κ by

$$\kappa = k(t_p)/k_{ts} \tag{6}$$

where k_{ts} is the TST reaction rate determined by $k(0)$.

The reactive flux correlation function for the *gauche–trans* isomerization reaction at the water–hexane interface is shown in Figure 3 and compared with $k(t)$ for the same reaction in bulk hexane, bulk water and at the liquid–vapor interface of water. The latter two have been published previously[8] and we will closely follow the analysis described in that paper.

In bulk media, $k(t)$ decreases rapidly in the first 0.1 ps and then reaches a plateau. This indicates that some non–TST trajectories advance slightly towards the product and then return to the reactant well. The early recrossings appear to be associated with the rapid intramolecular redistribution of energy between strongly coupled modes. If a rigid model of DCE is used this mechanism of energy coupling is no longer available and, consequently, the initial decrease in $k(t)$ is markedly reduced.[8]

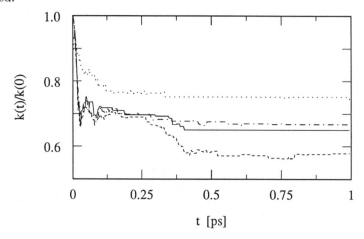

Figure 3. Reactive flux correlation functions for *gauche–trans* isomerization of DCE, $k(t)/k(0)$, at the water–hexane interface (solid line), at the liquid–vapor interface of water (dashed line), in bulk water (dot–dashed line) and in bulk hexane (dotted line).

For DCE at the interfaces, $k(t)$ closely follows the bulk media results for the first 0.3 ps. Then, for the next 0.2 ps, it continues to decrease before reaching a plateau. This behavior has been attributed to the existence of quasi–periodic trajectories, which pass through the product well, rebound off the wall confining the product, recross the transition state, and return to the reactant well without substantial energy exchange with the bath. This picture is supported by two arguments. First, if the trajectories are terminated when they reach the product well the values of κ in the bulk phases and at the interfaces are almost identical. Second, the behavior of $< \dot{\phi}^2(t) >$, which is a measure of the kinetic energy along the reaction coordinate, is periodic for at least 0.6 ps for trajectories exhibiting late recrossing. In contrast, other trajectories experience dissipation of energy into bath and loss of coherence between trajectories. The resulting value of $\kappa = 0.65$ at the water–hexane interface yields $\tau_i = 45$ ps.

The existence of inherently non–RRKM quasi–periodic trajectories at the water liquid–vapor interface is not entirely unexpected. Due to reduced density, the solvent is not able to damp these trajectories and the system can move considerable distances along the reaction coordinate without exchanging energy with the bath. It is, however, somewhat surprising that the same effect, although to a lesser extent, is also observed at the water-hexane interface. In this case, the system is expected to be in the high friction regime of the Kramers theory,[19] whereby the reaction coordinate is strongly coupled to the solvent. A small, additional correction to κ resulting from the quasi–periodic trajectories is probably of no practical significance but this effect is still interesting because it teaches us something about water–oil interfaces. In conjunction with the previously described results for the reversible work of inserting a cavity at the interface, we develop a picture in which the interactions between the two liquids in contact are quite small compared to water–water interactions and the liquids appear to "pull away" from each other.

C. Transfer across the interface

Probably the simplest, physically motivated, description of the transfer of DCE across the interface is by a one–dimensional diffusion equation in an external field,

$$\frac{\partial P(z,t)}{\partial t} = -\frac{\partial j(z,t)}{\partial z} \tag{7}$$

where $j(z,t)$ is the flux and $P(z,t)$ is the probability distribution of finding a particle (in this case the center of mass of a DCE molecule) between z and $z + \delta z$ at time t. This field is given by the calculated free energy $A(z)$. For a constant diffusion coefficient D, the flux is given by[20]

$$j(z,t) = -D\frac{\partial P(z,t)}{\partial z} - \beta D\frac{\partial A(z)}{\partial z}P(z,t) \tag{8}$$

Predictions of the model can be further tested by comparing the resulting $P(z,t)$ with the distributions obtained by generating MD trajectory with similar initial conditions and of the same duration. The same approach has been applied with some success to describe the transfer of ions across liquid–liquid interfaces.[21,22] Differences between results generated by MD and the diffusion model were attributed to non–equilibrium solvation and to difficulties in obtaining accurate $A(z)$ for the ion. These problems are probably less relevant to neutral solutes. On the other hand, we introduce additional approximation by ignoring lack of spherical symmetry of the diffusing molecules. If the probability distribution of the orientational degrees of freedom of the solute are functions of z, then the configurational space available for the exploration of these degrees of freedom will also depend on z. Similar problems have been recently addressed by Zwanzig.[23]

Solving Eq. (7) requires the knowledge of D for DCE, which was obtained from MD calculations in water, hexane and at the interface. The corresponding values were 0.18, 0.53 and 0.56 $\text{Å}^2\text{ps}^{-1}$ for the *trans* conformer and 0.18, 0.50 and 0.42 $\text{Å}^2\text{ps}^{-1}$ for the *gauche* conformer. Since the diffusion coefficient obviously changes along z it may be more appropriate to consider a diffusion equation with a spatially variable D[21,22]

$$j(z,t) = -\frac{\partial[D(z)P(z,t)]}{\partial z} - \beta D(z)\frac{\partial A(z)}{\partial z}P(z,t) \qquad (9)$$

There is really no unambiguous way to define $D(z)$. A reasonable approach is to identify this quantity with the lateral diffusion coefficient for the same value of z. We determined this coefficient for two narrow windows of z on the water side of the interface and obtained the full $D(z)$ by interpolation. Then, the diffusion equation was solved numerically by Monte Carlo, as described by Benjamin,[22] with initial conditions $P(z,0) = \delta(z - z_0)$ where z_0 was 5 and 6 Å from the interface for the *gauche* and the *trans* conformers, respectively. These initial values of z are close to points where $A(z)$ reaches a plateau, defining the borderline between bulk water and the interfacial region. The comparison between $P(z,t)$ obtained from the diffusion model and from MD trajectories is given in Figure 4. The overall agreement is good. Moreover, the results are not sensitive to the interpolation scheme chosen for $D(z)$. In fact, using a constant, intermediate value of $D = 0.4$ Å^{2}ps^{-1} also yields similar results.

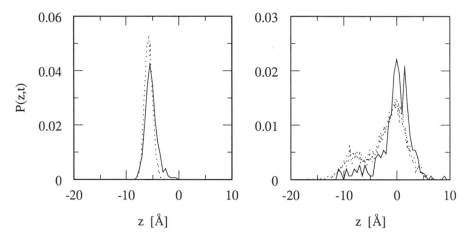

Figure 4.(a) Probability distributions, $P(z,t)$, of *trans*–DCE from MD (solid line) and diffusion model (dot–dashed) line after 1 ps; (b) as in (a), but after 20 ps.

Since the diffusion model proved to be successful in describing the transfer of DCE from hexane to the interface its application was extended to markedly longer time scales, for which MD trajectories are not available, and the lifetimes of the *gauche* and *trans* conformers at the interface were determined. The calculations yielded $\tau_t = 1.5$ ns and $\tau_g = 10.5$ ns. These values are much larger than τ_i, indicating that DCE molecules arriving at the interface from water will reach their new conformational equilibrium before escaping to hexane. Considering the large separation of time scales between isomerization and diffusion from the interface, it can be approximated that the transfer to hexane proceeds on the free energy surface statistically averaged between the two conformers. This yields a lifetime at the interface of 1.5 ns. Note that the considerations above would not hold if $A(z)$ was a decreasing function over the whole interfacial region. The time needed for half of the DCE molecules to reach the interface from water was only about 20 ps which is less than τ_i.

IV. ALANINE DIPEPTIDE AT THE WATER-HEXANE INTERFACE

Molecules with one torsional degree of freedom typically exhibit the same conformational minima in the gas phase and in different liquids. The polarity of the solvent influences only the relative stabilities of the conformational states and the heights of energy barriers separating these states. For molecules with more than one torsional degree of freedom, the situation may be, in general, markedly more complicated. In particular, some states preferred in the gas phase may become unstable in solution and new conformational states may emerge. Dialanine, which is a model of backbone conformational equilibria in peptides and has been studied theoretically, provides a simple example of this situation. The main conformational features of this molecule are determined by two angles ϕ and ψ. The remaining backbone torsions, peptide bond angles ω_1 and ω_2, are rigidly held at values close to 180°. In the gas phase, three stable conformations, C_5, C_{7eq} and C_{7ax} have been identified. All involve intramolecular hydrogen bonding between NH and CO groups leading to the formation of a 5-membered or 7-membered ring within the molecule. It is assumed that the same conformational preferences are preserved in nonpolar liquids. In contrast, at least some of these states are no longer stable in water, while several new conformational minima appear. These minima are characterized by the values of ϕ, ψ typical to α_R and α_L helices and β sheets in proteins. The differences in conformational preferences in nonpolar and aqueous solvents raise several interesting questions regarding behavior of dialanine at the interface between these two media. How does the molecule reach its new conformational equilibrium? What is the time scale of this process? In which region of space does it happen? Which conformational states are preferred at the interface? Before we address these questions we briefly summarize the results obtained for the bulk phases.

A. Alanine dipeptide in bulk solvents

Dialanine in water has been extensively studied using computer simulations[24-28] and integral equation theory.[29,30] The results of these studies are in qualitative agreement in the sense that they predict a significant solvent effect stabilizing open conformations (β, α_R and α_L) relative to the internally hydrogen bonded forms. This is a consequence of the competition between the hydration of polar groups in dialanine and the intramolecular N-H...O-C interactions. At the quantitative level, different studies yield somewhat different conformational free energies, due in part to different potential energy functions used. Our results resemble closest those from the most recent work of Tobias and Brooks,[28] probably because both studies employed potential functions for dialanine designed on the basis of a similar level ab initio quantum mechanical calculations. We focus mostly on four conformational states - β, α_R, C_{7eq} and C_5. The remaining states of potential interest (α_L and C_{7ax}) are markedly less stable in water.[28] Also, they are less relevant biologically.

In the ϕ, ψ region, studied by us in detail, only the β and α_R conformations are stable. Their free energies differ only by 0.3 kcal/mol, in favor of the β form. As shown in Figure 5, the two states are separated by a 3.0 kcal/mol energy barrier located at $\phi, \psi = 285, 40^o$. These values are in very good agreement with the results reported by Tobias and Brooks.[28] Note that C_{7eq} form, which is internally the most stable, is located close to the energy barrier in aqueous solution.

Conformational preferences for dialanine in hexane are quite similar to those in the gas phase. In this case, the two stable conformations in the same ϕ, ψ region are C_{7eq} and C_5, with the former being favored by 1.2 kcal/mol. The barrier of 2.6 kcal/mol between these stable conformations is located at $\phi, \psi = 250,130°$, near a typical β form. α_R is also not a stable state and its free energy is approximately 3.6 kcal/mol above that for C_{7eq}. The results for both bulk media are summarized in Table II.

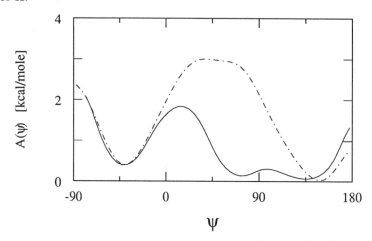

Figure 5. Potential of mean force, $A(\psi)$ for the $\alpha_R \leftrightarrow \beta$ isomerization reaction in dialanine at the water–hexane interface (solid line) and in bulk water (dashed line).

Table II. Energies (in kcal/mol) of Four Conformational States of Alanine Dipeptide in Water, Hexane and at the Water-Hexane Interface

conformation	ϕ, ψ	water	hexane	interface
α_R	(300,320)	0.4	3.6[a]	0.3
β	(300,150)	0.0	0.9[a]	0.0
C_{7eq}	(285,60)	2.9[a]	0.0	0.1
C_5	(205,160)	a,b	1.2	0.3

[a] not a stable state; [b] not calculated

B. The free energy surface

The first step in analyzing conformational transitions in dialanine near the water–hexane interface is to calculate the free energy profile, A(z), of transferring this molecule across the interface. This profile is shown in Figure 6 and is properly averaged over different conformational states. We will argue this point further in this section. As we can see, A(z) exhibits a deep minimum in the interfacial region.

Relative to the bulk water and hexane phases (end points in the profile), dialanine is stabilized at the interface by about 5.5 kcal/mol and 10.0 kcal/mol, respectively.

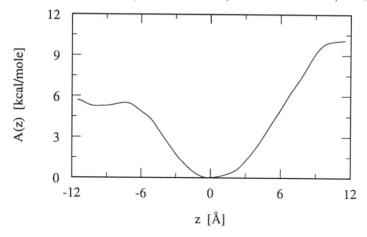

Figure 6. Free energy of dialanine as a function of the distance from the water–hexane interface. Water is located at $z < 0$ and hexane solvent is located at $z > 0$.

Considering the depth of the free energy minimum, dialanine at the interface can be regarded as a well defined species and its conformational preferences can be investigated without imposing any constraints on the z-coordinate. This was done by running four MD trajectories, each initiated at one of potentially stable conformational states. In each case several transitions between different states were observed over the course of simulations and the resulting probability distributions of ϕ and ψ were found to be similar. The cumulative distribution is shown in Figure 7.

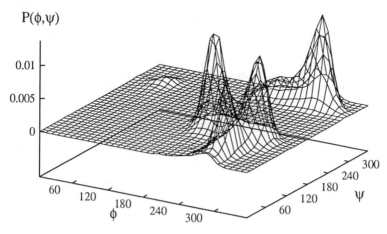

Figure 7. Probability distribution of finding dialanine at different values of ϕ and ψ at the interface.

This distribution exhibits three peaks of almost identical heights, corresponding to α_R, β and C_5 states and a smaller peak near C_{7eq}. The barrier between α_R and β was estimated at 1.4 kcal/mol and the barrier separating β from C_5 is only 0.6 kcal/mol. The barrier between β and C_{7eq} conformations is so low, compared to the thermal energy, that they should be considered as a single state. These results were confirmed by performing more precise calculations for the $\beta \leftrightarrow \alpha_R$ transition using umbrella sampling. The free energy for this transition as a function of ψ is shown in Figure 5. As in aqueous solution, β was found to be more stable than α_R by 0.3 kcal/mol. The barrier obtained from these calculations is 1.7 kcal/mol.

Two aspects of these results are worth special notice. First, all stable conformations in water and hexane correspond to the free energy minima at the interface. In this respect, the interface truly represents an intermediate case between the two bulk phases. Second, the energy barriers at the interface are markedly lowered, compared with both water and hexane. This fact, in conjunction with the deep interfacial free energy minimum, leads to a conclusion that dialanine molecules arriving at the interface reach the conformational equilibrium characteristic to this region before leaving for the bulk phases. In some respects, the interface acts as a catalyst to isomerization. It can bind the substrate and reduce the energy barrier for the reaction. The only missing element for the full catalytic behavior is an effective mechanism of product desorption.

C. Transfer between bulk hexane and the interface

The next issue that we address is how a new conformational equilibrium is established during the processes of desorption from and adsorption at the interface. We will first discuss desorption. Since all the energy barriers involved in conformational transitions are quite low, compared to the free energy difference of transferring dialanine from the interface to bulk hexane, it can be assumed that conformational equilibria are established in different regions of z. Thus, changes in populations of different states along the dialanine pathway simply reflect changes in the underlying ϕ, ψ free energy landscape. The corresponding changes in the probability distributions of ϕ and ψ are shown in Figure 8. As the molecule moves away from the interface, the peak at $\phi = 200°$, characteristic of C_5 state, progressively decreases and eventually reaches the value reflecting the population of this state in bulk hexane. For ψ, the peak around $300°$, representing α_R, quickly disappears and the peak near $150°$, corresponding to β and C_5 states, decreases as the former state disappears and the population of the latter is markedly reduced. Simultaneously, the height of the peak at $60°$, characteristic of C_{7eq}, greatly increases. Thus, the change in conformational equilibria happens progressively over a range of about 7 Å, roughly equal to the dimension of the solute molecule.

The adsorption of dialanine at the interface from hexane was studied by generating MD trajectories and solving the diffusion model in much the same fashion as described for DCE. The initial molecular conformation was almost entirely C_{7eq}, which dominates in hexane. In this case, the model gave the less satisfactory result of predicting faster evolution of $P(z,t)$ towards the interface than the MD calculations. One possible reason for this discrepancy may be that applying the one-dimensional diffusion equation to a strongly asymmetric molecule such as dialanine is no longer justified. Another reason has to do with using the equilibrium free energy profile, $A(z)$. This would be approximately correct if the distribution of conformational states for the trajectories which reached the interface corresponded to the equilibrium distribution found in this region. This is, however, not the case—the populations of β and C_5 are markedly depleted. It this context, it may seem surprising at first that the population of α_R, which is separated from C_{7eq} by the largest energy barrier is similar to that at the interfacial equilibrium. This, however, does not necessarily mean that the system actually had to transverse this barrier, located at $\psi = 40°$. We should remember that this barrier does not exist in hexane and the corresponding position on the ϕ, ψ map is a part of a broad C_{7eq} minimum. Thus, molecules in this state moving toward the interface which happen to have a low value of ψ can cross the growing barrier without significant energy requirements.

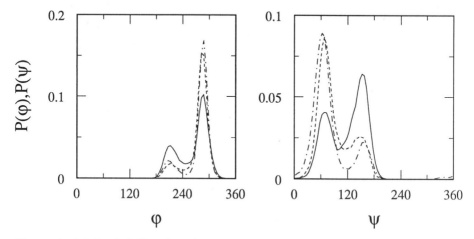

Figure 8. (a) Probability distribution of angle ϕ in dialanine located in hexane approximately 3-4 Å from the interface (solid line), 6-7 Å from the interface (dashed line) and in bulk hexane (dot-dashed line); (b) the same for angle ψ.

D. Desorption from the interface to water

Desorption from the interface towards water is characterized by two main features. First, the peaks in the probability distributions of ϕ and ψ corresponding to states with intramolecular hydrogen bonds disappear as the molecules in these states shift to the β form. Second, the number of transitions between α_R and β progressively decreases as the molecule moves away from the interface. This is due to the growing ridge in the free energy landscape separating these two states.

So far, we have ignored the fact that dialanine is not a spherical molecule and, therefore, its orientational degrees of freedom may be important in characterizing the process of transfer across anisotropic environment. We will describe the orientational preferences of dialanine with the angle θ, formed between the vector joining the C-terminal and N-terminal methyl groups and the interface normal. The probability distributions of this angle for different positions of the center of mass of dialanine along the z-coordinate are shown in Figure 9. At the interface, the probability is strongly peaked at $\theta = 70°$ indicating that the solute is most often oriented parallel to the interface. This result sheds light on the origins of dialanine's stability at the interface. The parallel orientation allows the molecule to simultaneously bury its hydrophobic methyl groups in hexane and expose its polar N-H and C-O groups to water. When dialanine is moved into hexane, the distribution simply broadens and eventually becomes approximately uniform, as expected in a bulk phase. On the aqueous side of the interface the situation is different. As can be seen in Figure 9, P(θ) initially shifts toward 180°, which corresponds to the molecular orientation perpendicular to the interface, before undergoing broadening. This phenomenon has already been observed in our study of amphiphilic molecules at the water liquid–vapor interface[7] and can be explained by the tendency to minimize the hydrophobic portion of the molecule exposed to water. In this case, dialanie adopts the orientation which allows for removing at least one terminal methyl group from the aqueous environment. Since we have not studied adsorption of dialanine at the interface from water, we cannot unambiguously answer the question whether the same orientational changes are sufficiently fast to occur during this process.

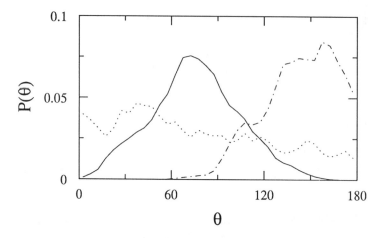

Figure 9. Probability distribution of angle θ between the vector joining C-terminal and N-terminal of dialanine and the interface normal for dialanine at the interface (solid line), in water 3-5 Å from the interface (dot-dashed line) and 9-12 Å from the interface (dotted line).

V. CONCLUSIONS

In this paper we have presented a study of two small, conformationally flexible solutes, DCE and dialanine, at the water–hexane interface. One interesting result from this work is that the free energies of transferring these molecules across the interface exhibit minima. This result, which is in disagreement with the solubility–diffusion model, would have been anticipated for amphiphilic molecules. In this respect, the case of DCE, which does not have a well defined hydrophobic and hydrophilic portions, is of particular interest. Its interfacial behavior does not appear to be an isolated example. Another substituted ethane, 1,1,2-trifluoroethane, which is a fairly potent anesthetic, was recently shown to have a similar free energy profile across the water–hexane interface.[31]

The shapes of the free energy profiles for DCE and dialanine imply that these molecules experience "resistance" during transport across the interface. The existence of this resistance is conventionally identified with an interfacial free energy barrier.[32] In these terms, the free energy minimum can be called "a barrier to exiting the interface". However, this is more than a subtle semantic issue. One potentially important consequence of the interfacial free energy minima is that the equilibrium concentrations of solutes at the interface are greatly enhanced. This is directly related, for example, to one of the leading hypotheses of anesthetic action, which assumes that sufficient concentration of anesthetic compounds at the water–membrane interface is required for their activity.[33]

For low concentration, non–equilibrium conditions across the interface, typical, for example, to drug delivery systems, the existence of interfacial free energy minima greatly prolongs the lifetimes of solute molecules at the interface. The typical relaxation times for isomerization reactions become much shorter than these lifetimes, and solutes reach conformational equilibria characteristic to the interface. This, however, would not necessarily be true if the free energy profile was monotonous across the whole interfacial region.

The distribution of conformational states at the interface for a simple molecule with only one torsional degree of freedom, such as DCE, can be anticipated based on the dielectric model arguments. However, even in that case, attention to molecular structure of the interface is required.[8] For a more complex molecule, dialanine, which has different stable conformational states in hexane and in water, both sets of states are populated at the interface. Moreover, the energy barriers between these states are markedly reduced and the molecule adopts orientation which allows for exposing its polar groups to the water and simultaneously burying the methyl groups in the oil. This result is probably directly related to the fact that some membrane active small peptides, such as neurotransmitters, toxins or ion transporters, which are disordered in water, adopt ordered structures at the membrane surface. [34]

These considerations raise a question to what extent the results obtained for the water–hexane interface are related to the situation in biologically relevant water–membrane systems. There is no doubt that these two systems have somewhat different properties regarding transport of nonelectrolytes, and some of these differences are well documented.[35] However, both systems share one basic feature that the solutes are transferred between a polar aqueous medium and a nonpolar oily environment. This similarity determines that 1,1,2-trifluoroethane exhibits a free energy minimum not only at the water–hexane interface but also in the water–membrane system.[31]

Acknowledgments. The authors express their thanks to Wendy Cornell for deriving potential energy functions for alanine dipeptide, to Ilan Benjamin and Karl Schweighofer for making available their program to solve the diffusion equation and to Lawrence Pratt for helpful discussions. This work was supported by NASA-UCSF Consortium Agreement No. NCC 2-772 and NASA-UCSF Joint Research Interchange No. NCA 2-792. Computer facilities were provided by National Cancer Institute and by the National Aerodynamics Simulator (NAS).

APPENDIX. Potential Function Parameters for Alanine Dipeptide

The parameters used in this work were taken from the new AMBER force field, currently under development. Partial charges and torsional parameters are not necessarily in their final form. All 1-4 van der Waals interactions were scaled by 0.5 and 1-4 electrostatic interactions were scaled by 1/1.2. For parameters not listed below see ref. 36.

Table III. Partial Charges on Atoms of Alanine Dipeptide

atom		type	q
C	(N-terminal CH$_3$)	cm	-0.3787
H	(N-terminal CH$_3$)	hm	0.1152
C		co	0.5902
O		oc	-0.5569
N		n	-0.4241
H		hn	0.2872
C	(α C)	ca	0.0103
H	(on α C)	ha	0.0894
C	(β C)	cb	-0.1781
H	(on β C)	hm	0.0607
C	(C-terminal CH$_3$)	cm	-0.1883
H	(C-terminal CH$_3$)	hm	0.1084

Table IV. Torsional Parameters for Alanine Dipeptide; all V$_n$ in kcal/mol

	atom types			V_n	n	θ_{eq}
*	co	ca	*	0.001	2	0.
*	co	cm	*	0.001	2	0.
*	ca	n	*	0.001	3	0.
*	cm	n	*	0.001	3	0.
*	ca	cb	*	0.144	3	0.
*	co	n	*	5.0	2	180.
n	ca	co	oc	0.034	3	180.
ha	ca	co	oc	0.034	3	180.
cb	ca	co	oc	0.034	3	180.
hm	cm	co	oc	0.034	3	180.
hn	n	co	oc	0.33	1	0.
hn	n	co	oc	1.25	2	180.
co	n	ca	co	0.8	2	180.
n	ca	co	n	0.8	2	180.
cb	ca	n	co	1.0	1	0.
cb	ca	co	n	0.6	2	0.
cb	ca	co	n	0.3	1	0.

Table V. Van der Waals Parameters for Atoms in Alanine Dipeptide

atom type	$\epsilon^{a,b}$	σ^b
ca	0.1094	1.7
cb	0.1094	1.7
cm	0.1094	1.7
co	0.0860	1.7
oc	0.21	1.48
n	0.17	1.625
hm	0.0157	1.325
ha	0.0157	1.325
hn	0.0157	0.535

^{a} ϵ is in kcal/mol, σ is in Å; ^{b} ϵ and σ are related to A_{ii} and B_{ii} in Eq. (1) by the relations $A_{ii} = 4\epsilon(\sigma)^{12}, B_{ii} = 4\epsilon(\sigma)^6$

REFERENCES

1. R. H. Guy and D. H. Honda (1984), "Solute transport resistance at the octanol–water interface", Int. J. Pharm. **19**, 129-137.
2. W. R. Lieb and W. D. Stein (1986), "Non-Stokesian Nature of Transverse Diffusion within Human Red Cell Membranes", J. Membrane Biol. **92**, 111-119.
3. M. A. Wilson and A. Pohorille, to be published, see also ref 16.
4. W. L. Jorgensen, J. Chandrasekhar, J. D. Madura, R. W. Impey, and M. L. Klein (1984), "Comparison of simple potential functions for simulating liquid water", J. Chem. Phys. **79**, 926-935.
5. M. A. Wilson, A. Pohorille, and L. R. Pratt (1987), "Molecular dynamics of the water liquid-vapor interface", J. Phys. Chem. **91**, 4873-4878.
6. M. Wilson and A. Pohorille (1991), "Interaction of monovalent ions with the liquid–vapor interface of water: a molecular dynamics study", J. Chem. Phys. **95** 6005-6013.
7. A. Pohorille and I. Benjamin (1991), "Molecular dynamics of phenol at the liquid–vapor interface of water", J. Chem. Phys. **94**, 5599-5605.
8. I. Benjamin and A. Pohorille (1993), "Isomerization reaction dynamics and equilibrium at the liquid–vapor interface of water. A molecular dynamics study", J. Chem. Phys. **98**, 236-242.
9. W. L. Jorgensen, J. D. Madura and C. J. Swenson (1984), "Optimized intermolecular potential functions for liquid hydrocarbons", J. Am. Chem. Soc. **106**, 6638-6646.
10. W. D. Cornell, private communication.
11. W.D. Cornell, P. Cieplak, C.I. Bayly, and P.A. Kollman (1993), "Applications of RESP charges to calculate conformational energies, hydrogen bond energies and free energies of solvation", J. Am. Chem. Soc., in press.
12. M. P. Allen and D. J. Tildesley (1987), *Computer Simulations of Liquids*, Clarendon Press, Oxford, p 78.
13. G. Ciccotti and J. P. Ryckaert (1986), "Molecular dynamics simulation of rigid molecules", Comput. Phys. Rep. **4**, 345-392.
14. D. Chandler (1987), *Introduction to Modern Statistical Mechanics*, Oxford Univ. Press, New York, ch 6.3.

15. A. Pohorille and I. Benjamin (1993), "Structure and energetics of model amphiphilic molecules at the Water liquid–vapor interface. A molecular dynamic study", J. Phys. Chem. **97**, 2664-2670.
16. A. Pohorille and M. A. Wilson (1993) "Viepoint: Molecular structure of aqueous interfaces", J. Mol. Struct. THEOCHEM, in press.
17. P. Pechukas (1976) in *Dynamics of Molecular Collisions; Part B*, edited by W. H. Miller, Plenum, New York.
18. D. Chandler (1978), "Statistical mechanics of isomerization dynamics in liquids and the transition state approximation", J. Chem. Phys. **68**, 2959-2970.
19. J. T. Hynes (1984), *The Theory of Chemical Reactions*, edited by M. Baer,, CRC Press, Boca Raton, FL, Vol 4.
20. N. G. van Kampen, (1985) *Stochastic Proceses in Physics and Chemistry*, North Holland, Amsterdam, ch. X.
21. K. J. Schweighofer and I. Benjamin (1993), "Dynamics of ion desorption from the liquid–vapor interface of water", Chem. Phys. Lett. **202**, 379-383.
22. I. Benjamin (1992), "Dynamics of ion transfer across a liquid–liquid" interface: A comparison between molecular dynamics and a diffusion model", J. Chem. Phys. **96**, 577-585.
23. R. Zwanzig (1992), "Diffusion past an entropy barrier", J. Phys. Chem., **96**, 3926-3930.
24. P. J. Rossky and M. Karplus (1979), "Solvation. A molecular dynamics study of a dipeptide in water", J. Am. Chem. Soc. **101**, 1913-1937.
25. J. Brady and M. Karplus (1985), "Configurational entropy of the alanine dipeptide in vacuum and in solution. A molecular dynamics study", J. Am. Chem. Soc. **107**, 6103.
26. G. Ravishanker, M. Mezei and D. L. Beveridge (1986), "Conformational stability of the Ala dipeptide in free space and water: Monte Carlo computer simulation studies", J. Comput. Chem. **7**, 345-348.
27. A. Anderson and J. Hermans (1988), "Microfolding: conformational probability map for the alanine dipeptide un water from molecular dynamics simulations", Proteins **3**, 262-273.
28. D. J. Tobias and C. L. Brooks (1992), "Conformational equilibrium in the alanine dipeptide in the gas phase and aqueous solution: A comparison of theoretical results", J. Phys. Chem. **96**, 3864-3870.
29. B. M. Pettitt and M. Karplus (1985), "The potential of mean force surface for the alanine dipeptide in aqueous solution: A theoretical approach", Chem. Phys. Lett. **121**, 194-201.
30. B. M. Pettitt and M. Karplus (1988), "Conformational free energy of hydration for the alanine dipeptide: Thermodynamic analysis", J. Phys. Chem **92**, 3994-3997.
31. A. Pohorille, M. A. Wilson and P. Cieplak, to be published.
32. R. B. Gennis (1989), **Biomembranes**, Springer, New York, p 242.
33. E. I. Eger, J. Liu, D. Koblin, M. Halsey and B. Chortkoff (1993), "Molecular properties of the "ideal" anesthetic", Anesthesia, October 1993, in press.
34. E. T. Kaiser and F. J. Kezdy (1987) "Peptides with affinity to membranes", Ann. Rev. of Biophys. Biophys. Chem. **16**, 561-582.
35. V. M. Knepp and R. H. Guy (1989), "Transport of steroids at model biomembranes surfaces and across organic liquid-aqueous phase interfaces", J. Phys. Chem. **93**, 6817-6823.
36. S. J. Weiner, P. A. Kollman, D. T. Nguyen and D. A. Case (1986), "An all atom force field for simulations of proteins and nucleic acids", J. Comput. Chem **7**, 230-252.

ULTRAFAST SOLVATION DYNAMICS BY DEGENERATE FOUR WAVE MIXING: A THEORETICAL AND EXPERIMENTAL STUDY

R. Richert[*], S.Y. Goldberg, B. Fainberg and D. Huppert
Beverly and Raymond Sackler Faculty of Exact Sciences

School of Chemistry, Tel-Aviv University, Tel-Aviv 69978, ISRAEL

Abstract

We developed theoretically and experimentally the principles of a new spectroscopical method based on four-wave mixing for quantitative description of solvation dynamics of excited large molecules in liquid solutions. We have found that the solvation dynamics of LDS 750 in methanol, ethanol and propanol solutions on a time scale of 1 ps is almost identical. The solvation dynamics in these solvents is biphasic where the long component decays exponentially with a 400 fs decay time. The fast solvation process is followed by the longtitudinal solvent relaxation with relaxation times of 5, 10 and 20 fs for methanol ethanol and propanol respectively.

Introduction

Ultrafast time-resolved spectroscopy has been applied to probe the dynamics of electronic spectra of molecules in solutions [1-4]. Typically, a fluorescent probe molecule is electronically excited and the fluorescence spectrum is monitored as a function of time. Relaxation of the solvent polarization around the newly created excited molecular state led to time dependent Stokes shift of the luminescence spectrum. Such investigations are aimed to study the mechanism of solvation effects on electron transfer processes, protons transfer, etc. [1-4].
Recently an interesting phenomenon has been observed: the existence of fast (subpicosecond) components in the solvation process [3-6].

In this paper we propose a new method for the observation of ultrafast solvation dynamics: the resonance transient grating spectroscopy (RTGS) [7-12]. This method is characterized by high time-resolution and provides additional spectroscopical information, in particular it also senses the dynamics in the ground electronic state [11] which is principally absent in luminescence measurements.

We shall show theoretically and experimentally that the RTGS is rather sensitive to the solvation dynamics and reflects its fine details.

*Permanent Address:Physikalische Chemie, Philipps-Universität, D-35043 Marburg, Germany

J. Jortner et al. (eds.), Reaction Dynamics in Clusters and Condensed Phases, 227–244.
© 1994 Kluwer Academic Publishers. Printed in the Netherlands.

Theoretical Background

Consider a molecule with two electronic states n=1 and 2 in a solvent described by the Hamiltonian

$$H_o = \sum_{n=1}^{2} |n> [E_n + W_n (Q)] <n|, \quad E_2 > E_1 \tag{1}$$

where E_n is the energy of state n, $W_n(Q)$ is the adiabatic Hamiltonian of a reservoir (the vibrational subsystems of a molecule and a solvent interacting with the two-level electron system under consideration in state n).

The molecule is affected by electromagnetic radiation

$$E(r,t)=E^+(r,t) + E^-(r,t) = \frac{1}{2} \sum_{m=1}^{3} \left\{ e_m \mathscr{E}_m(t) \exp[i(k_m r - \omega t) + c.c.] \right\}$$

Since we are interested in intermolecular relaxation processes we shall single out the solvent contributions to E_n and $W_n(Q)$,

$$E_n = E_n^0 + <V_n^{el}>, \tag{2}$$

$$W_n(Q) = W_{nM} + W_{so} + W_{ns}, \tag{3}$$

where W_{so} is the Hamiltonian governing the nuclear degrees of freedom of the solvent in the absence of the solute, W_{nM} is the Hamiltonian representing the nuclear degrees of freedom of a solute molecule, E_n^0 is the energy of state n of the isolated molecule, W_{ns} and V_n^{el} describe interactions between the solute and the nuclear and electronic degrees of freedom of the solvent, respectively [13]. It is possible to replace the operators V_n^{el} in the Hamiltonian by their expectation values $<V_n^{el}>$ [13].

In transient four-photon spectroscopy two pump pulses with wave vectors k_1 and k_2 create a light-induced grating in the sample under investigation with a wave vector $q_1 = k_1 - k_2$ (see Fig. 4 below). The grating effectiveness is measured by the diffraction of a time delayed probe pulse k_3 with the generation of a signal with a new wave vector $k_s = k_3 + (k_1 - k_2)$.

The signal intensity J_s can be calculated from the cubic polarization:

$$J_s(\tau) \sim \int_{-\infty}^{\infty} dt \, |P^{(3)+}(r,t)|^2 \tag{4}$$

where τ is the delay time of the probe pulse $\mathbf{k}_3$ with respect to the pump ones. We shall calculate $\mathbf{P}^{(3)}(r,t)$, using a general theory [11,14-16]:

$$\mathbf{P}^{(3)+}(\mathbf{r},t)= \sum_{mm'm''} \mathbf{B}_{mm'm''} \iiint_o^\tau d\tau_1 d\tau_2 d\tau_3 \exp\left\{-[i(\omega_{21}-\omega)+\gamma]\tau_1 - \frac{\tau_2}{T_1}\right\}\mathcal{E}_{m''}(t-\tau_1)\left\{\mathcal{E}_m(t-\tau_1-\tau_2)\right. \times$$

$$\mathcal{E}_m^*(t-\tau_1-\tau_2-\tau_3)\exp\left[\left[i(\omega_{21}-\omega)-\gamma\right]\tau_3\right]F_1(\tau_1,\tau_2,\tau_3)+\mathcal{E}_{m'}(t-\tau_1-\tau_2-\tau_3)\mathcal{E}_m^*(t_1-\tau_1-\tau_2) \times$$

$$\exp\left[-\left[i(\omega_{21}-\omega)+\gamma\right]\tau_3\right]F_2(\tau_1,\tau_2,\tau_3)\right\} \tag{5}$$

where

$$F_1(\tau_1,\tau_2,\tau_3) = K(0,\tau_3,\tau_1+\tau_2+\tau_3,\tau_2+\tau_3) + K(0,\tau_2+\tau_3,\tau_1+\tau_2+\tau_3,\tau_3), \tag{6a}$$

$$F_2(\tau_1,\tau_2,\tau_3) = K^*(0,\tau_3,\tau_2+\tau_3,\tau_1+\tau_2+\tau_3)+ K^*(0,\tau_1+\tau_2+\tau_3, \tau_2+\tau_3,\tau_3), \tag{6b}$$

$$K(0,t_1,t_2,t_3) = \langle\exp\left[\frac{i}{\hbar}\tilde{W}_2 t_1\right]\exp\left[\frac{i}{\hbar}W_1(t_2-t_1)\right]\exp\left[-\frac{i}{\hbar}\tilde{W}_2(t_2-t_3)\right]\exp\left[-\frac{i}{\hbar}W_1 t_3\right]\rangle \tag{7}$$

are four-time correlation functions [17]. The value $u = W_2-W_1-\langle W_2-W_1\rangle \equiv \tilde{W}_2-W_1$ represents both the perturbations of the molecular nuclear system and the solvent nuclear system respectively during the electronic transition. The angle brackets indicate thermal averaging over the variables of the vibrational subsystems in the ground electronic state of the molecule. We can subdivide the operator u to the intra(M) and inter(S) molecular contributions, $u = u_M + u_S$, where $u_{M,S} = W_{2M,S} - W_{1M,S} - \langle W_{2M,S} - W_{1M,S}\rangle$.

$$u(t) = \exp\left(\frac{i}{\hbar}W_1 t\right)u(Q)\exp\left(\frac{i}{\hbar}W_1 t\right)$$

$$\mathbf{B}_{mm'm''} = \frac{-iNL^4}{8\hbar^3}|D_{12}^{el}|^4 \langle\vec{\kappa}^*(\vec{\kappa} e_{m''})(\vec{\kappa}e_{m'})(\vec{\kappa}^* e_m)\rangle_{or} \exp\left[i\left((k_{m'}+k_{m''}-k_m)r-\omega t\right)\right],$$

$\vec{\kappa} D_{12}^{el}$ is a matrix element of the dipole-moment operator taken with respect to the electron wave function; $\langle...\rangle_{or}$ signifies averaging over various molecule orientations,

$T_1=(2\gamma)^{-1}$ is the lifetime of the excited state 2; $\omega_{21}=(E_2^O + \langle V_2^{el}\rangle - E_1^O - \langle V_1^{el}\rangle)/\hbar + \langle W_2-W_1\rangle/\hbar$ is the frequency of the $1 \to 2$ Franck-Condon transition; N is the system particle density and L is the Lorentz correction factor for a local field. The summation in Eq.(5) is carried out over all fields that satisfy the condition:

$$k_s = k_{m'} + k_{m''} - k_m .$$

We shall calculate $P^{(3)}$ for the following conditions:

1) We use a Gaussian approximation for the value of u. In this case the four-time correlation functions can be represented as follows [11,15-16]:

$$K(0,t_1,t_2,t_3) = \exp\left[g(t_3-t_2)+g(t_1)+g(t_2-t_1)-g(t_2)-g(t_3-t_1)+g(t_3)\right] \tag{8}$$

where $g(t) = -\hbar^{-2} \int_0^t dt'(t-t') \langle u(0)u(t')\rangle$ is the logarithm of the characteristic

function (Fourier transformation) of the spectrum of one-photon absorption after subtraction of a term which determines the first moment of the spectrum, $g(t) = g_M(t) + g_S(t)$. Note that the Gaussian approximation is valid for the description of intermolecular relaxation [13,18].

2) We consider molecules with broad structureless electronic spectra.

3)   The duration (>150 fs) of the laser pulses are longer than the reciprocal bandwidths of both the absorption and the luminescence spectra and longer than intramolecular vibrational relaxation times.

4) We shall consider the translational and the rotational motions of the liquid molecules as classical, at room temperature, as their characteristic frequencies are smaller than the thermal energy kT.   The intermolecular relaxation is described by the corresponding correlation function $\langle u_S(0)u_S(t)\rangle \equiv \hbar^2\sigma_{2S}S(t)$ where $\sigma_{2S} = \langle u_S^2(0)\rangle/\hbar^2$ is the contribution of the solvent to the second central moment of both the absorption and the luminescence spectra. The solvent contribution ω_{st} to the Stokes shift of the equilibrium spectra of the absorption and the emission is of the order of 1000 cm^{-1}. For the classical case we have [13,18-19]

$$\sigma_{2S} = \omega_{st} \frac{kT}{\hbar} \tag{9}$$

Bearing in mind conditions (1)-(4), we can significantly   simplify the expression for the nonlinear polarization (5).   First of all, it is apparent that the values $K(0,t_1,t_2,t_3)$ (formula (7)) can be represented in the form

$$K(0,t_1,t_2,t_3) = K_M(0,t_1,t_2,t_3) \cdot K_S(0,t_1,t_2,t_3) \tag{10}$$

due to the fact that the $u = u_M + u_S$. Let us denote by τ_s the characteristic time of the attenuation of the "intermolecular" correlation function $\langle u_s(0)u_s(t)\rangle$. In any case $\tau_s \gtrsim 10^{-13}$s [4]. Since $\sigma_{2s}^{-1/2} \sim 10^{14}$s, the parameter $\sigma_{2s}\tau_s^2 \gtrsim 10^2 \gg 1$. For this case we can write [11,15-16]

$$F_{1,2}=\left\{\exp\left[g_M(-\tau_1)\right]+\exp\left[g_M(\tau_1)+i\omega_{st}\left(1-S(\tau_2)\right)\right]\right\} \times$$

$$\exp\left[-\frac{\sigma_{2s}}{2}\left(\tau_1^2+\tau_3^2\mp2\tau_1\tau_3 S(\tau_2)\right)+g_M(\pm\tau_3)\right] \tag{11}$$

It follows from Eq.(11) that the intermolecular relaxation is described by the correlation function $S(\tau_2)$, and the upper boundary of the value of times τ_1 and $\tau_3 \sim \sigma_{2s}^{-1/2}$. Therefore, we can put $\tau_1 \approx \tau_3 \approx 0$ in the arguments of the field functions $\mathcal{E}_m$, $\mathcal{E}_{m'}$, $\mathcal{E}_{m''}$ in Eq.(5) [10-11]. In addition, the time τ_2 is of the order of the intermolecular relaxation time τ_s, the intramolecular functions $g_M(t)$, which depends on τ_2, will attenuate to zero in accordance with condition 3). Keeping this in mind, we can integrate the right hand side of formula (5) with respect to τ_1 and τ_3. Using Eqs.(4), (6), (8) and (10)-(11), we obtain for the signal excited by very short non-overlapping pulses:

$$J_s(\tau) \sim e^{-2\tau/T_1}\,|A(\tau)|^2 \tag{12}$$

The term $e^{-\tau/T_1}$ describes the attenuation of $P^{(3)}$ due to the destruction of the population grating.

The term $|A(\tau)|^2$ describes the contribution of the solvation dynamics to the time evolution of the signal. Let us consider the main physical processes, occurring in a solvating system under a laser excitation (fig.1).

The pump pulses of frequency ω create a hole (A) in the initial thermal distribution (C) relative to generalized solvation coordinate in the ground electronic state and, simultaneously, a spike (B) in the excited electronic state. Apparently, such formations have a space modulation $\sim \exp[-i(k_1-k_2)r]$. These distributions tend to the equilibrium point of the corresponding potentials over time, and are also broadened during their movements. These changes are measured by the probe pulse delayed by a time τ relative to the pump pulses.

Let us adduce at first the formula for $A(\tau)$ without taking into account the intramolecular degrees of freedom [11]:

$$A(\tau)\sim F^e_{s\alpha}(\omega-\omega_{21})\left\{F_{s\alpha}(\omega-\omega_\alpha,\tau)+F_{s\varphi}(\omega-\omega_\varphi,\tau)+i\frac{2}{\sqrt{\pi}}\left[X_\alpha(\omega-\omega_\alpha,\tau)+X_\varphi(\omega-\omega_\varphi,\tau)\right]\right\} \tag{13}$$

The formula completely corresponds to the physical processes taking place in solvation of the system considered before. The value of $A(\tau)$ depends on changes related to nonequilibrium solvation processes in both the $F_{\alpha,\varphi}$ absorption (α) and the emission (φ) spectra [11]

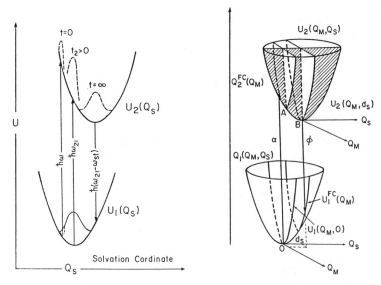

Fig.1. Potential surfaces of the ground and excited electronic states of a
        solute molecule in liquid.
a.   One dimensional potential surfaces   b. Two dimensional potential surfaces
     as a function of generalized   of the ground and electronic states.
     solvent polarization coordinate.

$$F_{\alpha,\varphi}(\omega-\omega_{\alpha,\varphi},\tau) = \frac{1}{\sqrt{2\pi\sigma(\tau)}} \exp\left[-\left(\omega-\omega_{\alpha,\varphi}(\tau)\right)^2/2\sigma(\tau)\right] \qquad (14)$$

at the active pulse frequency ω, as well as on the corresponding changes in both the spectra of the refraction index $X_{\alpha,\varphi}(\omega-\omega_{\alpha,\varphi},\tau)$, where "e" means the equilibrium value. $X_{\alpha,\varphi}$ are related to $F_{\alpha,\varphi}$ by the Kramers-Kronig formula, and have the following form [11]

$$X_{\alpha,\varphi}(\omega-\omega_{\alpha,\varphi},\tau) = F_{\alpha,\varphi}(\omega-\omega_{\alpha,\varphi},\tau) \; \text{Erfi}\left[\frac{\omega - \omega_{\alpha,\varphi}}{(2\sigma(\tau))^{1/2}}\right], \qquad (14a)$$

$$\text{Erfi}(x) = \int_{0}^{x} \exp(y^2)dy$$

As can be seen from Eq.(14), the changes in both spectra $F_{\alpha,\varphi}$ at each   instant   in time τ are Gaussian functions with time dependent width proportional to $\left[2\sigma(\tau)\right]^{1/2}$

$$\sigma(\tau) = \sigma_{2S}\left(1-S^2(\tau)\right) \tag{15}$$

Thus, as it follows from Eq.(15) the width of the light-induced changes in both spectra are small for small delay times τ $\left(S(\tau) \approx 1\right)$. The hole and the spike distribution broaden in time relative to the solvation coordinate (fig.1a).

The detunings $\omega - \omega_{\alpha,\varphi}(\tau)$ of $F_{\alpha,\varphi}$ are functions of the delay time τ [11]:

$$\omega_{\alpha}(\tau) = \omega_{21} + (\omega-\omega_{21})S(\tau),$$
$$\omega_{\varphi}(\tau) = (\omega_{21} - \omega_{st}) + (\omega-\omega_{21}+\omega_{st})S(\tau) \tag{16}$$

The detuning $\omega_{\alpha}(\tau)$ is connected with the motion of the hole in time, and the detuning $\omega_{\varphi}(\tau)$ is dependent on the motion of the spike (fig.1a).

The values $X_{\alpha,\varphi}(\omega-\omega_{\alpha,\varphi},\tau)$, that are related to $F_{\alpha,\varphi}(\omega-\omega_{\alpha,\varphi},\tau)$ by the Kramers-Kronig formula display the corresponding changes in the index of refraction.

Now, let us take into account the intramolecular vibrations. In this case the adiabatic potentials will be represented by hypersurfaces (Fig. 1b). And the corresponding absorption and emission spectra will be represented by the convolutions

$$F_{\alpha}^{e}(\omega-\omega_{21}) - \int d\omega' F_{M}(\omega')F_{s\alpha}(\omega-\omega_{21}-\omega') \tag{17a}$$

$$F_{\varphi}^{e}(\omega_{21}-\omega_{st}-\omega) = \int d\omega' \, F_{M}(\omega')F_{s\varphi}(\omega_{21}-\omega_{st}-\omega-\omega') \tag{17b}$$

The shape of the "intramolecular "spectrum $F_{M}(\omega')$ is determined schematically by the 1-D potentials $U_{1}(Q_{M},0)$ and $U_{2}^{FC}(Q_{M})$ (fig. 1b) that are obtained by the intersection of the hypersurfaces $U_{2}(Q_{M},Q_{S})$ and $U_{1}(Q_{M},Q_{S})$ by a vertical plane passing through the "molecular" coordinate Q_{M} :

$$F_{M}(\omega') = (2\pi)^{-1} \int_{-\infty}^{\infty} dt \exp\left[g_{M}(t) - i\omega't\right] \tag{18}$$

The "intermolecular" spectrum $F_{s\alpha}(\omega-\omega_{21}-\omega')$ is determined by "1-D" potentials that are obtained by the intersection of the hypersurfaces by a vertical plane passing through the solvation coordinate Q_S.

The situation is similar for the emission spectrum. Typical "intramolecular" and whole spectra are shown in fig. 2.

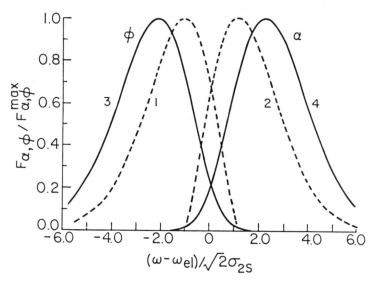

Fig. 2.   The shape of the "intramolecular" spectra $F_M(\omega')$. 1 and 2 are the equilibrium luminescence and absorption spectra of a molecule, respectively, when the contribution from the solvent is absent; 3 and 4 are the equilibrium spectra of a molecule in solution. The arrow shows the relative position of the excitation frequency ω for the four-photon signal calculation (Fig.3).

We now calculate $A(\tau)$ in the general "2-D" case where both the intramolecular and intermolecular contributions are taken into account.   The calculation is the generalization of the results [11] for the case of an arbitrary spectrum $F_M(\omega')$, corresponding to the reorganization of the ultrafast intramolecular degrees of freedom during the electron transition.

Bearing in mind Eq.(10) one might assume that the spectral changes in Eq.(13) must be changed by the corresponding convolutions with the "intramolecular" spectra. However, it is correct only for the longtime limit value of $A(\tau)$:

$$A(\tau \to \infty) = F_\alpha^e(\omega - \omega_{21})\left[F_\alpha^e(\omega - \omega_{21}) + F_\varphi^e(\omega_{21} - \omega_{st} - \omega) - i\Phi_\alpha^e(\omega - \omega_{21}) + i\Phi_\varphi^e(\omega_{21} - \omega_{st} - \omega)\right] (19)$$

Here $\Phi_{\alpha,\varphi}^e$ are the equilibrium spectra of the refraction index corresponding to the absorption (F_α^e) and the emission (F_φ^e) spectra:

$$\Phi^e_{\alpha,\phi}(\omega_1) = \frac{P}{\pi} \int d\omega' \frac{F^e_{\alpha,\phi}(\omega')}{\omega' - \omega_1} \quad , \tag{20}$$

where P is the symbol of the principal value.

In the general case the value of $A(\tau)$ is represented by the 2-D integral:

$$A(\tau) = \int\int d\omega' d\omega'' F_M(\omega') F_M(\omega'') F^e_{s\alpha}(\omega-\omega_{21}-\omega'') \text{ x}$$

$$\left\{ F_{s\alpha}(\omega-\omega_\alpha,\tau) + F_{s\phi}(\omega-\omega_\phi,\tau) + i \frac{2}{\sqrt{\pi}} \left[X_\alpha(\omega-\omega_\alpha,\tau) + X_\phi(\omega-\omega_\phi,\tau) \right] \right\}(21)$$

that does not reduce to the product of the one-dimensional integrals (according to Eq.(17)). The reason is that the frequencies $\omega_{\alpha,\phi}$ in Eq.(21) are functions of both ω' and ω'':

$$\left. \begin{array}{l} \omega_\alpha(\tau) = (\omega_{21} + \omega') + (\omega-\omega_{21}-\omega'') \ S(\tau) \\[2mm] \omega_\phi(\tau) = (\omega_{21}-\omega'-\omega_{st}) + (\omega-\omega_{21}-\omega'' + \omega_{st}) \ S(\tau) \end{array} \right\} \tag{22}$$

The physical reason for such a dependence is given as follows.

Let us return to fig.1a. The situation that is shown in this figure, is characteristic also for the "2-D" case, however, it is true only for the intersections of hypersurfaces by the vertical plane passing through the coordinate Q_s (fig. 1b). Therefore, any distribution shown in fig. 1a will be accompanied by the equilibrium distribution with respect to the "intramolecular" coordinate Q_M.

Let us consider for the definition only the processes, corresponding to the second and the fourth addends in Eq.(21). The pump pulses act along the transition α between the "plane" potentials $U_1(Q_M,0)$ and $U_2^{FC}(Q_M)$ (fig. 1b), bearing the spike of the distribution on the bottom of the Franck-Condon potential $U_2^{FC}(Q_M)$ (point A) due to the instantaneous intramolecular relaxation. If the delay time τ of the probe pulse is small with comparison to the relaxation time with respect to the coordinate Q_S, the probe pulse will act also between the potentials $U_2^{FC}(Q_M)$ and $U_1(Q_M,0)$ and, correspondingly the spectra $F_M(\omega')$ and $F_M(\omega'')$ in Eq.(13) will be strongly correlated. For large delays τ, the spike will relax to the equilibrium state (point B). Therefore, the pump pulse will act in the range of the pair of potentials: $U_2(Q_M,d_s)$ and $U_1(Q_M,0)$ (the transition ϕ). The corresponding spectra $F_M(\omega')$ and $F_M(\omega'')$ will not correlate. In this case the double-integration reduces to the product of 1-D integrals, i.e. to the product of the corresponding equilibrium spectra.

Figs. 3 illustrate the time behaviour of the signal $J_s(\tau)$ (Eq.(12)) that was calculated by formulae (12), (14)-(15), (18), (21)-(22). The shape of "intramolecular" spectrum $F_M(\omega')$ is modelled by a "smoothed" dependence of one optically active intramolecular vibration of frequency ω_0 [20-21]: $F_M(\omega') \sim \overline{S}^x/\Gamma(x+1)$ where $\Gamma(x+1)$ is the gamma-function, $x = (\omega'-\omega_{el})/\omega_0$, $\omega_{el} = (E_2^0 + <V_2^{el}> - E_1^{el} - <V_1^{el}>)/\hbar$ is the frequency of the purely electron transition of a molecule in solution. We used the following values for the parameters: $\omega_{st}(2\sigma_{2s})^{-1/2} = 2$, $\overline{S} = 1.5$, $\omega_0(2\sigma_{2s})^{-1/2} = 1.14$. The shape of the "intramolecular" spectrum $F_M(\omega')$ for these parameters is shown in Fig. 2 in the form of the equilibrium spectra F_α^e $(\omega-\omega_{el})$ and F_φ^e $(\omega_{el}-\omega)$ when the contribution from the solvent is absent. $F_\alpha^e(\omega-\omega_{el})$ and F_α^e $(\omega_{el}-\omega)$ are determined by formulae (17) for the substitutions $\omega_{21} \to \omega_{el}$, $\omega_{st} = 0$, $F_{s\alpha} \to \delta(\omega-\omega_{el}-\omega')$ and $F_{s\varphi} \to \delta(\omega_{el}-\omega-\omega')$, $\delta(x)$ is the δ-function of Dirac. The equilibrium spectra of the molecule in solution F_α^e $(\omega-\omega_{21})$ and F_φ^e $(\omega_{21}-\omega_{st}-\omega)$ are also shown in Fig.2.

It follows from Eqs.(12), (21), that the signal $J_s(\tau)$ depends on the excitation frequency ω. We chose $\omega = \omega_{el} + \omega_{st}/2$ which approximately corresponds to the experimental situation (see below). The excitation frequency ω is also shown in Fig.2.

We used two forms for the correlation function $S(t)$

$$S(t) = a_2 \exp(-a_3 t^2) + (1-a_2-a_4)\exp(-a_1 t) + a_4 \exp(-a_5 t) \qquad (23a)$$

and

$$S(t) = \exp(-\Gamma|t|) \left[\cos \Omega t + (\Gamma/\Omega)\sin \Omega |t|\right] \qquad (23b)$$

corresponding to a Brownian oscillator [4,15-16,22-24].

The first addend in expression (23a) for the first correlation function corresponds to a fast Gaussian component, observed in [3]. The second one corresponds to the relatively fast exponential component with an attenuation time of 200-400 fs observed in [3] and in our experiment (see below). The third component corresponds to a slower attenuation with a decay time of the longitudinal relaxation τ_L. It is worth noting that such a division by different contributions to the correlation function is purely formal, and is used here to impart the realistic form of the correlation function. As a matter of fact, both the short and the long time components of the correlation function are manifestations of one physical process.   We shall discuss this issue in more detail below.   We also showed in Figs. 3 the time dependence of the correlation functions $S(\tau)$, used for the calculation of corresponding signals $J_s(\tau)$.

One can see that the dependences $S(\tau)$ and $J_s(\tau)$ are very similar (but not identical), and the signal $J_s(\tau)$ reflects the fine details of $S(\tau)$.   Thus, the RTGS can be used for the ultrafast study of the solvation dynamics.

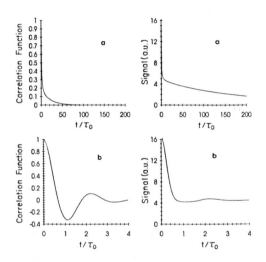

Fig.3.   Model calculations of the RTGS signal - a: the solvation correlation function consists of the Gaussian and exponential decays (formula (23a)), b: the correlation function corresponds to a Brownian oscillator model for the liquid behavior(formula (23b)); τ_0=200fs. a: (T_1/τ_0)=400, $a_3 \cdot \tau_0^2$=7.7016, $a_1 \cdot \tau_0$=0.5, $a_5 \cdot \tau_0$=0.05, a_2=0.34, a_4=0.19;   b. T_1/τ_0=∞, $\Gamma \cdot \tau_0$=1, $\Omega \cdot \tau_0$ = 2.83.

Experimental Details

The schematics of the laser system used for generation of intense ultrashort coherent pulses for the time-resolved four wave mixing measurements is shown in Fig. 4.   The laser source consists of a CW mode locked Nd: YAG laser (Coherent Antares) operating at 76 MHz.   A small portion of the 1.06 μm radiation (~20 mW) is used to   seed a CW Nd:YAG regenerative amplifier.   The input and output to the amplifier are switched by a LiNbO$_3$ Pockels cell controlled by a two-pulse driver (Medox Electro-Optic model DR 85-A) operating at 1 kHz and synchronized to the mode-locker driver. The regenerative amplifier output pulses of 1.1 mJ energy at 1.06 μm are doubled with a beta barium borate (BBO) crystal and reach an energy per pulse of 0.4 mJ at 532 nm. The doubled frequency output of the amplifier (70 ps full width half maximum) was used to amplify the ultrashort laser pulse 140 fs FWHM, 1 nJ generated by a synchronously pumped dye laser.

The synchronously pumped dye laser (Satori, Coherent) is pumped by the second harmonic of the CW model locked YAG (2 W average power). The dye laser design utilizes a saturable absorber in combination with group velocity dispersion compensation prisms to achieve a stable pulse width of the order of 140 fs. The dye amplifier consists of three flowing dye cells pumped by the regenerative amplifier second-harmonic pulse. With Kiton red dye the dye laser operates at 640 nm and the amplification is achieved by DCM dye to ~15 μJ with a pulse width comparable with the non-amplified pulse.

The four wave mixing optical setup is shown at the bottom of Fig. 4. The amplified (15 μJ) 140 fs laser pulse was split into three beams. Optical delay lines were used to overlap in time the pump beams and to control the time delay of the probe beam. The three beams were focused onto the sample by a single lens of 50 cm focal length.

In DFWM experiments the signal beam exit the sample at a unique direction $k_s = (k_1 - k_2) + k_3$ and therefore it is easily separated from the three generation beams.

LDS 750 (styryl 7) and LDS 821 (styryl 9) were purchased from Exciton and were used without further purification. The solvent used were either analytical or of a spectroscopical grade. Samples were circulated in a flowing cell of 1 mm pathlength.

Experimental Results

The time-resolved four-wave mixing signal was measured by the experimental setup shown in Fig.4. The time dependent four-wave mixing signals of LDS 750 in methanol and ethanol are shown in Fig.5. The signals were collected with a relatively low time resolution by scanning the probe beam delay stage at 0.5 ps steps. As seen from Fig.5 the signal decay curves for LDS 750 in methanol and ethanol are nonexponential and consist of several time domains. The long life time component exhibit an exponential decay law with corresponding life times of 120 ps and 200 ps for methanol and ethanol respectively. These lifetimes we attribute to the population grating attenuation. The fluorescence lifetimes of LDS 750 in methanol and ethanol are 240 ps and 400 ps respectively [25]. The factor of two between the electronically excited state lifetime measured by luminescence technique and the decay of the DFWM signal arises from the following argument. The signal in DFWM experiments is proportional to $|P^{(3)}|^2$ where $P^{(3)}$ is the cubic polarization. If population gratings are formed in such experiments then $P^{(3)}$ decay as $\exp(-t/T_1)$ where T_1 is the excited state lifetime. However, the DFWM signal decays as $\exp -2t/T_1$. Thus the decay rate of the population grating in a DFWM experiment is twice larger than the actual decay rate of the excited state population.

The shorter time components of the DFWM signal of LDS 750 in methanol, ethanol and propanol are seen on a shorter time scale with an expanded time resolution (50 fs time steps) in Fig. 6. Each of the decay curves shown in Fig. 6 consists of three time components. We attribute all these time components to the solvation dynamics of LDS 750.

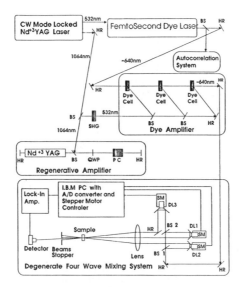

Fig. 4.   The optical setup for the generation of ultrashort laser pulses and the conduction of DFWM experiments.

The longest component can be approximately fitted to an exponential decay with decay times of 5,10 and 20 ps for methanol, ethanol and propanol respectively. These relaxation times correspond to the longitudinal dielectric relaxation time τ_L of the particular liquids. The dielectric relaxation properties of alcohols were studied quite extensively [26-27]. While the dielectric relaxation time τ_D is obtained for measurements that senses the orientational motion of the liquid molecules at constant field, the longitudinal relaxation time τ_L provides the liquid relaxation time at constant charge. The two relaxation times are related by a simple formula $\tau_L = \dfrac{\varepsilon_\infty}{\varepsilon_s}\tau_D$ where ε_s and ε_∞ are the low and high frequency dielectric constants respectively. The constant charge, longitudinal relaxation time is more appropriate to compare with the solvation dynamics of excited solute molecules [28-30].

The dielectric relaxation properties of neat normal primary alcohols present a complex behaviour. This complexity is attributed to the hydrogen bonding between adjacent molecules. The relatively long relaxation time is attributed to the breaking of hydrogen bonds in molecular aggregates followed by ROH Rotation. In addition to the long relaxation component, shorter relaxation times are observed in alcohols. Since the dielectric relaxation meausurements are frequency limited by the instrument response, the high frequency dielectric response obtained in these meausurements is inaccurate and often not available. Garg and Smyth [26] analyzed their data for propanol to dodecanol in terms of three different relaxation times for each alcohol. They explained the intermediate relaxation time as arising from rotation of a free monomeric molecule. The shortest relaxation time is that for the relaxation of the hydroxyl group by rotation around its C-O bond.

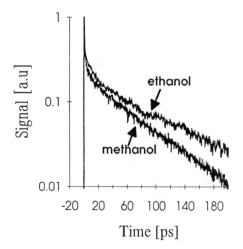

Fig. 5.   Degenerate four wave mixing signal intensities as a function of delay
           time between the pump pulses and the probe pulse for resonance
           excitation of LDS 750 in methanol and ethanol.

The short time components of the DFWM signals of LDS 750 in methanol and
acetonitrile solutions are shown in Fig. 7, using 10 fs time steps of the probe
beam delay stage. The DFWM signal for both solvents consists of an ultrashort
spike followed by a ~200 fs (acetonitrile) and a ~ 400 fs (methanol) decay. The
initial Gaussian shape spike is due to a contribution of two superimposed
components. A coherent contribution arises due to repumping of energy from the
pumping beams to the probe beam and is often found in DFWM experiments. The
coherent spike full width half maximum is determined by the laser pulse
correlation function and hence by the laser pulse width. The coherent spike
prevents us for the time being to resolve accurately the first ~150 fs of the
solvation dynamics.

The ultrafast solvation dynamics of LDS 750 in acetonitrile was studied by
Rosenthal et al. [3] using time resolved luminescence technique with ~125 fs fwhm
instrument response function. The solvation response consisted of two distinctive
parts. A fast initial decay accounted for ~ 80% of the amplitude was fit by a
Gaussian. The slower tail decayed exponentially with a decay time of 200 fs. In a
subsequent study, Cho et al. [4] meausured the time dependent non resonant optical
Kerr effect in neat acetonitrile liquid. Both experiments have shown the biphasic
character of the solvent response. A vibrational model was used to describe
quantitatively the solvation and the neat liquid dynamics[4]. A Brownian
oscillator with frequency distribution of the vibrational modes produce a very
good fit of both experimental data.

The short time scale dynamics of the DFWM signal in our experiments of LDS 750 in
acetonitrile is very similar to the experiments of the Chicago group [3-4]. The
shortest time component has a Gaussian shape (See Fig. 7) but can not be time
resolved since the coherent spike is superimposed on it. Also the pulse duration
in our experiment is longer than the predicted Gaussian component of the solvation

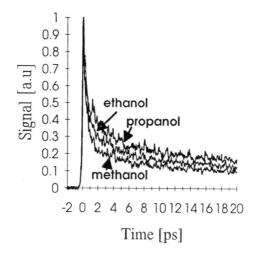

Time [ps]

Fig.6.   Time dependent DFWM signal for LDS 750 in methanol, ethanol and propanol.

dynamics. The second component of the solvation dynamics of LDS 750 in acetonitrile decays with a time constant of 200 fs as was found by the previous experiments.

We now wish to compare the solvation dynamics on the short time scale <1 ps of LDS 750 in methanol, ethanol and propanol. On this short time scale the solvation dynamics in all three solvents is quite similar. The relative height of the coherent spike superimposed on the Gaussian compound versus the subsequent fast exponential decay is the same in all three liquids. The decay time of the exponential component is ~400 fs in all liquids. This decay time is about twice as longer than in acetonitrile. It is interesting to note that while the long solvation component in these liquids is strongly dependent on the particular liquid, the ultrafast solvation dynamics is almost identical (within the S/N ratio of the experimental data).

As mentioned previously, the dielectric relaxation measurements indicated that in alcohols ultrafast dielectric relaxations of the order of ~2 ps exists. Our solvation dynamics experiments show that the time scale of the fastest solvation components are much shorter, it consists of a Gaussian component of ≤ 100 fs followed by a ~ 400 fs exponential decay. Molecular dynamics simulations of solvation dynamics in methanol [31] have shown that the solvation dynamics is biphasic. A Gaussian contribution with <100 fs is identified as arising from the inertial rotational motion of the solvent molecules. The second component is longer and approximately decays exponentially with 400 fs decay time. Classical molecular dynamics simulation of methanol [32] performed to longer times 10 ps than the solvation simulations [31] work shows the long solvation component of 5 ps (found in our experiment) as well as the short components. The ultrafast solvation components can be deduced from far infrared absorption measurements [33-34]. The Fourier transform of the far-infrared absorption line shape of neat acetonitrile and acetonitrile in other liquids [33] shows both the Gaussian and the exponential components of the solvent orientational correlation function.

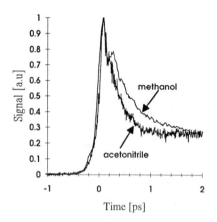

Fig.7.   The ultrashort time of the solvation response of LDS 750 in acetonitrile
         and methanol.

Discussion

In this work we developed theoretically and experimentally the principles of a new
method for the observation of ultrafast solvation dynamics: the resonance
transient grating spectroscopy.   Theoretical results reproduce the main properties
of the experimental curves.

The theory presented in the paper connects a four-photon signal with the
correlation function $S(t)$ that describes the fluctuations of the value $u_s(t)$ and
the transient Stokes shift of the luminescence spectrum ($\omega_\varphi(t)$). The analytical
form of $S(t)$ can be arbitrary in principal.   Its calculation is an independent
problem.

There are number of papers devoted to the calculation of $S(t)$.   There are computer
simulations [6,35]. $S(t)$ is also obtained from far-infrared spectroscopy [36].
$S(t)$ is obtained by Kubo's stochastic modulation theory [37], and, at last, the
modeling of $S(t)$ by a strongly overdamped Brownian oscillator [23,24] or by a
superposition of Brownian oscillators [4].   Concerning the methods of modeling
$S(t)$ [4,23-24,35-37], we shall note that the model of the strongly overdamped
Brownian oscillator [23,24] describes only the long time (exponential) behavior of
$S(t)$.   The use of a frequency distribution of the solvation modes in the form of
Brownian oscillators [4] describes both the fast and the slow components of $S(t)$.
However, such a description is rather characteristic for polar crystals than for
polar solvents, where the change of the polarization originates from the rotation
of the molecules.   The approach based on Kubo's stochastic theory [37] that
describes both the short and the long components, seems to us rather attractive.
However, an approach on the basis of the generalized Langevin equation [22,38] is
more consistent.   On the basis of this equation we proposed before the
non-Markovian model of an optically active oscillator for electron transitions in
molecules [22].   We shall use such an approach to describe the solvation
correlation function $S(t)$ elsewhere.

Acknowledgements

This work was supported by grants from the Ministry of Science and Technology, the James Franck Binational German-Israel Programme in Laser Matter Interaction. One of us (RR) gratefully acknowledges the financial support by the Minerva-Gessellschaft.

References

1. Kahlow, M.A., Jarzeba, W., DuBruil, T.P., and Barbara, P.F. (1988), Rev.Sci.Instrumen. **59**, 1098.
2. Maroncelli, M., and Fleming, G.R. (1988), J.Chem.Phys. **89**, 875.
3. Rosenthal,S.J., Xie,X., Du,M., and Fleming,G.R.(1991), J.Chem.Phys. **95**,4715.
4. Cho, M., Rosenthal, S.J., Scherer, N.F., Ziegler, L.D., and Fleming, J.R., (1992), J.Chem.Phys. **96**, 5033.
5. Bagchi, B., and Chandra, A. (1992), J.Chem.Phys. **97**, 5126.
6. Neria, E., and Nitzan, A. (1992), J.Chem.Phys. **96**, 5433.
7. Shen, Y.R. (1984) "The principles of nonlinear optics", Wiley, New York.
8. Ishida, Y., and Yajima, T. (1987). Rev.Phys.Appl. **22**, 1629.
9. Weiner,A.M., De Silvestri,S., and Ippen,E.P. (1985). J.Opt.Soc.Amer. **B2**, 654.
10. Fainberg, B.D. (1986). Opt.Spectry **60**, 120.
11. Fainberg, B.D. (1990). Opt.Spectry **68**, 305.
12. Goldberg, S.Y., Pines, D., Meltsin, A., Fainberg, B., and Huppert, D. (1993). Nonlinear Optics, **5** .
13. Loring, R.F., Yan, Y.J., Mukamel, S. (1987). J.Chem.Phys. **17**, 5840.
14. Fainberg, B.D. (1985). Opt.Spectry **58**, 323.
15. Fainberg, B.D., and Neporent, I.B. (1986). Opt.Spectry **61**, 31.
16. Fainberg, B.D., and Myakisheva, I.N. (1987). Sov.J.Quant.Electr. **17**, 1595.
17. Mukamel, S. (1985). Phys.Rev. **A28**, 3480; J.Phys.Chem. **89**, 1077.
18. Mazurenko, Yu. T. (1990). Opt.Spectry **48**, 388.
19. Fainberg, B.D., and Neporent, B.S. (1980). Opt.Spectry **48**, 393.
20. Lin, S.H. (1968). Theor.Chim.Acta **10**, 301.
21. Fainberg, B.D. (1980). Opt.Spectry **49**, 95.
22. Fainberg, B.D. (1987). Opt.Spektrosk **63**, 738 [Opt. Spectrosc. (USSR) **63** (1987) 436].
23. Yan, Y.J., and Mukamel, S. (1988). J.Chem.Phys. **89**, 5160.
24. Yan Y.J., and Mukamel, S. (1990). Phys.Rev. **A41**, 6485.
25. Castner, E.W., Maroncelli, M., and Fleming, G.R. (1987). J.Chem.Phys. **86**, 1090-1097
26. Garg, S.K., and Smyth, C.P. (1965). J.Phys.Chem. **69**, 1294.
27. Bertolini, D., Casettari, M., and Salvetti, G. (1983). J.Chem.Phys. **78**, 365.
28. Kosower, E.M., and Huppert, D. (1983). Chem.Phys.Letters. **96**, 483.
29. Kosower, E.M., and Huppert, D. (1986). Annual Reviews of Physical chemistry **37**, pp.122-156.
30. Barbara, P.F., and Jarzeba, W. (1990). Adv.Photochem. **15**, 1.
31. Fonseca, T., and Ladanyi, B.M. (1991). J.Phys.Chem. **95**, 2116-2119.
32. Bultmann, T., Kemeter, K., and Rusbuldt, Ch. (1993). Ph. A. Bopp and N.P. Ernstig in "Reaction Dynamics in Clusters and Condensed Phases", the 26th Jerusalem Symposium in Quantum Chemistry and Biochemistry, B. Pullman and J. Jortner, eds. Kluwer Academic. Publishers: Dordrecht (in press).

33. Van Aalst, P.M., Van der Elsken, J., Frenkel, D., and Wegdam, G.W. (1972). Faraday Disc.Chem.Soc. **6**, 94.
34. Rothschild, W.G. "Dynamics of Molecular Liquids",(Wiley-Interscience, New York, 1984).
35. Perera, L., and Berkowitz, M. (1992). J.Chem.Phys. **96**, 3092.
36. Roy, S., and Bagchi, B., in press.
37. Rips, I., and Jortner, J. (1987). J.Chem.Phys. **87**, 2090.
38. Mori, H. (1965). Progr.Theor.Phys. **33**, 423; Kubo, R.(1966). Rep.Progr.Phys. **29**, 255; Adelman, S.A., and Doll, J.D. (1976). J.Chem.Phys. **69**, 2375.

FEMTO-SECOND PHOTODISSOCIATION OF TRIIODIDE IN SOLUTION

U. Banin and S. Ruhman

the Department of Physical Chemistry and
the Farkas center for light-induced processes
the Hebrew University, Jerusalem 91904 Israel.

ABSTRACT. The photodissociation of triiodide in polar solvents is followed directly by femtosecond spectroscopy. In both ethanol and water solutions, diiodide fragment ions are formed within 300 fsec in a partially coherent vibrational state, as is discerned from modulations in the transient transmittance measurements. Substantial solvent effects are observed, regarding different dynamical processes. The primary geminate recombination yields and vibrational relaxation times on ground state triiodide are affected strongly by solvent. TRISRS (Transient Resonance Impulsive Stimulated Raman Scattering) is used for the first time to directly follow the vibrational relaxation stage of the diioidide fragment in ethanol solution. The feasibility of the TRISRS scheme is demonstrated and the results of this experiment corroborate the transient transmission measurements. From these complimentary experiments an ultrafast vibrational relaxation time on the order of a few picoseconds is deduced for nascent diiodide ions.

1. Introduction:

The technological advancement in ultra-short spectroscopy, has provided a new handle to study contiguously the detailed dynamics of chemical reactions in gas, solid and solution[1]. Femtosecond temporal resolution, enables direct following of the full course of a chemical reaction[2], and this is most valuable especially in the liquid phase where the solvent inherently induces rapid dephasing and relaxation processes. This causes quick loss of memory limiting the applicability of alternative methods in the frequency domain where information on the reactive stages is deduced after the fact.

Aside from the inherent short temporal resolution of femtosecond experiments, a new situation arises. The pulse length may be substantially shorter than vibrational periods of molecular modes. The interaction of the ultra short pulse with the molecular system can then create coherence in these modes in a process termed Impulsive Stimulated Raman Scattering (ISRS) when excitation is off resonance[3], or Resonant Impulsive Stimulated Raman Scattering (RISRS) when excitation is on resonance[4, 5]. Such coherence has been observed in transient absorption, emission and scattering, for

245

J. Jortner et al. (eds.), Reaction Dynamics in Clusters and Condensed Phases, 245–259.
© 1994 Kluwer Academic Publishers. Printed in the Netherlands.

many molecular systems in different environments[6, 7]. The RISRS process has also been studied theoretically by a perturbative approach in a density matrix formalism notably by Mathies and coworkers[8], and in a non perturbative approach and wavepacket formalism by Ruhman et. al[9]. When the excited state potential is fully dissociative, the coherence is induced exclusively in the ground state, and thus grossly speaking, Raman equivalent information is obtained on ground state vibrational modes.

In this paper we discuss the detailed dynamics of the photodissociation process of I_3^- in solution[10, 11];

$$I_3^- + h\nu \quad \text{-------->} \quad I_2^- + I \tag{1}$$

We have studied this reaction in different polar solvents using femtosecond time resolved transient absorption spectroscopy. Compared to a diatomic dissociation, both the richness and complexity of the above triatomic process is enhanced[12]. In addition to possible geminate recombination back to the ground state, excess energy from the excitation stage may be deposited in product degrees of freedom. This will be followed by relaxation processes involving the surrounding solvent molecules. Since this case falls within the strong coupling regime of solute - solvent interactions[13], substantial solvent effects on the observed dynamics are expected.

We present results of a new experimental method - TRISRS (Transient Resonant Impulsive Stimulated Raman Scattering), designed to directly follow the vibrational relaxation stage of the diiodide fragment ions formed in reaction 1. From the transient transmission results in ethanol, we have deduced that the vibrational relaxation of the diiodide ion is extremely fast - on the order of ~4 psec. This is indicated by a narrowing of the I_2^- IR absorption band[11]. The TRISRS experiment is designed to corroborate this assignment and to gain additional information on this very rapid vibrational relaxation. A three pulse sequence is involved in TRISRS: Pump - Push - Probe. The pump pulse, initiates the photdissociation process as in regular Pump-Probe spectroscopy. At a certain delay with respect to the photolysis, which can be varied in different experiments, the 'push' pulse sets in motion ground state coherence. This coherence is followed by measuring the modulations in the transient transmission by a variably delayed probe pulse. This free induction decay, carries information on the incoherent vibrational distribution at the instant of the push interaction.

The outline of the article is as following; In section 2 we present the experimental setup concentrating on the TRISRS experiment. In the following section, the main results for transient transmission measurements in ethanol are outlined, along with results in other solvents. Then the results of the TRISRS experiment are shown. In section 4, a discussion including the tentative interpretation of the transient transmission results is followed by an analysis of TRISRS results. Concluding remarks are brought in section 5.

2. Experimental:

The laser system has been described in detail elsewhere. Briefly, an anti-resonant ring synch-pumped dye laser constructed in lab is pumped by a mode locked Nd;YLF laser. The output is amplified in a dye amplifier at a repetition rate of 1 kHz. The dye cells are pumped by the doubled output of a Nd;YLF regenerative amplifier. The ultimate output consists of 65 fsec pulses, centered at 616 nm, containing 20-30 µJ of energy. In regular pump-probe experiments, One portion of this output is frequency doubled to produce the pump pulse at 308 nm. The probe pulse is obtained from a second portion, and can be continuously delayed relative to pump via a motorized translation stage. Spectrally filtered segments of white light continuum formed in the probe arm allow measurement of the transient transmission signal at different central frequencies. Another option used is probing with a secondary doubled pulse at 308 nm. The pump and probe pulses are overlapped in the flowing sample (flowing jet or quartz cell, path length of 200 µ) and the experimental signal is obtained by differential measurement of probe intensity in two photo-diodes, before and after the sample.

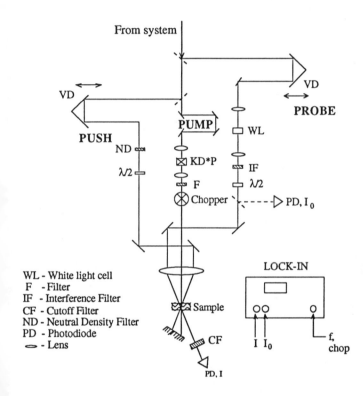

Figure 1: TRISRS experimental setup. The push is at 616 nm and can be variably delayed relative to the pump (at 308 nm). The three pulses are overlapped in the flowing sample. See text for details.

WL - White light cell
F - Filter
IF - Interference Filter
CF - Cutoff Filter
ND - Neutral Density Filter
PD - Photodiode
⬥ - Lens

The TRISRS experimental setup is shown schematically in figure 1. In addition to the pump and probe pulses, a third portion to comprise the "push" pulse at 616 nm is split off the pump arm. The push time can be variably delayed relative to pump by an additional motorized translation stage. Parallel polarization configuration was used to maximize the experimental signal. A chopper was introduced in the pump arm and the differential transmittance of the probe was detected by lock-in amplification (SR530), at the chop frequency. Samples of triiodide in all solvents studied, were prepared from I_2 and KI (Highest available purity). The solvents were GR grade, and the I_3^- concentrations were determined spectrophotometrically to be 1 - 5 mM.

3. Results:

3.1: Transient transmission results in ethanol and other solvents:

In figure 2, short time transient transmission results are presented at probe wavelengths within the I_2^- IR absorption band. In both solvents presented (water and ethanol), the overall appearance is similar. At time zero there is an immediate absorption, the rise of which is limited by our experimental temporal resolution. This is followed by a rapid fall off in the absorbance, on a time scale of 200 - 300 fsec. We wish to draw attention to the appearance of weak modulations superimposed on the overall transient transmission, appearing at about 400 fsec in both solvents. As the probe wavelength is changed from the blue side of λ_{max} of the I_2^- IR band (c.a. 620 nm), to a wavelength to the red of λ_{max} (c.a. 880 nm) - the phase of the modulations is shifted by 1π radians. In ethanol a fitting procedure was used to determine the frequency of the modulations to be ~95 cm^{-1}, and the decay time on the order of ~400 fsec.

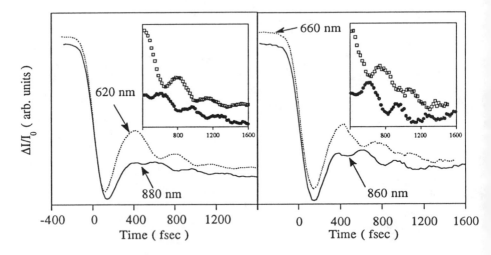

Figure 2: Transient Transmission measurements in the red at two probe wavelengths. The solvent is **ethanol** on the left and **water** on the right. In the insert, the region of modulations in the signal is enlarged.

The same overall features in this time scale are observed in other probe wavelengths in the region between 550 - 950 nm. The behavior at longer times and in additional wavelengths in the red and near IR is presented in figure 3 (ethanol as solvent). All scans were conducted in 'magic angle' relative polarization of pump and probe, thus eliminating any reorientational contributions in the transient transmittance measurements. The features observed constitute an overall narrowing of the absorption in the IR band, on a time scale of ~4 psec. At probe wavelengths at both sides of the absorption band - there is a gradual decay of the absorption (for example - 620 and 950 nm), and in the same time there is a growing in of absorption in wavelengths at the center of the band (760 nm).

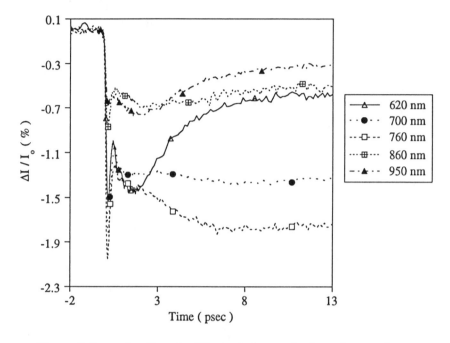

Figure 3: Long time Transient Transmission results for red spectral region. The solvent is ethanol. Note the overall narrowing of the absorption.

When the probe wavelength is tuned into the blue spectral region, the immediate absorption observed at time zero persists, but other features change. Modulations are superimposed on the overall transient transmission at short times, these oscillations initiate at time zero and become more pronounced when the probe is tuned into the region where I_3^- absorption is strong. Presented in figure 4 are short time scans with probing at 308 nm, the pump frequency. Results for water and ethanol are shown for comparison. Superimposed on the observed bleach, there are strong modulations. The frequency of the modulations in both solvents is ~110 cm^{-1}, but the characteristic decay time changes substantially from ~1.5 psec in ethanol, to ~0.7 psec in water.

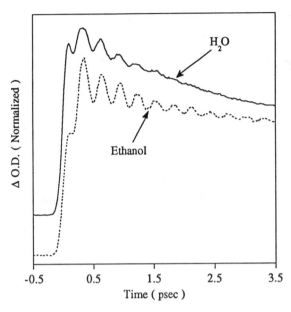

Figure 4: Short time results for 308 nm transient transmission. Pump and probe polarizations are parallel. The oscillation decay time is substantially shorter for water.

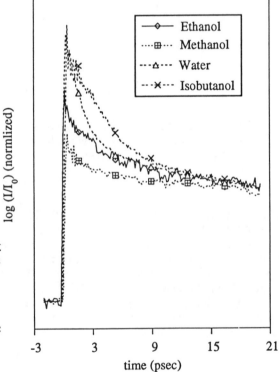

Figure 5: 308 nm transient transmission (magic angle relative pump-probe polarization). Shown are results for 3 alcohols and water. The scans were normalized from ~15 psec onwards.

This feature is only one imprint of solvent effects that are observed. The transient transmission at 308 nm, in longer times and in additional solvents is presented in figure 5. Modulations in short times are observed for all solvents, but we focus now on the overall bleach decay at 308 nm. This decay proceeds notably on two time scales, A fast component with a characteristic decay time of less then 10 psec depending on the solvent studied, and a much slower and less pronounced decay proceeding on a ~100 psec time scale. In a series of alcohol solvents; methanol - ethanol - propanol - isobutanol, there is a systematic behavior. The relative contribution of the fast decay component gradually rises as the alcohol becomes heavier. An interesting behavior is observed for water as solvent, the fast component contribution is pronounced and comparable to that observed for the heaviest alcohol. The decay time of this component is extremely rapid - on the order of 2 - 3 psec.

3.2) Trisrs results in ethanol:

The sequence of events for a typical TRISRS experiment is demonstrated in figure 6; presented, is a comparison of the transient transmission signal for 840 nm probe, with and without the 'push' pulse. The push delay in the case presented is 4 psec relative to pump. At this time - a bleach on the order of ~ 0.15 of the overall absorption signal is observed. At 13 picoseconds the bleach has not fully recovered back to the value observed without the push pulse.

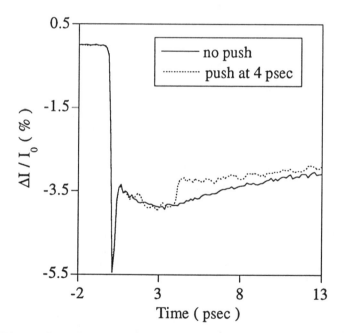

Figure 6: Transient transmission data at 840 nm with and without the secondary TRISRS push pulse at a delay of 4 psec. The solvent is ethanol.

The longer time features of the signal are clearly interesting on their own right, but here we focus on the coherent part. Results for short time probing with various push delays are shown in figure 7. The probe wavelength is 680 nm at a push delay of 2 psec, and 840 nm in the later delays. In all cases, a bleach is observed at time zero (zero of time taken at center of push pulse). Superimposed upon the decaying bleach, periodic modulations of the absorption are discerned. The differences of the oscillatory feature between push delays of 2, 2.7 and 4 psec is evident even without detailed numerical analyses of the results. Measurements at longer push delays show that the coherent part of the data does not change.

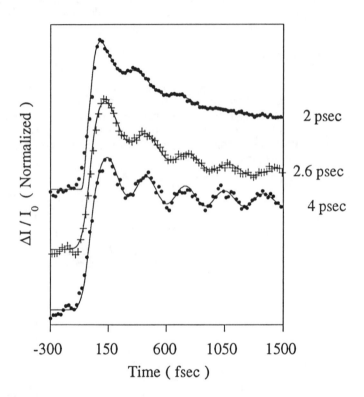

Figure 7: TRISRS data recorded at three push delays. The experimental points are presented along with the best fit (solid lines) after convolution with the experimental response. See text for details.

Along with the data, we show fits obtained by a non linear least squares fitting procedure. The system's Greens function G(t) is convoluted with the instrument response according to the following equation;

$$S(t) = \int_{-\infty}^{\infty} \int_{-\infty}^{t'} I_{push}(t'') \, G(t' - t'') \, I_{probe}(t' - t) \, dt' \, dt'' \qquad \text{eq. 2}$$

The push and probe pulses were taken as squared hyperbolic secant functions of 65 fsec FWHM, and the Green's function is given in equation 3;

$$G(t) = A + B \exp(- \Gamma t) + C \sin (\omega t + \phi) \exp (- t / \tau) \qquad \text{eq. 3}$$

The physically important parameters obtained from this procedure are given in the following table:

PUSH DELAY (psec)	ω (cm^{-1})	τ (psec)	ϕ (radians)
2.0	102 +/- 4	0.5 +/- .2	0.2 +/- 0.4
2.67	105 +/- 3	0.9 +/- .2	-1.2 +/- 0.4
4.0	112 +/- 2	1.2 +/- .3	-2.0 +/- 0.4

As the push delay rises from 2 psec on to 2.7 and 4 psec, it is seen that both the frequency and decay times of the oscillatory part gradually rise. After 4 psec, no change is discerned within the experimental error.

4) Discussion:

4.1) Interpretation of transient transmittance results:

The I_2^- fragment is characterized by an absorption band in the red and near IR where I_3^- ground state absorption is negligible. We thus start sorting out the results in this spectral region. A fuller account has been presented elsewhere[11] and here only a summary of the earlier reasoning is presented, along with a discussion of the additional results regarding different solvent effects. The initial absorption upon excitation is associated to the excited state accessed via the optical electronic transition. It must be stressed that this is obviously a population in transition, and during the finite excitation pulse it may have undergone considerable rearrangement on the excited state potential. The following decay in the absorption is associated with the bond cleavage stage, which occurs on a time scale of ~300 fsec. This assignment was corroborated by quantum mechanical simulations of the collinear triiodide dissociation process in gas phase[14].

Following this, there is the initiation of the anti - phased oscillations in the transient transmittance. We have attributed this to coherent vibrational motion in the nascent diiodide fragment ions. This assignment is based on the following points: (a) The oscillations initiate at ~300 fsec after excitation and are not retractable back to the zero of time. This should have been the case if the oscillations were a result of a RISRS type process, in the ground state or a bound excited state of triiodide. (b) Via photoselection experiments, very different reorientation times were measured for the 620 nm absorption (~5 psec), and the 308 nm transients (~15 psec). These substantial differences can be comprehended if the absorption at 620 nm is assigned to the product state that can indeed have a much shorter reorientation time in comparison to the ground or a bound excited state of triiodide. (c) Perhaps the most conclusive evidence for the above assignment is the anti phase observed in the oscillations upon changing the probe frequency. For diiodide, the excited state reached by probing is dissociative. We may now envision wavepacket motion in the ground state of diiodide; when the wavepacket is at the bound side of the potential, absorption will be maximal in the blue side of λ_{max}, simultaneously there is minimum absorption in the red side of λ_{max}. After a half of a vibrational period the wavepacket moves to the free side and then absorption is maximum in the red and minimum in the blue. This leads to a phase shift of 1 π radians as is indeed observed experimentally. Such phenomena of coherence being carried on to a product state was previously observed by Zewail and co workers in the photodissociation of IHgI in the gas phase[15]. Here we see evidence for such a process in solution, and the observation persists in several solvents. Recent MD calculations show such phenomena to be generally possible also in barrier descent dynamics[16].

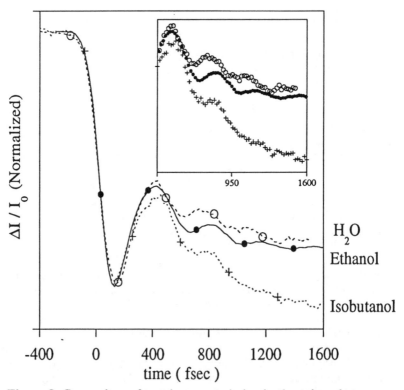

Figure 8: Comparison of transient transmission in short times between 3 solvents. The probe wavelength is 660 nm for water and isobutanol, and 620 nm for ethanol. The insert shows the oscillations resulting from coherence in product diiodide ions on an expanded scale.

It is instructive to compare more closely the short time results in the red observed in the two solvents presented earlier - water and ethanol, along with isobutanol (figure 8). All three stages discussed above are apparently seen in all solvents. The stage we have assigned to the fragment recoil is almost in perfect overlap. This is consistent with the picture obtained by MD calculations for a growing number of molecular systems[12,]. The very early dynamics of fragment recoil is dominated by the repulsive potential and the solvent hardly effects this stage. Thus the change of solvent is not expected to interfere in this process. In these MD calculations the solvent was a rare gas with only lennard - jones pair wise interactions between solvent and solute. In polar solvents the interaction consists also of long range coulombic forces. It is reasonable however to expect even in this 'strong coupling' case[13] that the recoil stage will be dominated by the repulsive potential as in gas phase. On the other hand, a closer inspection of the oscillatory part of the data assigned to coherence in the product state, shows some differences between the three solvents: For water the frequency of oscillations is lower, this implies that the coherent product ensemble formed in water emerges with less energy in the vibration in comparison to the situation in ethanol (assuming that ω_e and $\omega_e\chi_e$ do not differ substantially between these three solvents). In isobutanol the frequency is even higher then in ethanol. The dephasing times also differ. At present we cannot account for this behavior, but it proves the relevance and importance of this observable measurable exclusively in the time domain, for understanding the role of solvent in the dynamics.

From the frequency of the oscillations and published ω_e, $\omega_e\chi_e$[17, 18]; the excess energy in the coherent part of the I_2^- population in ethanol is estimated to be on the order of ~2000 cm^{-1}. This picture, is consistent conceptually with the Resonance Raman spectra for triiodide in solution which shows a long progression of the symmetric stretch vibration[19]. This implies initial motion on the excited state along the symmetric stretch coordinate, leading to possible vibrational excitation in the fragment. Quantum simulations on this system in gas phase, indeed show that the I_2^- population emerges with substantial vibrational excitation[14]. Classical MD simulations by Benjamin et. al.[20] including the solvent substantiates the above picture. The main conclusions are that the excess energy in both vibration and translation in the fragment relax on picoseconds timescale. Coherence in the product is also observable through anti-phased oscillations in the simulated absorption.

Probing in 308 nm in contrast to the visible, allows us to concentrate on the parent ion population. The oscillations observed are a signature of coherent excitation in the ground state symmetric stretch vibration of I_3^-. This is an example of a RISRS type process which can be visualized as following[21]: The impulsive pump pulse preferentially depletes probability density from populated ground state geometries that meet the Franck resonance condition. This 'coherent depletion' and impulsive transfer of momentum involves all ground state populated vibrational levels which carry oscillator strength within the pump pulse spectrum, and will result in coherent redistribution of population into a number of levels above and below each such vibronic state. Following this interaction, the coherent superposition vibrates and dephases, giving rise to time dependent decaying spectral modulations in the transient absorption of a variably delayed weak probing pulse of similar duration. As is observed, changing the solvent from ethanol to water shortens the dephasing time considerably by a factor of nearly 2.

Equivalent information can be obtained from line shape analysis of the Raman or Resonance Raman spectra[22], but here it is demonstrated that the detailed information can be extracted also directly in the time domain. The above description lays the foundation for the idea behind the TRISRS scheme. By using RISRS in a Transient way, one can obtain Raman equivalent information for rapidly evolving species in the fastest conceivable way.

The other features and the solvent effects at 308 nm can be explained as following: The fast component is assigned to primary geminate recombination followed by rapid vibrational relaxation of ground state I_3^-. Thus, in the series of alcoholic solvents the relative contribution of this component becomes more significant as the solvent mass increases, providing more efficient caging. Similar dependence was observed by Barbara and coworkers[23] for I_2^- photodissociation. The increase in mass for the series of alcoholic solvents is accompanied by a reduction of polarity, it seems though that the caging efficiency is determined in this case by the solvent mass rather than the long range attractive interactions with the solute. For the study of iodine photodissociation it was indeed determined that the geminate recombination process occurs very early in the dynamics[24], still in the region which is mostly governed by the kinematics. A single or very few collisions with solvent relax the translational energy rapidly, enabling the geminate recombination process to occur via curve crossings at the asymptote of the dissociative potential. Thus it is indeed reasonable that the mass effect will be more important then polarity in this stage. The observed time scale for the decay of the fast component is determined by the vibrational relaxation time on the triiodide ground state. The tightly bound local structure of water also causes very efficient caging and the vibrational relaxation time that is deduced from the decay is extremely fast. A fuller study of the above process is currently underway.

4.2) Interpretation of TRISRS results:

The inherent lack of fine structure in liquid phase electronic absorption, especially when the species are vibrationally (or electronically) excited, makes it advantageous to supplement transient transmission measurements with transient Raman spectroscopy[25]. The vibrational relaxation stage of the fragment diioidide ions tenders a special challenge in this respect. First, the observed narrowing of the IR band may result from processes other then vibrational relaxation, for example solvation effects may also change the spectral appearance. Second, the vibrational relaxation is a multilevel process and one may hope that vibrational spectral information will enable to obtain a more detailed picture of this complex process. The overall relaxation time deduced from the transient spectra is on the order of 4 psec only about 10 to 15 times the vibrational period in this case. The above considerations and the ultrafast relaxation time are the motivation for the TRISRS experiment that can serve to corroborate the above assignment.

The push pulse is centered at 616 nm, this frequency lies within the I_2^- IR absorption band and is well separated from triiodide ground state absorption. Eigenstates above v ~ 2 of dioidide have significant oscillator strength at this wavelength. The low vibrational frequency of diiodide enables RISRS excitation of ground state coherence by the push pulse. At all push delays studied, oscillations in the transient transmission are indeed observed. The oscillations initiate at time zero as is expected from a RISRS type excitation. The frequency of oscillations reaches an asymptotic value of ~112 cm^{-1}

within 4 psec push delay. This value equals the published diiodide vibrational frequency[11] within the experimental error. In light of the above reasoning, we can unequivocally attribute the observed oscillations to ground state TRISRS excited vibrational coherences in the I_2^- fragments. The oscillation phase and its change with the variation of probe frequency agree qualitatively with this assignment. At a push delay of 2 psec both push and probe pulses are to the blue of the center of the I_2^- absorption band, and the oscillatory contribution to the transmission should resemble a cosine, whereas in the later delays, a negative cosine contribution would be expected. While a full change of 1π in phase is not observed upon the variation of probe frequency, and the initial phase is not exact, the trend of change is correct. Considering the uncertainty in our determination of time zero, and the possibility of small contributions of self phase modulation coherent artifacts, this may be within our margin of error. The question of the exact phase change that is to be expected will be addressed theoretically in a future publication[21].

Within this conceptual framework, interpretation of the TRISRS results is straight forward. The first push delay immediately supersedes the decay of spectral oscillations due to bond fission, and the fragment distribution must be broad, involving highly excited vibrational levels. It may however have undergone substantial cooling since its inception about 200 fsec after the UV excitation, when the mean vibrational quantum level of the coherent vibrations are experimentally[11] and theoretically[20] determined to be at $v \approx 20$. Using the documented vibrational constants of I_2^-, the coherence frequency of 102-103 cm^{-1} indicates that the earliest TRISRS induced superposition is made up of levels centered around $v \approx 13$. The ultrafast rate of coherence dephasing at this push delay cannot be due to anharmonicity alone, and must involve substantial homogeneous contributions such as pure dephasing. Inhomogeneity of the solvent following the violent act of bond fission may also contribute to rapid vibrational dephasing.

As the push delay is increased, the population of fragment ions has relaxed further, leading to a reduction of both the first and second moments of the vibrational distribution. Accordingly, the frequency and the timescale of RISRS coherence dephasing should increase. The frequency will increase due to the larger level spacings closer to the bottom of the potential well. The dephasing is prolonged because of a reduction in the rate of pure dephasing at low v levels[26], and a reduced contribution of anharmonicity which is proportional to $\omega_e\chi_e$ times the second moment of the distribution. These trends are in fact observed. Within 4 psec of push delay, the modulations have already reached the asymptotic vibrational frequency, within our experimental error. The accumulating evidence from this experiment, along with the previous transient transmission data, show clearly that the highly vibrationally excited population of diiodide ions does in fact lose most of its excess vibrational energy within a few picoseconds. This finding agrees with a growing body of work on vibrational relaxation of molecular ions in polar solvents[27, 28, 29]. Despite the fact that at longer probe delays reorientation will influence these all parallel pulse polarization experiments, this effect will not qualitatively influence these trends, and the quantitative contribution to the dephasing times is minor.

A closer examination of the transient transmission results, shows that after 4 picoseconds there is still spectral evolution of the diiodide IR band, while the coherence observed in the TRISRS experiment does not change. This demonstrates that the above method is in this case complementary to the transient transmission measurements, being more sensitive to highly vibrationally excited populations. For such populations, the absorption spectra is very wide but both frequency and dephasing times of the coherence differ substantially from the thermal values. On the other hand, For the final relaxation stages, the coherence created is less sensitive to the small changes, and then transient transmission can be used.

5) Concluding Remarks:

In this study we have used femtosecond spectroscopy to directly follow the detailed course of a chemical reaction in solution. Information relevant to the effect of solvation is deduced, and substantial solvent effects are observed. Primary geminate recombination yields, vibrational relaxation times, coherence transfer to products, dephasing times, are amongst the observables that are found to be solvent dependent. TRISRS has been demonstrated as a method to study the vibrational evolution of short lived species. It is used for the first time to follow the vibrational relaxation of the diiodide fragments, and the results fully corroborate the transient transmission measurements. The TRISRS scheme has the potential to evolve into a valuable method to obtain Raman equivalent spectral information in the fastest conceivable way. In the above case it is uniquely capable of obtaining this information.

Acknowledgments: We wish to thank Prof. R. Kosloff for helpful discussions. A. Waldman and Dr. E. Mastov are thanked for technical assistance. This work was supported by the Israel Science Foundation.

References:

1) Fleming G. R. , Chemical Applications of Ultrafast Spectroscopy, Oxford University Press (1986).

2) Zewail A. H., Faraday Discuss. Chem. Soc., **91**, 207 (1991).

3) Yan, Y. -X.; Cheng, L.T.; Nelson, K. A. in Advances in Nonlinear Sepecroscopy, R. G. H. Clarke, R. E. Hester ed., Wiley, New York (1987).

4) Chesnoy J. and Mokhtari A., Phys. Rev. A **38**, 3566 (1988).

5) Walmsley I. A.,Wise F. W. and Tang C. L., Chem. Phys. Lett. **154**, 315 (1989).

6) Dexheimer S. L., Wang Q., Peteanu L. A., Pollard W. T., Mathies R. A., and Shank C. V., Chem. Phys. Lett. **188**, 61 (1992).

7) Weiner A. M., Leaird D. E., Wiederrecht G. P. and K. A. Nelson, Science **247**, 1317 (1990).

8) Pollard W. T., and Mathies R. A., Annu. Rev. Phys. Chem., **43**, 497 (1992).

9) Hartke B., Kosloff R. and Ruhman S., Chem. Phys. Lett. **158**, 238 (1988).

10) Banin U., Waldman A., and Ruhman S., J. Chem. Phys. **96**, 2416 (1993).

11) Banin U., and Ruhman S., J. Chem. Phys. **98**, 4391 (1993).

12) Benjamin I. and Wilson K.R., J. Chem. Phys. **90**, 4176 (1989).

13) Gertner B. J., Whitnell R. M., Wilson K. R., and Hynes J. T., J. Am. Chem. Soc.

113, 74 (1991).

14) Banin U., Kosloff R. and Ruhman S., Isr. J. Chem., in press.

15) Dantus M., Bowman R. M., Gruebele M., and Zewail A. H., J. Chem. Phys. **91**, 7437 (1989).

16) Ben-Nun M., Levine R.D., Chem. Phys. Lett. **203**, 450 (1993).

17) Tripathi G. N. R., Schuler R. H. and R. W. Fessenden, Chem. Phys. Lett. **113**, 563 (1985).

18) Chen E. C. M. and Wentworth W. E., J. Phys. Chem. **89**, 4099 (1985).

19) a)Kaya K., Mikami N. and Ito M., Chem. Phys. Lett. **16**, 151 (1972).
b) Keifer W. and Bernstein H. J., Chem. Phys. Lett. **16**, 5 (1972).

20) Benjamin I., Banin U. and Ruhman S., J. Chem. Phys., in press.

21) Banin U., Bartana A., Kosloff R. and Ruhman S., in preparation.

22) Rothschild W. G., Dynamics of molecular liquids, Wiley, New York (1984).

23) Johnson A. E., Levinger N. E. and P. F. Barbara, J. Phys. Chem. **96**, 7841 (1992).

24) Harris A. L., Brown J. K., and Harris C. B., Ann. Rev. Phys. Chem. **39**, 341 (1988).

25) Xu X., Yu S. C., Lingle R., Zhu H. and Hopkins J.B., J. Chem. Phys. **95**, 2445 (1991).

26) Kosloff R., and Rice S. A., J. Chem. Phys. ,**72**, 4591 (1980).

27) Kliner D. A. V., Alfano J. C. and Barbara P. F., J. Chem. Phys. **98**, 5375 (1993).

28) Li M., Owrutsky J., Sarisky M., Culver J. P., Yodh A. and Hochstrasser R. M., J. Chem. Phys. **98**, 5499 (1993).

29) Whitnell R. M., Wilson K.R. and Hynes J.T., J. Phys. Chem. **96**, 5354 (1992).

SOLVENT POLARITY EFFECTS ON CIS-STILBENE PHOTOCHEMISTRY FROM THE FIRST FEMTOSECONDS TO THE FINAL PRODUCTS

Anne B. Myers,* Jon-Marc Rodier,* and David L. Phillips[#]
*Department of Chemistry, University of Rochester, Rochester, NY 14627-0216
#Department of Chemistry, University of Hong Kong, Pokfulam Road, Hong Kong

1. Abstract

The ultrafast torsional isomerization and ring closure reactions of photoexcited *cis*-stilbene have been examined in hexane, cyclohexane, methanol, and acetonitrile solvents through a combination of ground state resonance Raman intensities, two-color uv picosecond anti-Stokes resonance Raman scattering, and product quantum yield measurements. The quantum yields for ring closure to ground state dihydrophenanthrene are factors of two to three higher in the nonpolar solvents than in the polar ones, while the yield of *trans*-stilbene is slightly higher in the polar solvents. Vibrationally hot ground-state *trans*-stilbene is formed within the 10 ps time resolution of the time-resolved Raman experiments in both solvents, but the subsequent vibrational cooling is about a factor of two faster in methanol than in cyclohexane. The resonance Raman spectra of the *cis*-stilbene ground state exhibit slightly greater relative intensities in modes involving primarily out-of-plane and/or hydrogen wagging or rocking motions in the polar solvents than in the hydrocarbons. These observations, together with the femtosecond transient absorption and fluorescence data from other groups, indicate that increasing solvent polarity causes vertically excited *cis*-stilbene to distort more rapidly along the torsional isomerization coordinate, shortening the excited state lifetime and decreasing the quantum yield for the competing ring closure reaction.

2. Introduction

The photoinduced interconversions of the stilbene isomers (Figure 1) are among the most widely studied of all photochemical reactions.[1] Virtually every existing technique for exploring molecular dynamics, energetics, and structure has been applied, with at least some degree of success, to the stilbene system. While *trans*-stilbene undergoes only one important primary photoreaction, the *trans-cis* isomerization, *cis*-stilbene undergoes not only *cis-trans* photoisomerization but also a photochemical ring closure reaction to form 4a,4b-dihydrophenanthrene (DHP). Studies using traditional photochemical methods established the identity of the major products and the dominant excited singlet state pathway for the photoreactions of both stilbene isomers.[2-8] The photoisomerization of *trans*-stilbene was also one of the earliest reactions to be studied through picosecond time-resolved absorption and fluorescence techniques, and dependence of the isomerization rates on solvent viscosity and polarity have been probed in detail in this way.[1,9-24] The structure and dynamics of the S_1 excited state of *trans*-stilbene have also been examined via pump-probe resonance Raman techniques.[25-30] Due to the very short lifetime of *cis*-stilbene's excited state, direct time-resolved studies of its photochemical dynamics have only recently become feasible, but by now a number of such experiments utilizing both transient ab-

261

J. Jortner et al. (eds.), Reaction Dynamics in Clusters and Condensed Phases, 261–278.
© 1994 *Kluwer Academic Publishers. Printed in the Netherlands.*

sorption and time-resolved fluorescence techniques have been reported.[31-42] Ground-state resonance Raman intensities,[43] steady-state fluorescence,[44,45] and spectra in supersonic jet expansions[46-48] are other techniques that have shed light on the dynamics of *cis*-stilbene's ultrafast
photochemistry.

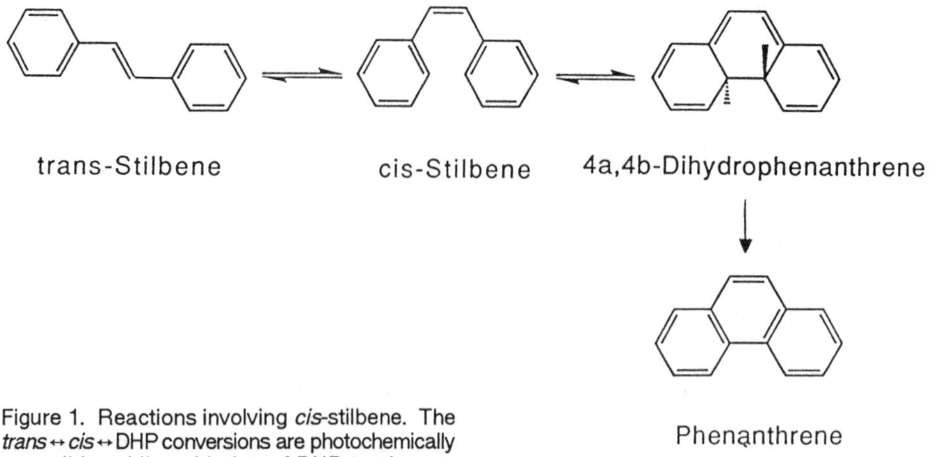

Figure 1. Reactions involving *cis*-stilbene. The
trans ↔ *cis* ↔ DHP conversions are photochemically
reversible, while oxidation of DHP to phenanthrene is essentially irreversible.

Both *trans*- and *cis*-stilbene exhibit a significant solvent viscosity effect on their excited state
lifetimes, attributable to the hindering effect of environmental friction on the large-amplitude torsional isomerization. *Trans*-stilbene also exhibits generally shorter excited state lifetimes in polar
solvents than in nonpolar ones having comparable viscosities, and attempts to extract the barrier
to excited-state isomerization from temperature and viscosity dependences have generally concluded that the barrier is greatly reduced (relative to its value in the isolated molecule or in nonpolar solvents) or even eliminated in polar solvents. This is consistent with the idea that the $90°$
twisted excited state, presumed to be a global minimum on the S_1 surface, is highly polarizable
and/or dipolar (the "sudden polarization" effect)[49] and is thus stabilized by polar solvents, although there are complicating issues having to do with the time scale on which the solvent repo-
larization can occur.[1] *Cis*-stilbene also exhibits a shortening of its excited state lifetime in polar
solvents compared with hydrocarbons of similar viscosity. However, in the *cis* isomer the situation is complicated by the presence of two important reaction channels. Does solvent polarity affect only the torsional isomerization channel, or the ring closure pathway as well? We have chosen to examine the solvent dependence of *cis*-stilbene's photochemistry through three distinct experimental approaches which, together, are sensitive to the dynamics on time scales from a few
femtoseconds to hours.

First, we employ steady-state resonance Raman spectroscopy on the *cis*-stilbene ground state
as a function of solvent and interpret the intensities within the context of the time-dependent
wavepacket picture, which shows that resonance Raman intensities for large molecules in solution reflect the initial dynamics of excited-state motion along specific ground-state normal coordinates.[43,50-52] A quantitative analysis of *cis*-stilbene's excited state dynamics in one solvent (cyclohexane) has already been reported.[43] In view of *cis*-stilbene's very short excited state lifetime, it seems likely that the lifetime as well as the overall photochemical product yields and/or

branching ratios may be strongly influenced by the initial dynamics of nuclear motion near the vertically excited geometry.

Second, we utilize two-color picosecond pump-probe anti-Stokes resonance Raman spectroscopy to obtain vibrational spectra of the *trans*-stilbene product formed immediately following excitation of *cis*-stilbene, and to follow its subsequent conformational and/or vibrational energy relaxation. This is, to our knowledge, the first application of time-resolved vibrational spectroscopy to *cis*-stilbene's photoreactions. A pump pulse at 295 nm excites *cis*-stilbene, and a variably delayed probe pulse at 278 nm excites anti-Stokes Raman scattering from the *trans*-stilbene product, which is formed with a high degree of vibrational excitation. The time-resolved vibrational spectra of the products should provide information regarding the conformation of the photoisomerized product when it first appears on the ground state surface (planar vs. twisted), how the excess vibrational energy is partitioned among the various vibrational modes, the intra- and intermolecular relaxation of the excess vibrational energy, and whether any of these dynamics depend on solvent polarity.

Finally, we carry out careful traditional photochemical yield measurements to evaluate the effect of solvent polarity on the quantum yields for *cis*-stilbene's two reaction pathways. The yields for the *cis*→*trans* and *cis*→DHP reactions are determined in two polar and two nonpolar solvents by irradiation of ground-state *cis*-stilbene with a monochromatized cw lamp followed by quantitation of the photoproduct concentrations by absorption spectroscopy. In this way we explore the connection between the femtosecond and picosecond vibrational dynamics and the net photochemistry for these ultrafast reactions.

3. Experimental

3.1 GROUND-STATE RESONANCE RAMAN INTENSITIES

Cis-stilbene (Aldrich) was used as received, as were the hexane, cyclohexane, acetonitrile, and methanol solvents, all HPLC or spectroscopic grade. Resonance Raman spectra of ground state *cis*-stilbene were obtained in all four solvents with 266 nm excitation (100 µJ/pulse) provided by the fourth harmonic of a 20 Hz Nd:YAG laser. The sampling arrangement used 45° backscattering from an open flowing jet, and the detection system used reflective collection optics, a 0.75-m single spectrograph, and an intensified diode array multichannel detector. The apparatus and methods have been described in detail previously.[53,54] Sample concentrations were approximately 4.6 mM. The spectra were calibrated in wavenumber using the Raman bands of cyclohexane and were corrected for reabsorption and detection sensitivity and the integrated peak areas determined as described elsewhere.[53,54]

3.2 PICOSECOND ANTISTOKES RAMAN SPECTROSCOPY

Picosecond time-resolved transmission and Raman experiments were carried out with an amplified, synchronously-pumped dye laser system utilizing frequency doubling and mixing to generate the uv pump and probe frequencies. A small piece of the 1053 nm output from a Coherent Antares mode-locked Nd:YLF laser was used to seed a cw-pumped Nd:YLF regenerative amplifier (Quantronix 117/Medox) operating at 500 Hz, while the remainder was frequency doubled in a heated 5 mm long KTP crystal and used to pump a home-built, three-mirror dye laser with Rhodamine 590 in ethylene glycol as the gain medium, tuned to 590 nm with a two-plate birefringent filter. The dye laser output was amplified in a three-stage longitudinally pumped dye chain pumped by the frequency doubled (5 mm BBO) output of the regenerative amplifier (0.8 mJ/pulse at 527 nm) with Rhodamine 610 in ethanol as the amplifier dye. The amplified output was typically ≈20 µJ/pulse with a pulse duration of ≈6 ps. Pump pulses at 295 nm (3-4 µJ/pulse) were obtained by frequency doubling the amplified dye laser, while anti-Stokes probe pulses at 278 nm (0.8-1.2 µJ) were generated by mixing the amplified dye laser with 10% of the 527 nm output from the regenerative amplifier. Both the doubling and mixing crystals were 2 mm long KDP.

The probe beam was variably delayed relative to the pump by traversing a motorized delay line, and pump and probe beams were focused through a 5 cm lens and overlapped within the sample flowing in an open jet. Raman scattering was collected in a backscattering geometry from the front face of the jet with a fused silica condenser lens and detected with the same spectrograph and detector used for the steady-state resonance Raman experiments. The picosecond Raman experiments were performed with pump and probe beams having the same polarization. Spatial and temporal overlap of pump and probe beams was typically performed with a sample of *trans*-stilbene in the jet by detecting the pump-induced transient bleaching with a photodiode intercepting the transmitted probe beam. The instrument response function was obtained by differentiating the rising edge of the *trans*-stilbene bleaching signal, and typically had a full width at half maximum of 9-10 ps.

3.3 PHOTOCHEMICAL QUANTUM YIELDS

Samples of 3 ml of *cis*-stilbene in each solvent were degassed to 10^{-6} Torr by repeated freeze-pump-thaw cycles on a vacuum line and were then sealed in their 1 cm path length fused silica cuvettes. The samples, thermostatted at 27°C, were irradiated with a cw Hg-Xe lamp transmitted through a monochromator centered at 280 nm with a 4 nm bandpass. The total number of photons absorbed was monitored with an actinometer using actinochrome 1R. Uv-visible absorption spectra were recorded with Hewlett-Packard 8451A diode array or IBM scanning spectrophotometers to monitor the photoreactions.

Four types of experiments were performed. The yields for DHP production were determined by irradiating samples of 10 mM concentration, which allowed easily detectable concentrations of the visible-absorbing DHP to build up before converting a significant fraction of the *cis*-stilbene starting material. Visible absorption spectra were recorded after absorption of 8.9×10^{-7} Einsteins, corresponding to absorption of a photon by approximately 3% of the *cis*-stilbene molecules, and the DHP concentrations were calculated from the absorbances at the λ_{max} (445-450 nm) assuming a solvent-independent molar extinction coefficient of 6750 M^{-1} cm^{-1}.[6] In contrast, since both stilbene isomers absorb only in the uv, the yields for *trans*-stilbene production had to be determined using samples of low concentration (0.055 mM). Spectra were recorded after 4.7×10^{-8} Einsteins absorbed, corresponding to absorption of a photon by ≈28% of the *cis*-stilbene molecules. The uv absorption spectra in the 250-330 nm range, consisting of a mixture of contributions from *cis*-stilbene, *trans*-stilbene, and DHP, were then fit to a weighted sum of the spectra of all three species with the concentration of DHP constrained to the value determined from its visible absorbance. The DHP uv absorption spectrum was estimated from that published in ref. 6. Third, the *trans*-stilbene and DHP yields were determined in a separate experiment by irradiating 0.055 mM *cis*-stilbene samples until 4.4×10^{-8} Einsteins had been absorbed, and then exposing the samples to air to oxidize the DHP to phenanthrene. The concentrations were determined by fitting the resulting uv absorption spectra to a sum of the spectra of *cis*-stilbene, *trans*-stilbene, and phenanthrene. This approach eliminates the need to use the uv spectrum of DHP, a species that has not been isolated in pure form and whose spectrum is not known very accurately. Finally, the rate of purely thermal decay of DHP at 27°C was determined by first generating DHP photochemically and then monitoring the decay of its visible absorbance at 1 min intervals.

4. Results

4.1   GROUND-STATE RESONANCE RAMAN INTENSITIES

Figure 2 shows the absorption spectra of *cis*-stilbene in hexane, cyclohexane, methanol, and acetonitrile. The spectra in all four solvents are extremely similar in shape, intensity, and position. The spectrum in cyclohexane is slightly red-shifted relative to the other solvents, consistent with the somewhat higher refractive index of this solvent. The molar extinction coefficient appears to be slightly lower in acetonitrile than in the other three solvents, a result for which we have obvious explanation. The similar bandshapes of the absorption spectra in different solvents imply that the initial dynamics of nuclear motion in the vertically excited electronic state are quite similar.[50-52,55] However, when the absorption spectra exhibit no vibrational structure, the resonance Raman intensities are much more diagnostic of the dynamics along specific vibrational coordinates.[50-52]

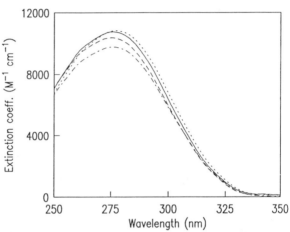

Figure 2. Uv absorption spectra of *cis*-stilbene in hexane (solid curve), cyclohexane (short dashes), methanol (long dashes), and acetonitrile (dot-dashed).

Figure 3 shows the resonance Raman spectra of *cis*-stilbene in the same four solvents. Again the spectra are qualitatively very similar. They are dominated by the ethylenic and ring C=C stretches near 1600 cm^{-1}, with significant intensity also appearing in the "in-plane" CH rocking and C-C stretching modes near 1190 and 1330 cm^{-1}, the "in-plane" ring deformation near 1000 cm^{-1}, the hydrogen out-of-plane wagging mode near 970 cm^{-1}, and modes containing large components of ethylenic torsion near 400 and 560 cm^{-1}.[43] However, careful examination of both the vibrational frequencies and the integrated band intensities shows that there are systematic differences among solvents, as summarized in Table I. In particular, the 560 and 970 cm^{-1} modes, which have large components of motion along the presumed reaction coordinate for isomerization, have higher ground-state frequencies and higher relative intensities in the polar solvents than in the nonpolar ones. The frequency differences indicate a small solvent effect on the ground state equilibrium geometry and/or electron distribution, while the intensity differences suggest slightly faster distortion along the isomerization coordinate in the polar solvents.[56] The nominally CH rocking modes at 1186 and 1329 cm^{-1} also exhibit more intensity in the polar solvents. This indicates that while there are differences in the initial nuclear dynamics from one solvent to another, these differences are complex and may involve motions along the ring closure coordinate as well.

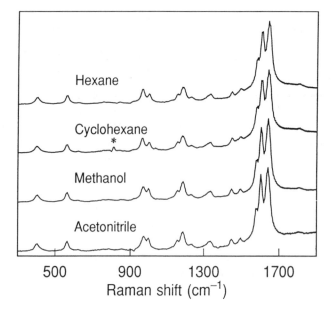

Figure 3. Resonance Raman spectra of *cis*-stilbene with 266 nm excitation in the indicated solvents. The pure solvent spectra have been subtracted; an asterisk labels a subtraction artifact in the cyclohexane spectrum.

Table I. Resonance Raman intensities of *cis*-stilbene.[a]

Methanol		Acetonitrile		Hexane		Cyclohexane	
Freq.	Rel. int.	Freq.	Rel. int.	Freq.	Rel. int.	Freq.	Rel. int.
405	0.059	405	0.066	403	0.056	403	0.063
563	0.057	563	0.060	561	0.052	561	0.051
969	0.12	971	0.11	964	0.098	963	0.087
999	0.039	999	0.047	1000	0.038	1000	0.040
1187	0.14	1187	0.15	1185	0.12	1185	0.11
1329	0.065	1330	0.057	1329	0.044	1326	0.054

[a]Frequencies in cm^{-1}, intensities in terms of integrated areas relative to ethylenic C=C stretch region (1500-1700 cm^{-1}).

4.2   TIME-RESOLVED ANTISTOKES RESONANCE RAMAN SPECTRA

Figure 4 shows representative anti-Stokes resonance Raman spectra of *cis*-stilbene's photoproducts in both cyclohexane and methanol solvents. The probe-only spectra are dominated by solvent bands and by the low-frequency modes of *cis*-stilbene. Since the intensity of these *cis*-stilbene bands relative to solvent is nearly the same in the probe-only, zero delay, and long time delay spectra, these bands must be attributed mainly to thermal Boltzmann population in the unpumped ground state *cis*-stilbene molecules. On the other hand, the bands in the 1550-1650 cm^{-1}, 1150-1200 cm^{-1}, and 950-1000 cm^{-1} regions clearly increase in intensity near time zero and then decay at longer delay times, so they must be attributed to vibrationally hot molecules formed subsequent to absorption of a photon from the pump pulse. Within the signal-to-noise ratio of these data, there are no dramatic time-dependent changes in the relative intensities of the bands assignable to *trans*-stilbene, nor do any new bands possibly arising from highly conformationally distorted molecules appear.

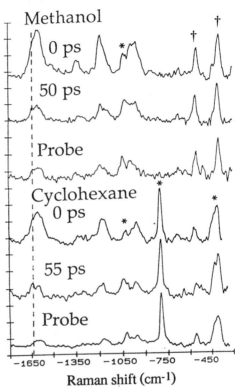

Figure 4. Picosecond anti-Stokes resonance Raman spectra (295 nm pump, 278 nm probe) of *cis*-stilbene's photoproducts. "Probe" is the probe-only spectrum. Asterisks mark solvent bands, and daggers label bands unique to *cis*-stilbene.

Since photoexcited *cis*-stilbene can decay to the hot ground state of the parent species, to *trans*-stilbene, or to DHP, the transient anti-Stokes spectra could in principle contain contributions from all three species. However, at our probe wavelength of 278 nm nearly all of the observed anti-Stokes scattering appears to arise from *trans*-stilbene. *Cis*- and *trans*-stilbene have similar resonance Raman spectra at frequencies above 600 cm^{-1},[43,57] but, as mentioned above, the nearly unchanged intensity of the low-frequency *cis*-stilbene bands between probe-only and pump-probe spectra indicates that the *cis* isomer makes little contribution to the pump-induced spectra. DHP has a very different Raman spectrum from that of the stilbene isomers,[56] and does not appear to contribute to our pump-probe spectra. At room temperature, the molar extinction coefficient of *trans*-stilbene at 278 nm is about 1.7 times larger than that of *cis*-stilbene and about 9 times larger than that of DHP, and resonance Raman cross sections scale roughly as the square of the absorption coefficient. The room temperature Stokes resonance Raman cross sections for the C=C stretching region have been measured to be about five times larger for *trans*-stilbene than for *cis* near 278 nm.[58] Thus we expect *trans*-stilbene to dominate the transient anti-Stokes Raman spectra, although it must be recognized that the anti-Stokes cross sections of vibrationally hot molecules could differ substantially from the Stokes cross sections at room temperature.

Because the excited state lifetime of *cis*-stilbene is shorter than the pump pulse duration, and the *trans*-stilbene product absorbs strongly at the pump wavelength, the vibrationally excited *cis*-, or, more likely, *trans*-stilbene molecules created by the pump pulse could absorb a second pump photon. We have evaluated the possible contribution of such double-pumped molecules to the signal by examining the dependence of the time zero anti-Stokes intensity on pump energy. The anti-Stokes intensity was found to be nearly linear with pump pulse energy over the range of energies used in our experiments, as shown in Figure 5. This indicates that the ground state *cis*-stilbene population is not seriously depleted under our experimental conditions.

Figure 5. Dependence of C=C stretching anti-Stokes Raman intensity on pump pulse energy at constant probe energy.

Figure 6 plots the anti-Stokes intensity in the C=C stretching region (1500-1700 cm^{-1}) relative to the strongest solvent band as a function of delay time in both solvents. The probe-only intensity has been subtracted. These curves clearly show that the anti-Stokes intensity decays faster in methanol than in cyclohexane. In these experiments using pump and probe pulses with parallel polarizations, the anti-Stokes resonance Raman intensity decay will have contributions from rotational motion of the hot stilbene product as well as from vibrational cooling. In order to account for the rotational contribution, the data of Figure 6 were fit to the function[59]

$I(t)=I(0)\exp(-t/\tau_{pop})\{2r(0)\exp(-t/\tau_{rot})+1\}$ convolved with a hyperbolic secant squared instrument function. Here τ_{pop} and τ_{rot} are the vibrational population decay and rotational reorientation times, respectively, and $r(0)$ is the initial anisotropy of the *trans*-stilbene product formed from *cis*-stilbene. The value of $r(0)$ in both solvents was taken to be 0.20 as determined by Sension *et al.* from data in hexane and hexadecane,[41] while the rotational diffusion time τ_{rot} was taken to be 16 ps in methanol (measured directly)[36] and 39 ps in cyclohexane (estimated as that expected for a linear hydrocarbon having the same viscosity as cyclohexane, $\eta = 1.0$ cP).[17] The vibrational population decay time τ_{pop} was left as the single variable in a nonlinear least-squares fitting procedure. The resulting best fit theoretical curves are compared with experiment in Figure 6. The decay time of the anti-Stokes population is found to be approximately a factor of two faster in methanol (17 ps) than in cyclohexane (39 ps). These values were not very sensitive (±10% or less) to varying the assumed value of $r(0)$ over the range from 0 to 0.4. We conclude that there is a significant difference in vibrational cooling rates between the two solvents that cannot be attributed merely to the faster anisotropy decay in methanol.

Figure 6. C=C stretching anti-Stokes Raman intensity of the *trans*-stilbene product as a function of delay time in cyclohexane (circles, solid curve) and in methanol (squares, dashed curve). The points are the experimental data and the curves are best fits to a single-exponential vibrational population decay with a time constant of 39 ps in cyclohexane and 17 ps in methanol (see text).

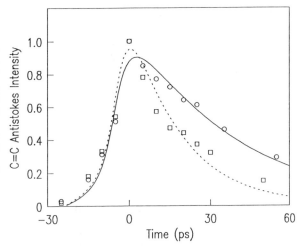

The Stokes resonance Raman spectrum of room temperature *trans*-stilbene exhibits two strong bands at 1638 and 1599 cm^{-1} which correspond to the fundamentals of normal modes that are primarily the ethylenic C=C stretch and a C=C ring stretch, respectively.[57,60] At the longest time delays the transient anti-Stokes spectra in both solvents exhibit two resolved peaks near these frequencies, but at shorter times the C=C stretches are broad, poorly resolved, and shifted to lower frequencies. Figure 7 plots the C=C stretching peak position (either the maximum of the unresolved band at short time delays or the average of the two resolved peaks at longer delays) as a function of time. In both solvents the average frequency shifts by nearly 20 cm^{-1} between 0 and 25 ps, by which time it is approximately equal to the room-temperature Stokes value. This indicates that by 25 ps nearly all of the anti-Stokes intensity originates from molecules that are excited by only one C=C stretching quantum, while the shift at earlier delay times reflects contributions from molecules that either have more than one quantum in the C=C stretch or are excited in both the stretch and other modes that are anharmonically coupled to it. The diagonal anharmonicity of the C=C stretching modes can be determined by examining the Stokes overtone spectrum of room-temperature *trans*-stilbene (Figure 8). By curve fitting the bands up to the 0→3 transition region, we find 0→1, 1→2, and 2→3 intervals of 1638, 1629, and 1619 cm^{-1}, respectively, in the higher-frequency mode, and 1599, 1591, and 1582 cm^{-1} in the lower-frequency mode. Thus the observed anti-Stokes frequency shift of about 20 cm^{-1} at short times would require excitation of 3-4 quanta in the C=C stretches if the shift were due to anharmonicity in the stretches alone. Coupling of the ethylenic-localized C=C stretch with lower-frequency modes, particularly torsional modes, is also expected to be significant since the ethylenic stretching frequency should depend strongly on bond order and be sensitive to out-of-plane distortions.

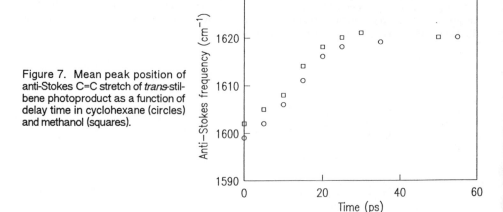

Figure 7. Mean peak position of anti-Stokes C=C stretch of *trans*-stilbene photoproduct as a function of delay time in cyclohexane (circles) and methanol (squares).

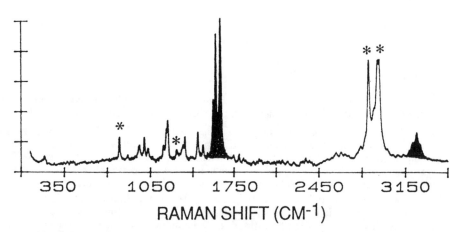

RAMAN SHIFT (CM⁻¹)

Figure 8. Ground state resonance Raman spectrum of *trans*-stilbene in cyclohexane. Stilbene C=C stretch fundamental and overtone regions are shaded; solvent bands are labeled with asterisks.

4.3 PHOTOCHEMICAL QUANTUM YIELDS

The thermal decay of DHP was fit to a first order rate constant of approximately 7×10^{-5} s^{-1} in all four solvents, amounting to a loss of less than 10% of the DHP initially formed during a 20 min experiment. The thermal decay was ignored in our subsequent analysis of the photochemical yields.

Table II summarizes our photochemical quantum yields for the *cis*-stilbene→DHP and *cis*-stilbene→ *trans*-stilbene reactions. The quantum yields for both reactions determined by the different methods are in reasonably good agreement. The DHP yields are roughly a factor of three higher in the nonpolar solvents than in the polar ones, as we reported in a previous preliminary study.[56] The widely quoted literature value of ϕ_{DHP} = 0.1, measured in methylcyclohexane/isohexane at 0°C with 313 nm excitation,[6] falls between our polar and nonpolar solvent values. We find the yields for *trans*-stilbene formation to be slightly solvent dependent (higher in the polar solvents), bracketing the literature value of ϕ_{TS} = 0.35 reported in both methylcyclohexane/isohexane and ethanol/methanol at 25°C with 313 nm excitation.[61]

Table II. Photochemical quantum yields for *cis*-stilbene.[a]

Solvent	$\phi_{cis \rightarrow trans}$		$\phi_{cis \rightarrow DHP}$	
	Method 2[b]	Method 3[c]	Method 1[d]	Method 3[c]
Methanol	0.37	0.38	0.07	0.08
Acetonitrile	0.35	0.39	0.06	0.05
Hexane	0.33	0.32	0.19	0.16
Cyclohexane	0.35	0.35	0.18	0.17

[a]Temperature 27°C, irradiation wavelength 280 nm for all experiments.
[b]By fitting the uv spectra of oxygen-free samples to a sum of *cis*-stilbene, *trans*-stilbene, and DHP spectra.
[c]By air oxidation of irradiated samples followed by fitting the uv spectra to a sum of *cis*-stilbene, *trans*-stilbene, and phenanthrene spectra.
[d]By direct measurement of DHP visible absorbance.

5. Discussion

While the *dynamics* of *trans*-stilbene's excited state isomerization are clearly sensitive to solvent polarity, there is little evidence that the quantum yields are; while the presumed 90° twisted excited state minimum is reached more quickly in polar solvents, the partitioning from this state into the *cis* and *trans* ground states appears to be very nearly equal in different solvents,[11,61] and the fluorescence yields near room temperature are generally low enough that the amount of ground-state *trans* formed radiatively is quite a minor perturbation on the overall product yields. *Cis*-stilbene is a more interesting case in that it has two important photoproducts, and our photochemical measurements show clearly that hydrocarbon solvents give a much higher yield for ring closure and a slightly lower yield for isomerization as compared with the polar solvents. Together with the slightly higher resonance Raman intensities for modes involving out-of-plane motion in the polar solvents, this suggests that the principal effect of polar solvation is to speed up the motion along the twisting coordinate in the excited state, reducing the lifetime and also reducing the yield of the competing ring closure reaction.

The quantum yield data of Table II can be described nearly quantitatively with a very simple model similar to that proposed by Todd and Fleming.[42] We assume that the total rate of excited state decay is the sum of the rates of torsional isomerization to form the 90° twisted state and of ring closure to form electronically excited DHP. We further consider that the 90° twisted geometry partitions equally between the *cis* and *trans* ground states, and that F_{DHP}, the fraction of electronically excited DHP molecules that partition to ground state DHP, is independent of solvent. The quantum yield for *trans*-stilbene formation is then given by $0.5k_{twist}\tau$ where τ is the excited state lifetime and k_{twist} is the rate constant for formation of the 90° twisted state, while the quantum yield for ground state DHP formation is given by $F_{DHP}k_{rc}\tau$ where k_{rc} is the ring closure rate constant.[62] This model together with literature values for the excited state lifetimes[37,39,41,42] gives rate constants for twisting that vary by more than a factor of four across our solvent series (higher in the polar solvents), while the rate constants for ring closure are the same to within experimental uncertainty.[62] These results imply that the dynamics of excited state ring closure are not strongly affected by the nature of the solvent. Nikowa *et al.*, on the other hand, presented a model based on their pressure-dependent lifetime data in which the rate constants for both twisting and ring closure are strongly solvent-dependent.[39] They measured excited state lifetimes only, and while their model fits their lifetime data well, it does not account for our product quantum yields.

The effect of solvent polarity on the initial dynamics of nuclear motion on the excited-state surface, as revealed through the ground-state resonance Raman intensities together with the absorption spectra, is relatively minor. This is reasonable in view of the expectation that the vertically excited state is still relatively nonpolar, with significant charge separation occurring only as the 90° twisted structure is approached. (Recent time-resolved photocalorimetric[63] and microwave conductivity measurements[64] provide strong evidence for the zwitterionic nature of the much longer-lived perpendicularly twisted state of tetraphenylethylene.) The solvent polarity effect on the excited state surface is expected to become increasingly pronounced as the ethylenic twist develops. An additional complication of the resonance Raman intensity data is the fact that polar solvents also enhance some modes that are best characterized as hydrogen rocking modes having little or no out-of-plane or torsional component. These are modes that might have significant projections along the ring-closure coordinate, suggesting that motion along this reaction pathway is also enhanced by polar solvents. Petek and co-workers have argued strongly for the importance of motion along the ring closure coordinate in the radiationless decay of *cis*-stilbene.[47,65] However, if the *initial* motion along this coordinate as well as along the torsional coordinate is indeed enhanced in polar solvents, it does not show up in the overall rate constant for formation of the ground state of DHP. It is possible that there is also a solvent effect on the quantity F_{DHP} describing the partitioning from DHP* to *cis*-stilbene and ground state DHP, but this explanation would require a fortuitous cancellation of solvent effects and appears unlikely. (A similar explanation could also be used to rationalize the model of ref. 39 with our data.)

In ref. 43, the resonance Raman intensities of *cis*-stilbene in cyclohexane were combined with a ground state normal mode description to develop a picture of the rms motions of the individual atoms during the first tens of femtoseconds following electronic excitation. A similar analysis of the present intensity data might provide a clearer picture of the differences in the initial vibrational dynamics among solvents. However, the existing ground-state normal mode analyses, which have not been tested against any isotopic data, are of unknown accuracy. We therefore choose to discuss our present resonance Raman intensity data only qualitatively.

Our picosecond time-resolved anti-Stokes resonance Raman spectra reveal a strong solvent dependence of the vibrational population decay of the highly vibrationally excited *trans*-stilbene product. Several controversies and surprises regarding this vibrational relaxation have recently come to light. It had been generally believed that the solvent-induced friction that accompanies motion from either the *cis* or the *trans* side to the twisted excited-state minimum dissipates enough energy to trap the molecule at the twisted geometry, but the Saltiel group's recent reports of fluorescence from the excited state of *trans* following photoexcitation of *cis*[44,45] suggest that this idea needs to be reevaluated. Fleming *et al.*[66] also concluded, through modeling the absorption and fluorescence spectra, that the observed room-temperature fluorescence of *cis*-stilbene is largely vibrationally unrelaxed. Intramolecular vibrational relaxation is normally thought to be fast compared with loss of the vibrational energy to solvent, but Hochstrasser and co-workers have concluded, based on modeling the transient absorption spectra of the *trans*-stilbene product, that most of the energy placed through photoexcitation into the Franck-Condon active modes of *cis*-stilbene remains in the corresponding modes of the *trans* product on a long (>6 ps) time scale.[38,41] Unfortunately, the modest signal-to-noise ratio of our data combined with a variety of technical reasons discussed in detail in ref. 67 prevent us from being able to draw conclusions about the nature of the vibrational energy distribution probed in our experiments. The "vibrational population decay time" we obtain by fitting the experimental intensities to a simple exponential decay cannot be directly equated to the population of excited vibrational levels because the anti-Stokes resonance Raman cross sections depend on the changing populations of all of the vibrational modes of the molecule. However, since both the room-temperature absorption spectra and the Stokes resonance Raman spectra of *trans*-stilbene are only slightly solvent dependent,[68] it seems safe to conclude that the populations of the Franck-Condon active vibrations do decay considerably faster in methanol than in cyclohexane. It should be noted that our results in only two solvents do not demonstrate that polarity is the salient difference between cyclohexane and methanol. Factors not directly related to solvent polarity, such as the solvent's mass, its internal vibrational frequencies, and the local solvation structure, may also be important determinants of vibrational relaxation times.[69-73] Additionally, since we can monitor only a few of stilbene's 72 vibrations, we cannot determine whether the vibrational cooling we observe corresponds to an intra- or an intermolecular process.

The anti-Stokes decay time we observe in methanol agrees well with the vibrational cooling time of 17.5 ps inferred by Sension and co-workers from red-edge absorption measurements on the *trans*-stilbene product,[41] but they also observed comparable decay times in hexane and hexadecane. Our anti-Stokes decay times in cyclohexane are considerably longer than those observed by Reid and co-workers for the triene products of photochemical ring-opening reactions in the same solvent.[74] The cooling of highly vibrationally excited azulene following internal conversion, monitored by red-edge absorption, is also found to be about 15 ps in hydrocarbon solvents,[69] although slower cooling on the 20-50 ps scale is observed in halogenated solvents.[69,72,75] It is not clear why the vibrational cooling of stilbene in cyclohexane should be a factor of two to four slower than in the smaller trienes and azulene, but it should be stressed that the resonant anti-Stokes intensities do not correspond directly to vibrational populations, nor are the decays fit particularly well by a single exponential in either solvent.

It is tempting to suggest that the faster anti-Stokes intensity decay we observe for *trans*-stilbene in methanol relative to cyclohexane reflects less excess vibrational energy in the nascent ground state photoproduct in methanol due to polar solvent stabilization of the twisted excited state. However, this idea seems inconsistent with the very similar time-dependent shifts in the C=C stretching band positions in the two solvents (Figure 7). The anti-Stokes intensity in the C=C stretching region approximately reflects the total population in states that are excited to at least one quantum in a C=C stretching mode, while the bandshape and frequencies reflect the fraction of the total population that is in levels higher than n=1, and/or the population in other modes that are anharmonically coupled to the C=C stretch. This suggests that the difference between the vibrational relaxation process in methanol and in cyclohexane may lie mostly in the final step (loss of the last vibrational quantum in a C=C stretch) rather than in the initial cooling of a molecule that is excited to many quanta in several modes. We conclude that while the anti-Stokes Raman results are consistent with the general importance of the solvent environment in determining stilbene's photoisomerization dynamics, we do not yet have a clear interpretation of these data.

Figure 9 summarizes the hypothesized solvent polarity dependence of the excited state potential surfaces of *cis*- and *trans*-stilbene. This static picture accounts qualitatively for a large number of experimental observations on the photochemical dynamics and quantum yields of both stilbene isomers. However, the photochemical interconversions are clearly also affected by dynamic solvents effects such as friction and dielectric relaxation. More studies addressing these aspects of solvation, as well as experiments that are more directly sensitive to the potential surface and dynamics in the region near the 90° twisted geometry, will contribute to a more complete picture of this prototypical unimolecular reaction.

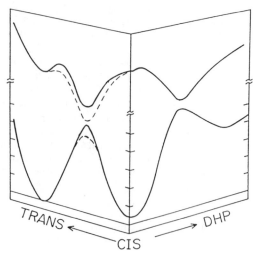

Figure 9. Hypothetical potential energy surfaces for *cis*-stilbene, *trans*-stilbene, and dihydrophenanthrene in nonpolar solvents (solid) and in polar solvents (dashed).

6. Acknowledgements

This work was supported in part by NIH grant GM 39724 and NSF grant CHE-9020844. A.B.M. is the recipient of a Packard Fellowship in Science and Engineering, a Sloan Research Fellowship, an NSF Presidential Young Investigator Award, and a Dreyfus Teacher-Scholar Award. We are indebted to Professor R. J. Dwayne Miller and members of his research group, particularly Dr. Luis Gomez-Jahn and Pierre Basseras, for their assistance in designing and setting up our picosecond laser system, and to Professor David Whitten and David Lawrence for the use of their facilities for carrying out the photochemical quantum yield measurements.

7. References

1.   Waldeck, D. H. (1991) "Photoisomerization dynamics of stilbenes", Chem. Rev. 91, 415-436.
2.   Moore, William M., Morgan, David D., and Stermitz, Frank R. (1963) "The photochemical conversion of stilbene to phenanthrene. The nature of the intermediate", J. Am. Chem. Soc. 85, 829-830.
3.   Malkin, Shmuel, and Fischer, Ernst (1964) "Temperature dependence of photoisomerization. III. Direct and sensitized photoisomerization of stilbenes", J. Phys. Chem. 68, 1153-1163.
4.   Mallory, Frank B., Wood, Clelia S., and Gordon, Janice T. (1964) "Photochemistry of stilbenes. III. Some aspects of the mechanism of photocyclization to phenanthrene", J. Am. Chem. Soc. 86, 3094-3102.
5.   Saltiel, Jack (1967) "Perdeuteriostilbene. The role of phantom states in the *cis-trans* photoisomerization of stilbenes", J. Am. Chem. Soc. 89, 1036-1037.
6.   Muszkat, K. A., and Fischer, E. (1967) "Structure, spectra, photochemistry, and thermal reactions of the 4a,4b-dihydrophenanthrenes", J. Chem. Soc. B 662-678.

7. Muszkat, K. A., and Schmidt, W. (1971) "Energy profiles for ring formation- and ring opening-processes in the *cis*-stilbene-4a, 4b-dihydrophenanthrene system. An example of the feasibility of a process forbidden by the rules of orbital symmetry conservation", Helv. Chim. Acta **54**, 1195-1207.

8. Birks, J. B. (1976) "The photoisomerization of stilbene", Chem. Phys. Lett. **38**, 437-440.

9. Rothenberger, G., Negus, D. K., and Hochrasser, R. M. (1983) "Solvent influence on photoisomerization dynamics", J. Chem. Phys. **79**, 5360-5367.

10. Syage, J. A., Felker, P. M., and Zewail, A. H. (1984) "Picosecond dynamics and photoisomerization of stilbene in supersonic beams. II. Reaction rates and potential energy surface", J. Chem. Phys. **81**, 4706-4723.

11. Sundström, Villy, and Gillbro, Tomas (1984) "Dynamics of the isomerization of *trans*-stilbene in n-alcohols studied by ultraviolet picosecond absorption recovery", Chem. Phys. Lett. **109**, 538-543.

12. Sundström, Villy, and Gillbro, Tomas (1985) "Dynamics of *trans-cis* photoisomerization of stilbene in hydrocarbon solutions", Ber. Bunsenges. Phys. Chem. **89**, 222-226.

13. Lee, Minyung, Holtom, G. R., and Hochrasser, R. M. (1985) "Observation of the Kramers turnover region in the isomerism of *trans*-stilbene in fluid ethane", Chem. Phys. Lett. **118**, 359-363.

14. Lee, M., Bain, A. J., McCarthy, P. J., Han, C. H., Haseltine, J. N., Smith, A. B., III, and Hochrasser, R. M. (1986) "Picosecond photoisomerization and rotational reorientation dynamics in solution", J. Chem. Phys. **85**, 4341-4347.

15. Balk, Michael W., and Fleming, Graham R. (1986) "Unimolecular reactions in isolated and collisional systems: Is the transition-state rate an upper limit for the isomerization of stilbene?", J. Phys. Chem. **90**, 3975-3983.

16. Hicks, J.. M., Vandersall, M. T., Sitzmann, E. V., and Eisenthal, K. B. (1987) "Polarity-dependent barriers and the photoisomerization dynamics of molecules in solution", Chem. Phys. Lett. **135**, 413-420.

17. Kim, Seong K., and Fleming, Graham R. (1988) "Reorientation and isomerization of *trans*-stilbene in alkane solutions", J. Phys. Chem. **92**, 2168-2172.

18. Courtney, Scott H., Balk, Michael W., Philips, Laura A., Webb, Steven P., Yang, Ding, Levy, Donald H., and Fleming, Graham R. (1988) "Unimolecular reactions in isolated and collisional systems: deuterium isotope effect in the photoisomerization of stilbene", J. Chem. Phys. **89**, 6697-6707.

19. Kim, Seong K., Courtney, Scott H., and Fleming, Graham R. (1989) "Isomerization of *t*-stilbene in alcohols", Chem. Phys. Lett. **159**, 543-548.

20. Lee, Minyung, Haseltine, John N., Smith, Amos B., III, and Hochrasser, Robin M. (1989) "Isomerization processes of electronically excited stilbene and diphenylbutadiene in liquids: Are they one-dimensional?", J. Am. Chem. Soc. **111**, 5044-5051.

21. Sivakumar, N., Hoburg, E. A., and Waldeck, D. H. (1989) "Solvent dielectric effects on isomerization dynamics: Investigation of the photoisomerization of 4,4'-dimethoxystilbene and *t*-stilbene in n-alkyl nitriles", J. Chem. Phys. **90**, 2305-2316.

22. Park, N. S., and Waldeck, D. H. (1990) "Influence of polar solvents on reaction dynamics. Photoisomerization studies of dihydroxystilbene", J. Phys. Chem. **94**, 662-669.

23. Park, N. S., and Waldeck, D. H. (1990) "On the dimensionality of stilbene isomerization", Chem. Phys. Lett. **168**, 379-384.

24. Schroeder, J., Schwarzer, D., Troe, J., and Voss, F. (1990) "Cluster and barrier effects in the temperature and pressure dependence of the photoisomerization of *trans*-stilbene", J. Chem. Phys. **93**, 2393-2404.

25. Gustafson, T. L., Roberts, D. M., and Chernoff, D. A. (1983) "Picosecond transient Raman spectroscopy: The photoisomerization of *trans*-stilbene", J. Chem. Phys. **79**, 1559-1564.

26. Hamaguchi, Hiro-o, Kato, Chihiro, and Tasumi, Mitsuo (1983) "Observation of transient resonance Raman spectra of the S_1 state of *trans*-stilbene", Chem. Phys. Lett. **100**, 3-7.

27. Gustafson, Terry L., Roberts, Dale M., and Chernoff, Donald A. (1984) "The structure of electronic excited states in *trans*-stilbene: picosecond transient Stokes and anti-Stokes Raman spectra", J. Chem. Phys. **81**, 3438-3443.

28. Weaver, William L., Huston, Lisa A., Iwata, Koichi, and Gustafson, Terry L. (1992) "Solvent/solute interactions probed by picosecond transient Raman spectroscopy: mode specific vibrational dynamics in S_1 *trans*-stilbene", J. Phys. Chem. **96**, 8956-8963.

29. Butler, Roger M., Lynn, Matthew A., and Gustafson, Terry L. (1993) "Solvent-solute interactions probed by picosecond transient Raman spectroscopy: Band assignments and vibrational dynamics of S_1 *trans*-4,4'-diphenylstilbene", J. Phys. Chem. **97**, 2609-2617.

30. Iwata, Koichi, Toleutaev, Bulat, and Hamaguchi, Hiro-o (1993) "Anomalous lifetime shortening of S_1 *trans*-stilbene in carbon tetrachloride as revealed by picosecond time-resolved Raman, absorption, and fluorescence spectroscopies", J. Chem. Phys. submitted.

31. Greene, B. I., and Scott, T. W. (1984) "Time-resolved multiphoton ionization in the organic condensed phase: picosecond conformational dynamics of *cis*-stilbene and tetraphenylethylene", Chem. Phys. Lett. **106**, 399-402.

32. Doany, F. E., Hochstrasser, R. M., Greene, B. I., and Millard, R. R. (1985) "Femtosecond-resolved ground-state recovery of *cis*-stilbene in solution", Chem. Phys. Lett. **118**, 1-5.

33. Sension, R.J., Repinec, S.T., and Hochstrasser, R.M. (1990) "Femtosecond laser study of energy disposal in the solution phase isomerization of stilbene", J. Chem. Phys. **93**, 9185-9188.

34. Abrash, S., Repinec, S., and Hochstrasser, R.M. (1990) "The viscosity dependence and reaction coordinate for isomerization of *cis*-stilbene", J. Chem. Phys. **93**, 1041-1053.

35. Todd, David C., Jean, John M., Rosenthal, Sandra J., Ruggiero, Anthony J., Yang, Ding, and Fleming, Graham R. (1990) "Fluorescence upconversion study of *cis*-stilbene isomerization", J. Chem Phys. **93**, 8658-8668.

36. Repinec, Stephen T., Sension, Roseanne J., Szarka, Arpad Z., and Hochstrasser, Robin M. (1991) "Femtosecond laser studies of the *cis*-stilbene photoisomerization reactions. The *cis*-stilbene to dihydrophenanthrene reaction", J. Phys. Chem. **95**, 10380-10385.

37. Rice, Jane K., and Baronavski, A. P. (1992) "Ultrafast studies of solvent effects in the isomerization of *cis*-stilbene", J. Phys. Chem. **96**, 3359-3366.

38. Sension, Roseanne J., Szarka, Arpad Z., and Hochstrasser, Robin M. (1992) "Vibrational energy redistribution and relaxation in the photoisomerization of *cis*-stilbene", J. Chem. Phys. **97**, 5239-5242.

39. Nikowa, L., Schwarzer, D., Troe, J., and Schroeder, J. (1992) "Viscosity and solvent dependence of low barrier processes: Photoisomerization of *cis*-stilbene in compressed liquid solvents", J. Chem. Phys. **97**, 4827-4835.

40. Pedersen, S., Bañares, L., and Zewail, A. H. (1992) "Femtosecond vibrational transition-state dynamics in a chemical reaction", J. Chem. Phys. **97**, 8801-8804.

41. Sension, Roseanne J., Repinec, Stephen T., Szarka, Arpad Z., and Hochstrasser, Robin M. (1993) "Femtosecond laser studies of the *cis*-stilbene photoisomerization reactions", J. Chem. Phys. **98**, 6291-6315.

42. Todd, David C., and Fleming, Graham R. (1993) "*Cis*-stilbene isomerization: Temperature dependence and the role of mechanical friction", J. Chem. Phys. **98**, 269-279.

43. Myers, Anne B., and Mathies, Richard A. (1984) "Excited-state torsional dynamics of *cis*-stilbene from resonance Raman intensities", J. Chem. Phys. **81**, 1552-1558.

44. Saltiel, Jack, Waller, Andrew, Sun, Ya-Ping, and Sears, Donald F., Jr. (1990) "*Cis*-stilbene fluorescence in solution. Adiabatic $^1c^* \rightarrow {}^1t^*$ conversion", J. Am. Chem. Soc. **112**, 4580-4581.

45. Saltiel, Jack, Waller, Andrew S., and Sears, Donald F., Jr. (1993) "The temperature and medium dependencies of *cis*-stilbene fluorescence. The energetics for twisting in the lowest excited singlet state", J. Am. Chem. Soc. **115**, 2453-2465.

46. Petek, Hrvoje, Fujiwara, Yoshihisa, Kim, Dongho, and Yoshihara, Keitaro (1988) "Observation of a local minimum on the S_1 surface of *cis*-stilbene solvated in inert gas clusters", J. Am. Chem. Soc. **110**, 6269-6270.

47. Petek, Hrvoje, Yoshihara, Keitaro, Fujiwara, Yoshihisa, Lin, Zhe, Penn, John H., and Frederick, John H. (1990) "Is the nonradiative decay of S_1 *cis*-stilbene due to the dihydrophenanthrene isomerization channel? Suggestive evidence from photophysical measurements on 1,2-diphenylcycloalkenes", J. Phys. Chem. **94**, 7539-7543.

48. Petek, Hrvoje, Yoshihara, Keitaro, Fujiwara, Yoshihisa, and Frey, Jeremy G. (1990) "Isomerization of *cis*-stilbene in rare-gas clusters: Direct measurements of *trans*-stilbene formation rates on a picosecond time scale", J. Opt. Soc. Am. B **7**, 1540-1544.

49. Bonacic-Koutecky, Vlasta, Bruckmann, Peter, Hiberty, Philippe, Koutecky, Jaroslav, Leforestier, Claude, and Salem, Lionel (1975) "Sudden polarization in the zwitterionic Z_1 excited states of organic intermediates. Photochemical implications", Angew. Chem., Int. Ed. Engl. **14**, 575-576.

50. Heller, Eric J., Sundberg, Robert L., and Tannor, David (1982) "Simple aspects of Raman scattering", J. Phys. Chem. **86**, 1822-1833.

51. Myers, Anne B., and Mathies, Richard A. (1987) "Resonance Raman intensities: A probe of excited-state structure and dynamics", in T. G. Spiro (ed.), Biological Applications of Raman Spectroscopy, Wiley, New York, pp. 1-58.

52. Myers, Anne B. (1990) "Femtosecond molecular dynamics probed through resonance Raman intensities", J. Opt. Soc. Am. B **7**, 1665-1672.

53. Myers, Anne B., Li, Bulang, and Ci, Xiaopei (1988) "A resonance Raman intensity study of electronic spectral broadening mechanisms in CS_2/cyclohexane", J. Chem. Phys. **89**, 1876-1886.

54. Ci, Xiaopei, Pereira, Marco A., and Myers, Anne B. (1990) "Resonance Raman spectra of *trans*-1,3,5-hexatriene in solution: Evidence for solvent effects on excited-state torsional motion", J. Chem. Phys. **92**, 4708-4714.

55. Heller, Eric J. (1981) "The semiclassical way to molecular spectroscopy", Acc. Chem. Res. **14**, 368-375.

56. Rodier, Jon-Marc, Ci, Xiaopei, and Myers, Anne B. (1991) "Resonance Raman spectra of 4a,4b-dihydrophenanthrene, the photocyclization product of *cis*-stilbene", Chem. Phys. Lett. **183**, 55-62.

57. Myers, Anne B., Trulson, Mark O., and Mathies, Richard A. (1985) "Quantitation of homogeneous and inhomogeneous broadening mechanisms in *trans*-stilbene using absolute resonance Raman intensities", J. Chem. Phys. **83**, 5000-5006.

58. Myers, Anne B. (1984) Ph.D. Thesis, University of California, Berkeley.

59. Myers, Anne B., and Hochstrasser, Robin M. (1986) "Comparison of four-wave mixing techniques for studying orientational relaxation", IEEE J. Quantum Electron. **QE-22**, 1482-1492.

60. Baranovic, Goran, Meic, Zlatko, Güsten, Hans, Mink, Janos, and Keresztury, Gabor (1990) "Intramolecular vibrational coupling in the ground electronic state (S_0) of *trans*-stilbene", J. Phys. Chem. **94**, 2833-2843.

61. Gegiou, Dina, Muszkat, K. A., and Fischer, Ernst (1968) "Temperature dependence of photoisomerization. VI. The viscosity effect", J. Am. Chem. Soc. **90**, 12-18.

62. Rodier, Jon-Marc, and Myers, Anne B. (1993) "*Cis*-stilbene photochemistry: Solvent effects on the initial dynamics and quantum yields", J. Am. Chem. Soc. submitted.

63. Morais, Joseph, Ma, Jongseok, and Zimmt, Matthew B. (1991) "Solvent dependence of the twisted excited state energy of tetraphenylethylene: Evidence for a zwitterionic state from picosecond optical calorimetry", J. Phys. Chem. **95**, 3885-3888.

64. Schuddeboom, Wouter, Jonker, Stephan A., Warman, John M., de Haas, Matthijs P., Vermeulen, Martien J. W., Jager, Wolter F., de Lange, Ben, Feringa, Ben L., and Fessenden, Richard W. (1993) "Sudden polarization in the twisted, phantom state of

tetraphenylethylene detected by time-resolved microwave conductivity", <u>J. Am. Chem. Soc.</u> **115**, 3286-3290.

65. Frederick, John H., Fujiwara, Y., Penn, John H., Yoshihara, Keitaro, and Petek, Hrvoje (1991) "Models for stilbene photoisomerization: experimental and theoretical studies of the excited-state dynamics of 1,2-diphenylcycloalkenes", <u>J. Phys. Chem.</u> **95**, 2845-2858.

66. Todd, David C., Fleming, Graham R., and Jean, John M. (1992) "Calculations of absorption and emission spectra: A study of *cis*-stilbene", <u>J. Chem. Phys.</u> **97**, 8915-8925.

67. Phillips, David L., Rodier, Jon-Marc, and Myers, Anne B. (1993) "*Cis*-stilbene photochemistry: Direct observation of product formation and relaxation through two-color uv pump-probe Raman spectroscopy", <u>Chem. Phys.</u> in press.

68. Ci, Xiaopei, and Myers, Anne B. (1989) "A resonance Raman study of solvent effects on the excited state potential surface of *trans*-stilbene", <u>Chem. Phys. Lett.</u> **158**, 263-270.

69. Sukowski, U., Seilmeier, A., Elsaesser, T., and Fischer, S. F. (1990) "Picosecond energy transfer of vibrationally hot molecules in solution: Experimental studies and theoretical analysis", <u>J. Chem. Phys.</u> **93**, 4094-4101.

70. Johnson, Alan E., Levinger, Nancy E., and Barbara, Paul F. (1992) "Photodissociation and recombination dynamics of I_2^- in water and alcohols", <u>J. Phys. Chem.</u> **96**, 7841-7844.

71. Xu, Xiaobing, Yu, Soo-Chang, Lingle, Robert, Jr., Zhu, Huiping, and Hopkins, J. B. (1991) "Ultrafast transient Raman investigation of geminate recombination and vibrational energy relaxation in iodine: The role of energy relaxation pathways to solvent vibrations", <u>J. Chem. Phys.</u> **95**, 2445-2457.

72. Schultz, Karen E., Russell, D. J., and Harris, Charles B. (1992) "The applicability of binary collision theories to complex molecules in simple liquids", <u>J. Chem. Phys.</u> **97**, 5431-5438.

73. Elsaesser, Thomas, and Kaiser, Wolfgang (1991) "Vibrational and vibronic relaxation of large polyatomic molecules in liquids", <u>Ann. Rev. Phys. Chem.</u> **42**, 83-107.

74. Reid, Philip J., Doig, Stephen J., Wickam, Steven D., and Mathies, Richard A. (1993) "Photochemical ring-opening reactions are complete in picoseconds. A time-resolved UV resonance Raman study of 1,3-cyclohexadiene", <u>J. Am. Chem. Soc.</u> submitted.

75. Wild, W., Seilmeier, A., Gottfried, N. H., and Kaiser, W. (1985) "Ultrafast investigations of vibrationally hot molecules after internal conversion in solution", <u>Chem. Phys. Lett.</u> **119**, 259-263.

SIMULATION OF ELECTRONIC SPECTROSCOPY AND RELAXATION IN AQUEOUS SOLUTION

Peter J. Rossky, Tim H. Murphrey, and Wen-Shyan Sheu
Department of Chemistry and Biochemistry
The Univeristy of Texas at Austin
Austin, Texas 78712-1167
U. S. A.

Abstract

Computer simulation of quantum systems in solution are allowing the direct observation of electronic dynamics of solutes at a molecular level on the same timescale as that probed by ultrafast transient spectroscopy. Here, we describe some of our recent theoretical approaches to the analysis of electronic spectroscopy and relaxation dynamics in solution, and outline some of the recent results obtained for the experimentally probed cases of energetic excess electrons in liquid water and for an aqueous halide ion.

1. Introduction

The fact that the majority of chemical reactions occur in solution continues to stimulate a major research effort aimed at elucidating the details of chemical dynamics in liquids. Parallel recent progress in laser technology and in theoretical methods is enabling direct assessment of dynamical behavior of chemical systems on the time scale of the fundamental events. In particular, new algorithms now allow for both classical and quantum mechanical dynamical simulation of phenomena on time scales accessible by these new experimental techniques. These efforts are resulting in an increasing ability to directly connect theoretical and experimental results.

In this article, we first outline the methods now available for quantum simulation of dynamics in solution, including state to state tranistions. We then briefly discuss some recent theoretical results for the hydrated electron and aqueous halide ions, both of which have been investigated experimentally.

2. Methods

The elements of molecular-level simulation are, for the most part, common to both classical and quantum mechanical simulation. In particular, in the studies discussed here, employing an aqueous solvent, we have used the SPC (Simple Point Charge) model for water, and the corresponding flexible generalization, which are the simplest interaction site models available and reproduce a wide variety of properties[1]. In quantum simulations, with explicit treatment of one or more electrons, the interaction of solute electrons with solvent employs a pseudopotential for water, which

J. Jortner et al. (eds.), Reaction Dynamics in Clusters and Condensed Phases, 279–288.

treats the electrons associated with solvent molecules implicitly[2]. Several related pseudopotentials for electron-solvent interactions for water, ammonia, and methanol are now available.[3-6]

Two basic approaches have evolved for representation of the quantum distribution. The first is the so-called Feynman path integral representation[7], and considerable progress is being made in applying this formalism to the computational study of quantum rate processes, an important parallel development[8-10]. Our focus has been on a direct description in terms of wavefunctions, a particularly convenient approach in a spectroscopic context. Given a simulated ensemble of solvent configurations and the respective eigenstate manifolds, the vertical excitation energies and the corresponding dipole transition matrix elements provide direct access to steady state and transient spectra. An explicit eigenstate description is also convenient for the treatment of time dependent phenomena, in both equilibrium and non-equilibrium situations. Neglecting corrections to the Born-Oppenheimer approximation, the system can be described adiabatically, and the set of forces on the nuclei required to propagate the solvent nuclear positions in time can be evaluated via the Hellmann-Feynman theorem.[11-14]

The algorithm we have developed for non-adiabatic dynamics[14] prescribes a semiclassical simulation of the electronic and solvent nuclear dynamics via a sequence of time steps, each of which is composed of a time propagation of the wavefunction from an initial time t_0 to a final time t, a branching among adiabatic electronic eigenstates at time t, and a time propagation of the nuclear coordinates subject to a total force that recognizes the nature of both the initial and final quantum state. This is accomplished via a novel combination of the surface hopping idea of Tully and Preston[15] and the non-adiabatic scattering formalism of Pechukas[16]. For short times, the quantum system evolves correctly as a mixed state and the bath evolves with retention of coherence. For longer times, we have to date made a specific (but not *required*) assumption in implementation of our algorithm in condensed phases : coherence is dropped for times longer than the step size[14]. In recent calculations, Neria et al.[17] have explicitly estimated the coherence effects associated with a non-adiabatic electronic transition in aqueous solution and found that the characteristic decay time is in fact only a few femtoseconds providing some empirical support for our type of approach. Tully[18] has provide an alternative algorithm with related goals, and it has been applied to the relaxation of an electron in helium by Space and Coker[19], using the numerical methods of Webster et al.[14]. The optimal treatment of bath coherence, not accounted for in the explicitly classical treatment of the solvent in either algorithm[18], remains an important topic for investigation.

3. Results

In this section we outline selected results obtained by the methods described above, considering first the hydrated electron[20-23] and then some very recent results on halide excited states in water[24-26].

3.1.   THE HYDRATED ELECTRON

The hydrated electron is ubiquitous in aqueous photochemistry[27]. However, a deeper understanding of the process of energetic relaxation and electron localization has been the subject of much recent progress. Until recently, the prevailing view assigned dominant control to solvent configurational rearrangement[28]. However, the original ultrafast experiments in water[29,30] manifested spectroscopically an apparently stepwise solvation process. The transient spectra do *not* manifest a continuous wavelength shift, in contrast to available theories[28].

Using quantum simulation, the steady state spectrum of the hydrated electron was originally investigated some time ago[21]. The broad intense equilibrium spectrum, from a more recent evaluation[23] with a flexible solvent model[1], is shown in Fig. 1.

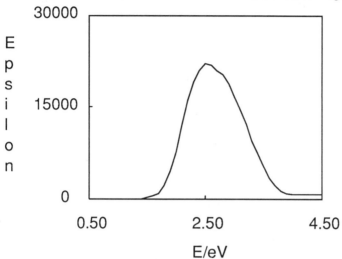

Figure 1. Simulated steady-state equilibrium optical absorption spectrum of an excess electron in liquid water.

The absorption was shown[21] to be predominantly comprised of the superposition of three dominant s-like to p-like transitions, and these excited states have been implicated in the relaxation dynamics as well, as discussed below. Adiabatic dynamics[22] showed that *ground state* solvation of the electron would be evidenced spectroscopically on a time scale of only 10's of femtoseconds, in contrast to the times on the order of 200 fs characterizing the experimental data, leading to the conjecture that, in fact, the electron is solvated in an effective competition with non-radiative energy loss, so that the electron could evolve into an intermediate solvated excited state. Assuming the existence of such a state, Long et al.[31] have carried out further experimental studies on the relaxation dynamics and have reported an analysis yielding an absorption spectrum characteristic of the intermediate state.

Nonadiabatic dynamical simulation of the relaxation of an initially energetic excess electron in water[20,23] has led to results that are consistent in many respects with the hypotheses and the data described above. Using an internally flexible water model, the electronic relaxation has been studied[23] at two different initial excess energies (approximately 2 and 2.5 eV excess kinetic energy), both of which correspond in the model to states in the continuum. In Fig. 2 and Fig. 3, we show results manifesting the most important characteristic behaviors observed in relaxation trajectories for the lower of these energies. These correspond closely[23] to the results for an internally rigid solvent model[21], except for a substantial reduction in the excited state lifetime with internal flexibility.

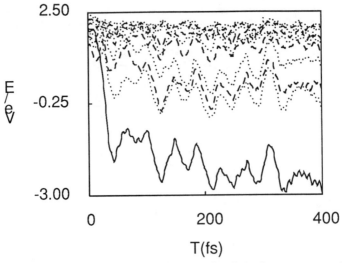

Figure 2. Dynamical history of adiabatic state energies for a
trajectory in which the electron first localizes in the ground state.
The bold solid line denotes the occupied eigenstate.

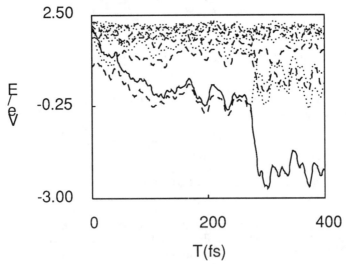

Figure 3. Dynamical history of adiabatic state energies for a
trajectory in which the electron first localizes in the first excited
state. The bold solid line denotes the occupied eigenstate.

These results show relaxation to occur either via a cascade directly to the ground state
(Fig. 2) or via intermediate trapping in a p-like excited state[20] (Fig. 3). The lifetime of this
excited state has been computed to be about 160 fs[23], in reasonable agreement with

inferences from experiment[29-31] and with recent alternative calculations of the excited state lifetime[17].

The fact that the experiment appears interpretable without invoking the channel shown in Fig. 2[29-31] remains under study, but may be associated with the necessarily simple form assumed in the kinetic analysis of the data[32].

A critical feature of simulated data is the ability to unequivocally separate spectral contributions. In particular, the published experimental analysis[31] has reported a spectrum for the intermediate excited state, derived under an assumed kinetic analysis. In Fig. 4, we show the computed spectrum of the excited state based on our dynamical results[23].

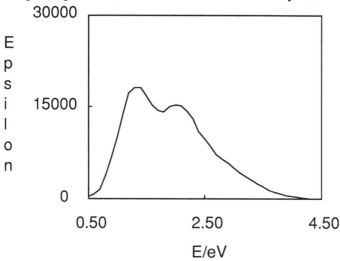

Figure 4. Simulated optical absorption spectrum of the solvated first excited state.

A comparison of these results to those reported in Ref. 31 show that, although the computed ground and excited state spectra are both blue shifted with respect to experiment, the relative contributions are remarkably similar. Thus, the conjecture that this excited state and its lifetime play an important role in the relaxation[22] appears supported.

3.2. AQUEOUS HALIDES

Related work has been carried out in our laboratory on the problem of photoexcited aqueous halides.[24-26] The unique role of the solvent on the spectral behavior of such simple negative ions has long attracted attention from physical chemists : these ions exhibit strong, broad UV absorption spectra in solution but not in the gas phase, and solvated electrons can result from photoexcitation[27,33-36]. The associated ionic excited states have been termed charge-transfer-to-solvent (CTTS) states. However, due to the limited tools available, the theoretical description of these states predated our studies by about thiry years[35,36]. Femtosecond spectroscopic data have also recently been reported[30,37,38]. In particular, Long et al.[38] have carried out transient multiphoton absorption experiments on an aqueous iodide ion and concluded that the relaxation proceeded via an initial CTTS state.

Using a pseudopotential model for the halide, quantum simulation techniques have now been applied to this problem as well[24-26]. First, the properties of the solute ground and excited electronic states have been analyzed with respect to their energies and approximate symmetries[24]. The CTTS spectra are found to have a complex structure with several overlapping subbands, with an overall lineshape resulting from the separate response of each subband to the field exerted by the solvent surrounding the ionic cavity. This structure is strongly reminiscent of the structure identified for the hydrated electron[21,22]. The results for one model of a solvated halide[24] are shown in Fig. 5.

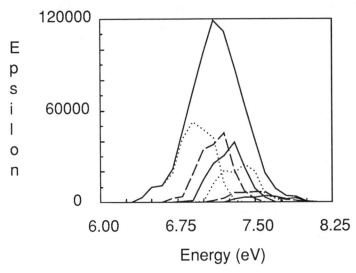

Figure 5 : Computed ground state absorption spectrum of an aqueous iodide . The solid line denotes the envelope corresponding to the full spectrum. Also shown are the subbands associated with the transitions of the p-type valence electron to the six one-electron states of mixed s and d character identified in the calculation as important in the principal CTTS absorption band.

The result corresponds reasonably to the observed intense ultraviolet absorption of aqueous halides, with a featureless overall bandshape. Further analysis[24] has led to the conclusion that the results reflect a very significant influence on each state of the field of both the atomic halogen atom as well as the solvent. Earlier models[35,36] emphasized only the atomic core or solvent for each state, so that the electronic characteristics of the excited states have been found to be considerably more complex then previously anticipated.

In further analysis[25,26], we have examined possible dynamical pathways at early time following the experimental two photon excitation of such halides[37,38]. By comparison of the calculated absorption spectra in aqueous solution[25] of the ground state and of the excited electronic states of the halide immediately following excitation, with the experimental observations[37,38], it has been possible to conclude[25] that the initially excited state is located above the principal CTTS absorption and was not directly observed; the

CTTS state absorption initially observed is the result of an ultrafast relaxation process occurring within instrumental resolution.

Most recently, direct dynamical simulation of the processes resulting following two photon excitation of the halide ion has led to evidence for a rich dynamical behavior.[26] These excited states can, in some cases, exhibit direct, nonadiabatic, electron transfer to preexisting solvent cavities. However, the predominant excited state relaxation channel observed does exhibit the anticipated ultrafast relaxation to the lowest state of the CTTS band. This is followed, sequentially, by adiabatic electron detachment and subsequent ultrafast geminate recombination. The entire process occurs on a time scale of a few picoseconds. These results are consistent with both classic photochemistry experiments[33,34] and modern ultrafast measurements[37,38], but provide considerable new details. It is important to address the generality of such alternative relaxation and ionization channels in the numerous known examples of molecular photoionization in solution[27] in future studies.

4. Conclusions

The emerging methods of mixed quantum-classical simulation are providing details of solution spectroscopy and dynamics that complements the details emerging from experimental analysis. The prospects for comparably detailed investigations of increasingly more complex systems in the future are very good. In the short term, it is of great interest to extend the studies to the important areas of electron and proton transfer reactions. Via applications of these simulation methods and new developments that are certain to arise, the anticipated progress is virtually certain to be rapid.

Acknowledgements

The work reported here has been supported by grants from the Robert A. Welch Foundation (F-0761) and from the Office of Naval Research. The computational support of the Center for High Performance Computing of The University of Texas System is also gratefully acknowledged.

References

1.   The SPC, SPC/E and the flexible "SPC/F" models and their properties are summarized in: K. Watanabe and M. L. Klein, "Effective Pair Potentials and the Properties of Water", Chem. Phys. **131**,157 (1989).

2.   J. Schnitker and P.J. Rossky, "An Electron-Water Pseudopotential for Condensed Phase Simulation", J. Chem. Phys. **86**, 3462 (1987).

3.   G. J. Martyna and M.L. Klein, "The electronic states of lithium atoms in ammonia clusters and solution", J. Chem. Phys. **96**, 7662 (1992).

4.   A. Wallqvist, D. Thirumalai, and B. J. Berne, "Path integral Monte Carlo study of the hydrated electron", J. Chem. Phys. **86,** 6404 (1987).

5.   R. N. Barnett, U. Landman, C. L. Cleveland and J. Jortner, "Electron localization in water clusters. I. Electron-water pseudopotential", J. Chem. Phys. **88**, 4421 (1988).

6. M. Hilczer, W. M. Bartczak, and M. Sopek, "An Excess Electron-Methanol Pseudopotential", J. Phys. Chem. **96**, 2736 (1992).

7. B. J. Berne and D. Thirumalai, "On the Simulation of Quantum Systems: Path Integral Methods", Ann. Rev. Phys. Chem. **37**, 401 (1986).

8. J. S. Bader, R. A. Kuharski, and D. Chandler, "Role of nuclear tunneling in aqueous ferrous-ferric electron transfer", J. Chem. Phys. **93**, 230 (1990).

9. J.K. Hwang, Z. T. Chu, A. Yadav, and A. Warshel, "Simulations of Quantum Mechanical Corrections for Rate Constants of Hydride-Transfer Reactions in Enzymes and Solutions", J. Phys. Chem. **95**, 8445 (1991).

10. J. Lobaugh and G. A. Voth, "Calculation of quantum activation free energies for proton transfer reactions in polar solvents", Chem. Phys. Letters **198**, 311 (1992).

11. D. Thirumalai, E. J.Bruskin, and B. J.Berne, "On the use of semiclassical dynamics in determining electronic spectra of Br_2 in an Ar matrix", J. Chem. Phys. **83**, 230 (1985).

12. A. Selloni, P. Carnevali, R. Car, M. Parrinello, "Localization, Hopping, and Diffusion of Electrons in Molten Salts", Phys. Rev. Letters **59**, 823 (1987).

13. R. N. Barnett, U. Landman, and A. Nitzan, "Dynamics and spectra of a solvated electron in water clusters", J. Chem. Phys. **89**, 2242 (1988).

14. F. Webster, P. J. Rossky, and R. A. Friesner, "Nonadiabatic Processes in Condensed Matter: Semi-Classical Theory and Implementation", Comput. Phys. Commun. **63**, 494 (1991).

15. J.C. Tully and R.K. Preston, "Trajectory Surface Hopping Approach to Nonadiabatic Molecular Collisions : The Reaction of H^+ with D_2", J. Chem. Phys. **55**, 562 (1971).

16. P. Pechukas, "Time-Dependent Semiclassical Scattering Theory. I. Potential Scattering", Phys.Rev. **181**, 166 (1969); "Time-Dependent Semiclassical Scattering Theory. II. Atomic Collisions", Phys.Rev. **181**, 174 (1969).

17. E. Neria, A. Nitzan, R. N. Barnett, and U. Landman, "Quantum Dynamical Simulations of Nonadiabatic Processes : Solvation Dynamics of the Hydrated Electron", Phys. Rev. Lett. **67**,1011 (1991); E. Neria and A. Nitzan, "Semiclassical Evaluation of Non-Adiabatic Rates in Condensed Phases", J. Chem. Phys. (in press). The second of these works provides revised values for the excited state lifetimes reported in the first.

18. J.C. Tully, "Molecular dynamics with electronic transitions", J. Chem. Phys. **93**, 1061 (1990).

19. B. Space and D.F. Coker, "Nonadiabatic dynamics of excited excess electrons in simple fluids", J. Chem. Phys. **94**, 1976 (1991); "Dynamics of trapping and localization of excess electrons in simple fluids", *ibid* **96**, 652 (1992).

20. F. J. Webster, J. Schnitker, M. S. Friedrichs, R. A. Friesner, and P. J. Rossky, "Solvation Dynamics of the Hydrated Electron: A Nonadiabatic Quantum Simulation", Phys. Rev. Lett. **66**, 3172 (1991).

21. J.S. Schnitker, A.K. Motakabbir, P.J. Rossky, and R.A. Friesner, "An *A Priori* Calculation of the Optical Absorption Spectrum of the Hydrated Electron", Phys. Rev. Letters **60**, 456 (1988).

22. P.J. Rossky and J.S. Schnitker, "The Hydrated Electron: Quantum Simulation of Structure, Spectroscopy, and Dynamics", J. Phys. Chem. **92**, 4277 (1988).

23. T. H. Murphrey and P. J. Rossky, "The Role of Solvent Intramolecular Modes in Excess Electron Solvation Dynamics", J. Chem. Phys. (in press).

24. W.-S. Sheu and P. J. Rossky, "Charge-Transfer-to-Solvent Spectra of an Aqueous Halide Revisited via Computer Simulation", J. Am. Chem. Soc. (in press).

25. W.-S. Sheu and P. J. Rossky, "The Electronic Dynamics of Photoexcited Aqueous Iodide", Chem. Phys. Lett. **302**, 186 (1993).

26. W.-S. Sheu and P. J. Rossky, "Dynamics of Electron Photodetachment from an Aqueous Halide Ion", Phys. Rev. Lett. (submitted for publication).

27. see, for example, J. C. Mialocq, "La Formation de L'Electron Solvate en Photochimie", J. Chim. Phys. **85**, 31 (1988).

28. see, for example, D. F. Calef and P. G. Wolynes, "Smoluchowski-Vlasov theory of charge solvation dynamics", J. Chem. Phys. **78**, 4145 (1983).

29. A. Migus, Y. Gauduel, J. L. Martin, and A. Antonetti, "Excess Electron in Liquid Water : First Evidence of a Prehydrated State with Femtosecond Lifetime", Phys. Rev. Lett. **58**, 1559 (1987).

30. Summaries of a number of recent measurements are provided in : H. Lu, F. H. Long, and K. B. Eisenthal, "Femtosecond Studies of Electrons in Liquids", J. Opt. Soc. Am. B **7**, 1511 (1990); Y. Gauduel, S. Pommeret, A. Migus, N. Yamada, and A. Antonetti, "Femtosecond investigation of single electron transfer and radical reactions in aqueous media and bioaggregate mimetic systems", *ibid.* **7**, 1528 (1990).

31. F. H. Long, H. Lu, and K. B. Eisenthal, "Femtosecond Studies of the Presolvated Electron : An Excited State of the Solvated Electron?", Phys. Rev. Letters **64**, 1469 (1990).

32. E. Keszei, S. Nagy, T. H. Murphrey, and P. J. Rossky, "Kinetic Analysis of Computer Experiments on Electron Hydration Dynamics", J. Chem. Phys. (submitted for publication).

33. J. Jortner, M. Ottolenghi, and G. Stein, "On the Photochemistry of Aqueous Solutions of Chloride, Bromide and Iodide Ions", J. Phys. Chem. **68**, 247 (1964).

34. F. S. Dainton and S. R. Logan, "Primary processes in the photolysis of the iodide ion in aqueous solution", Proc. Roy. Soc. (London) **A287**, 281 (1965).

35 T. R. Griffiths and M. C. R. Symons, "Solvation Spectra 3. - Further studies of the effect of environmental changes on the ultra-violet spectrum of iodide ions", Trans. Faraday Soc. **56**, 1125 (1960).

36. J. Jortner, and A. Treinin, "Intensities of the Absorption Bands of Halide Ions in Solution", Trans. Faraday Soc. **56**, 1503 (1961).

37. F. H. Long, H. Lu, X. Shi, and K. B. Eisenthal, "Femtosecond studies of electron photodetachment from an iodide ion in solution : the trapped electron", Chem. Phys. Lett. **169**, 165 (1990).

38. F. H. Long, Ph. D. Dissertation, Columbia University, 1991; F. H. Long, H. Lu, X. Shi, and K. B. Eisenthal, "Photodetachment from halide ions in solution" (to be published).

ELECTRONIC STRUCTURE AND CHEMICAL REACTIONS IN SOLUTION

James T. Hynes, Hyung J. Kim,[a] Jeffery R. Mathis, Roberto Bianco,[b] Koji Ando, and Bradley J. Gertner

Department of Chemistry and Biochemistry, University of Colorado, Boulder, CO 80309-0215, USA. a) Present address: Department of Chemistry, Carnegie Mellon University, Pittsburgh, PA 15213-3890; b) also associated with Dipartimento di Chimica e Chimica Industriale, Università di Pisa, I-56126 Pisa, Italy

ABSTRACT

A review is given of a recent theoretical approach to the issue of the title. Novel aspects of the theory include incorporation of nonequilibrium solvation, attention to time scales of the solvent electronic polarization and construction of reaction paths and free energetics. The theory is illustrated with examples of an S_N2 nucleophilic displacement, S_N1 ionization, and an ion-radical recombination reaction.

1. INTRODUCTION

Our understanding of the classical, <u>nuclear</u> aspects of solution reactions has reached a certain stage of maturity. In particular, the theory of Grote and Hynes [1,2] for assessing the influence of solvent dynamical effects on reducing the reaction rate constant below its Transition State Theory value has proven to be highly successful when compared to detailed microscopic molecular dynamics computer simulations for a wide variety of reaction types [3]. It has also proven useful in interpreting experimental results [4]. Several reviews [2,5] may be consulted for a survey.

However, our understanding of, and ability of prediction for, solute <u>electronic</u> structural aspects of chemical reactions in solution is comparatively much less developed. These features, and their alteration by the solvent, will have a significant impact on solution reaction rates: they will influence the reaction free energy barrier height. By contrast, the dynamical effects referred to above will generally influence only the

289

J. Jortner et al. (eds.), Reaction Dynamics in Clusters and Condensed Phases, 289–309.

nonexponential prefactor in the reaction rate constant. Quantum chemical electronic features will be especially important in reactions involving charge transfer or shift in polar solvents, due to strong Coulombic solute-solvent interactions.

In fact, some special aspects of the solute electronic structure problem in solution have long been studied. One example is activated electron transfers [6], in which the barrier often is completely or significantly determined by the solvent. There are also numerous studies employing self-consistent [7] and direct [8] reaction field methods to determine solute electronic structure in the special case when the solvent is equilibrated to the solute.

Here we present a brief review of some of our own efforts to understand, and construct theories for, solute electronic structure in solution in a chemical reaction context [9]. Compared to past efforts, some novel aspects include a) allowance for the solvent to be out of equilibrium with the solute, i.e., nonequilibrium solvation; b) explicit incorporation of the time scale of the solvent electronic polarization, which in part determines whether a self-consistent or Born-Oppenheimer or some intermediate description is appropriate; and c) a focus on reaction energetic profiles and paths, over the entire range of solute electronic coupling.

The outline of this review is as follows: In Sec. 2, we briefly sketch the issues and the theoretical framework. The remaining sections deal with the highlights of the results for differing reaction systems: an S_N2 reaction (Sec. 3), S_N1 ionization (Sec. 4), and radical-ion recombination (Sec. 5). We only give a brief overview and a few highlights in each case, and refer the reader to the original papers for details.

2. THEORY

A central issue of electronic structure in solution is that of delocalization versus localization. Consider as an example [9] two degenerate diabatic valence bond states Ψ_{A^-A} and Ψ_{AA^-} for a symmetric electron transfer (ET) system $D^-A \rightarrow DA^-$, with D=A. The electronic or resonance coupling between them will favor the formation of a delocalized, electronically adiabatic symmetric ground state $\Psi_S = \frac{1}{\sqrt{2}}(\Psi_{A^-A} + \Psi_{AA^-})$, while instead the Coulombic interactions of either diabatic structure with a polar and polarizable solvent will favor that localized structure. Although not usually discussed this way, the activation barriers for activated, electronically adiabatic ET reactions can be comprehended in this manner [9]. One important consequence of this competition is that the solute electronic structure will be altered by the solvent-an especially dramatic example of this is the S_N1 ionization (Sec. 4).

While a molecular description of the solvent is desirable (and under construction), our efforts to date have adopted a dielectric continuum [10] solvent characterization as a useful first approach. The solvent is thereby characterized by an electronic polarization field $\underline{P}_{el}(\underline{x})$ at any point $\underline{x}$ in the solvent, related to the electronic polarizability of the solvent molecules, and an 'orientational' polarization field $\underline{P}_{or}(\underline{x})$, related to the nuclear orientation (as well as translational and vibrational degrees of freedom) of the solvent molecules, e.g., their permanent dipole moments. The electric susceptibilities χ_i—the proportionality constant between $\underline{P}_i(\underline{x})$ and the local electric field for these (when equilibrium holds)—are

$$\chi_{el} = \frac{1}{4\pi} (\varepsilon_\infty - 1) \; ; \quad \chi_{or} = \frac{1}{4\pi} (\varepsilon_0 - \varepsilon_\infty) \; , \tag{1}$$

where ε_∞ is the high-frequency (optical), and ε_0 is the low-frequency or static, dielectric constant.

Beginning from a general expression for the nonequilibrium free energy of the system, treating $\underline{P}_{el}(\underline{x})$ in a quantum mechanical fashion sketched below, and with the solute wave function ψ, represented by the expansion over two diabatic valence bond states,

$$\Psi = c_1\Psi_1 + c_2\Psi_2 \; , \tag{2}$$

the free energy of the nonequilibrium solute-solvent system is [9]

$$
\begin{aligned}
G[\Psi, \underline{P}_{or}] \quad &= \quad (1-f) \, G_{BO}[\Psi, \underline{P}_{or}] + f \, G_{SC}[\Psi, \underline{P}_{or}]; \\
G_{BO,SC} \quad &= \quad \langle \Psi | H^0 | \Psi \rangle - \frac{1}{8\pi} (1 - \frac{1}{\varepsilon_\infty}) \int dx \; A_{BO,SC} \\
&\quad - \frac{1}{\varepsilon_\infty} \int dx \; \langle \Psi | \underline{E} | \Psi \rangle \cdot \underline{P}_{or} \\
&\quad + \frac{2\pi\varepsilon_0}{\varepsilon_\infty(\varepsilon_0 - \varepsilon_\infty)} \int dx \; \underline{P}_{or} \cdot \underline{P}_{or}; \\
A_{BO} \quad &= \quad \langle \Psi | \underline{E}^2 | \Psi \rangle; \; A_{SC} = \langle \Psi | \underline{E} | \Psi \rangle^2 \tag{3}
\end{aligned}
$$

Here G_{BO} and G_{SC} are the free energies in the Born-Oppenheimer (BO) and self-consistent (SC) approximations, described in some detail below. $\underline{E}$ is the (bare) electric field arising from the solute, and the spatial integrals are over the solvent outside of a cavity or cavities encasing the solute. The fraction

$$f = \rho(2c_1c_2 + \rho)^{-1} = (\tau_{el} + \tau_{tr})^{-1} \tau_{el} \quad , \tag{4}$$

where the ratio ρ involving the electronic coupling β between the diabatic states and the frequency associated with the electronic polarization is

$$\rho = \frac{2\beta}{\hbar\omega_{el}} \quad , \tag{5}$$

is related to the time scales of the electronic polarization $\tau_{el} = 2\pi/\omega_{el}$ and of the transferring electron, $\tau_{tr} = 2\pi\hbar c_1 c_2/\beta$. As discussed below, $\hbar\omega_e \geq 2$ eV for typical solvents, while β can range up to several eV, depending on the reaction.

In the SC approximation $\rho \to \infty$, $f \to 1$, the electronic polarization is sufficiently slow that $\underline{P}_{el}$ is equilibrated to the average solute field. In the opposite BO limit $\rho, f \to 0$, it is fast enough that the equilibrated $\underline{P}_{el}$ "sees" the fields of the individual VB components in the solute charge distribution. The latter approximation has been the standard one for weak to modest electronic coupling ET reactions [6]; the former approximation has been the usual one in equilibrium reaction field studies [7] (although some of these formulations use the BO approximation [8]).

The variational principle optimization of G with respect to ψ yields the nonlinear Schrödinger equation [9]

$$H[\Psi]\Psi = E\Psi \quad . \tag{6}$$

In general, when f is finite, the Hamiltonian depends through the equilibrated $\underline{P}_{el}(\underline{x})$ on ψ, i.e., in the Hamiltonian H whose explicit form we do not need, the transferring electron interacts with $\underline{P}_{el}(\underline{x})$, which itself depends upon the solute wave function.

The nonlinear Eq. (6) can be solved for the coefficients $\{c_i\}$, which are functions of $\underline{P}_{or}$. With Eq. (3), this gives the nonequilibrium free energy for any given nonequilibrium or equilibrium $\underline{P}_{or}$. We refer the reader to Ref. 12a-c for the details and to Ref. 12d for a generalization to multiple VB states, including attention to solute cavity boundary effects.

In the case of full equilibrium, in which both the polarizations $\underline{P}_{el}$ and $\underline{P}_{or}$ are equilibrated to the solute, one has

$$P_{\sim or}^{eq} = \frac{1}{4\pi} (\frac{1}{\epsilon_\infty} - \frac{1}{\epsilon_0}) \, \epsilon_\infty \, \langle \Psi \mid \underline{E} \mid \Psi \rangle \, , \tag{7}$$

and the equilibrium version [9] of the theory sketched above gives the electronic structure and free energy when there is equilibrium solvation. The predictions of this theory reduce to the predictions of SC [7] or direct (BO) [8] equilibrium reaction field methods in the appropriate ρ limits, but in general they differ from those predictions by the influence of the P_{el} and transferring electron time scales via the factor f.

Eqs. (3) and (6) were derived [9] via a multiconfiguration self-consistent approach in which, for given P_{or} , the solute and P_{el} are treated quantum mechanically,

$$\mid \Phi \rangle = c_1 \mid 1 \rangle \mid P_{el,1} \rangle + c_2 \mid 2 \rangle \mid P_{el,2} \rangle \tag{8}$$

where the $\mid P_{el,i} \rangle$ are treated [9] as coherent states [11]. The quantum description of P_{el} is necessitated by its high characteristic frequency ω_{el}, which is approximately the UV-visible absorption frequency of the solvent: $\hbar\omega_{el} \geq 2$ eV [9].

According to Eqs. (3) and (4), if the solute is localized, $c_1 c_2 \to 0$, the general result reduces to the SC limit which is identical to the BO description in this limit. For any degree of delocalization, however, the situation depends on the value of ρ, Eq. (5), and thus f; for a given solvent, the dependence is on the value of the electronic coupling β, which can vary widely between reaction classes, and within a given reaction class. For activated ET reactions, β is often $\leq k_B T$ and though the differences are not large in this limit [9,12a], the BO approximation is more appropriate [9,13,14] than is an SC treatment [12,15]. (In our original work which initiated the construction of an electronic structure-nonequilibrium solvation theory over the entire coupling range [12a,b], the SC limit was inappropriately [9,13,14] applied in this weak coupling regime.) For higher coupling ET reactions, neither limit [12-15] is appropriate [9]. The situation for S_N2, S_N1, and ion-radical recombination reactions, where the coupling is $O(1$ eV$)$ or greater, is discussed below.

3. S_N2 REACTIONS

We first examine how the theory sketched in Sec. 2 can be applied to the S_N2 reaction class [16]

$$N^- + RX \to NR + X^- \, , \tag{9}$$

where N⁻ is an incoming nucleophile, R is a primary or secondary alkyl group [17], and X is the leaving group. S_N2 reaction rates in vacuum and solution differ considerably. In vacuum, the rates are often so fast as to be experimentally inaccessible [18]. In solution, however, the point-like charge distribution is solvated to a much greater extent than the delocalized transition state intermediate, which results in an activation barrier often in the range of 15-30 kcal/mol, and means as much as a 10^{10} fold decrease in rate.

The symmetric chloride-methylchloride reaction system

$$Cl^- + CH_3Cl \rightarrow ClCH_3 + Cl^- \tag{10}$$

is a common example of an S_N2 reaction, and one for which an accurate *ab initio* vacuum potential energy surface has been computed [19]. To begin the analysis, we use a two VB state description for the reactants and products according to [16]

$$\psi_1 = [Cl^- / CH_3Cl] \quad ; \quad \psi_2 = [ClCH_3 / Cl^-] . \tag{11}$$

This two VB state is akin to that used by Pross and Shaik [20a] in their extensive investigations of S_N2 reactions in both vacuum and solution. However, in the simple application discussed, the source of the CH_3Cl dipole moment is assumed to arise from purely covalent sources [21], and does not result from a pre-mixing of the ionic state [20a]

$$\psi_3 = [Cl^- CH_3^+ Cl^-] \tag{12}$$

into ψ_1 and ψ_2. We leave to a future publication further discussion of the involvement of the third state ψ_3 in S_N2 reactions, as a three-state model is considerably more complicated.

The diabatic energies H^0_{11} and H^0_{22} for the states ψ_1 and ψ_2 respectively, can be suitably obtained [16], and are displayed schematically in Fig. 1 along the reaction coordinate, taken to be the [Cl-CH₃-Cl]⁻ asymmetric stretch [18-20a]. The adiabatic vacuum potential energy surface E for Eq. (10) is then given by

$$E = \frac{1}{2}[H^0_{11} + H^0_{22}] - \frac{1}{2}[(H^0_{22} - H^0_{11})^2 + 4\beta^2]^{1/2} \tag{13}$$

where $-\beta = \langle \psi_1 | H | \psi_2 \rangle$ is the electronic coupling between the two states, and is treated as an adjustable parameter. It is found that in order to reproduce both the accepted gas-phase activation energy of 13.7 kcal/mol [19], as measured from the ion-dipole complex and the ion-dipole complex well depth, a coordinate dependence must be given to β. This is of interest in that β is usually assumed to be constant [20a]. The value of β necessary to reproduce the vacuum activation barrier is found to be 31.2 kcal/mol.

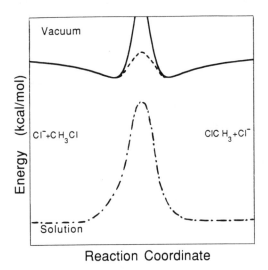

Fig. 1   Schematic VB diagram for the $Cl^- \longrightarrow CH_3Cl$ S_N2 reaction system along the reaction coordinate, defined here as the $[Cl-CH_3-Cl]^-$ asymmetric stretch. —— vacuum diabatic energies; - - - - vacuum adiabatic energy; — • — adiabatic profile in solution. Preferential solvent stabilization of the reactants results in a much larger activation barrier in solution as compared to that in vacuum.

As mentioned above, a much larger activation barrier is present in solution compared to that in vacuum, due to the greater solvent stabilization of the reactants than the transition state, as illustrated schematically in Fig. 1. However, the magnitude of the transistion state stabilization in solution depends strongly on how the electronic polarization is treated. For example, we find that in water solvent, the BO activation free energy calculated according to the procedure given in Ref. 16 is 22.2 kcal/mol, while the SC activation free energy is 29.9 kcal/mol. This is a significant difference given the exponential sensitivity of the reaction rate constant to this quantity. Using the exact theory outlined in Sec. 2 gives an intermediate result: 24.2 kcal/mol, and is closest to other estimates (26.6 kcal/mol [22b]) based on other experimentally and empirically grounded methods. This result indicates that one should be very cautious in applying the standard reaction field methods [22]. The reader is referred to Ref. 16 for a full disclosure of the procedure used to obtain these activation free energies and its discussion.

4. S_N1 IONIZATION

The S_N1 unimolecular ionization $RX \rightarrow R^+ + X^-$ is another fundamental reaction class in organic chemistry [17]. It is also one for which the solvent is critical. As Fig. 2 illustrates, the vacuum valence bond covalent and ionic state energetics are typically such that no thermal ionization is to be expected in the gas phase; it is only the solvent stabilization of the ionic state that allows the ionization to occur. This key feature was pointed out years ago by Evans, Polanyi, Ogg and their coworkers [23]. If one were to electronically couple the two VB states to generate an electronically adiabatic surface in vacuum, it would be purely covalent at large RX separations. The subsequent solvation of this state would not produce any stable ionic configurations; instead, the electronic coupling and solvation effects must be allowed to operate simultaneously, as in the theory sketched in Sec. 2.

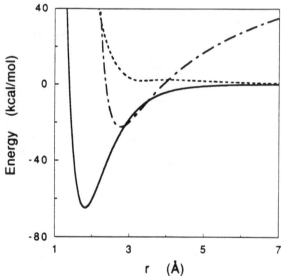

Fig. 2   The gas phase diabatic energies and coupling for t-BuCl. —— and — • — denote the vacuum covalent and ionic curves, respectively, while • • • represents the electronic coupling β.

This ionization for a model of t-BuCl—the most well-known S_N1 reactant—has been treated [24] via the theory sketched in Sec. 2, with the additional feature that the RX separation r is taken into account. Thus the solute electronic structure is described by the expansion

$$\Psi(r,s) = c_C(r,s)\Psi_C(r) + c_I(r,s)\Psi_I(r) \tag{14}$$

over (orthogonalized [24]) covalent and ionic valence bond states. The electronic coupling between these states is strong, $\beta \sim 0.5 - 0.75$ eV near the solution reaction transition states. The coefficients here depend both on r and a solvent coordinate s, which gauges the extent of the solvent orientational polarization via the definition

$$\underset{\sim or}{P} = s\frac{1}{4\pi}(\frac{1}{\epsilon_\infty} - \frac{1}{\epsilon_0})\,\epsilon_\infty\,\langle\,\Psi_I\,|\,\underset{\sim}{E}\,|\,\Psi_I\,\rangle\ , \tag{15}$$

where it is assumed for simplicity that the covalent state has no dipole moment.

With this formalism, two dimensional nonequilibrium free energy surfaces for the S_N1 ionization can be constructed. Fig. 3 gives an example for t-BuCl in acetonitrile, where two reaction paths are displayed. The first is the equilibrium solvation path (ESP), along which the solvent is imagined to always be equilibrated, at any r, to the solute: $\partial G/\partial s = 0$. This is the path that would be implied by considering the equilibrium potential of mean force [2,25]—the typical object of equilibrium statistical mechanical calculation, and is closely associated with standard, equilibrium solvation transition state theory [26]. The second path is the solution reaction path (SRP). This is the generalization to solution, by Lee and Hynes [27], of the familiar intrinsic reaction path in the gas phase, introduced by Fukui [28] and often studied [29]. Fig. 3 shows that near the transition state, which is the saddle point on the surface, the SRP differs markedly from the ESP. Passage over the barrier is so rapid that there is not sufficient time for the solvent to equilibrate to the rapidly changing solute charge distribution, i.e., for the ESP to be followed. One consequence of this is that the transmission coefficient

$$\kappa = \frac{k}{k^{TST}} \tag{16}$$

which is a measure of the effect of nonequilibrium solvation through the ratio of the actual rate constant to its equilibrium solvation TST value [26], is noticeably less than unity. For example, it is estimated [24b] that $\kappa = 0.65$ for t-BuCl in CH3CN.

The basic characteristics of the reaction path and the deviation from TST have been confirmed, via an application of a variant of the formulation of Ref. 24a, in a Molecular Dynamics computer simulation of a model of the t-BuCl ionization in water [30]. In this simulation, $\kappa = 0.53$. This value is reproduced [30] by Grote-Hynes Theory [1,2] and its nonadiabatic solvation limit [25]; by contrast, the Kramers Theory [31] prediction is far below this: $\kappa_{KR} = 0.02$.

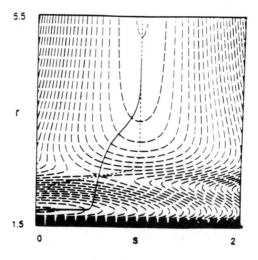

Fig. 3 Free energy contour map for t-BuCl in CH3CN. — — represent the equi-free energy lines in units of 0.1 eV. The free energy value for the contour in the upper center of the map is -3.1 eV. r is in Å. —— and • • • denote the solution reaction path (SRP) and equilibrium solvation path (ESP), respectively.

But perhaps the most novel results of the S_N1 study [24] concern the transition state location and free energetics. It is well known experimentally that the activation free energy $\Delta G^{\ddagger}$ decreases for t-BuCl (and other) ionizations with increasing solvent polarity [17,32]. The conventional explanation of this is due to Hughes and Ingold [17]: since the transition state has some ionic character $\{R^{\delta+} X^{\delta-}\}^{\ddagger}$ while the reactant RX has very little, increasing solvent polarity will preferentially stabilize the transition state and lower the activation barrier. But, a quite different picture emerges from the S_N1 study [24a]: the decrease of $\Delta G^{\ddagger}$ for t-BuCl arises from the separation r-variation of the large electronic coupling (~13-18 kcal/mol) between the covalent and ionic valence bond states. The transition state stabilization actually <u>decreases</u> with increasing solvent polarity. A simplified explanation of these striking features is the following [24a]. Fig. 4 illustrates the solvated diabatic curves. For increasing solvent polarity, the curve crossing—which is an approximation to the TS location—occurs at more reactant-like geometries, in accord with the Hammond postulate [33]. The TS ionic character also becomes more reactant-like, i.e., <u>less</u> ionic, in a Hammond fashion. (Actually this does not follow directly from the diabatic curve crossing location, and a detailed analysis [24a] is required to establish it). It is the diminishing TS ionic character that leads to a less favorable TS solvent stabilization; the tighter TS location in r

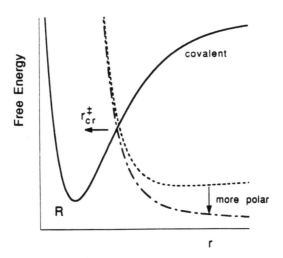

Fig. 4 Hammond postulate diagram. The equilibrated ionic curves for CH3CN and C6H5Cl are noted by — • — and • • • respectively, while the equilibrated covalent curve is represented by ——. With increasing solvent polarity, the reaction becomes less endothermic as indicated by the vertical arrow; the crossing point moves in towards the reactant state *R*. The transition state becomes more reactant-like with growing solvent polarity.

leads to an increased electronic coupling (approximately exponential in r), and it is this increased coupling which increasingly suppresses the activation barrier (cf. Fig. 5). (For why the solvent stabilization nonetheless applies in the SN2 example of Sec. 3, see Ref. 16.)

It is interesting to note that a number of computational results for a model of hydride transfers with a reaction field designed to be appropriate for an enzyme active site [34] show a number of the features sketched here and may independently support our theoretical perspective.

A possible experimental avenue to test these unconventional predictions is [20a] via the Bronsted coefficient α connecting the activation and reaction free energies

$$\alpha = \frac{d\Delta G^{\ddagger}}{d\Delta G_{rxn}} \quad .$$

$$(17)$$

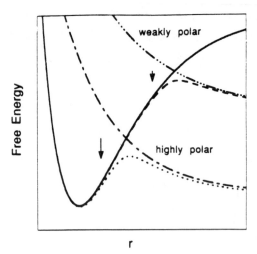

Fig. 5 Schematic diagram for electronic coupling contribution to the activation free energy. The covalent curve is denoted by ——, while the (equilibrated) ionic curves for weakly and highly polar solvents are represented by — • • — and — • — , respectively. The corresponding adiabatic states are denoted by — — — and • • •. The magnitude of the coupling stabilization of the transition state is indicated by an arrow; as solvent polarity increases, the transition state separation decreases (i.e., an earlier crossing) and the coupling stabilization increases.

A smaller α signifies [24a] a more reactant-like TS. Experiment should indicate a trend like that displayed in Fig. 6.

Other S_N1 ionizations have been studied as well. In the series t-BuCl, t-BuBr and t-BuI the basic features noted above for t-BuCl are repeated [24c]. The t-BuI example, however, displays extra interesting features. In this case, the electronic coupling at the TS is sufficiently low ($\approx$ 4 kcal/mol), that there is a barrier in the solvent coordinate for the ionization; the contrast with the t-BuCl case is sketched in Fig. 7. However, the solvent barrier is quite low, ~ 1 kcal/mol. Another system in which a larger solvent barrier (~ 4.5 kcal/mol) is calculated is isopropyl iodide [24c]. In each case, the TS ionic character still decreases with increasing solvent polarity and the TS solvent stabilization decreases. But now the electronic coupling r variation is small and is not responsible for the lowering of $\Delta G^{\ddagger}$. Instead it is the r variation of the vacuum diabatic energy difference [24c].

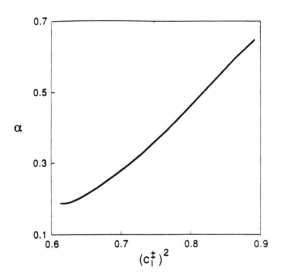

Fig. 6 The Brønsted coefficient α—transition state ionic character $(c_I^{\ddagger 2})$ plot. An almost linear correlation between the two holds for a wide solvent polarity range $(2.5 \lesssim \varepsilon_0 \lesssim 80,$ $\varepsilon_\infty = 2)$. Since the TS ionic character is found to diminish with increasing solvent polarity in model theoretical study [24], the solvent polarity dependence of α can serve as an experimental probe of this novel aspect.

All the S_N1 results discussed above were produced via either a perturbation method [9,24a,b] or an exact numerical analysis [9,24c]. A more "quantum chemical" sort of approach to the numerical calculations is one introduced in Ref. [9] and commented upon in Sec. 3. Here one returns to the model two-state Hamiltonian of the solute-solvent system in which the quantum electronic polarization of the solvent and its conjugate momentum appear as operators [9]. This may then be diagonalized using a suitably orthogonalized basis set constructed from the four states [35]

$$| C \rangle | P_{\tilde{el},C}^{BO} \rangle, \; | C \rangle | P_{\tilde{el},I}^{BO} \rangle, \; | I \rangle | P_{\tilde{el},I}^{BO} \rangle, \; | I \rangle | P_{\tilde{el},C}^{BO} \rangle . \tag{18}$$

The first and the third states here would combine to give the ground state in the BO approximation. The second and fourth would be involved in excited states in a BO picture. All these states contribute at finite ρ and f. The results of the numerical diagonalization [35] agree extremely well with those obtained via the perturbation [9,24a,b] or the exact numerical method [9,24c]. An advantage of this general method is that it can be easily used to generate actual electronically excited states of a solute-solvent system. But, since it is a numerical diagonalization procedure, it does not provide an analytical framework and such a framework was crucial in unravelling the source of the S_N1 TS patterns described above.

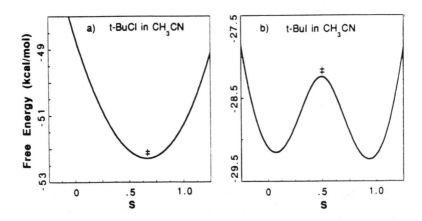

Fig. 7 Free energy profiles in CH3CN along the solvent cordinate s at fixed transition state separation $r^{\ddagger}$ for a) t-BuCl and b) t-BuI. For t-BuCl, the coupling is large enough ~ 18 kcal mol^{-1}) so that no solvent barrier is present. In the case of t-BuI, however, the coupling is small enough (~ 4 kcal mol^{-1}), to allow for the presence of a solvent barrier.

5. Molecular Ion-Radical Combination Reactions

As our final illustration, we turn to the example [36] of the ion-radical combination reaction of I$^-$ and I to form the stable complex I$_2^-$ in solution: this system,

$$I + I^- \rightarrow I_2^-$$

(19)

which differs fundamentally from the much studied I + I $\rightarrow$ I$_2$ system [37] due to the coulombic solute solvent interactions, is now under considerable experimental and computational scrutiny [38].

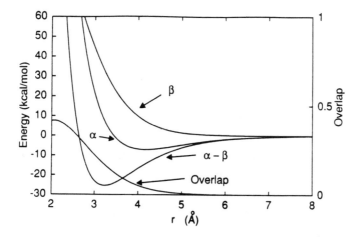

Fig. 8 Various vacuum quantities calculated for the I_2^- problem. (Here α denotes the diagonal VB energy.)

If we again adopt a two VB state point of view, here using ψ_{I-I} and ψ_{II}-, the system at small, near equilibrium internuclear separation r will be [9,14,36] a delocalized complex due to the strong resonance coupling $\beta(r)$. At larger separations, where the coupling between these two VB states has diminished, the localizing electrostatic interactions of a sufficiently polar solvent will produce two localized states separated by a barrier in a solvent coordinate. Figure 8 illustrates the calculated [36] vacuum potential energy curves for the degenerate diabatic states (plus the electronically adiabatic curve for reference) and the vacuum electronic coupling. The nonorthogonal VB wave functions are taken to be of the form $\psi_{I-I} = |\, I_A \bar{I}_A I_B\, |$ and $\psi_{II}\text{-} = |\, I_A I_A \bar{I}_B\, |$, where e.g

$$|\, I_A \bar{I}_A I_B\, | \;=\; \{\, I_A^c\ I_B^c\ I_A(1)\ \alpha(1)\ I_A(2)\ \beta(2)\ I_B(3)\ \alpha I(3)\, \} \tag{20}$$

so that three valence electrons are treated explicitly, with the remainder assigend to a core (c). These states, their energies, overlap and coupling are then determined [36] via the semi-empirical VB method of Zeiri and Shapiro [39]. the symmetric orthogonalized states ψ_{I-I}- and are then constructed as in Ref. 12, and it is to these states to which Fig. 8 and the ensuing discussion refer.

The computed [36] two dimensional free energy surface G(r,s) for I_2^- in (dielectric continuum) acetonitrile CH3CN solvent is displayed in Fig. 9. Both the single well delocalized minima at larger r are clearly apparent. The barrier in the solvent coordinate disappears rather "late" in the combination process, at r ≈ ψÅ.

The ESP and SRP are also illustrated in Fig. 9. The former is particularly misleading for the process. Along the ESP, the solvent would equilibrate instantly to the rapidly changing electronic structure in the transistion region where the solvent barrier is disappearing. Then, at a fixed solvent equilibrated to the delocalized structure, the path to ultimate equilibrium at the well bottom is exclusively in the solute nuclear coordinate r. By contrast, the SRP reflects the solvent inertia in that the solvent lags both the r motion and the changing electronic structure through, and after, the transition region. This point is exposed more explicitly in Fig. 10, where the delocalization variable $y = 2c_{I^-I}\,C_{II^-}$ is displayed.

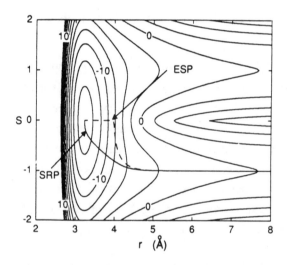

Fig. 9 Calculated free energy surface for I_2^- in acetonitrile solvent. s is the solvent coordinate such that s = ± 1 corresponds to solvent equilibrium with ψ_{II^-}, ψ_{I^-I}, while s = 0 corresponds to solvent equilibrium with the delocalized electronic structure. This and Fig. 10 are calculations in the full KH treatment.

The changing electronic structure of the combining solute leads to an extra "polarization" force which would be absent if the electronic structure were fixed. If for example, we refer the BO system free energy to its vacuum value in the delocalized state, then the polarization free energy is [36]

$$G_P = \beta y - \frac{1}{2}[(\Delta G)^2 + 4\beta^2]^{1/2} . \tag{21}$$

The polarization free energy G_P depends explicitly upon the evolving solute electronic stucture and vanishes in the delocalized state; it has not been accounted for in any MD simulation to date. The associated polarization force on the diatomic coordinate

$$F_P = -\partial G_P/\partial r \tag{22}$$

can be calculated [36]. Figure 10 shows that this force peaks in the transition state region, and there it is dominated [36] by the electronic coupling variation contribution in Eq. (22), and is larger than the solvent force arising from the fixed electronic structure term. The force is repulsive since the solvent would "prefer" a more localized charge distribution to maximize solute-solvent stabilization.

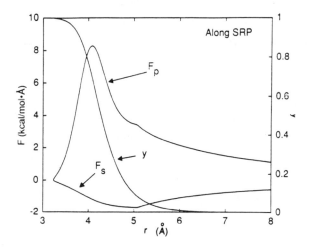

Fig. 10   The polarization force F_P and solvent force F_S (see the text) calculated along the solution reaction path in Fig. 9. $y = 2c_{I^-}c_{II^-}$ is the delocalization variable.

It seems likely that this force could play a significant role in the vibrational relaxation stabilization of the incipient I_2^-. In particular, this force should be significant in the beginning to intermediate stages of the process - where all to more than say half of (solution phase) vibrational energy resides in the solute molecular ion and the solvent has not yet shifted much from its initial s=1 neighborhood.

The analytical description of the vibrational relaxation presents an interesting challenge. For example, Fig. 10 shows that Fp is a strongly nonlinear function of the I_2^- coordinate r, and linearization approximations often employed in relaxation problems will fail. Nonetheless, the problem can alternatively be studied by MD computer simulation. In particular, a molecular Hamiltonian with the structure of Eq. (21) can be constructed which incorporates the evolving electronic structure [36].

6. CONCLUDING REMARKS

Here we have given a brief overview of some of our recent efforts to elucidate the issues of electronic structure in solution and thier impact on dynamical reactive processes for several chemical reaction types. Other applications to electron transfer [9], proton transfer [40] and excited electronic state twisted intramolecular charge transfer (TICT) reactions [41] have also been carried out. For all these reaction types, electronic stucture issues prove to be key in comprehending reaction paths and rates, and we expect that this will also be true, probably in surprising ways, as the range of reaction classes studied is expanded.

ACKNOWLEDGMENTS

This work was supported in part by NIH grant GM41332 and a Shannon award, NSF grant CHE88-07852, and grants of CRAY YMP time from the Pittsburgh Supercomputer Center. The late Teresa Fonseca played a vital and inspirational role in the creation of the theory described in Sec. 2.

REFERENCES

1. R. F. Grote and J. T. Hynes, *J. Chem. Phys.*, 73 (1980) 2715.
2. J. T. Hynes, in M. Baer (Editor),*The Theory of Chemical Reaction Dynamics*, Vol. 4, CRC Press, Boca Raton, 1985, Ch.4, p.171.
3. (a) J. P. Bergsma, J. R. Reimers, K. R. Wilson and J. T. Hynes, *J. Chem. Phys.*, 85 (1986) 5625; (b) J. P. Bergsma, B. Gertner, K. R. Wilson and J. T. Hynes, *J. Chem. Phys.*, 86 (1987) 1356; B. J. Gertner, J. P. Bergsma, K. R. Wilson, S. Lee and J. T. Hynes, *J. Chem. Phys.*, 86 (1987) 1377; B. J. Gertner, K. R. Wilson and

J. T. Hynes, *J. Chem. Phys.*, 90 (1989) 3537; (c) G. Ciccotti, M. Ferrario, J. T. Hynes and R. Kapral, *J. Chem. Phys.*, 93 (1990) 7137; (d) D. A. Zichi, G. Ciccotti, J. T. Hynes and M. Ferrario, *J. Phys. Chem.*, 89 (1989) 2093; (e) S. B. Zhu, J. Lee and G. W. Robinson, *J. Phys. Chem.*, 92 (1988) 2401; B. J. Berne, M. Borkovec and J. E. Straub, *J. Phys. Chem.*, 92 (1988) 3711; B. Roux and M. Karplus, *J. Phys. Chem.*, 95 (1991) 4856; R. Rey and E. Guardia, *J. Phys. Chem.*, 96 (1992) 4712.

4. B. Bagchi and D. W. Oxtoby, *J. Chem. Phys.*, 78 (1983) 2735; J. Ashcroft, M. Besnard, V. Aquada and J. Jonas, *Chem. Phys. Lett.*, 110 (1984) 430; D. M. Zeglinski and D. H. Waldeck, *J. Phys. Chem.*, 92 (1988) 692; N. Sivakumar, E. A. Hoburg and D. H.Waldeck, *J. Chem. Phys.*, 90 (1989) 2305; N. S. Park and D. H. Waldeck, *J. Phys. Chem.*, in press; G. E. McManis and M. J. Weaver, *J. Chem. Phys.*, 90 (1989) 1720.

5. D. G. Truhlar, W. L. Hase and J. T. Hynes, *J. Phys. Chem.*, 87 (1983) 2664; J. T. Hynes, *Ann. Rev. Phys. Chem.*, 36 (1985) 573; B. J. Berne, M. Borkovec and J. E. Straub, *J. Phys. Chem.*, 92 (1998) 3711; P. Hänggi, T. Talkner and M. Borkovec, *Rev. Mod. Phys.*, 62 (1990) 251; R. M. Whitnell and K. R. Wilson, *Rev. Comp. Chem.*, Vol. 4, in press.

6. See, e.g., the reviews by M. D. Newton and N. Sutin, *Ann. Rev. Phys. Chem.*, 35 (1984) 437, R. A. Marcus and N. Sutin, *Biochim. Biophys. Acta*, 811 (1985) 265; and R. D. Cannon and J. F. Endicott, in M. V. Twigg (Editor),*Mechanisms of Inorganic and Organometallic Reactions*, Vol. 6, Plenum, New York, 1989, Ch.1, p.3.

7. (a) H. Beens and A. Weller, *Chem. Phys. Lett.*, 3 (1969) 666; H. Beens and A. Weller in J. B. Birk (Editor), *Organic Molecular Photophysics*, Vol. 2, Wiley, London, 1975,Ch.4, p.159; (b) M. D. Newton, *J. Chem. Phys.*, 58 (1973) 5833; M. D. Newton, *J. Phys. Chem.*, 79 (1975) 2795; (c) S. Yomosa, *J. Phys. Soc. Jpn.*, 35 (1973) 1738; S. Yomosa, *J. Phys. Soc. Jpn.*, 36 (1974) 1655; S. Yomosa, *J. Phys. Soc. Jpn.*, 44 (1978) 602; (d) O. Tapia and O. Goscinski, *Mol. Phys.*, 29 (1975) 1653; O. Tapia in R. Daudel, A. Pullman, L. Salem and A. Veillard (Editors), *Quantum Theory of Chemical Reactions*, Vol. 2, Reidel, Dordrecht, 1980, p.25; O. Tapia, *J. Mol. Struc.*, 226 (1991) 59; (e) J. -L. Rivail and D. Rinaldi, *Chem. Phys.*, 18 (1976) 233; J. -L. Rivail in M. Moreau and P. Turq (Editors), *Chemical Reactivity in Liquids*, Plenum, New York, 1988; (f) C. Ghio, E. Scrocco and J. Tomasi, in B. Pullman (Editor), *Environmental Effects on Molecular Structure and Properties*, Reidel, Dordrecht, 1976, p.329; S. Miertuÿ, E. Scrocco and J. Tomasi, *Chem. Phys.*, 55 (1981) 117; S. Miertuÿ, E. Scrocco and J. Tomasi, *Chem. Phys.*, 65 (1982) 239; (g) G. Karlström, *J. Phys. Chem.*, 92 (1988) 1315; (h) K. V. Mikkelsen, E. Dalgaard and P. Swanstrøm, *J. Phys. Chem.* 91 (1987) 3081; K. V. Mikkelsen and M. A. Ratner, *Int. J. Quantum Chem. Quantum Chem. Symposium*, 22 (1988) 707; (i) M. Karelson, T. Tamm, A. R. Katritzky, M. Szafran and M. C. Zerner, *Int. J. Quantum Chem.*, 37 (1990) 1; (j) M. W. Wong, K. B. Wiberg and M. Frisch, *J. Chem. Phys.*, 95 (1991) 8991; M. W. Wong, M. Frisch and K. B. Wiberg, *J. Am. Chem. Soc.*, 113 (1991) 4776.

8. (a) J. Hylton, R. E. Christoffersen and G. G. Hall, *Chem. Phys. Lett.*, 26 (1974) 501; J. Hylton McCreery, R. E. Christoffersen and G. G. Hall, *J. Am. Chem. Soc.*, 98 (1976) 7191; (b) B. T. Thole and P. T. van Duijnen, *Theoret. Chim. Acta*, 55 (1980) 307; B. T. Thole and P. T. van Duijnen, *Chem. Phys.*, 71 (1981) 211; (c) For an extension, see J. G. Ångyán and G. Jansen, *Chem. Phys. Lett.*, 175 (1990) 313.

9. H. J. Kim and J. T. Hynes, *J. Chem. Phys.*, 96 (1992) 5088.
10. See, e.g., H. Fröhlich, *Theory of Dielectrics*, 2nd ed., Oxford, Cambridge, 1958.
11. R. J. Glauber, *Phys. Rev.* 131 (1963) 2766; T. W. Kibble, in S. M. Kay and A. Maitland (Editors), *Quantum Optics*, Academic, London, 1970, p.11.
12. (a) H. J. Kim and J. T. Hynes, *J. Phys. Chem.*, 94 (1990) 2736; (b) H. J. Kim and J. T. Hynes, *J. Chem. Phys.*, 93 (1990) 5194; (c) H. J. Kim and J. T. Hynes, *J. Chem. Phys.*, 93 (1990) 5211; (d) R. Bianco and J.T. Hynes, to be submitted,
13. J. Gehlen, D. Chandler, H. J. Kim and J. T. Hynes, *J. Phys. Chem.* 96 (1992) 1748.
14. R. A. Marcus, *J. Phys. Chem.*, 96 (1992) 1753. For a recent discussion from a more quantum chemical perspective, see R.H. Young, *J. Chem. Phys.* 97, 5261 (1992).
15. R. A. Marcus, *Faraday Symp. Chem. Soc.*, 10 (1975) 60.
16. J. R. Mathis, R. Bianco and J. T. Hynes, *J. Mol. Liq.* (1993), in press.
17. E. D. Hughes and C. K. Ingold, *Trans. Faraday Soc.*, 37 (1941) 657; C. K. Ingold, *Structure and Mechanism in Organic Chemistry*, Cornell University, Ithaca, 2nd ed., 1969.
18. W. N. Olmstead and J. I. Brauman, *J. Am. Chem. Soc.* 99, 4219 (1977); S. E. Barlow, J. M. Van Doren and V. M. Bierbaum, *J. Am. Chem. Soc.* 110, 7240 (1988); C. H. DePuy, S. Gronert, A. Mullin and V. M. Bierbaum, *J. Am. Chem. Soc.* 112, 8650 (1990); B. D. Wladkowski and J. I. Brauman, *J. Phys. Chem.* (1993), in press.
19. S. C. Tucker and D. G. Truhlar, *J. Am. Chem. Soc.* 93, 8138 (1989); S. C. Tucker and D. G. Truhlar, *J. Am. Chem. Soc.* 112, 3338 (1990); S. C. Tucker and D. G. Truhlar, *J. Am. Chem. Soc.* 112, 3347 (1990).
20. (a) For a general survey of this approach, see, e.g. S. S. Shaik, H. B. Schlegel and S. Wolfe, *Theoretical Aspects of Physical Organic Chemistry: The S_N2 Mechanism*, (John Wiley and Sons, New York, 1992) and references therein; (b) D. J. McLennan, *Aust. J. Chem.* 31, 1897 (1978); W. J. Albery and M. M. Kreevoy, *Adv. Phys. Org. Chem.* 16, 87 (1978); W. J. Albery, *Ann. Rev. Phys. Chem.* 31, 227 (1980).
21. See, e.g. C. A. Coulson, *Valence*, 2nd Ed. (Oxford Univ. Press, London, 1961).
22. See, e.g., M.V. Basilevsky, G.E. Chudinov and D.V. Napolov, *J. Phys. Chem.* 97 3270 (1993).
23. R. A. Ogg, Jr. and M. Polanyi, *Trans Faraday Soc.*, 31 (1935) 604; E. C. Baughan, M. G. Evans and M. Polanyi, *Trans Faraday Soc.*, 37 (1941) 377; A. G. Evans, *Trans Faraday Soc.*, 42 (1946) 719.
24. (a) H. J. Kim and J. T. Hynes, *J. Am. Chem. Soc.* 114, 10508 (1992); H.J. Kim and J.T. Hynes, *ibid.*, 10529 (1992); J.R. Mathis, H.J. Kim and J.T. Hynes, *ibid.*, in press.
25. G. van der Zwan and J. T. Hynes, *J. Chem. Phys.*, 76 (1982) 2993; G. van der Zwan and J. T. Hynes, *J. Chem. Phys.*,78 (1983) 4174; G. van der Zwan and J. T. Hynes, *Chem. Phys.*, 90 (1984) 21.
26. S. Glasstone, K. J. Laidler and H. Eyring, *The Theory of Rate Processes*, McGraw-Hill, New York, 1941. For a modern exposition, see, e.g., J. I. Steinfeld, J. S. Francisco and W. L. Hase, *Chemical Kinetics and Dynamics*, Prentice-Hall, Englewood Cliffs, 1989.
27. S. Lee and J. T. Hynes, *J. Chem. Phys.*, 88 (1988) 6853, 6863.
28. K. Fukui, *J. Phys. Chem.*, 74 (1970) 4161; K. Fukui, *Acc. Chem. Res.*, 14 (1981) 363.

29. W. H. Miller, N. C. Handy and J. E. Adams, *J. Chem. Phys.*, 72 (1980) 99; W. H. Miller, *J. Phys. Chem.*, 87 (1983) 3811. For a recent exposition, see, e.g., V. Melissas, D. G. Truhlar and B. Garrett, *J. Chem. Phys.*, 96 (1992) 5758 and references therein.
30. W. Keirstead, K. R. Wilson and J. T. Hynes, *J. Chem. Phys.*, 95 (1991) 5256.
31. H. A. Kramers, *Physica*, 7 (1940) 284.
32. (a) E. Grunwald and S. Winstein, *J. Am. Chem. Soc.*, 70 (1948) 846; A. H. Fainberg and S. Winstein, *J. Am. Chem. Soc.*, 78 (1956) 2770; (b) I. A. Koppel and V. A. Palm, *Org. React. (Tartu)*, 6 (1969) 213; (c) M. H. Abraham, *J. Chem. Soc. Perkin Trans. II*, (1972) 1343; (d) For a recent review, see G. F. Dvorko, E. A. Ponomareva and N. I. Kulik, *Russ. Chem. Rev.*, 53 (1984) 547.
33. G. S. Hammond, *J. Am. Chem. Soc.*, 77 (1955) 334. See also R. P. Bell, *Proc. Roy. Soc. (London) A*, 154 (1936) 414, and M. G. Evans and M. Polanyi, *Trans. Faraday Soc.*, 34 (1938) 11.
34. O. Tapia, J. Andres, J. M. Aullo and R. Cardenas, *J. Mol. Structure*, 167 (1988) 395.
35. H. J. Kim, R. Bianco, B. J. Gertner and J. T. Hynes, *J. Phys. Chem.* 97, 1723 (1993).
36. K. Ando, B. J. Gertner and J. T. Hynes, to be submitted.
37. A. L. Harris, J. K. Brown and C. B. Harris, *Ann. Rev. Phys. Chem.* 39, 341 (1988).
38. D. Ray, N. E. Levinger, J. M. Papanikolas and W. C. Lineberger, *J. Chem. Phys.* 91, 6533 (1989); J. M. Papanikolas, J. R Gord, N. E. Levinger, D. Ray, V. Vorsa and W. C. Lineberger, *J. Phys. Chem.* 95, 8028 (1991); A. E. Johnson, N. E. Levinger and P. F. Barbara, *J. Phys. Chem.* 96, 7841 (1992); I. Benjamin and R. M. Whitnell, *Chem. Phys. Lett.* 204, 45 (1993).
39. Y. Zeiri and M. Shapiro, *Chem. Phys.* 31, 217 (1978); M. Shapiro and Y. Zeiri, *J. Chem. Phys.* 70, 5264 (1978); Y. Zeiri and M. Shapiro, *J. Chem. Phys.* 75, 1170 (1981).
40. J. Juanós i Timoneda and J. T. Hynes, *J. Phys. Chem.*, 95 (1991) 10431.
41. T. Fonseca, H. J. Kim and J. T. Hynes, J. Mol. Liq., (1993) ,in press

SUPPRESSION OF ACTIVATED RATE PROCESSES INDUCED BY SPACE DEPENDENT, TIME DEPENDENT AND ANISOTROPIC FRICTION

Alexander M. Berezhkovskii[a], Anatoli M. Frishman and Eli Pollak
Department of Chemical Physics
Weizmann Institute of Science
Rehovot 76100, Israel

[a] Permanent address: Karpov Institute of Physical Chemistry, ul. Obukha 10, 103064, Moscow K-64, Russia

Abstract

This paper summarises recent work on rate theory of activated proceses in condensed phases. Various studies have shown that nonlinearities in the potential of the reacting particle or its interaction with the bath may lead to substantial reductions of the rate constant relative to predictions of the standard theories. It is shown that the optimized planar dividing surface variational transition state theory can account correctly for all these observed suppressions.

1. Introduction

The theory of activated rate process in condensed phases was introduced by Kramers in his paper of 1940.[1] He considered the escape rate of a particle with coordinate q, moving under the influence of a potential $V(q)$, a frictional force characterized by a damping constant γ and an external Gaussian Markoffian random force. Initially the particle is trapped in the well of the potential and it escapes by crossing a potential barrier. When the damping is weak, Kramers showed that the escape rate is limited by the rate of transfer of energy to the particle and so is proportional to the damping. When the damping is moderate or strong, the process is limited by the spatial rate of diffusion of the particle across the barrier. In this paper we will concentrate primarily on the spatial diffusion limited dynamics, although many of the ideas can be and are being extended to account for the energy diffusion limited process also.

In the spatial diffusion case, Kramers provided two limits. When the damping is very strong, one can reduce the dynamics to a one dimensional Smoluchowski equation for which practically everything is known. In the moderate damping limit, a solution was found by assuming that the process is limited only to the immediate vicinity of the barrier, such that one can consider only the parabolic barrier. When the barrier height is large enough (in units of $k_B T$) this approximation becomes

311

J. Jortner et al. (eds.), Reaction Dynamics in Clusters and Condensed Phases, 311–329.
© 1994 Kluwer Academic Publishers. Printed in the Netherlands.

exact.

Kramers' approach was generalized by Langer[2] to the multidimensional case. Here, the particle is moving on a multidimensional potential energy surface, characterized by a reaction coordinate q and nonreactive modes $\underline{z}$. Each of the modes is coupled separately to a bath and thus feels its own frictional and random force. By considering the dynamics only in the vicinity of the saddle point of the multidimensional potential energy surface $V(q, \underline{z})$ Langer was able to generalize Kramers' expression for the rate in the spatial diffusion limit.

A further generalization of the problem came with introduction of memory friction. Instead of the Langevin equation of motion, one may consider a generalized Langevin equation (GLE) in which the friction function is no longer Markoffian. The effect of memory friction on activated rate processes has been studied extensively during the past decade.[3] Grote and Hynes[4] realized that Kramers' theory[1] for the rate in the spatial diffusion limit may be generalized. They pointed out that the rate for spatial diffusion across a barrier is a function of frequency dependent friction and is determined by the component of the friction at the barrier frequency rather than the static friction as in Kramers' original theory.

A further refinement of Kramers theory may be obtained by introducing also a space dependent friction. The equation of motion, as derived first by Lindenberg and coworkers[5,6] is substantially more complex looking than the GLE. Although some theoretical developments were made by Carmeli and Nitzan for this problem,[7] they were primarily limited to the energy diffusion limited regime.

This sums up the state of affairs as of the middle of the 1980's. The dominant theoretical approach was based primarily on the parabolic barrier approximation. During the past few years it has become evident that the Kramers-Langer-Grote-Hynes theory may be insufficient. The first major surprise came when Agmon,[8-10] Berezhkovskii and Zitserman,[11-14] and others[15,16] started considering the effect of anisotropic friction on the reaction rate. By anisotropic friction one means that the ratio of damping constants γ_q, γ_z describing the frictional force on the reaction coordinate and a perpendicular mode respectively is either much larger or much smaller than one. Consider for simplicity the case when $\gamma_z \gg \gamma_q \gg 1$. Since the diffusion along the z coordinate will be the slowest, one might expect that the rate will be inversely proportional to γ_z. In fact, Berezhkovskii and Zitserman[11-14] demonstrated that if the coupling between the reaction coordinate q and the perpendicular mode z is moderate, then the Langer theory will predict a $\frac{1}{\gamma_q}$ dependence for the rate. A more careful analysis shows that the rate is determined by the z diffusion and goes as $\frac{1}{\gamma_z}$.[14] The reason for the failure of the Langer theory comes because the nonlinearity of the potential plays a crucial role in determining the rate.

A second failing is related to memory friction. In the presence of a finite memory time τ, if the damping $(\gamma \equiv \hat{\gamma}(0)$, cf Eq. 3.3 below) is very weak one expects that the rate limiting step will be the energy diffusion process. If the damping γ is very strong one expects that the time t_{diff} it takes the particle to diffuse across the barrier will be much longer than the memory time τ, so that the parabolic barrier Grote-Hynes estimate for the rate will be valid. But for finite long memory, there will be an intermediate range of damping values, for which the motion across the barrier occurs on a time scale which is similar to the memory time. When this occurs, and the nonlinearity of the potential is not negligibly small, the particle will 'remember' the nonlinearity and one may observe rate suppression.[17,18] The magnitude of the nonlinearity is controlled by the temperature in terms of the reduced barrier height $\frac{V^{\ddagger}}{k_B T}$.

A third, but important mechanism which may lead to deviations from the parabolic barrier limit has to do with spatial dependent friction. Consider first a memoryless but spatial dependent friction. If the friction is much larger away from the barrier then one should expect that the effective friction is larger than used in the standard theory and the rate would be suppressed. In the presence of memory friction the effect may be much stronger, since as may already be realised from the discussion above, the particle will 'remember' the larger damping strength exerted upon it as it moves along the barrier. Strong deviations from the Kramers-Grote-Hynes limit induced by memory and space dependent friction have been recently observed in numerical simulations by Voth and coworkers.[19,20]

The real challenge at this point is to present an extension of the Kramers-Langer-Grote-Hynes theory which is capable of treating uniformly all three seemingly different effects. Here, we demonstrate that variational transition state theory (VTST)[21] using optimized planar dividing surfaces[22] meets this challenge. In section II, we review briefly the optimized planar dividing surface VTST for the generalized Langevin equation. Results are then presented in section III for memory, anisotropic and space dependent induced suppression of the rate for some specific models.

2. Optimal planar dividing surfaces.

a. *One dimensional systems*

In this section we will review briefly our recent generalization of the Kramers-Langer-Grote-Hynes theory for the rate in the spatial diffusion limit. Instead of presenting all details, most of which have already been published, we will review the essential ideas by concentrating on the one dimensional case and memory friction. The generalization to multidimensional systems and space dependent friction is really technical in nature, the central idea remains the same.

The GLE for a one dimensional system is of the form:

$$\ddot{q} + \frac{dV(q)}{dq} + \int_0^t d\tau \gamma(t - \tau)\dot{q}(\tau) = \xi(t). \qquad (2.1)$$

Here, q is the (mass weighted) system coordinate and $V(q)$ is the system potential. The Gaussian random force $\xi(t)$ is related to the friction kernel $\gamma(t)$ through
the second fluctuation dissipation theorem: $< \xi(t)\xi(0) > = \frac{1}{\beta}\gamma(t)$ and we use the
notation $\beta \equiv \frac{1}{k_B T}$ throughout this paper.

It is well known that the dynamics of the GLE (Eq. 2.1) is equivalent to the
dynamics of the Hamiltonian[23,24]

$$H = \frac{p_q^2}{2} + V(q) + \sum_j \frac{1}{2}[p_{x_j}^2 + (\omega_j x_j - \frac{c_j q}{\omega_j})^2] \qquad (2.2)$$

where the system coordinate q is coupled bilinearly to a bath of harmonic oscillators with frequencies ω_j. The summation is in principle over an infinite set of bath
oscillators which tends towards a continuum. The bath coordinates x_j are mass
weighted. By explicit solution for the time dependence of each of the bath coordinates, one can show that Hamilton's equation of motion for the system coordinate
q reduces to the GLE (Eq. (2.1)), with the identification that

$$\gamma(t) = \sum_j \frac{c_j^2}{\omega_j^2} \cos(\omega_j t). \qquad (2.3)$$

The TST expression for the escape rate is:[25-27]

$$\Gamma = \frac{\int dp_q dq \prod_j dp_{x_j} dx_j \delta(f)(\nabla f \cdot \mathbf{p})\theta(\nabla f \cdot \mathbf{p})e^{-\beta H}}{\int dp_q dq \prod_j dp_{x_j} dx_j \theta(-f)e^{-\beta H}}. \qquad (2.4)$$

The Dirac delta function $\delta(f)$ localizes the integration onto the dividing surface $f = 0$. The gradient of the surface (∇f) is in the full phase space, $\mathbf{p}$ is the generalized
velocity vector in phase space with components $\dot{q}, \dot{p}_q, [(\dot{x}_j, \dot{p}_{x_j}), j = 1, ..., N]$ and
$\theta(y)$ is the unit step function which chooses the flux in one direction only. The
term $\nabla f \cdot \mathbf{p}$ is proportional to the velocity perpendicular to the dividing surface.
The numerator is the reactive flux and the denominator is the partition function
of reactants. The TST expression is an *upper bound* for the rate.[25-28] VTST is
obtained by varying the dividing surface f looking for that dividing surface which
gives the least upper bound.

The choice for the transition state implicit in Kramers' paper[1] is the barrier
top ($q = 0$) of the potential $V(q)$. In this case the dividing surface takes the
form $f = q = 0$ and the rate expression (Eq. 2.3) reduces to the well known one
dimensional result:

$$\Gamma_{1D} = (2\pi\beta)^{-\frac{1}{2}} \frac{e^{-\beta w(0)}}{\int dq \theta(-q)e^{-\beta w(q)}} \simeq \frac{\omega_a}{2\pi} e^{-\beta V^{\ddagger}}. \qquad (2.5)$$

where ω_a is the frequency at the bottom of the wells.

The Kramers-Grote-Hynes expression for the rate may be derived from the TST formulation by noting that for a purely parabolic barrier $(V(q) = -\frac{1}{2}\omega^{\ddagger 2} q^2)$ the Hamiltonian (Eq. 2.2) is a bilinear form which may be diagonalized using a normal mode transformation.[29] In the diagonal form, one finds one unstable mode, denoted ρ, with associated barrier frequency, denoted $\lambda_\infty^\ddagger$ and N stable modes. The ∞ subscript will serve to remind us that this is the solution for the purely parabolic barrier, or equivalently for an infinte reduced barrier height. The normal mode barrier frequency $(\lambda_\infty^\ddagger)$ is the solution of the equation:

$$\omega^{\ddagger 2} = \lambda_\infty^{\ddagger 2}(1 + \frac{\hat{\gamma}(\lambda_\infty^\ddagger)}{\lambda_\infty^\ddagger}). \tag{2.6}$$

where $\hat{\gamma}(s)$ denotes the Laplace transform of the friction kernel with frequency s. The rate may now be obtained by choosing the dividing surface $f = \rho = 0$. The result is the usual Kramers-Grote-Hynes result for the spatial diffusion limit:

$$\Gamma_\infty = \frac{\lambda_\infty^\ddagger}{\omega^\ddagger}\Gamma_{1D}. \tag{2.7}$$

Note, that the Kramers-Grote-Hynes solution has been obtained by replacing the one dimensional dividing surface $f = q = 0$ by a dividing surface in the full space of system and bath, $f = \rho = u_{00}q + \sum_j u_{0j}x_j = 0$, where the u_{ij}'s are elements of the orthogonal normal mode transformation matrix. To obtain a generalization of this approach, in the presence of a finite reduced barrier height $(\beta V^\ddagger)$ we pose the question: *what is the optimal* **planar** *dividing surface?* The most general planar dividing surface may be written as $f = a_0 q + \sum_j a_j x_j - \rho_0 = 0$ where ρ_0 denotes the distance of the dividing surface from the origin. A generalization of the Kramers-Grote-Hynes theory, is obtained by minimizing the TST expression for the rate with respect to the coefficients $a_0, a_j, j = 1, ..., N$ and the shift ρ_0. The details are given explicitly in Ref. 22, here we summarise the results needed to apply the theory.

The optimized estimate for the rate is:

$$P_0 \equiv \frac{\Gamma}{\Gamma_{1D}} = (\frac{\beta A^2}{2\pi})^{\frac{1}{2}} \int_{-\infty}^{\infty} dq\, exp(-\beta[\frac{1}{2}A^2(Cq - \rho_0)^2 + V(q) - V(0)]). \tag{2.8}$$

The optimized effective frequency A, coupling constant C and coefficient a_0 are expressed in terms of the temeprature dependent effective barrier frequency $\lambda^\ddagger$ through the following relations:

$$a_0^2 = [1 + \frac{1}{2}(\frac{\hat{\gamma}(\lambda^\ddagger)}{\lambda^\ddagger} + \frac{\partial\hat{\gamma}(s)}{\partial s}|_{s=\lambda^\ddagger})]^{-1}. \tag{2.9}$$

$$C = a_0(1 + \frac{\hat{\gamma}(\lambda^{\ddagger})}{\lambda^{\ddagger}}) \tag{2.10}$$

$$A^2 = \frac{\lambda^{\ddagger^2}}{a_0 C - 1} \tag{2.11}$$

The effective barrier frequency is determined by a generalization of the Kramers-Grote-Hynes equation (2.6) for the barrier frequency:

$$A^2 C^2 - \frac{1}{\beta < (q - <q>)^2 >} = \lambda^{\ddagger^2}(1 + \frac{\hat{\gamma}(\lambda^{\ddagger})}{\lambda^{\ddagger}}). \tag{2.12}$$

where we have used the notation:

$$< q^n > \equiv \frac{P_n}{P_0}$$

$$P_n \equiv (\frac{\beta A^2}{2\pi})^{\frac{1}{2}} \int_{-\infty}^{\infty} dq q^n exp(-\beta[\frac{1}{2}A^2(Cq - \rho_0)^2 + V(q) - V(0)]) \tag{2.13}$$

Finally the shift parameter is determined by the equation:

$$\rho_0 = C < q > \tag{2.14}$$

from which it is clear that for a symmetric potential a solution for the shift parameter is $\rho_0 = 0$.

Equation 2.12 must be solved for the barrier frequency $\lambda^{\ddagger}$. Given the barrier frequency, one can find all the parameters (A, C, a_0, ρ_0) necessary to evaluate the optimized rate. In practice, since the dependence of the three parameters A, C, a_0 on the barrier frequency parameter $\lambda^{\ddagger}$ is known, one may think of the transmission probability P_0 given in Eq. 2.8 as a function of the two independent variables $\lambda^{\ddagger}, \rho_0$ and minimize the function numerically. For a symmetric potential this procedure is even simpler, since the shift $\rho_0 = 0$, the transmission probability is a function of one variable only $(\lambda^{\ddagger})$, and one need only find its lowest minimum.

b. *Two dimensional systems.*

The dynamics is assumed to be described without loss of generality in terms of two coupled GLE's having the form:

$$\ddot{q} = -\frac{\partial V(q,z)}{\partial q} - \int_0^t d\tau \gamma_q(t - \tau)\dot{q}(\tau) + \xi_q(t). \tag{2.15}$$

$$\ddot{z} = -\frac{\partial V(q,z)}{\partial z} - \int_0^t d\tau \gamma_z(t - \tau)\dot{z}(\tau) + \xi_z(t). \tag{2.16}$$

The two Gaussian random forces $\dot{\xi}_q, \dot{\xi}_z$ have zero mean, are taken as independent and obey respective fluctuation dissipation relations $< \xi_j(t)\xi_j(\tau) > = \frac{1}{\beta}\gamma_j(t-\tau), j = z, q$.

The potential energy surface $V(q, z)$ is assumed to have a saddle point at $q = z = 0$ and a well at $q = q_a, z = z_a$ and may be written without loss of generality as:

$$V(q, z) = \frac{1}{2}(V_{qq}q^2 + 2V_{qz}qz + V_{zz}z^2) + V_1(q, z) \tag{2.17}$$

This form defines the nonlinear part of the potential $V_1(q, z)$ which is of course negligible in the vicinity of the saddle point. The normal modes of the surface at the saddle point will be denoted q', z', such that q' will be the 'reaction coordinate' (for the frictionless dynamics) associated with the normal mode barrier frequency $\omega^{\ddagger}$ and z' denotes the stable mode with associated normal mode frequency $\omega_{z'}$. The potential may then also be written as:

$$V(q', z') = \frac{1}{2}(-\omega^{\ddagger^2}q'^2 + \omega_{z'}^2 z'^2) + V_1(q', z'). \tag{2.18}$$

With these definitions one has the identity:

$$V_{qq}V_{zz} - V_{qz}^2 = -\omega^{\ddagger^2}\omega_{z'}^2 \equiv det(\mathbf{K}^{\ddagger}), \tag{2.19}$$

where $\mathbf{K}^{\ddagger}$ denotes the force constant matrix at the saddle point. A potential of mean force along the reaction coordinate q' is defined as:

$$w(q') = -\frac{1}{\beta}ln[\frac{1}{L}\int dz' e^{-\beta V(q',z')}]. \tag{2.20}$$

The length scale L is defined as:

$$L \equiv \int dq' e^{-\beta w(q')}. \tag{2.21}$$

To apply VTST to this system it is first necessary to write down the equivalent Hamiltonian which now has the form:[22]

$$H = \frac{p_q^2}{2} + \frac{p_z^2}{2} + V(q, z) + \sum_j \frac{1}{2}[p_{x_j}^2 + (\omega_j x_j - \frac{c_j q}{\omega_j})^2 + p_{y_j}^2 + (\omega_j y_j - \frac{d_j z}{\omega_j})^2] \tag{2.22}$$

Again, explicit solution for each of the bath coordinates leads to the identifications:

$$\gamma_q(t) = \sum_j \frac{c_j^2}{\omega_j^2}cos(\omega_j t). \tag{2.23}$$

$$\gamma_z(t) = \sum_j \frac{d_j^2}{\omega_j^2} cos(\omega_j t). \tag{2.24}$$

The general expression for the rate as given by transition state theory is identical in form to Eq. 2.4 except that both in the numerator and the denominator one must add an integration over the additional degree of freedom z and its conjugate momentum p_z as well as over all coordinates (y_j) and momenta (p_{y_j}) of the bath coupled to z. 'Conventional' TST is based on the choice of the simple dividing surface $f = q' = 0$. This leads to a one dimensional rate expression as in Eq. 2.5.

The Kramers-Langer theory as derived by Langer[2,3], assumes that in the intermediate to strong damping limits the rate is determined by the dynamics in the close vicinity of the barrier. Therefore one must expand only up to quadratic terms about the saddle point. This means that the potential has the general form:

$$V(q, z) = \frac{1}{2}(V_{qq}q^2 + 2V_{qz}qz + V_{zz}z^2). \tag{2.25}$$

and again the Hamiltonian as given in Eq. 2.22 is a quadratic form which may be diagonalized. The result of this diagonalization is that now the barrier frequency associated with the unstable normal mode is the single positive root of the equation:

$$det[\lambda^{\ddagger^2}\mathbf{I} + \lambda^{\ddagger}\hat{\underline{\gamma}}(\lambda^{\ddagger}) + \mathbf{K}^{\ddagger}] = 0. \tag{2.26}$$

The Laplace transformed friction matrix $\hat{\underline{\gamma}}$ is diagonal with the elements $\hat{\gamma}_q(\lambda^{\ddagger}), \hat{\gamma}_z(\lambda^{\ddagger}$ the force constant matrix $\mathbf{K}^{\ddagger}$ has been defined previously (Eq. 2.19) and $\mathbf{I}$ is the unit matrix. The rate expression takes the form:

$$\Gamma_\infty \simeq \frac{1}{2\pi}(\frac{det[\mathbf{K}_a]}{|det[\mathbf{K}^{\ddagger}]|})^{\frac{1}{2}}\lambda^{\ddagger}e^{-\beta V^{\ddagger}} \tag{2.27}$$

where $\mathbf{K}_a$ denotes the force constant matrix at the reagents well.

This is just the generalization of the Kramers-Gote-Hynes formula for the rate to the multidimensional case. As in the one dimensional problem, it is based on a quadratic expansion about the top of the barrier. Our goal again will be to find the optimal planar dividing surface which now has the general form:

$$f = a_0 q + \sum_j a_j x_j + b_0 z + \sum_j b_j y_j - \rho_0 = 0. \tag{2.28}$$

where ρ_0 is the distance of the optimal planar dividing surface from the origin. Using this form for the dividing surface, it is a matter of straightforward Gaussian integrals to show that the transmission coefficient can be expressed as:

$$P_{00} \equiv \frac{\Gamma_a}{\Gamma_{a_{1D}}} = (\frac{\beta A^2}{2\pi L^2})^{\frac{1}{2}} \int_{-\infty}^{\infty} dq dz exp(-\beta[\frac{1}{2}A^2(Cq + Dz - \rho_0)^2 + V(q, z) - w(0)]).$$
$$\tag{2.29}$$

After minimizing this expression with respect to the coefficients defining the optimal planar dividing surface one finds the following explicit expressions for the various parameters appearing in Eq. 2.29:

$$C = a_0(1 + \frac{\hat{\gamma}_q(\lambda^{\ddagger})}{\lambda^{\ddagger}});$$ (2.30)

$$D = b_0(1 + \frac{\hat{\gamma}_z(\lambda^{\ddagger})}{\lambda^{\ddagger}});$$ (2.31)

and the frequency A^2 is given by the expression:

$$A^2 = \frac{\lambda^{\ddagger^2}}{a_0 C + b_0 D - 1}$$ (2.32)

The shift parameter is just:

$$\rho_0 = C < q > + D < z >$$ (2.33)

where the average is defined as in Eqs 2.13, except that one has an integration over both the q and the z coordinates.

The remaining three unknown quantities which are the coefficients a_0, b_0 and the barrier frequency $\lambda^{\ddagger}$ are given by the following relations:

$$1 = a_0^2[1 + \sum_j \frac{c_j^2}{(\omega_j^2 + \lambda^{\ddagger^2})^2}] + b_0^2[1 + \sum_j \frac{d_j^2}{(\omega_j^2 + \lambda^{\ddagger^2})^2}]$$

$$= a_0^2[1 + \frac{1}{2}(\frac{\hat{\gamma}_q(\lambda^{\ddagger})}{\lambda^{\ddagger}} + \frac{\partial \hat{\gamma}_q(s)}{\partial s}\Big|_{s=\lambda^{\ddagger}})] + b_0^2[1 + \frac{1}{2}(\frac{\hat{\gamma}_z(\lambda^{\ddagger})}{\lambda^{\ddagger}} + \frac{\partial \hat{\gamma}_z(s)}{\partial s}\Big|_{s=\lambda^{\ddagger}})];$$

(2.34)

$$\beta A^2[C < (q - < q >)^2 > + D < (qz - < q >< z >) >] = a_0;$$ (2.35)

$$\beta A^2[C < (zq - < z >< q >) > + D < (z - < z >)^2 >] = b_0.$$ (2.36)

As in the one dimensional case, instead of solving these equations one may again think of the transmission probability P_{00} (cf Eq. 2.29) as a function of the two variables $\lambda^{\ddagger}, \rho_0$ and minimize numerically with respect to these two independent parameters. If the potential is symmetric $[V(q, z) = V(-q, -z)]$ then the shift parameter is $\rho_0 = 0$ and one is left with a minimization with respect to the barrier frequency only.

c. *Space dependent friction.*

A formally consistent approach to space and time dependent friction has been introduced by Lindenberg and coworkers[5,6]. They introduce the space dependent friction by allowing for a nonlinear coupling between the particle and the oscillator bath in the Hamiltonian equivalent:

$$H = \frac{1}{2}p_q^2 + V(q) + \frac{1}{2}\sum_{j=1}^{N}p_{x_j}^2 + \frac{1}{2}\sum_{j=1}^{N}\omega_j^2\left(x_j - \frac{c_j f(q)}{\omega_j^2}\right)^2 . \tag{2.37}$$

The function $f(q)$ couples the bath modes nonlinearly to the reaction coordinate. The GLE (2.1) is regained by assuming that $f(q) = q$. Lindenberg and coworkers[5,6] (see also Carmeli and Nitzan[7]) proceed to derive a stochastic dynamical equation which is a generalization of the GLE (2.1) to the case of space and time dependent friction:

$$\ddot{q} = -\frac{dV(q)}{dq} - \int^t dt'\left(\frac{df[q(t)]}{dq(t)}\gamma(t-t')\frac{df[q(t')]}{dq(t')}\right)\dot{q}(t') + \frac{df[q(t)]}{dq(t)}\xi(t) . \tag{2.38}$$

The random force $\xi(t)$ is Gaussian, with zero mean and correlated as usual to the time dependent friction $\gamma(t)$ (as given in Eq. 2.3).

As in the one dimensional GLE, one may again solve for the optimal planar dividing surface which minimizes the flux.[30] Even though the coupling to the bath modes is now nonlinear, the bath modes themselves are harmonic and the dividing surface is by definition linear in the bath modes. Therefore as in the GLE, one may explicitly integrate over all bath modes. The resulting expression for the transmission coefficient is formally very similar to the same expression derived for the GLE (cf Eq. 2.8):[30]

$$P_0 \equiv \frac{\Gamma}{\Gamma_{1D}} = \left(\frac{\beta A^2}{2\pi}\right)^{\frac{1}{2}}\int_{-\infty}^{\infty}dq\,exp(-\beta[\frac{1}{2}A^2(C(q)q - \rho_0)^2 + V(q) - V(0)]). \tag{2.39}$$

and

$$C(q) = \left(1 + \xi^2 B_+\right)^{-1/2}\left[\tilde{f}(q)\xi\,\hat{\gamma}(\lambda^{\ddagger})/\lambda^{\ddagger} + 1\right] . \tag{2.41}$$

The quantities $B_\pm$ are given by

$$B_\pm = \frac{1}{2}\left[\frac{\hat{\gamma}(\lambda^{\ddagger})}{\lambda^{\ddagger}} \pm \left(\frac{\partial\hat{\gamma}(s)}{\partial s}\right)_{s=\lambda^{\ddagger}}\right] , \tag{2.42}$$

and $\tilde{f}(q) \equiv f(q)/q$ [cf. eq. (10)]. The two quantities ρ_0 and ξ in the above equations are given by the averages

$$\rho_0 = \langle q\, C(q)\rangle \quad , \tag{2.43}$$

and

$$\xi = \frac{\langle q\tilde{f}(q)[C(q)q - \rho_0]\rangle}{\langle q[C(q)q - \rho_0]\rangle} \tag{2.44}$$

where the averages are performed with the (normalized) distribution function defined by the integrand in eq. (2.39). The expression for the transmission probability (2.39) depends on the three independent parameters $\lambda^{\ddagger}$, ξ, and ρ_0. The variational TST solution for the transmission coefficient P_0 is found by minimizing Eq. 2.39 with respect to those three parameters.

In principle, one can carry out this formal variation and obtain three transcendental equations (of which eq. (2.43) is one). However, as in the previous cases, solution of the three transcendental equations is not simpler than direct minimization of eq. (2.39). For a symmetric system where both the potential $V(q)$ and the spatial dependence of the friction $\tilde{f}(q)$ are symmetric about $q = 0$, the variation simplifies considerably. In this case, the solution for the shift parameter is $\rho_0 = 0$ and one is left with only a two parameter minimization problem.

3. Rate Suppression

a. Memory friction

The system potential $V(q)$ is taken throughout this paper as the symmetric quartic form:

$$V(q) = -\frac{1}{2}\omega^{\ddagger 2} q^2 + \frac{a}{4}q^4. \tag{3.1}$$

The length scale for this potential is $q_0 = \dfrac{\omega^{\ddagger}}{(2a)^{\frac{1}{2}}}$, the well is located at $\sqrt{2}q_0$, the barrier at $q = 0$, and the barrier height is $V^{\ddagger} = \frac{1}{2}\omega^{\ddagger 2} q_0^2$.

The memory friction used is the exponential form:

$$\gamma(t) = \frac{\omega^{\ddagger 2}\gamma}{\tau}e^{-\frac{\omega^{\ddagger}t}{\tau}}, \tag{3.2}$$

identifying the parameter γ as the reduced static friction while τ is the reduced memory time. The Laplace transform of the friction kernel is:

$$\hat{\gamma}(s) \equiv \int_0^{\infty} dt\, e^{-st}\gamma(t) = \frac{\omega^{\ddagger 2}\gamma}{\omega^{\ddagger} + s\tau}. \tag{3.3}$$

To demonstrate the suppression of the rate with respect to the Kramers-Grote-Hynes expression it is useful to define a 'suppression coefficient':

$$P \equiv \frac{\Gamma}{\Gamma_\infty} = \frac{P_0}{\lambda_\infty^\ddagger / \omega^\ddagger}. \tag{3.4}$$

The positivity of the nonlinearity $(\frac{a}{4}q^4)$ assures[31] that the transmission coefficient is less or equal unity. The deviation from unity is a clear measure of the memory induced suppression.

A detailed description of all results is given in Ref. 18. Here we bring only one figure, obtained with a reduced memory time of $\tau = 100$ and a reduced barrier height of $\beta V^\ddagger = 5$. The suppression coefficient is plotted vs the parameter $\frac{\gamma}{\tau}$ in Figure 1. The solid line is the prediction of VTST based on the optimized planar

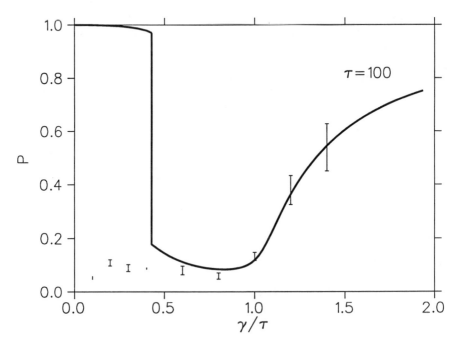

Figure 1. Memory induced suppression of activated rate processes. The solid line is the suppression probability. Bars denote the numerical results. Deviations from the predicted suppression coefficient in the weak damping limit are due to an energy diffusion limited process.

dividing surface. The bars denote results of numerical simulations. Note the two branches. For low values of the damping, the optimized planar dividing surface is practically identical to the Kramers-Grote-Hynes estimate, but vastly overestimates the rate. This is because in this region the rate is dominated by energy diffusion. Inclusion of a term which accounts for the energy diffusion process gives satisfactory results as shown in Ref. 18.

The more interesting results in the context of the present paper which is limited to the spatial diffusion regime, are for $\frac{\gamma}{\tau} \geq .5$. Here the optimal planar dividing surface departs radically from the normal mode and accounts quantitatively for the memory suppression of the rate. We have shown in Ref. 18 that the suppression is associated with a rotation of the optimal planar dividing surface such that in the suppression region it is almost perpendicular to the normal mode dividing surface.

Finally, note that for larger damping values, the rate is again well predicted by the parabolic barrier estimate. As noted in the Introduction, when the damping is siufficiently large, the rate of spatial diffusion becomes slower than the memory time and the nonlinearity in the potential becomes again unimportant.

b. *Anisotropic friction*

As already described briefly in the Introduction, in a multidimensional activated rate process, one may find strong deviations from the Langer limit. To be more specific, we study a potential energy surface of the form:

$$V(q, z) = V(q) + \frac{1}{2}\Gamma(z - q)^2 \qquad (3.5)$$

where $V(q)$ denotes the potential along the reaction coordinate q and will be taken in the following specific example to be the quartic double well potential (Eq. 3.1). The parameter Γ couples the harmonic perpendicular mode z to the reaction coordinate. The friction acting on the two degrees of freedom will be taken as Ohmic and δ correlated such that $\gamma_j(t) = 2\gamma_j\delta(t); j = q, z$. The Langer estimate for the rate Γ_∞ is given by Eqs. 2.26 and 2.27.

In the limit of large damping ($\gamma_q \gg 1$) and large anisotropy $\frac{\gamma_z}{\gamma_q} \gg 1$, if the coupling between the two modes is intermediate $\frac{\Gamma}{\omega^{\ddagger^2}} <\simeq 1$ the Langer estimate is inversely proportional to the damping along the reaction coordinate. Berezhkovskii and Zitserman demonstrated[11-14] that when $\frac{\Gamma}{\omega^{\ddagger^2}} < 1$ the Langer result is wrong and the rate is inversely proportional to the friction along the perpendicular coordinate z.[14] Here, we will demonstrate through a numerical example, not only the limit, but the transition from the Langer limit to the anisotropic friction limit by studying the rate as a function of the anisotropy.

The suppression coefficient analogous to the one defined in the previous subsection (Eq. 3.4), which gives a measure of the deviation of the rate from the Langer limit is defined in this case as:

$$P = \frac{P_{00}}{\Gamma_\infty/\Gamma_{1D}} \qquad (3.6)$$

where Γ_{1D} is the one dimensional estimate for the rate as obtained from Eq. 2.5 using the potential of mean force as in Eq. 2.20. The suppression coefficient is bounded by unity and any deviation from unity demonstrates the deviation from the Langer estimate for the rate.

A detailed study of the suppression coefficient will be given in Ref. 32, here we provide a typical example, choosing the following set of parameters: $\Gamma = \frac{1}{4}\omega^{\ddagger 2}$, $\gamma_q = \omega^{\ddagger}$ and $\beta V^{\ddagger} = 10$. In Figure 2 we plot the suppression coefficient vs the dimensionless

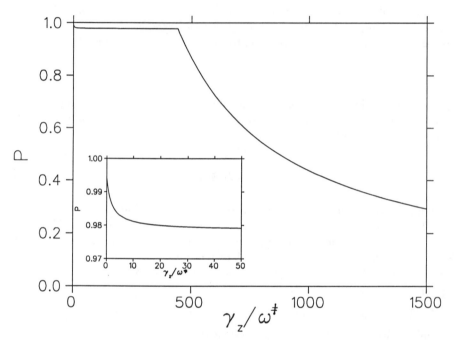

Figure 2. Anistropy induced suppression of activated rate processes. The inset shows the suppression coefficient for low values of γ_z. For further details see text.

anisotropy parameter $\tilde{\gamma}_z = \frac{\gamma_z}{\gamma_q}$ at the fixed value of γ_q. The suppression coefficient is close to unity as long as the anisotropy is not too high. It then deviates and tends to 0 as $\frac{1}{\tilde{\gamma}_z}$ exactly as predicted by the Berezhkovskii-Zitserman theory.

We stress that this behaviour is typical for moderate coupling between the two modes, or in our case, as long as $\Gamma \leq \omega^{\ddagger^2}$. For stronger coupling, the suppression is negligible and the coefficient is close to unity over the whole anisotropy range.

c. *Space and memory dependent friction*

Voth and coworkers[20] have studied numerically the reaction rate for the quartic double well potential (cf Eq. 3.1) in the presence of space and time dependent friction. The memory term was taken to be exponential (cf Eq. 3.2). The nonlinear coupling function between the system and the bath was chosen to have the form:

$$f(q) = q\left[1 + \epsilon s(q)\right] \qquad (3.7)$$

where the function $s(q)$ is:

$$s(q) = 1 - exp[-q^2/2\Delta q^2]. \qquad (3.8)$$

The reduced timescale of the friction was chosen to be equal to unity ($\tau = 1$), the reduced barrier height was chosen as $\beta V^{\ddagger} = 10$. The lengthscale Δq of the function $s(q)$ was taken to be $0.2\,|q_{min}|$, where $q_{min} = \pm\sqrt{2}q_0$ are the minima of the symmetric double well. The magnitude of the nonlinearity parameter ϵ was taken to be 2. This model gives rise to a spatially dependent friction which is 9 times larger in the wells than at the barrier. The exact value of the classical transmission coefficient was calculated using the well-known reactive flux correlation function method[3]. All results were calculated as a function of the dimensionless damping parameter of the exponential memory γ.

The results of numerical simulation are compared with the optimal planar dividing surface based theory in Fig. 3. The open circles are the numerical results for the case of nonlinear coupling and spatially dependent friction, while the open squares are for the linear coupling, spatially independent friction limit. The dashed and dot-dashed lines are the results of the analytic theory for the nonlinear and linear cases, respectively. In the weak damping limit (below the turnover) a correction has been added to the transmission coefficient which accounts for the energy diffusion limited process. But for all values above turnover, the rate is determined only by the spatial diffusion and the theoretical lines reflect the optimal planar dividing surface for both cases respectively. The theory is seen to be in excellent agreement with the simulation results over the entire range of friction.

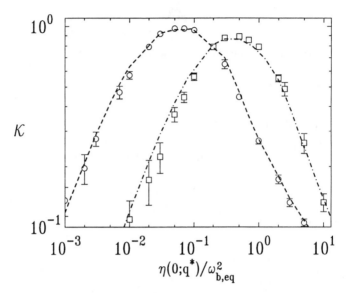

Figure 3. Space dependent friction induced suppression of activated rate processes. The open circles are numerically exact results for the case of nonlinear coupling and spatially dependent friction, the open squares are for linear coupling, spatially independent friction. The solid line is the predicted theoretical transmission coefficient ($\kappa \equiv P_0$, cf Eq. 2.39) for both case respectively. In the weak dampinlg limit, it has been corrected to account also for the energy diffusion limited process.

As is evident from the figure, the space dependent friction induces a substantial suppression of the rate relative to the what would be obtained from a naive application of the Kramers-Grote-Hynes theory. The difference between the nonlinear and linear dissipation results can be explained by an examination of the expression for the multidimensional planar dividing surface that is employed in the derivation of P_0 in Eq. 2.39. The dividing surface is defined in the discrete limit by the condition

$$f = a_0 q + \sum_{i=1}^{N} a_i x_i - \rho_0 = 0, \tag{3.9}$$

where the dividing surface coefficients can be shown to satisfy the relationship

$$a_i = -\xi \frac{c_i}{\lambda^{\ddagger 2} + \omega_i^2} a_0 \quad , \tag{3.10}$$

and ξ is given by Eq. (2.44). In the case of spatially *independent* friction, ξ equals unity in the above equation[22]. Therefore, by defining scaled coupling coefficients as $\bar{c}_i \equiv c_i \xi$ in eq. (3.10), and by noting the definition of the friction in Eq. (2.3), the effect of the spatially dependent friction is seen to be essentially equivalent to an *effective spatially independent friction* scaled by the factor ξ^2. In the spatially dependent friction case, it is found that $\xi^2 \approx 5$. Thus, it can be concluded that the trajectories passing through the barrier region "feel" an effective friction which is five times larger than the friction at the very top of the barrier.

IV. Discussion

This paper serves two major purposes. On the one hand we have brought together all known deviations from the Kramers-Langer-Grote-Hynes estimate for the rate in the spatial diffusion limit. We have also shown how all these may be accounted for using an optimized planar dividing surface VTST. As may be seen from the different applications, the rate suppression is always associated with a substantial rotation of the optimized dividing surface from the one used in the standard parabolic barrier theories. The 'true' bottleneck is rotated, hence the lower flux and rate.

In a sense though, this result is somewhat abstract. Elsewheres we will show[32] that there is an intimate dynamical relationship between the three forms of suppression presented in this paper. It is well known[33] that the GLE is equivalent to two coupled equations of motion, in which the reaction coordinate is bilinearly coupled to a perpendicular mode and that this mode is subjected to Ohmic friction. Thus, the memory friction equation of motion may be thought of as a special case of anisotropic friction.

Moreover, we have recently shown[34] that also the space dependent generalization of the GLE, as given in Eq. 2.38 may be reformulated in terms of two coupled equations of motion in which the reaction coordinate is nonlinearly coupled to a perpendicular harmonic mode which is subjected to Ohmic friction. Thus, all three effects may be derived from a single set of coupled equations of motion in the presence of Ohmic friction. The optimized planar dividing surface approach is then just the VTST solution for the rate of these two coupled equations of motion.

We have seen that nonlinearities can induce strong deviation from the parabolic barrier estimates. This would seem to imply that the same nonlinearities should induce suppression also in a quantum mechanical context. Although VTST is a powerful theory, it is classical in nature. It remains a challenge to the future to understand how the mechanism of suppression operates in the quantum world.

Acknowledgment

This work has been supported by grants from the Minerva foundation, the Israeli Ministry for Absorption of Immigrant Scientists and the Einstein center at the the Dept. of Physics of the Weizmann Institute of Science.

References

[1] Kramers, H.A., *Physica* **7**, 284 (1940).

[2] J.S.Langer, Ann. Phys. (N.Y.) **54**, 258 (1969).

[3] P. Hänggi, P. Talkner and M. Borkovec, *Rev. Mod. Phys.* **62**, 251 (1990).

[4] R.F. Grote and J.T. Hynes, J. Chem. Phys. **73**, 2715 (1980).

[5] K. Lindenberg and V. Seshadri, Physica A **109**, 483 (1981).

[6] K. Lindenberg and E. Cortés, Physica A **126**, 489 (1984).

[7] B. Carmeli and A. Nitzan, Chem. Phys. Lett. **102**, 517 (1983).

[8] N. Agmon and J.J. Hopfield, J. Chem. Phys. **78**, 6947 (1983).

[9] N. Agmon and R. Kosloff, J. Phys. Chem. **91**, 1988 (1987).

[10] N. Agmon and S. Rabinovich, Ber. Bunsenges. Phys. Chem. **95**, 278 (1991).

[11] A. M. Berezhkovskii, L.M. Berezhkovskii and V. Yu. Zitserman, Chem. Phys. **130**, 55 (1989).

[12] A. M. Berezhkovskii and V. Yu. Zitserman, Chem. Phys. Lett. **158**, 369 (1989).

[13] A. M. Berezhkovskii and V. Yu. Zitserman, Physica A **166**, 585 (1990).

[14] A. M. Berezhkovskii and V. Yu. Zitserman, Chem. Phys. **157**, 141 (1991).

[15] M.M. Klosek, B.M. Hoffman, B.J. Matkowsky, A. Nitzan, M. Ratner and Z. Schuss, J. Chem. Phys. **90**, 1141 (1989).

[16] M.M. Klosek, B.J. Matkowsky, and Z. Schuss, Ber. Bunsenges. Phys. Chem. **95**, 331 (1991).

[17] A. M. Frishman and E. Pollak, J. Chem. Phys. **96**, 8877 (1992).

[18] A. M. Frishman and E. Pollak, J. Chem. Phys., in press.

[19] J. B. Straus and G. A. Voth, J. Chem. Phys. **96** 5460 (1992).

[20] J. B. Straus, J. M. Gomez Llorente, and G. A. Voth, J. Chem. Phys. **98** XXX (1993).

[21] E. Pollak, J. Chem. Phys. **95**, 533 (1991).

[22] A.M. Berezhkovskii, E. Pollak and V. Yu. Zitserman, J. Chem. Phys. **97**, 2422 (1992).

[23] R. Zwanzig, J. Stat. Phys. **9**, 215 (1973).

[24] A.O. Caldeira and A.J. Leggett, Phys. Rev. Lett. **46**, 211 (1981); Ann. Phys. (NY) **149**, 374 (1983).

[25] J.C. Keck, Adv. Chem. Phys. **13**, 85 (1967).

[26] P. Pechukas in *Dynamics of Molecular Collisions, Part B*, ed. W.H. Miller (Plenum Press, NY, 1976), p. 269.

[27] W.H. Miller, J. Chem. Phys. **61**, 1823 (1974).

[28] E.P. Wigner, Trans. Far. Soc. **34**, 29 (1938).

[29] E. Pollak, J. Chem. Phys. **85**, 865 (1986).

[30] G. R. Haynes, G. A. Voth, and E. Pollak, Chem. Phys. Lett., in press.

[31] E. Pollak, S. C. Tucker, and B. J. Berne, Phys. Rev. Lett. **65**, 1399 (1990).

[32] A.M. Berezhkovskii, A.M. Frishman and E. Pollak, to be published.

[33] J.E. Straub, M. Borkovec and B.J. Berne, J. Chem. Phys. **84**, 1788 (1986).

[34] E. Pollak and A.M. Berezhkvoskii, J. Chem. Phys., in press.

Multidimensional Kramers and the Onset of Protein Relaxation

Noam Agmon

Department of Physical Chemistry and the Fritz Haber Research Center, The Hebrew University, Jerusalem 91904, Israel

Theory and computation of non-equilibrium barrier crossing in two dimensions indicate that the onset of relaxation perpendicular to the reaction coordinate occurs at the first maximum of the logarithmic derivative of the survival probability. This is verified by multi-pulse experiments on CO binding to horse-myoglobin. A model potential surface provides semi-quantitative estimates for features of the decay curves resulting from the protein relaxation process.

I. INTRODUCTION

A well known treatment of chemical reactions in solution [1] deals with diffusive motion of the reactants over a one dimensional (1d) potential barrier. The Kramers model has found numerous applications in chemistry and physics [2]. When chemical reactions occur in complex environments, such as inside heme proteins [3], more than one degree of freedom need be considered [4]. Here, two dimensional (2d) diffusive barrier crossing is considered on the model potential shown in Fig. 1. Both degrees of freedom, the iron-CO distance and the "protein coordinate", are considered to be diffusive: The ligand motion within the protein is diffusive due to collisions and interactions with the protein interior [5]. The protein coordinate represents a large amplitude, "soft" mode which fluctuates out of equilibrium for sufficiently long times to affect the binding kinetics.

Since the two modes are very different, one anticipates different response to variations in the environment. Thus at very low temperatures protein motion is essentially frozen while the ligand may still bind from within the "heme pocket" [3]. Raising the temperature is expected to speed-up protein motion much more than that of the ligand. This leads to a picture of anisotropic diffusion on a multi-dimensional potential energy surface. Several studies of the anisotropy dependence of the reaction rate coefficient for such problems have appeared [6–11]. These refer to the long-time asymptotic characteristics of the reactants' survival probability. Ligand binding to heme proteins is a non-equilibrium barrier crossing problem [12]: Dissociation of the heme-CO complex prepares a geminate pair, with the CO located within the heme pocket and the protein having the equilibrium *bound* state structure, which is removed from its equilibrium deoxy conformation. The intermediate time behavior is non-exponential due to increase of the effective barrier height with protein relaxation.

We have recently investigated the intermediate time, non-exponential phase of 2d barrier crossing in some detail [12,13]. Precise propagations of the 2d Smolu-

J. Jortner et al. (eds.), Reaction Dynamics in Clusters and Condensed Phases, 331–342.
© 1994 *Kluwer Academic Publishers. Printed in the Netherlands.*

chowski equation, describing diffusion in a potential field, have been carried out. From these, the survival probability has been calculated over a large range of diffusion anisotropies. The calculation has been compared with a number of approximate solutions for the multi-dimensional problem, each potentially valid in a different time-window and anisotropy regime. While some of our results are reviewed below, the present contribution focuses on the question of determining the onset of relaxation in the (slower) protein mode from the time course of the survival probability. The conclusion from the theoretical treatment is that the *most* non-exponential part of the decay namely, the first maximum in the logarithmic derivative of the survival probability, signals the onset of protein relaxation. The test is applied to multi-pulse data of CO rebinding to horse-myoglobin [14,15]. A more complete analysis of the experimental data is under preparation [16].

II. THEORY

We consider [12] 2d diffusion in a (dimensionless) potential field, $V(x,y) = U(x,y)/k_BT$, defined on a rectangular coordinate domain $[x_m, x_M; y_m, y_M]$. $U(x,y)$ is the potential surface of Fig. 1, with x being the Fe-CO distance and y a "protein coordinate" which is strongly coupled to the reactive process. k_BT is the thermal energy. Reflective boundary conditions are imposed along the perimeter of the rectangle. V is assumed to be a double-well potential in x (but not in y). A ridgeline on the potential surface separates the domain into reactants' and products' regions, denoted by $\mathcal{R}$ and $\mathcal{P}$, respectively. For simplicity, we assume a constant, diagonal diffusion tensor, (D_x, D_y). The diffusion anisotropy is then defined as $\eta \equiv D_y/D_x$.

The time evolution of the probability density, $p(x,y,t)$, for observing the system at (x,y) by time t is described by the Smoluchowski equation

$$\partial p(x,y,t)/\partial t = [D_x\mathcal{L}_x + D_y\mathcal{L}_y]\, p(x,y,t). \tag{2.1}$$

In Eq. (2.1), $\mathcal{L}_x$ and $\mathcal{L}_y$ are Smoluchowski operators in the two coordinates

$$\mathcal{L}_z \equiv \frac{\partial}{\partial z} e^{-V(x,y)} \frac{\partial}{\partial z} e^{V(x,y)}, \quad z = x, y. \tag{2.2}$$

They include the full form of the potential energy surface. The present work concentrates on the onset of relaxation rather than on the effect of an initially wide inhomogeneous distribution. Therefore a delta-function initial condition,

$$p(x,y,0) = \delta(x - x_0)\,\delta(y - y_0), \tag{2.3}$$

will generally be appropriate. Since during the fast photodissociation process the protein has no time to alter its conformation, y_0 will be taken as the equilibrium value for protein coordinate of the bound state, y_{eq}.

The fundamental property for comparison with kinetic experiments is the "survival probability",

$$S(t) \equiv \iint_{\mathcal{R}} p(x, y, t) \, dx \, dy \,, \tag{2.4}$$

obtained from the 2d density function by integration over the reactants' region. It is implicitly assumed that experimental observables, such as spectral changes occurring upon ligand binding, correspond to the same division into reactants' and products' regions as induced by the ridgeline on the surface.

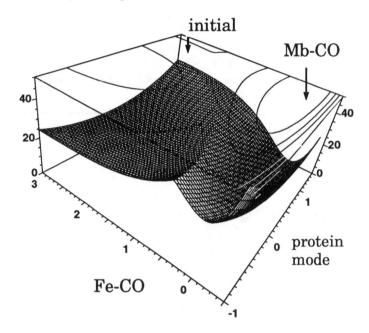

FIG. 1. The potential surface for MbCO.

In the anisotropy range where the y motion is slow as compared to the "reaction coordinate" (x motion), one may apply a series of approximations to Eq. (2.1) and obtain an analytical expression for the survival probability. The first step is to assume [17, Chap. 8.3], in the spirit of the Born-Oppenheimer approximation, that the fast x-coordinate dynamics are governed by the lowest eigenvalue, $\kappa(y) > 0$, where

$$\mathcal{L}_x \, p(x, y, t) \ = -\kappa(y) \, p(x, y, t) \,. \tag{2.5}$$

Therefore $k(y) \equiv D_x \kappa(y)$ is essentially a 1d Kramers' rate coefficient for a cut in the potential at a constant value of y. If, near some $(y, x_R) \in \mathcal{R}$, the potential is locally separable

$$V(x, y) = V_1(x) + V_2(y), \tag{2.6}$$

the density function will factor in that region of space. One may then integrate over x to obtain the 1d sink Smoluchowski equation [12,9]

$$\partial \overline{p}(y, t)/\partial t = [D_y \mathcal{L}_2 - k(y)] \overline{p}(y, t), \tag{2.7a}$$

$$\mathcal{L}_2 \equiv \frac{\partial}{\partial y} e^{-V_2(y)} \frac{\partial}{\partial y} e^{V_2(y)}, \tag{2.7b}$$

$$\overline{p}(y, t) \equiv \int_{x_m}^{x_M} p(x, y, t) \chi_R(x, y) \, dx. \tag{2.7c}$$

The characteristic function, $\chi_R(x, y)$, equals 1 if $(x, y) \in \mathcal{R}$ and 0 otherwise. It is noted that the potential in Fig. 1 indeed obeys the local separability condition (2.6) in the reactants and products regimes, where it is given [4] as a sum of a Morse potential in x and a parabolic potential in y. Equation (2.7a) is the starting point in the treatment of Ref. [4].

The next step is to convert the effective 1d diffusion equation, (2.7a), to a rate equation with a time-dependent rate coefficient. By integrating it over y, using the reflective boundary conditions in the form

$$\int_{y_m}^{y_M} dy \, \mathcal{L}_2 \, \overline{p}(y, t) = 0, \tag{2.8}$$

and the initial condition, $S(0) = 1$, one obtains

$$S(t) = \exp \left(- \int_0^t \langle k \rangle_{t'} \, dt' \right), \tag{2.9a}$$

$$\langle k \rangle_t \equiv \int_{y_m}^{y_M} k(y) \overline{p}(y, t) \, dy \, / S(t). \tag{2.9b}$$

This result is formal, since calculation of the time-dependent rate coefficient, $\langle k \rangle_t$, still requires the solution of Eq. (2.7a). However, it may serve as a basis for approximations. For what follows, it suffices to consider the approximation

$$\langle k(y) \rangle_t \approx k(\langle y \rangle_t). \tag{2.10}$$

This inversion of the order of the averaging has been found [12] to give semi-quantitative agreement with the exact propagation results in the anisotropy regime $D_y < D_x$.

The above results can be used to characterize intermediate-time power-law regime(s) in the decay of the survival probability. If $S(t) = A/t^\alpha$, then $\alpha = -d \ln S/d \ln t$. Therefore one considers the logarithmic derivative of the survival probability

$$B(t) \equiv -d \ln S(t)/d \ln t \approx t \, k(\langle y \rangle_t). \tag{2.11}$$

A power-law regime is hence *defined* as a regime of constant $B(t)$. Prior to the "onset" of y-motion relaxation one has $\langle y \rangle_t = y_0$. Therefore, at short times $B(t) \approx tk(y_0)$ is linear in t and $S(t)$ decays exponentially with the rate coefficient $k(y_0)$. After the "termination" of the relaxation process, $\langle y \rangle_t = \langle y \rangle_\infty$, so that at long times $B(t) \approx tk(\langle y \rangle_\infty)$ and $S(t)$ once again decays exponentially, but with the *smaller* rate coefficient $k(\langle y \rangle_\infty)$. The decrease of the effective rate coefficient with time follows from the shape of the potential surface in Fig. 1: The larger the relaxation in the y direction, the higher the climbing back to the ridgeline along x.

Two conceivable interpolations between the linear asymptotic behaviors of $B(t)$ are: *(i)* Monotonic increase or *(ii)* through a maximum and a minimum. In our investigations [12,13], we always find behavior of type *(ii)* on this family of potentials. Thus *two* power-law phases are predicted, corresponding to the maximum and minimum in $B(t)$. Each power, α_i, is an extremal value of $B(t)$, occurring at some time t_i. Thus

$$\alpha_i = B(t_i), \tag{2.12a}$$

$$(dB(t)/dt)|_{t=t_i} = 0, \tag{2.12b}$$

$i = 1, 2$ denote the first maximum and minimum of $B(t)$, so that $t_1 \equiv t_{max}$ and $t_2 \equiv t_{min}$. For $t < t_1$ the protein is completely inhomogeneous, showing no conformational interconversion, so that $\langle y \rangle_t = y_0$. Near t_1, $\langle y \rangle_t$ starts to decrease and so does $k(\langle y \rangle_t)$. This gives rise to the maximum in $B(t)$. For $t > t_2$ the relaxation is essentially complete and $\langle y \rangle_t = \langle y \rangle_\infty$. Therefore t_2 marks the termination of the y-relaxation process. At longer times one expects the protein conformations to behave homogeneously.

The protein relaxation rate coefficient, k_{rel}, should be intermediate between the two characteristic times, $t_1 < 1/k_{rel} < t_2$. Assuming that $\langle y \rangle_t$ in Eq. (2.11) depends on time only thorough the dimensionless variable $\tau = k_{rel}\, t$, the condition for an extremum in $B(t)$ becomes

$$B'(t) \approx k(\langle y \rangle_\tau) + \tau\, k'(\langle y \rangle_\tau) \langle y \rangle_\tau' = 0. \tag{2.13}$$

A prime denotes differentiation with respect to the appropriately designated variable. Denoting the solutions of Eq. (2.13) by τ_i, one concludes that

$$k_{rel} = \tau_i/t_i. \tag{2.14}$$

This shows, under quite general conditions, that the first maximum and minimum in $B(t)$ are inversely proportional to the protein relaxation rate coefficient.

A more quantitative estimate for the time-constants can be obtained with the aid of a specific model. For the potential surface shown in Fig. 1, the following relations approximately hold [4]

$$k(y) = k_0 \exp(ay), \tag{2.15a}$$

$$\langle y \rangle_\tau = y_0 \exp(-\tau). \tag{2.15b}$$

The first relation is a Kramers'-like approximation for the fast ligand motion when the activation energy depends linearly on the protein conformation, y. Equation (2.15b) is the solution for diffusion in a harmonic potential ("the Ornstein-Uhlenbeck process"). It is approximate due to the neglect of the reaction along x. The above equations show that as $\langle y \rangle_t$ relaxes from y_0 to the equilibrium deoxy conformation, $y = 0$, $k(\langle y \rangle_t)$ decreases. The effect can be significant, depending on the value of the parameter a.

The parameters in Eq. (2.15) are related to the parameters of the potential as follows [4]:

$$k_0 = A \exp[-(\Delta + f y_{eq}^2/2)/(3k_B T)], \tag{2.16a}$$

$$a = f y_{eq}/(3k_B T), \tag{2.16b}$$

$$k_{rel} = D_y f / k_B T. \tag{2.16c}$$

Here f is the force constant in the protein coordinate while Δ determines the height of the barrier in the ligand coordinate. A is a constant pre-exponential factor. y_0 is the protein conformation at which the initial distribution peaks, while y_{eq} (denoted by x_0 in Ref. [4]) is the equilibrium y value of the bound heme-CO. See Eqs. (1) and (6) and Table I in Ref. [4].

Substituting the model-specific Eqs. (2.15) into Eq. (2.13) yields

$$\tau \exp(-\tau) = 1/(a y_0). \tag{2.17}$$

Knowledge of a for the specific heme protein, Eq. (2.16b), allows for a graphical solution of Eq. (2.17), as demonstrated in Fig. 2. For $a y_0 < e$, two solutions are indeed obtained, τ_1 and τ_2. By considering second derivatives, one may show that the smaller value, τ_1, corresponds to a maximum in $B(t)$ while τ_2 corresponds to the minimum in $B(t)$. The τ_i's depend only on the dimensionless potential, $V(x,y)$, but not on the two diffusion coefficients.

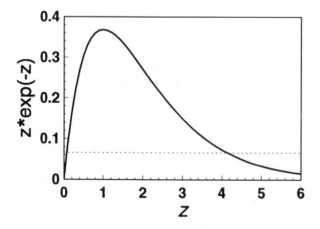

FIG. 2. Determinations of the roots of Eq. (2.17).

Since D_x and D_y cannot currently be determined by independent experiments, a quantitative comparison between model and experiment must rely on those quantities

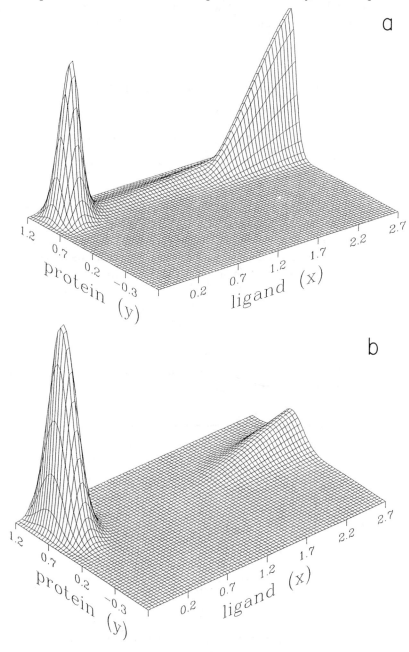

FIG. 3. The density function for α-heme CO at (a) $t = 0.5$ and (b) $t = 2$.

which are determinable from the parameters of the potential surface. The functional form of the potential is adjusted to reproduce the very low temperature (say, $T < 120K$) rebinding data [4]. The test then is whether this form can be used to extrapolate to higher temperatures. Appropriate quantities are identified as follows.

First, from Eq. (2.14) one gets

$$t_2/t_1 = \tau_2/\tau_1, \qquad (2.18)$$

Similarly, by using Eqs. (2.12a), (2.15a) and (2.17) in Eq. (2.11) one obtains

$$\alpha_i/t_i = k(y_0/a\tau_i) = k_0 \exp(1/\tau_i). \qquad (2.19)$$

These are again independent of the diffusion coefficients, D_x and D_y.

III. COMPUTATIONAL RESULTS

The survival probability obtained from 2d propagations on a α-heme potential [4] has been compared with various approximations in Ref. [12]. The α-heme potential involves lower barriers than that of Mb, therefore propagations to very long times are feasible. Qualitatively, one expects similar behavior on both surfaces.

The probability density $p(x, y, t)$, obtained from the numerical solution of Eq. (2.1), is shown in Fig. 3 at two different times. The initial density is a delta function at the upper right corner and time units are such that $D_x = 1$. Here $D_y = 0.01$, hence also $\eta = 0.01$. At $t = 0.5$, y motion is still frozen while at $t = 2$ the density function starts relaxing towards smaller y values. From the survival probability obtained by integrating it we have found that the maximum of $B(t)$ occurs at $t_1 = 1.6$ (see Table I in Ref. [12], point #1). Thus t_1 corresponds to what one may term "the onset of protein relaxation".

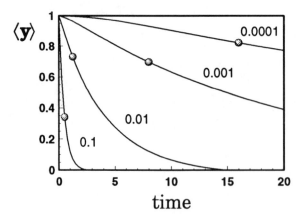

FIG. 4. Relaxation of the average protein conformation at various anisotropy values (curves), with corresponding t_1 values denoted by spheres.

A more quantitative measure is obtained from a calculation of the time dependence of the relaxing protein coordinate, $\langle y \rangle_t$. The calculation shown in Fig. 4 is based on the exact numerical solution of the effective 1d Smoluchowski equation (2.7a), cf. Fig. 7 of Ref [12]. Superimposed on the curves are the corresponding t_1 values characterizing the first maximum in $B(t)$ as obtained from the 2d propagations (e.g., Table I in Ref. [12]). For a large range of anisotropies, $\eta < 0.1$, $\langle y \rangle_{t_1} \approx 0.8$ namely, t_1 marks the beginning of the relaxation process.

IV. EXPERIMENTAL RESULTS

Multi-pulse data of Doster and coworkers [15] for CO binding to horse-myoglobin (Mb) at 185K in 75% (v) glycerol-water are shown in Fig. 5. The bold curve (labeled 1) is the usual kinetic trace following photodissociation of MbCO with a nsec laser and obtained by monitoring in the Soret band. A n-pulse kinetic trace is obtained by monitoring the transient absorption following a train of n such pulses at 10Hz (i.e., an inter-pulse separation of 0.1sec).

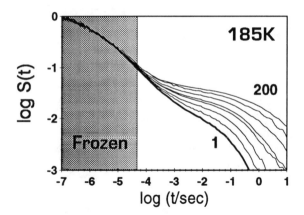

FIG. 5. Multi-pulse kinetics ($n = 1, 2, 5, 10, 20, 50$ and 200) for CO binding to horse myoglobin [15]. Inter-pulse separation is 0.1sec.

At short times the protein is completely inhomogeneous, each conformation reacts as if it was a distinct chemical species and it is impossible to "pump" the system into slower binding states. Thus the inhomogeneous time-regime is characterized by a universal (n-independent) $S(t)$. At longer times the protein is no longer inhomogeneous so that the curves diverge. The time where the curves begin to diverge should correspond to t_1, the "onset of protein relaxation".

To determine t_1, the data was fitted (on the log-log scale) to a polynomial of order 8 which was then differentiated analytically. The logarithmic derivatives thus obtained are shown in Fig. 6. These have the characteristic S shape predicted by theory. The maximum in $B(t)$ for $n = 1$ is indicated by the edge of the gray rectangle. Returning to Fig. 5 one notes that this time (t_1) is indeed where the multi-pulse data start to diverge.

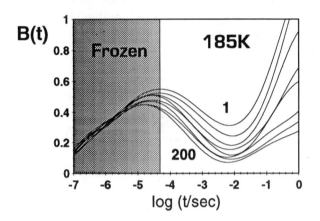

FIG. 6. The logarithmic derivative obtained from the experimental data of Fig. 5

As long as the inter-pulse separation is larger than t_1, the multi-pulse data always diverge at the same time. When it is smaller than t_1, the protein does not start relaxing and therefore the whole kinetics after a second pulse is expected to be identical to that after the first pulse. In Ref. [15] a second pulse was applied after 0.1msec, which is just slightly longer than t_1. A small degree of slowing down was observed which, however, began at the same t_1 value as in the 0.1sec case.

A more quantitative comparison with theory is suggested by Eqs. (2.18) and (2.19). Equation (2.18) implies that the separation between the maximum and minimum of $B(t)$ should be independent of both binding and relaxation rates and depend only on temperature and the potential surface. The potential shown in Fig. 1 was fitted 10 years ago [4] to data [3] of CO binding to sperm-whale Mb. A horse-Mb surface is yet unavailable, but the similarity of the low T binding data of the two Mb proteins suggests that the surfaces should be similar. From the parameters in Table I of Ref. [4], $y_0 = y_{eq} = 1.1au$ and $a = 14au^{-1}$ (see Eq. (2.16b)). The graphical solution of Eq. (2.17) gives $\tau_1 = 0.07$ and $\tau_2 = 4.1$, see Fig. 2. Therefore $\tau_2/\tau_1 = 60$, in comparison with the experimental single-pulse value of $t_2/t_1 = 155$. Given that the experiment spans some eight orders of magnitude in time, the hierarchy of approximations that have led to Eq. (2.18) and the inaccuracies in the potential as well as in the experimental data, the agreement should be considered remarkable.

Turning to Eq. (2.19), one first finds from Eq. (2.16a) that $k_0 = 0.1$sec. Therefore, for the maximum in $B(t)$ one gets $\alpha_1/t_1 = 16 \cdot 10^4$/sec, as compared with an experimental value of $1.2 \cdot 10^4$/sec. This again is good agreement. In contrast, for the minimum in $B(t)$ one gets $\alpha_2/t_2 = 0.13$/sec compared with an experimental value of

44/sec. The model produces α_2 values which are too small as compared with experiment. Therefore, to obtain the protein relaxation rate coefficient from Eq. (2.14) it seems preferable to use τ_1/t_1. This gives $k_{rel}(185K) = 1.6 \cdot 10^3$sec, which indeed lies between $1/t_1$ and $1/t_2$.

The "onset of relaxation" may be extrapolated to room temperature using an Arrhenius plot [16]. The extrapolation gives $t_1(300K)$ =150nsec. For sperm-whale Mb in the same solvent the half-lifetime for the second SVD component at room temperature was found to be about 30ns [18]. It is interesting to consider whether the relaxation of the multi-wavelength transient absorption data corresponds to the same relaxation process as deduced from the $B(t)$ function.

V. CONCLUSION

Measured survival probabilities for ligand binding to myoglobin span a huge range of temperatures and times. At low temperatures (say, $T < 120K$) the data decay smoothly with time in an almost power-law fashion, interpreted [3] as distributed kinetics of an inhomogeneous ensemble of protein molecules. At higher temperatures the data show "kinks" and "bends" which were initially attributed [3] to the sequential migration of the ligand from the heme pocket, through sites in the "protein matrix" to solution. Subsequently it became clear that part of the sequential kinetics may be replaced by "parallel kinetics", in which different heme conformations constitute parallel channels for rebinding. The inevitable conclusion from this picture, as reflected in Fig. 1, is that following ligand dissociation at ambient temperatures the tertiary protein relaxation slows down ligand rebinding. This tertiary-level autocatalytic effect [19] could complement the quaternary level cooperativity in hemoglobin.

The above point of view has been partially adopted in several recent publications [20,21]. However, it was claimed [21] that the relaxation occurs on a different "tier" of protein conformations, so that the "tier" which controls the heme reactivity remains inhomogeneous throughout the relaxation process. The present analysis shows that the first maximum in the logarithmic derivative of $S(t)$ is best interpreted as the time when the relaxation process begins. As demonstrated by the multi-pulse kinetics, at about the same time protein conformations cease to behave inhomogeneously. This agrees with a multi-dimensional non-equilibrium Kramers picture, in which relaxation occurs in degrees of freedom which are "perpendicular" to the reaction coordinate rather than in a hierarchal manifold of substates. It would be interesting to check this conclusion using independent spectroscopic markers, such as the transient evolution in the near-IR band of Mb-CO. While such experiments are much more difficult to perform than in the Soret regime, they are becoming feasible [22].

Consider the question of applicability of similar effects to the ultrafast room-temperature rebinding of NO to Mb [20]. Assuming, to a first approximation, that the protein relaxation is independent of the ligand, extrapolation to $300K$ gives a Mb relaxation rate slower than $t_1(300K) \approx 150$nsec in a 75% glycerol-water solution.

Since most of the NO kinetics occurs on the psec time scale, the interpretation [5] that the observed nonexponentiality in the data is due to NO diffusion within the protein matrix is not inconsistent with the present analysis.

Biophysical theory should correlate different observations and enable the prediction of new ones. Thus the qualitative understanding of the origin of natural phenomena takes precedence over the generation of quantitative fitting formulae with adjustable parameters. The 2d non-equilibrium Kramers model predicted [12,13] that $B(t)$ for MbCO should show a maximum and a minimum, that $\log(t_{min}/t_{max}) \approx 1.8$, and that the protein is inhomogeneous up to t_{max} and homogeneous as of t_{min}. These predictions were found to be in qualitative agreement with experiment.

ACKNOWLEDGMENTS. I thank Wolfgang Doster for the experimental data and its discussion. Work supported by the Zevi Hermann Schapira Research Fund. The Fritz Haber Research Center is supported by the Minerva Gesellschaft für die Forschung, München, FRG.

[1] H. A. Kramers, Physica **7**, 284 (1940).

[2] P. Hänggi and J. Troe, Eds., Ber. Bunsen-Ges. Phys. Chem. **95**, (1991).

[3] R. H. Austin *et al.*, Biochem. **14**, 5355 (1975).

[4] N. Agmon and J. J. Hopfield, J. Chem. Phys. **79**, 2042 (1983).

[5] Q. H. Gibson *et al.*, J. Biol. Chem. **267**, 22022 (1992).

[6] M. M. Kłosek-Dygas *et al.*, J. Chem. Phys. **90**, 1141 (1989).

[7] A. M. Berezhkovskii and V. Y. Zitserman, Physica A **166**, 585 (1990).

[8] A. M. Berezhkovskii and V. Y. Zitserman, Chem. Phys. **157**, 141 (1991).

[9] A. M. Berezhkovskii and V. Y. Zitserman, Physica A **187**, 519 (1992).

[10] N. Agmon and R. Kosloff, J. Phys. Chem. **91**, 1988 (1987).

[11] N. Agmon and S. Rabinovich, Ber. Bunsen-Ges. Phys. Chem. **95**, 278 (1991).

[12] N. Agmon and S. Rabinovich, J. Chem. Phys. **97**, 7270 (1992).

[13] S. Rabinovich and N. Agmon, Phys. Rev. E **47**, 3717 (1993).

[14] W. Doster *et al.*, J. Biol. Phys. **17**, 281 (1990).

[15] F. Post, W. Doster, G. Karvounis, and M. Settles, Biophys. J. **xx**, xxxx (1993).

[16] N. Agmon, W. Doster, and F. Post, in preparation.

[17] H. Risken, *The Fokker-Planck Equation* (Springer-Verlag, Berlin, 1984).

[18] A. Ansari *et al.*, Science **256**, 1796 (1992).

[19] N. Agmon, in *Tunneling: 19th Jerusalem Symposium on Quantum Chemistry and Biochemistry*, B. Pullman and J. Jortner, Eds. (Reidel, Dordrecht, 1986), pp. 373–381.

[20] J. W. Petrich *et al.*, Biochem. **30**, 3975 (1991).

[21] P. J. Steinbach *et al.*, Biochem. **30**, 3988 (1991).

[22] M. Lim, T. A. Jackson, and P. A. Anfinrud, in *Proceedings of the 8th International Symposium on Ultrafast Phenomena* (Springer-Verlag, Berlin, 1993).

NONBINARY BIMOLECULAR RELAXATION IN SOLUTIONS

A.I. Burshtein[1]
Weizmann Institute of Science
76100 Rehovot, Israel

Abstract

The angular momentum relaxation in dense gases and the hopping quenching in solid solutions are considered in integral and differential versions of binary kinetic theory. Results of the two methods are compared with each other and with the exact solutions when available. The differential theory is shown to have some advantages when the binary results are extrapolated to higher concentrations.

1 Introduction

There are two kinds of kinetic problems: the random modulation of the frequency $\omega = \omega_0 + \delta\omega(t)$ and the rate of dissipative process $W(t)$. In the former case, $\overline{\delta\omega(t)} = 0$; otherwise, the mean value of $\delta\omega$ may be added to ω and excluded from consideration. In the second case $W > 0$ and $\overline{W}$ is an upper limit of the rate reached at the ultrafast modulation limit (kinetic control reaction). This is definitely not zero but may be infinite in the models with the point particles. The further difference is that in the frequency modulation problem one is interested in the correlation function $K(t)$ whereas in the decay problem the survival probability $N(t)$ is usually calculated:

$$K(t) = \langle e^{i\int_0^t \delta\omega(t')dt'}\rangle \qquad N(t) = \langle e^{-\int_0^t W(t')dt'}\rangle . \qquad (1.1)$$

The random modulation is assumed to be stationary in both cases.

Although different these problems are similarly treated by two alternative methods that lead to integral and differential kinetic equations correspondingly. The integral theory results from the Fano-Zwanzig projection operator technique [1], [2] and becomes conventional in the spectral line shape theory [3], [4]. The differential theory actually appeared even earlier as a theory of diffusion-controlled reactions

---

[1] Meyerhoff Fellow. On leave from Institute of Chemical Kinetics and Combustion, Novosibirsk, USSR.

J. Jortner et al. (eds.), Reaction Dynamics in Clusters and Condensed Phases, 343–359.

[5, 6] but was recognized as the rate modulation problem much later [7, 8]. It was extended to a frequency modulation problems [9] to a more general random walk process [10] in so called "encounter theory". Sometimes the first approach is called COP (chronological ordering prescription) while the second is POP (partial ordering prescription) [11]. We prefer to call them the "integral" and "differential" theories as it was first done in [12].

In principle both approaches are exact but their applications need some additional approximations that break their equivalency. For example, the impact theory of gases and the encounter theory of liquid and solid solutions are both binary with respect to concentration of the particles c that produce the frequency shift or the decay rate. Occasionally one or another may be find to be exact at some particular case giving an etalon for checking the approximate results. Of course the accuracy of the binary approximation may be investigated straightforwardly by calculation of the ternary corrections. However, it is much simpler to find what is beyond this approximation by comparison the results of the two methods with each other and with the exact solution if it is available.

In the present article we will analyze from this point of view the linear molecule rotational relaxation due to a random variation of angular momentum (rotational speed) and the hopping quenching of excitation due to a random walk to the energy sinks. We will show that even at the lowest concentration of buffer atoms or sinks the binary approximation is limited in time and the very last stage of relaxation is collective in origin. It will be shown also that the differential theory has some advantages when the low concentration results are extrapolated to higher concentrations.

2   Angular momentum relaxation

The rotational relaxation is characterized by a correlation function of angular momentum $\vec{J}$:

$$K_J = <\vec{J}(t')\vec{J}(t'+t)> = K_J(t).$$

The alternative kinetic approaches are based on the following equations:

$$a)\quad \dot{K}_J = -\int_0^\infty R(t-t')\,K_J(t')\,dt' \qquad\qquad b)\quad \dot{K}_J = -c\,k(t)\,K_J \qquad (2.1)$$

where the kernel $R(t)$ and the time dependent reaction rate $k(t)$ are the functions of c and t that are not known in phenomenological theories . If the integral equation is considered as the first equation of the Mori chain than $R(c,t)$ is defined by the next and so on. However, there are no regular procedure to find it this way. The chain is usually truncated at the first step and the choice of $R(t)$ is arbitrary. The most popular choice for $R(t)$ is an exponential model $R = R_0 \exp(-t/\tau_c)$ that leaves $R_0(c)$ dependence unknown. The only alternative to such phenomenology is a binary collisional theory which Markovian version is known as impact approximation. In principle the impact theory gives the recipes for calculating either R or k in the lowest order with respect to c. However, in this section we will use the phenomenological approach to compare the alternative non-Markovian theories.

The important point is that in binary approximation there is a fundamental relation between R and k:

$$R(t) = c\dot{k}(t). \tag{2.2}$$

It follows from Eq.(2.1) when the "collision time"

$$\tau_c = \int_0^\infty \frac{R(t)}{R(0)} dt \tag{2.3}$$

is much shorter than the "free rotation time"

$$\tau_J = \int_0^\infty \frac{K_J(t)}{K_J(0)} dt. \tag{2.4}$$

The binary parameter is actually

$$\kappa = \frac{\tau_c}{\tau_J}. \tag{2.5}$$

When it is small enough $K_J(t') \approx K_J(t)$ may be factored outside the integral sign reducing Eq.(2.1a) to (2.1b) with $k(t)$ defined in Eq.(2.3).

Instead of exponential model for $R(t)$ let us use the following

$$R(t) = c\frac{\gamma}{\tau_c(1 + t^2/\tau_c^2)^{3/2}} \qquad k(t) = \frac{\gamma t}{\tau_c(1 + t^2/\tau_c^2)^{1/2}}. \tag{2.6}$$

This model is in accordance with the necessary condition (2.2) and the physical meaning of $R(t)$ as the correlation function of the torque switched on and off by collision. The asymptotic reaction constant in differential theory

$$\gamma = \lim_{t \to \infty} k(t) = \sigma\bar{v} \tag{2.7}$$

determines the rate of collisions $1/\tau = c\gamma$ with average velocity $\bar{v}$ and the cross-section σ. This is the rate of the exponential decay

$$K_J = <J^2> e^{-t/\tau} \qquad <J^2> = 2IkT \tag{2.8}$$

that is inherent to impact theory as well as to any other Markovian theory. In a non-Markovian differential theory such a decay is just the final stage of rotational relaxation that is :

$$K_J = <J^2> e^{-\tau_c/\tau} e^{-t/\tau} \qquad \text{at} \quad t \gg \tau_c. \tag{2.9}$$

However, it is not a case in integral theory. For the chosen model $K_J(t)$ becomes sign alternating at the very end [13]. Even at the lowest

$$\kappa = c\gamma\tau_c \ll 1$$

there is a negative loop at $t > t_f$ shown in **Fig.1**. The qualitative difference between the results of two binary theories is an evidence that the higher order corrections in the particle density play an important role , i.e., the long-time tail of $K_J(t)$ is of multiparticle origin.

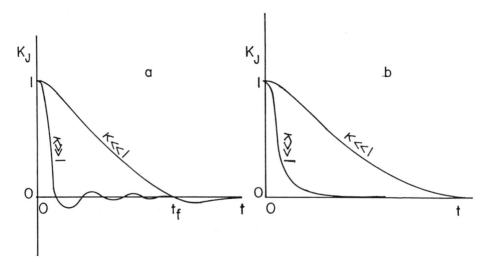

Fig.1 Angular momentum relaxation in integral (a) and differential (b) theories in binary approximation ($\kappa \ll 1$) and in the high density limit ($\kappa \gg 1$)

With increase of density $t_f \rightarrow \tau_c$ and the difference between integral and differential results becomes much stronger. The intermediate quasi-exponential asymptotic that takes place in integral theory at $\tau_c \ll t \ll t_f$ is ousted and the relaxation becomes quasi-periodical. Nothing like this happens to differential theory: the relaxation getting faster does not change its shape. The difference remains the same even at exponential kernel although sign-alteration appears only at $\kappa \gg 1$ [14].

The change of sign in $K_J(t)$ is of principle importance: the oscillations look like rotational vibration in the collective potential of surrounding particles. Its frequency is represented in light absorption and scattering spectra by the satellite lines or by superabsorption at the wing of the dielectric spectrum discovered by Poley [15, 16]. The integral theories have been used a few times to explain this way the peculiarities of the spectra [16, 17, 18], but it is unclear why they are preferable than differential theory. The latter does not predict any oscillations and all the effects following from them. Moreover, no binary theory can be extrapolated to such high densities ($\kappa \gg 1$) where the oscillation may appear unless it is known to coincide with exact solution. Unfortunately, the only example discussed in the literature shows that only the differential (and not integral theory) may be valid outside the limits of binary approximation.

This becomes possible in a rather special system perturbed by the point particles whose motion is independent of one another at any density. An example was first given for the atom line broadening by Anderson and Talman [19] and extended to purely nonadiabatic rotational relaxation (J-diffusion) in [13]. It was shown that in the system of independently moving perturbers $k(t)$ is the same for any densities. Hence the solution of Eq.(2.2a)

$$K_J = < J^2 > e^{-c \int_0^t k(t')dt'} \tag{2.10}$$

is actually an exact result. By integrating it in Eq.(2.5) we find for our model:

$$\tau_J = \tau_c e^{\kappa} K_1(\kappa) \tag{2.11}$$

where K_1 is a modified Bessel function and $\kappa = c \gamma \tau_c$. The effective frequency of collisions $1/\tau_J$ is evidently nonlinear function of density unlike that found from Laplace transform of integral equation (2.2a):

$$\frac{1}{\tau_J} = \int_o^{\infty} R(t)dt = R(0)\tau_c = \frac{\kappa}{\tau_c} \tag{2.12}$$

This is a rather general result inherent to integral theory as well as the oscillations. Both differential and integral estimations of $1/\tau_J$ are unacceptable outside the region where they coincide. The exact result shows that an increase in collision rate slows down with density because the collective potential is more smooth than the individual one. The integral theory does not reproduce this effect because it is approximate outside the binary limits. However, both results are far from the truth at $\kappa \gg 1$ (Fig.2). In reality the rate of collision increases nonlinearly with density due to a lack of the free volume in the system of hard spheres or real particles which are not points and hinder the relative motion. It means that the real $R(c)$ dependence may not be linear as it is in Eq.(2.7). As far as it is unknown one has to be very careful in extrapolating results of the integral theory to high κ region, outside the binary limits.

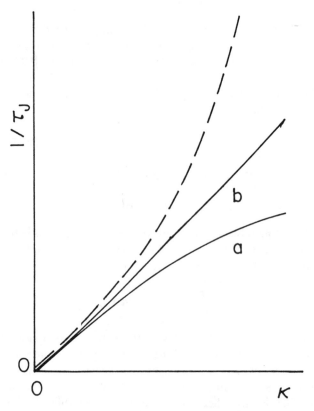

Fig.2 The rate of rotational relaxation in differential (a) and integral (b) theories. The dashed line stands for actual dependence for non-point particles.

3   Hopping quenching

The irreversible energy transfer from an excited donor molecule (D) to acceptors (A) may be described by the kinetic scheme

$$D^* + A \longrightarrow D + A,$$

which is a popular model of many bimolecular reactions in solutions either liquid or solid. The quenching process together with a natural decay of the excited state with time τ_D result in the luminescence intensity decrease in time after pulse excitation of the system. This may be presented as

$$I(t)/I(0) = P(t) \tag{3.1}$$

where $P(t)$ is a survival probability of excitation at time t. Since the kinetics of this process is usually non-exponential the effective excitation life-time is defined as in Eq.(2.5):

$$\tau_L = \int_0^\infty P(t)dt \tag{3.2}$$

The excitation encounter with acceptors may be accelerated by the particle diffusion (in liquid) or by a resonance energy transfer between the donors (in solid). At low acceptor concentration c the migration accelerated quenching (MAQ) may be described by encounter theory which is binary in acceptors [12]. In the frame of this theory the alternative kinetic approaches are either "integral" or "differential", as before:

$$a) \quad \dot{P}(t) = -c \int_0^t \sigma(t-t')P(t')dt' - P/\tau_D \qquad b) \quad \dot{P}(t) = -ck(t)P(t) - P/\tau_D \tag{3.3}$$

Neither the kernel $\sigma(t)$ nor the time dependent reaction constant $k(t)$ are now phenomenological. The encounter theory contains the ultimate definitions of them via the pair distribution function donor-acceptor $n(r,t)$:

$$a) \quad \sigma(t) = -\frac{\partial^2}{\partial t^2} \int [1 - n(r,t)]d^3r \qquad b) \quad k(t) = \int w(r)n(r,t)d^3r, \tag{3.4}$$

where $w(r)$ is a distant dependent rate of quenching. These definitions reduce the multi-particle problem presented by Eqs.(3.3) to a pair problem that is common for both integral and differential theories:

$$\dot{n}(r,t) = -w(r)n(r,t) + \hat{L}n(r,t) \qquad n(r,0) = 1, \qquad \frac{\partial n}{\partial r}\Big|_R = 0. \tag{3.5}$$

Here R is the distance of closest approach and $\hat{L}$ is an operator of the donor-acceptor relative motion. Actually only one of the functions (3.4) has to be find with a solution of Eq.(3.5) since there is a relation between them similar to Eq.(2.3):

$$\sigma(t) = \dot{k}(t) + \delta(t)\int w(r)d^3r. \tag{3.6}$$

Although both approaches are approximate (binary) the important particular case is well known when the differential theory is exact, i.e., valid at any c. It is similar to that discussed in the preceeding section. When donor is immobile and acceptors are point particles their motion is not correlated and the averaging over all trajectories can be performed rigorously [20]. In reality, this situation may occur in liquid solution when the heavy molecule is excited and quenched by the light particles like oxygen which diffuses much faster. In other cases it is simply a solvable model that helps to recognize what happens outside the low concentration limit.

When the relative motion of the particles is considered as a Markovian random walk in isotropic space the operator of motion is determined by the frequency of jumps $1/\tau_0$ and the free path distribution $\Phi(r)$ [10]:

$$\hat{L}n = -\frac{1}{\tau_0}\left[n - \int_0^\infty \frac{\Phi(|r-r'|) - \Phi(r+r')}{2r}n(r')r'dr'\right] = \begin{cases} D\Delta n & for \quad \lambda_0 \ll R_Q \\ \frac{1}{\tau_o}[1-n] & for \quad \lambda_0 \gg R_Q \end{cases}$$

$$(3.7)$$

Here $D = \lambda_0^2/\tau_0$ is the diffusion coefficient and $\lambda_0^2 = \frac{1}{2}\int_0^\infty r^2\Phi(r)dr$ defines the mean square size of the random steps. The alternative exhibited in Eq.(3.7) is related to a well-known division of the MAQ processes on diffusional and hopping quenching [21]. The random walk may be considered as the continuous diffusion to acceptor if λ_0 is much smaller than the effective quenching radius R_Q which determines the size of the reaction zone around acceptor . In the opposite limit when the step length is larger than R_Q either donor or acceptor enter into reaction zone by a single jump . This is the alternative "jumping" or "hopping" quenching mechanism introduced in [22] and then widely used as energy transfer theory for incoherent excitation in disordered systems either solid or liquid [21].

The encounter theory gives a unique opportunity to bridge between diffusional and hopping limits and it was actually done in [23]. However the discrimination between integral and differential description of quenching kinetics is a problem that must be solved separately in each limit. Here we will concentrate on the hopping quenching for the following reasons. First of all it is much simpler because the pair equation (3.5) with $\hat{L}n = \frac{1}{\tau_o}[1-n]$ has the general solution

$$n(r,t) = e^{-(w+\frac{1}{\tau_o})t} + \frac{1}{1+w\tau_o}\left[e^{-(w+\frac{1}{\tau_o})t} - 1\right].$$

$$(3.8)$$

independent of the width and shape of $\Phi(r)$. Inserting this into Eqs.(3.4) we immediately find two binary solutions of the problem. In particular

$$k(t) = \dot{Q}(t)e^{-t/\tau_o} + \frac{1}{\tau_o}\int_0^t \dot{Q}'(t')e^{-t'/\tau_o}dt'$$

$$(3.9)$$

where

$$Q(t) = \int[1 - e^{-w(r)t}]d^3r.$$

As has been mentioned already, the differential theory with $k(t)$ from Eq.(3.9) gives an exact solution for the immobile excitation quenched by point particles performing Markovian random walks which are hops over a regular lattice. Although useful, this is rather artificial situation. Usually the excitation jumps over regular or disordered donor system toward the immobile acceptors. The advantage of hopping limit is that there is a special, non-binary hopping theory also for this situation. It is known as "Burshtein's model" [22] that leads to a multiparticle quenching equation for $N(t)$ defined in (1.1)

$$N(t) = N_o(t)R_o(t) - \int_0^t N_o(t-t')\dot{R}_o(t-t')N(t')dt'.$$

$$(3.10)$$

Here

$$N_0(t) = \langle e^{-Wt} \rangle = \langle e^{-\sum_i w(r_i)t} \rangle = e^{-cQ(t)} \tag{3.11}$$

presents the static quenching of excitation which motion is frozen and

$$R_0(t) = \langle e^{-\sum_k u(r_k)t} \rangle \tag{3.12}$$

is so called "emigration kinetics" from one donor to the rest when there are no acceptors in a system. Originally the Burshtein equation was proposed for a markovian migration which is a random walk in a real or equivalent regular lattice. In this case averaging over the donor distribution is superfluous and

$$R = e^{-t/\tau_0}$$

where $1/\tau_0 = \sum_k u(r_k)$ is a lattice sum. For Markovian random walk the Burshtein model is equivalent to coherent potential approximation [24], [25] and was shown to be exact solution for MAQ in hopping quenching limit [26]. There is nothing like this in diffusional limit.

To compare binary theories between themselves and both of them with Burshtein's model we will use the relative quantum yield of the luminescence as an integral characteristic of MAQ:

$$\eta = I(c,t)/I(0,t) = \frac{\tau_L}{\tau_D} = \int_0^\infty P(t)dt/\tau_D = \int_0^\infty N(t)e^{-t/\tau_D}dt/\tau_D. \tag{3.13}$$

The Laplace transformation of the integral kinetic equation (3.3a) leads us straight to the famous Stern-Volmer relation [27]

$$\frac{1}{\eta} = 1 + cq \tag{3.14}$$

where the Stern-Volmer constant

$$q = \tau_D \int_0^\infty \sigma(t)dt = k\tau_D \tag{3.15}$$

and

$$k = k(\infty) = 4\pi R_Q D = \frac{4\pi R_w^3}{3\tau_0}$$

is the stationary reaction constant. The reproduction of the linear Stern-Volmer law (3.14) is in fact a weak point of the integral theory. The significant deviations from linearity was found experimentally in a high density region [28, 29] The higher order corrections in c must appear in Eq.(3.14) with increase of acceptor concentration that are of multiparticle origin. We will see what they are from the other approaches that are non-binary in two alternative cases: when either donors are immobile (differential theory) or vice versa (Burshtein's model).

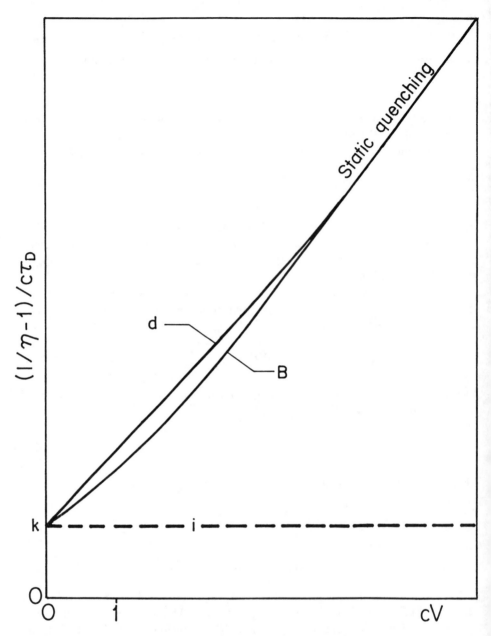

Fig.3 Deviations from Stern-Volmer law in differential (d) and Burshtein's model (B). The dashed line obtained in the integral theory (i) is the conventional Stern-Volmer result .

From now on we will consider a particular case of dipole-dipole quenching that has well-known static quenching kinetic [30]:

$$w = \frac{C_{DA}}{r^6} \qquad N_0 = e^{-2\Delta\sqrt{t}} \qquad (3.16)$$

where

$$\Delta = \frac{2}{3}\pi^{3/2}\sqrt{C_{DA}}\ c. \qquad (3.17)$$

In differential theory the quenching kinetics is expressed through an incomplete gamma function $\gamma(a, x)$:

$$N_{dif}(t) = exp\left\{-\frac{\Delta\sqrt{\tau_0}}{2}\left[\gamma\left(-\frac{1}{2};\frac{t}{\tau_0}\right)\frac{t}{\tau_0} - \gamma\left(\frac{1}{2};\frac{t}{\tau_0}\right)\right]\right\} = \begin{cases} \exp(-2\Delta\sqrt{t}) & t \ll \tau_0 \\ e^{-cV}\exp(-ckt) & t \gg \tau_0 \end{cases}$$
$$(3.18)$$

It starts from the static quenching and finishes exponentially with a hoping rate ck, where

$$k = \frac{2\pi^2}{3}\sqrt{\frac{C_{DA}}{\tau_0}} \qquad \text{and} \qquad V = \frac{\pi^2}{3}\sqrt{C_{DA}\tau_0}. \qquad (3.19)$$

To calculate the quantum yield at low density limit it is sufficient to use only the long-time asymptotics of N_{dif} in Eq.(3.13). Then at small c

$$\frac{1}{\eta} = (1 + ck\tau_D)e^{cV} = 1 + cq + c^2 kV\tau_D. \qquad (3.20)$$

The Stern-Volmer constant

$$q = k\tau_D\left(1 + \frac{\tau_0}{2\tau_D}\right) \qquad (3.21)$$

contains a correction $\tau_o/2\tau_D$ for the initial nonstationarity of the quenching. This is the hopping analogy of the well-known rate nonstationarity of diffusion-controlled reactions $(R/\sqrt{D\tau_D})$. Such corrections do not depend on concentration and are of no interest here. To avoid any complications we identify below κ with $k\tau_D$ assuming $\tau_o/2\tau_D \ll 1$.

The nonlinear correction to Stern-Volmer law in Eq.(3.20) has a different nature. The term

$$c^2 kV = c^2\left[\frac{2\pi^4}{9}C_{DA}\right] = \frac{\pi}{2}\Delta^2$$

does not depend on τ_o and differs only numerically from the static quenching rate $4\Delta^2$. However, this correction for static quenching does not appear in the integral theory that keeps only linear in c term in Eq.(3.15). This is a disadvantage of a binary theory avoided here. On the other hand the popular equation (3.20) widely used for explanation of any deviations from Stern-Volmer law is exact only for immobile donors.

This is not a case for immobile acceptors. To describe the resonance energy transfer to them one has to use the Burshtein equation (3.10) instead of differential theory. The Laplace transform of this equation leads to the following result:

$$\eta = \tilde{N}(\frac{1}{\tau_D})\frac{1}{\tau_D} = \frac{1 - f(x)}{1 + \frac{\tau_D}{\tau_o}f(x)} \qquad (3.22)$$

where

$$f(x) = \sqrt{\pi}x e^{x^2} erfc(x) \qquad x = \frac{\Delta}{\sqrt{1/\tau_0 + 1/\tau_D}}. \qquad (3.23)$$

For low density we find from Eq.(3.22)

$$1/\eta = 1 + \Delta\sqrt{\frac{\pi}{\tau_0}}[1 + \Delta\sqrt{\pi\tau_0}(1 - 2/\pi)] = 1 + ck\tau_D[1 + 0.723cV] \qquad (3.24)$$

Here the non-linear correction to the Stern-Volmer law is very close to that found in Eq.(3.20) by means of the differential theory.

At very high c the difference between Burshtein's model and differential theory disappears. Both of them (unlike integral theory) reproduce the static quenching which determines the quantum yield in the high concentration limit. At long enough τ_D any of them lead to

$$\lim_{\tau_0 \to \infty} \frac{1}{\eta} = \frac{\tau_D}{\int N_0 dt} = 2\Delta^2\tau_D = \frac{4}{\pi}c^2kV\tau_D \qquad cV \gg 1. \qquad (3.25)$$

In Fig.3 the horizontal line represent the Stern-Volmer relation ideally reproduced in integral theory. The slope of other curves at $c = 0$ is determined by ternary corrections that differ in Burshtein's and differential theory by only 30%. Therefore they first diverge but very soon converge to a common static limit (3.25). They are actually the upper and lower limits for the general case when both particles are mobile. The difference between these limits is so small that any of them can serve as an estimate of high order corrections to Stern-Volmer. This conclusion justifies the wide and successful application of differential theory to diffusional quenching independently of what is mobile: donor or acceptor. The relation (3.20) with diffusional rate constant $k = 4\pi RD$ gives rather satisfactory explanation of $\eta(c)$ dependence assuming $D = D_D + D_A$ although it is rigorous only for $D_D = 0$. An example may be found in [28].

An additional advantage of the differential theory is that it was shown to be a lower bound for $N(t)$ either for diffusional [31, 32] or hopping quenching [33]: $N_{dif} \leq N(t)$. The equality is related to immobile donors, while in all other cases the actual decay is slower. There is also another lower bound found in [34, 35]: $N_0(t)R_0(t) \leq N(t)$. Considered together they establish the following limits for quenching kinetics:

$$N_{dif}(t) , N_0(t)R_0(t) \leq N(t) \leq N_0(t) \qquad (3.26)$$

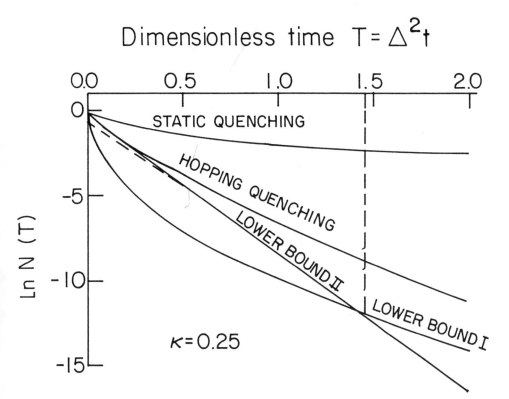

Fig.4 Kinetics of dipole-dipole hopping quenching together with its upper bound (static quenching) and two lower bounds:
Lower bound I: Emigration kinetics.
Lower bound II: Differential theory.
Hopping quenching: numerical solution of the Burshtein equation.

where the upper bound is put by the static quenching that is the slowest decay. The lower bound II established by differential theory is initially stronger than the other one but weaker at long times (Fig.4). Moreover, the asymptotics of differential theory is exponential up to the end while the other one is qualitatively different. In important case of disordered donor system the emigration kinetics is similar to that of static quenching [21, 36]

$$R_0(t) = e^{-\sqrt{\frac{6t}{\tau_0}}} \qquad \frac{1}{\tau_0} = \frac{8\pi^3}{27}C_{DD}\,\rho^2 \qquad (3.27)$$

where ρ is a concentration of donors and C_{DD}/r^6 is the resonance transfer rate. As is seen from Eqs (3.16) and (3.27) the lower bound I is

$$N_0(t)R_0(t) = \exp[-\sqrt{(4\Delta^2 + 6/\tau_0)t}\,]. \qquad (3.28)$$

It is much slower than an exponential decay in Eq.(3.18). The latter is not changed qualitatively if not Markovian but continuous time random walk model is used to describe properly the energy migration over disordered system [37]. The cross-point of two bounds shown by the vertical dashed line in Fig.4 put the upper time limit for the binary description of the quenching kinetics even at low acceptor concentrations. This time is similar to t_f appeared in Fig.1. At longer times the quenching is not exponential any more but approaches the "square root dependence" peculiar to the lower bound (3.28). The latter is multiparticle and analogous with the "fluctuation kinetics" of diffusion-controlled reactions known from the pioneering work of Balagurov and Vaks [38]. The migration-accelerated quenching is well described by binary differential theory at

$$\kappa = cV \ll 1 \tag{3.29}$$

but up to $t = t_f$ only. To get a more complete description of the process one has to use Burshtein's equation which is not binary and therefore capable to reproduce the fluctuation asymptotics [33]. The solution of this equation initially reproduces quasi-exponential MAQ kinetics but finally approaches the fluctuation asymptotics (Fig.4).

When migration is fast, the initial static stage is binary, short and not profound. Only those excitations that were created inside reaction spheres are quenched by direct (static) energy transfer to a neighboring acceptor. According to Eq.(3.29) they are not numerous. The remaining excitations must migrate from donor to donor to finally enter the reaction sphere. The common belief is that most of them are quenched during the migration-accelerated stage. However, there are some that were born far away from any sink, in fluctuation regions, occasionally empty of acceptors. These excitations must migrate out of the fluctuations to the ordinary places of the sample where they are quenched last. The fluctuation stage limited by emigration is always multiparticle in contrast with the two preceeding stages of quenching.

4 Concluding remarks

In both examples discussed above we found that the differential theory coincides in some special cases with the rigorous results valid at any concentrations. For the rest it may be considered as a reasonable interpolation between law density binary theory and the high density static limit. For the integral theory this limit is unattainable and the results are binary in principle. The exact linearity of Eqs.(2.12) and (3.14) may not be changed unless the kernels themselves are made nonlinear in concentration. It was actually done with a modified propagator approximation in [25] but the improved integral theory though nonlinear overestimates the ternary corrections and does not reproduce the static limit.

We do not mean to minimize the importance of the integral approach in some other respects. As was seen the sign-alternating relaxation of angular momentum confirmed by MD simulations of liquids is reproducible only in the frame of the integral theories. On the other hand some principle difficulties were also met in the differential theory applied to the back energy transfer problem. It is not widely known but well established in [39] that the quasi-stationary rate constant k diverges when the excited donor decay is faster than that of acceptor. These difficulties were avoided within the framework of the integral theory that does not imply the long-time decay to be necessarily exponential as it always is in the differential formalism. The further development of either integral or differential theory is necessary to combine the advantages of both approaches.

References

[1] U.FanoPhys.Rev.**131**, 259 (1963).

[2] R.Zwanzig Physica**30**, 1109, (1964).

[3] A.Ben-Reuven Advan.Chem.Phys. **33**, 235, (1975).

[4] A.Abraham The principles of nuclear magnetism (Oxford University Press, London, 1961).

[5] M.V.Smoluchovski, Z. Physik. Chem. **92**, 129, (1917).

[6] F.C.Collins, G.E.Kimball, J.Colloid Sci. **4**, 425, (1949)

[7] I.Z.Steinberg, E.Kachalsky J.Chem.Phys., **48**, 2404, (1968).

[8] G.Wilemski, M.Fixman, J.Chem.Phys., 4009 (1973).

[9] A.B.Doktorov, A.I.Burshtein ZhETF **68**, 1349, (1975) [Sov.Phys.JETP., **41**, 671, (1975)].

[10] A.B.Doktorov, A.A.Kiprianov, A.I.Burshtein ZhETF, **74**, 1184, (1978) [Sov.Phys.JETP **47**, 623 (1978)].

[11] S.Mukamel Chem.Phys. **37**, 33, (1979).

[12] A.A.Kiprianov, A.B.Doktorov, A.I.Burshtein Chem.Phys.**76**, 149,(1983).

[13] A.I.Burshtein, A.V.Storozhev Chem.Phys., **164**, 47, (1992)

[14] A.I.Burshtein, S.I.Temkin Spectroscopy of Molecular Rotation in Gases and Liquids, Nauka, Novosibirsk,(1981). English translation: Cambridge University Press (in press).

[15] B.M.Abdrakhmanov, A.I.Burshtein, S.I.Temkin Chem.Phys. **143**, 297, (1990).

[16] A.I.Burshtein, J.R.McConnel Physica A **157**, 933 (1989).

[17] P.Madden, D.Kivelson Advan. Chem.Phys. **56**, 467, (1984).

[18] B.Guillot, S.Bratos Phys.Rev., **16**, 424, (1977).

[19] P.W.Anderson, J.D.Talman in: Proceedings of the Conference on Broadening of Spectral Lines, University of Pittsburg (1955), Bell telephone Syst. Monogr. p.3117.

[20] A.Blumen, J.Manz J.Chem.Phys., **71**, 4694, (1979).

[21] A.I.Burshtein Usp.Fiz.Nauk, **146**'572, (1985) [Sov.Phys.Usp. **28**, 636, (1985)].

[22] A.I.Burshtein ZhETF, **62**, 1695, (1972) [Sov.Phys.JETP **35**, 882 (1972)].

[23] A.I.Burshtein, A.B.Doktorov, A.A.Kiprianov, V.A.Morozov, S.G.Fedorenko ZhETF, **88**, 878, (1985) [Sov.Phys.JETP **61**, 516, (1985).

[24] L.D.Huber Phys.Rev., **B 20**,2307, 5333, (1979).

[25] S.G. Fedorenko, A.I.Burshtein J.Chem.Phys. **97**, 8223 (1992).

[26] A.A.Kipriyanov, I.V.Gopich, A.B.Doktorov (in press).

[27] O.Stern, M.Volmer, Phys. Z. **20**, 183, (1919).

[28] M.R.Eftink, C.A.Ghiron J.Phys.Chem., **80**, 486, (1976).

[29] T.L.Nemzek, W.R.Ware J.Chem.Phys., **62**, 477, (1975).

[30] Th.Förster Disc.Farad. Soc. **27**,7, (1959).

[31] S.F.Burlatsky, A.A.Ovchinnikov ZhETF **92**, 1618, (1987) [Sov.Phys.JETP., **65**, 908, (1987)].

[32] A.M.Berezhkowskii, Yu.A.Makhnovskkii, R.A.Suris J.Stat.Phys. **65**, 1025, (1991).

[33] S.G.Fedorenko, A.I.Burshtein, A.A.Kipriyanov (in press).

[34] V.P.Sakun Fiz.Tverd. Tela **21**, 662,(1979).[Sov.Phys.Sol. State 21,230, (1979).

[35] S.F.Burlatsky, G.S.Oshanin, A.A.Ovchinnikov Phys. Rev. Lett., **A 139**, 241, (1989).

[36] A.I.Burshtein J.Luminescence **34**, 167, (1985).

[37] S.G.Fedorenko, A.I.Burshtein Chem.Phys.,**128**, 185, (1988)

[38] B.Balagurov, V.Vaks ZhETP **65**, 1939, (1973).

[39] N.N.Lukzen, A.B.Doktorov, A.I.Burshtein Chem.Phys. **102**, 289, (1986).

PRESSURE DEPENDENCE OF SOLVENT EFFECTS IN ELEMENTARY REACTIONS IN DENSE MEDIA

Jörg SCHROEDER and Jürgen TROE
Institut für Physikalische Chemie
Universität Göttingen
Tammannstraße 6
D-37077 Göttingen, Germany

ABSTRACT. The investigation of photoinduced conformational changes in cis- and trans-stilbene in dense solvents provides a test for models of reactions in condensed phase. Studies of the pressure- and temperature-dependences of the dynamics of transient absorption spectra allow, to some extent, to separate dynamic and static reagent-solvent interactions. In the high-damping regime, diffusion-controlled barrier crossing rates often are found to be inversely proportional to the solvent viscosity, ruling out major contributions from frequency-dependent friction.

1. INTRODUCTION

Elementary chemical reactions in fluid phase are not well understood. Their rate coefficients k by kineticists often are analyzed in terms of transition state theory (TST), i. e. using the relation

$$(1) \qquad k_{TST} = kT/h \, \exp(-\Delta G^{\neq}/RT) \, .$$

Absolute values and temperature- and pressure-dependences of k lead to the formal activation enthalpy $\Delta H^{\neq}$, activation entropy $\Delta S^{\neq}$, and activation volume $\Delta V^{\neq}$. The effect of the solvent can be identified particularly well, when $\Delta S^{\neq}$ and $\Delta H^{\neq}$ values of the gas phase reaction are compared with the corresponding values in dense fluid phase, and when the effect of increasing pressure on these quantities is expressed by the activation volume

$$(2) \qquad \Delta V^{\neq} = - RT\partial lnk/\partial P \, .$$

In contrast to this analysis, at the side of statistical molecular mechanics, the treatment often is based on a generalized Langevin equation of the type

$$(3) \qquad \mu\ddot{q} + \partial V(q)/\partial q + \mu \int_{o}^{t}dt'\gamma(t\text{-}t')\dot{q}(t') = \xi(t)$$

J. Jortner et al. (eds.), Reaction Dynamics in Clusters and Condensed Phases, 361–381.
© 1994 Kluwer Academic Publishers. Printed in the Netherlands.

with a friction kernel $\gamma(t)$ and a random force $\xi(t)$. $\gamma(t)$ and $\xi(t)$ are linked through the fluctuation-dissipation theorem, $<\xi(t)\xi(0)> = \mu k T \gamma(t)$. Eq. (3) allows to inlcude friction-memory effects. In the simplest case of the Markovian limit $\gamma(t) = 2\gamma\delta(t)$, Kramers' equation is obtained [1]. The general solution was treated in [2],[3] (for a discussion of the general field, see [4]). Kramers' solution of eq. (3) leads to the rate coefficient [1],[5]

$$(4) \qquad k = k_{TST}\{[1+(\beta/2\omega_B)^2]^{1/2} - \beta/2\omega_B\}$$

which, in the high-damping limit ($\beta/2\omega_B >> 1$), coincides with the Smoluchowski expression

$$(5) \qquad k = k_{TST}\{\omega_B/\beta\}\ .$$

The parameter ω_B, the "imaginary barrier frequency", is related to the barrier curvature which is expressed by a barrier potential of the form $V(r) \approx E_0 - 1/2\mu\omega_B^2(r-r_B)^2$. The dynamical friction coefficient β is linked with γ in eq. (3); if Stokes' law is valid, β is expressed by the bulk viscosity η through $\beta = 6\pi\eta a/\mu$ with an effective radius a of the diffusing species.

The comparison of eqs. (1) and (5) shows that the aspect of friction-control is underrepresented in the transition state treatment of eq. (1). On the other hand, effective solvent-modifications of the intramolecular potential energy, which are accounted for in eq. (1), generally are not emphasized in eqs. (3)-(5). The question arises whether friction-memory effects, e. g. of the Grote-Hynes type [6], are of practical relevance and, if they are, under which conditions they become apparent. The experimentalist, therefore, would like to have some guideline from theoreticians for the right experiments to do; likewise, he would like to have some simple basic expressions which are practically useful and allow for an interpretation of experiments such that experiments can be used as a diagnostic tool. Unfortunately, communication between theoreticians and experimentalists in this field has not been particularly good such that theories are designed which have often only little to do with reality and experimentalists have only little insight into the basic physical ideas behind complicated theoretical treatments. This situation, of course, is inacceptable and the present discussion, as well as earlier articles by the authors (e. g. [7]-[10]), insist in the necessity to reconcile the various descriptions of the rate coefficients. In the following, two representative examples from the authors' laboratory are shown which, on the basis of eqs. (1)-(5), suggest a number of conclusions, not withstanding the necessity of reinterpretations if the applied models turn out to be oversimplified.

The experimental observables considered in the following are thermal rate coefficients k as a function of temperature, solvent viscosity, and nature of the solvent. Whenever possible, experiments with isolated molecules, or isolated molecule-solvent clusters, under selective excitation conditions are also considered. Particular emphasis is given to the viscosity dependence of the rate coefficients which in the present experiments is produced through the variation of the solvent pressure. In this way, variable strengths of the reagent-solvent interaction and various magnitudes of transport control can be realized in the same reagent-solvent combination.

2. VISCOSITY DEPENDENCE OF A LOW-BARRIER PROCESS

Our best studied example of a low-barrier process in fluid phase is the isomerization of electronically excited cis-stilbene [11]. A schematic energy diagram of some of the possible chemical transformations is given in fig. 1. Excitation of the S_1-state of cis-stilbene gives rise to transient absorption spectra in the red spectral range which decay on the pico- to subpico-second time scale. The interpretation of the transient absorption spectra on this time scale is not without problems: the absorption samples distributions in configuration space, weighted by Franck-Condon factors. It is by no means clear that these signals can be related to "reagent concentrations" in the language of a kinetic rate law. Nevertheless, the observed signals and their strictly exponential decay (see fig. 2) suggest (i) that the absorption samples a region of configuration space which is separated from other regions by a bottle-neck, (ii) that other regions of configuration space do not contribute to the absorption signals, and (iii) that, within the probed part of configuration space, intramolecular vibrational redistribution (IVR) and/or collisional energy transfer leading to thermalization within the S_1-electronic state are over. As the time constants of the process are on the 0.3 - 1 ps scale, these assumptions are all not trivial and it even appears surprising that simple exponential decays of the absorption signals are observed. Such findings are in part in conflict with previous assumptions that the dynamics on the S_1 surface corresponds to a barrierless down-hill process (e. g. [11]-[14]) towards the photochemical funnel where internal conversion to the electronic ground state takes place or, if the twist-conformation is passed where internal conversion probably occurs, towards the trans-conformation of electronically excited stilbene, because in this case nonexponential absorption-time profiles would most probably be expected.

If the dynamics of cis-stilbene in the S_1-state at 1 bar is studied in various solvents with different viscosities, the picture of fig. 3 arises. A plot of k as a function of the inverse viscosity of the solvent suggests the appearance of an intercept at $\eta \to \infty$. Apart from this, the k - $1/\eta$ dependence is linear in alcohols, whereas there is only a weak dependence of k on $1/\eta$ in nonpolar solvents. Studies of the process in solid PMMA matrices showed multiexponential behavior with values of k embracing the intercepts of the k - $1/\eta$ plots of fig. 3.

The weak dependence of k on $1/\eta$ in fig. 3 is very misleading such as demonstrated by experiments in which, for a single solvent, the viscosity is varied by variation of the solvent pressure. At first, experimentally one recognizes complications which result in large residual absorptions, see fig. 4. These arise from the accumulation of the photoproduct trans-stilbene during the highly repetitive pulse irradiation experiment. By varying the flow rate of the high pressure liquid reaction mixture through the reaction cell, the residual absorption can be reduced to some extent but not completely suppressed. However, because the time dependence of trans-stilbene absorption has well been investigated separately (see below), the long-time component of the signal can easily be separated from the cis-stilbene signal (e. g. the short-time signal of fig. 4 has a time constant of 3.3 ps whereas the long-time signal has a time constant of 179 ps).

The increase of pressure and of solvent viscosity leads to an increase of the time constant of cis-stilbene dynamics. Fig. 5 demonstrates this behavior. Instead of the much weaker viscosity dependence of the 1 bar results of nonpolar solvents, now

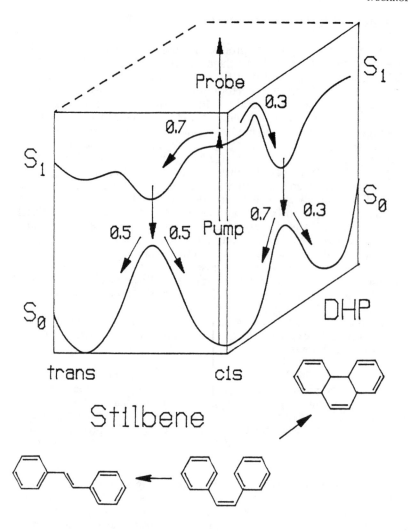

Fig. 1   Reaction scheme for photoisomerization of cis-stilbene
              (adapted from [13]).

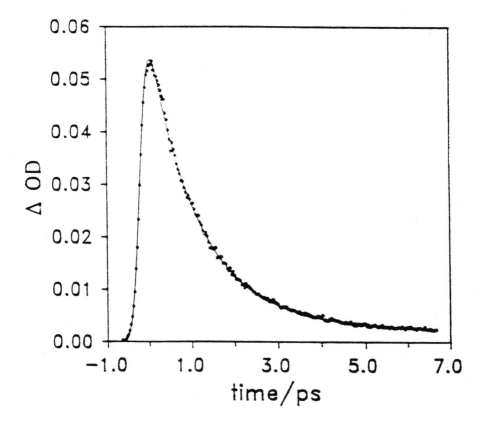

Fig. 2   Transient absorption signal of cis-stilbene at 612 nm after excitation
              at 306 nm. Experiment in n-octane at 1 bar and 295 K [11].

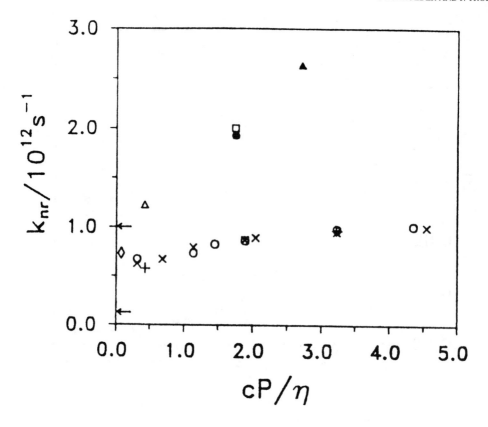

Fig. 3   First order rate coefficients k_{nr} of cis-stilbene decay in nonpolar (lower group) and polar solvents (upper group) at 1 bar (measurements from [11]-[13], arrows: measurements in PMMA matrices from [11]).

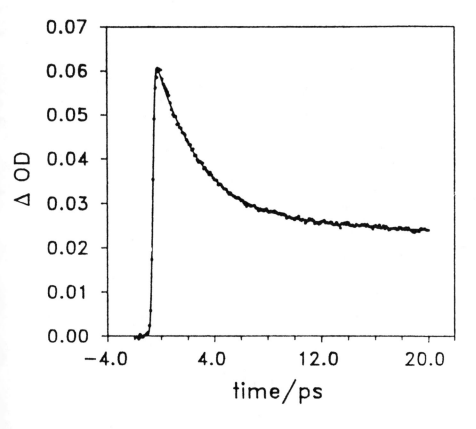

Fig. 4   Transient absorption signal of cis-stilbene at 612 nm after excitation
                at 306 nm. Experiment in n-octane at 3530 bar and 295 K [11]
                (residual absorption due to trans-stilbene photoproducts).

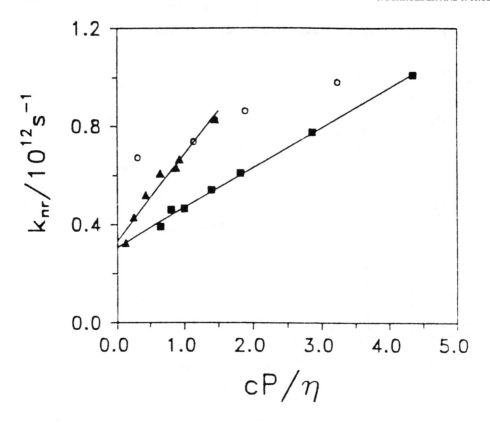

Fig. 5   Viscosity dependence of cis-stilbene isomerization rate coefficients k_{nr}
              in n-pentane (squares) and n-nonane (triangles) at 295 K [11]
              (1 bar measurements in n-alkanes (circles) are given for comparison,
              see fig. 3).

linear dependences of k on $1/\eta$ with non-zero intercepts are observed. This viscosity dependence is clearly missed if only measurements at 1 bar are performed. The first order rate coefficients k, such as derived from the dynamics of the cis-stilbene absorption signals, can well be represented by

(6) $k = k_I + k_{II}$

with

(7) $k_{II} = A/\eta$.

The parameters A depend on the nature of the solvent, being largest for polar solvents where they apparently do not vary too much from one solvent to the other, see fig. 6. The next property to inspect is the temperature dependence of k. Fig. 7 shows the k - $1/\eta$ plot for n-hexane at 295 and 390 K. Whereas the k_I component of k increases with T, the parameter A apparently is only very weakly temperature dependent. For n-hexane, k_I is represented by

(8) $k_I = 6.3 \times 10^{12} \exp(- 8.5 \text{ kJ mol}^{-1}/RT) \text{ s}^{-1}$.

The question arises whether the fairly clear experimental results can be rationalized by simple theoretical models such as sketched in the introduction or whether alternative or more complicated formulations are required. From investigations of the photoproducts of the reaction, one knows that excited cis-stilbene may also, by ring-closure, react to form dihydrophenanthrene (DHP), see fig. 1. The quantum yields indicated in the figure are representative for 1 bar experiments (see also [16], and the contribution by A. B. Myers at this symposium). It appears very reasonable to attribute the k_I-component of k to the DHP-pathway. The derived Arrhenius expression of k_I given by eq. (8) has a form which corresponds well to the TST relation of eq. (1), with $\Delta S^{\neq}$ close to zero such as this is typical for high barrier crossing processes. A linearity of the k - $1/\eta$ plots up to pressures of 5 kbar suggests that the DHP-process remains in its low-damping regime up to this pressure. In terms of the Kramers' equation (4), this would require that $\beta > \omega_B$ is still valid up 5 kbar. For a "sharp" high barrier, this is well possible: it would require values of ω_B larger than 10^{13} s^{-1} [11]. If this interpretation is correct, studies of the pressure dependence of k have allowed to separate the two reactive processes of excited cis-stilbene in a simple manner. However, more systematic investigations of the pressure and temperature dependences of the quantum yields for DHP formation would be desirable as well.

The interpretation of the component $k_{II} = A/\eta$ in eqs. (6) and (7) poses a more serious problem. On the one hand, the absorption signals follow strictly exponential time laws such as they are typical for barrier crossing processes involving bottle-neck situations. On the other hand, fig. 7 indicates that there is no major temperature dependence of the parameter A which would indicate the absence of a substantial barrier unless there is a compensation of the temperature dependences of a Boltzmann factor and other temperature dependences contained in A. The existence of a small barrier might be consistent with gas phase observations of cis-stilbene fluorescence in clusters [14]. The alternative interpretation of the term $k_{II} = A/\eta$ would follow the assumption of a barrierless process [15]. In this case, non-exponential time laws

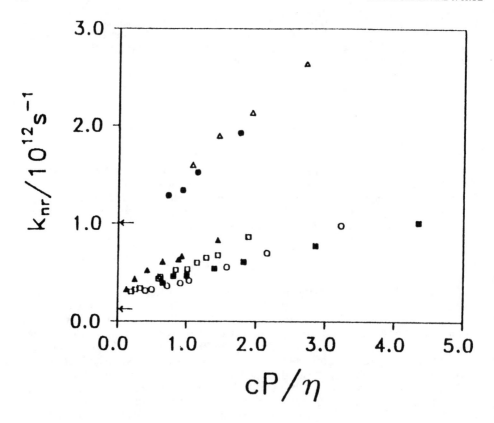

Fig. 6   Viscosity dependence of cis-stilbene isomerization rate coefficients k_{nr}
                in nonpolar (lower group) and polar solvents (upper group) [11]
                (viscosity variation by pressure variation between 1 and 4000 bar,
                arrows: measurements in PMMA matrices).

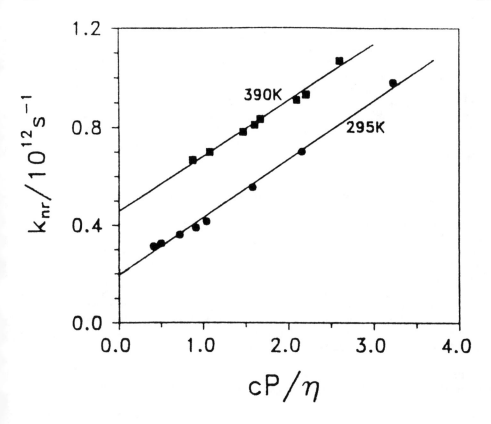

Fig. 7   Viscosity dependence of cis-stilbene isomerization in n-hexane at various temperatures [11].

would be expected unless the internal conversion through the photochemical funnel becomes rate determining. It appears premature to suggest which interpretation applies. An analysis of the data by the Smoluchowski limit of the Langevin equation, as applied to a low-barrier process, was given in [11], assuming solvent-dependent modifications of the small barrier. However, unless there are compensations of temperature dependent factors in the parameter A, this interpretation does not give an explanation for the equal slopes of the k - 1/η plots in fig. 7. Independent of the remaining uncertainties, our experiments show that, without studies of the combined pressure and temperature dependences of the process in various solvents, an interpretation of the transient absorption signals of cis-stilbene photoisomerization would have been impossible.

3. VISCOSITY DEPENDENCE OF A HIGH-BARRIER PROCESS

In the following we show some results on the isomerization of trans-stilbene in the S_1-state. This pathway is also included in fig. 1. Like the DHP pathway, it involves a high barrier on the way from the trans- to the twist-conformation. However, unlike the DHP pathway, this barrier probably is not sharp but flat. It appears of crucial importance to study the dependence of the rate of this process again in various solvents at varying pressures and temperatures. In addition, the process can be studied in isolated molecules. Fig. 8 shows measurements of specific rate constants k(E) of the process in jet-cooled isolated trans-stilbene molecules [17]. The results in all detail correspond to a "normal" unimolecular isomerization reaction which can be interpretated by suitable adaptations of RRKM theory [18]. The advantage of having isolated molecule measurements consists in the possibility of transforming these results into thermally averaged rate constants. Such rate constants would be expected for the liquid phase in the hypothetical absence of solvent-reagent interactions.

Measurements of the viscosity dependence of the trans-stilbene photoisomerization rate in liquid solvents again are most informative. If experiments only at 1 bar considered, the results are misleading. Fig. 9 shows a log k - log η plot of 1 bar results in a series of homologous n-alkanes. A dependence of k proportional to $\eta^{-0.32}$ is found. A careless interpretation of the η-exponent in terms of friction memory effects again would be in error such as illustrated in the following. Varying the viscosity not by varying the solvent (at 1 bar) but by increasing the pressure, the results of fig. 10 are obtained. Measurements in high pressure gas phase CH_4 and C_2H_6, as well as in compressed liquid C_2H_6, show the appearance of a "Kramers turnover", i. e. of the transition between a process controlled by collisional activation in the low pressure - low damping regime to a process controlled by diffusive barrier crossing in the high pressure - high damping regime. In the latter diffusion-controlled regime, k approaches a linear dependence on 1/η, quite in contrast to the suggestions based on fig. 9 (if measurements in the gas-liquid transition range are considered, a k - 1/η representation is replaced by a k - 1/D representation where D denotes the self-diffusion coefficient of the solvent). The weak dependence of k on 1/η in fig. 9, analogous to that observed in cis-stilbene (fig. 3), provides evidence for a superposition of static and dynamic solvent-reagent interactions. The static interactions appear unexpectedly strong even for nonpolar solvents, such as illustrated in fig. 9. Based on the isolated molecule measurements of fig. 8, one may design a "Kramers turnover" curve which, in the low pressure range, corresponds to the falloff

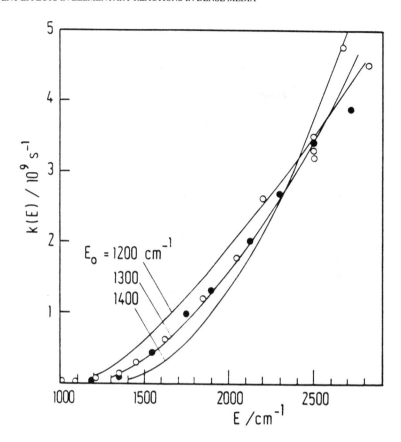

Fig. 8   Specific rate constants k(E) of trans-stilbene isomerization in the S_1-state (circles: experiments with jet-cooled molecules [17]; lines: RRKM fits with different threshold energies E_0 [18]).

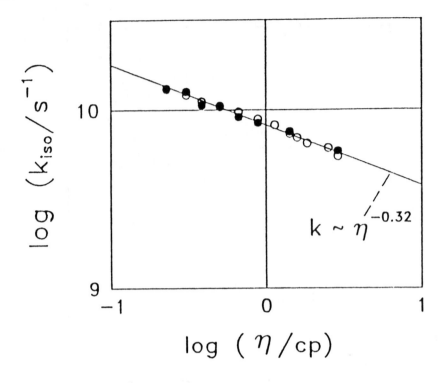

Fig. 9   Viscosity dependence of photoisomerization rate coefficients k_{iso} of
              trans-stilbene in the S_1-state in various alkane solvents at 1 bar
              (circles: experiments from [19],[20]).

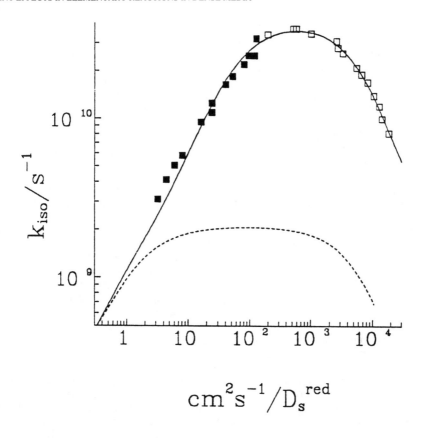

Fig. 10   Photoisomerization rate coefficients k_{iso} of trans-stilbene in the S_1 state in compressed methane (full squares) and ethane (open squares) as a function of a scaled solvent self-diffusion coefficient D_s^{red} (--: modelling with isolated molecule data, ——: modelling with a reduction of the energy barrier E_0 from 1250 cm^{-1} in the isolated molecule to 700 cm^{-1} in the clustered molecule [10]).

curve of a unimolecular isomerization and, in the high viscosity range, merges with a Kramers-Smoluchowski expression. Fig. 10 includes this hypothetical k - 1/D plot which strongly deviates from the real observations. The increase of the maximum rate constant by a factor of 40 over the hypothetical value can only be explained by very strong static solvent-reagant interactions which set in in the gas phase at pressures not far above 1 bar. We have attributed this effect to a lowering of the energy barrier in van der Waals clusters. It appears surprising that a proportionality between k and $1/\eta$ is approached at all, as this signifies that the solvent shift of the barrier is fully established in the clustering stage (from about 1250 cm^{-1} in the isolated molecule to 700 cm^{-1} in the clustered molecule), whereas no further solvent shift occurs in the compressed liquid.

Fig. 11 shows a series of log k - log 1/D plots for trans-stilbene photoisomerization. The figure shows that a linear k - 1/D relation is approached for n-alkane solvents up to n-butane. For larger n-alkanes such as n-hexane, the 1/D-dependence becomes weaker, see fig. 12. There is no unique interpretation of the different behavior of smaller and larger alkane solvents. One possible explanation might be that smaller alkanes "fit into lock-type sites" of trans-stilbene, thus modifying the barrier by conformational changes in the first solvation stage. Larger alkanes, on the other hand, may not fit easily into these lock-type sites such that their modifying effects of the barrier are developed more gradually, possible by influencing outer solvation shells; these static solvent effects apparently occur together with the diffusive action: Because of the stronger interaction, however, the barrier shifts are becoming more pronounced than with simpler alkanes. In alcohol solvents, the barrier shifts are even stronger, further reducing the barrier beyond the large alkane effects.

In summary, small alkane solvents in the investigated high-damping range do not show any sign of deviations from a linear k - $1/\eta$ dependence. At least for these solvents, there is no evidence for non-Markovian friction such that the Kramers description, using bulk viscosity data, appears adequate and no Grote-Hynes extension of this model appears necessary. On the other hand, there are obviously very pronounced "static" solvent-reagant interactions modifying the effective reaction barrier. The experiments with trans-stilbene suggest that there are at least two types of solvent-reagant interactions: (i) for small alkanes, there appears to be lock-type interactions which lower the barrier upon clustering; this effect becomes stronger with increasing molecular size; (ii) further lowerings of the barrier are observed with larger alkanes and, more strongly, with polar solvents. The two types of interactions are of different character, the first being saturable, whereas the second continues to strengthen with increasing density.

Another important set of experiments consists in studies of the temperature dependence of k in the various density regimes. In these experiments, surprisingly strong shifts of the turnover range of the k - 1/D plots are observed, see fig. 12 for n-hexane: with increasing temperature, the decrease of k with increasing viscosity sets in at lower values of η. The effect could be attributed to various origins: the effective barrier frequency ω_B might be excitation - and, hence, temperature-dependent or the barrier height might have different effective density dependences at different temperatures; the observations might be manifestations of multidimensional potential energy effects in the Kramers-Smoluchowski analysis [22]-[24]. The described

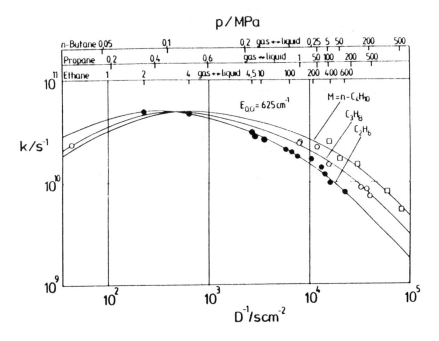

Fig. 11   Photoisomerization rate coefficients k of trans-stilbene in the S_1-state in compressed ethane, propane and n-butane [22] as a function of the self-diffusion coefficient D of the solvent.

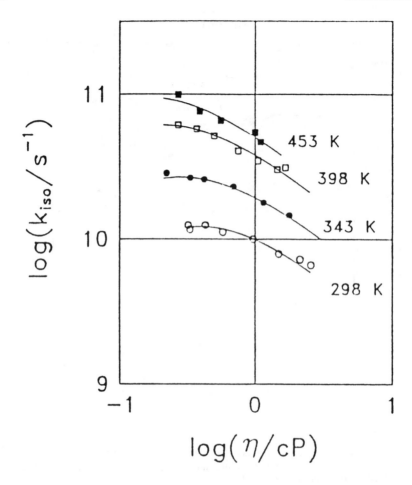

Fig. 12   Photoisomerization rate coefficient k_{iso} of trans-stilbene in the S_1-state
            in compressed n-hexane at various temperatures (experimental points
            and modelling with density-dependent threshold energies [20]).

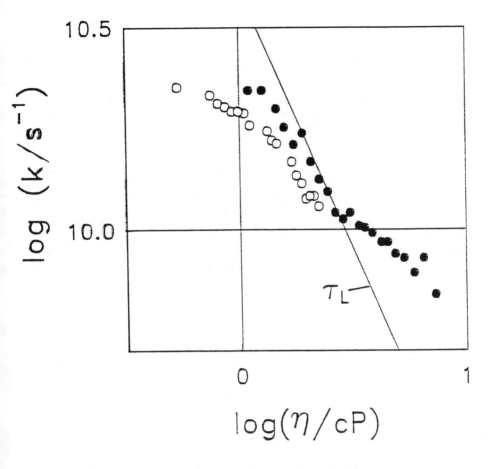

Fig. 13   Photoisomerization rate coefficients k of trans-stilbene in compressed liquid methanol (open circles) and ethanol (full circles) (comparison with longitudinal dielectric relaxation time τ_L [20]).

observations at present cannot be explained conclusively and the given interpretation, suggested by visual inspection of the experiments, remain highly speculative.

A final observation concerns the pressure dependence of trans-stilbene isomerization rates in compressed alcohols. Fig. 13 shows log k - log η plots in compressed liquid methanol and ethanol. In this case, an intermediate range is observed where k follows the dielectric relaxation of the solvent; at larger viscosities, the reaction is faster than solvent relaxation; at smaller viscosities, it is slower. In this case, the strong solvent-reagant interaction with a marked solvent shift of the barrier, at least in a certain density range, may serve as an indicator of the solvent dynamics.

In summary, the described experiments on the pressure and temperature dependence of simple photoisomerization reactions present a remarkly multi-facet picture. Whereas dynamic solvent effects apparently are fairly "uninteresting" in following bulk viscosity properties, static solvent effects modifying reaction barriers show very specific properties, possibly reflecting different types of solvent sites around the reaction center of the molecule. The separation of static and dynamic solvent effects, of course, is not unambiguous, although the pronounced experimental effects appear fairly suggestive. More theoretical work is required, particularly on static interaction models.

Acknowledgment

Financial support of this work by the Deutsche Forschungsgemeinschaft (SFB 357 "Molekulare Mechanismen unimolekularer Prozesse") is gratefully acknowledged.

REFERENCES

[1]   H. A. Kramers, Physica 7, 284 (1940).
[2]   E. Pollak, H. Grabert und P. Hänggi, J. Chem. Phys. 91, 4073 (1989).
[3]   J. E. Straub, M. Borkovec, and B. J. Berne, J. Chem. Phys. 83, 3172 (1985); 84, 1788 (1986).
[4]   P. Hänggi and J. Troe, Eds., "Rate Processes in Dissipative Systems: 50 Years after Kramers", Ber. Bunsenges. Phys. Chem. 95, 225 - 442 (1991).
[5]   S. Chandrasekhar, Rev. Mod. Phys. 15, 1 (1943).
[6]   R. F. Grote and J. T. Hynes, J. Chem. Phys. 73, 2715 (1980).
[7]   J. Troe, in "High Pressure Chemistry" (Ed. H. Kelm, Reidel, Dordrecht 1978).
[8]   J. Schroeder and J. Troe, Chem. Phys. Lett. 116, 453 (1985).
[9]   J. Schroeder and J. troe, Ann. Rev. Phys. Chem. 38, 163 (1987).
[10]   J. Schroeder and J. Troe, in "The Barrier Crossing Problem", (Eds. G. R. Fleming and P. Hänggi, World Scientific, 1993) p. 489.
[11]   L. Nikowa, D. Schwarzer, J. Troe, and J. Schroeder, J. Chem. Phys. 97, 4872 (1992).
[12]   D.C. Todd, J. M. Jean, S. J. Rosenthal, A. J. Ruggiero, D. Yang, G. R. Fleming, J. Chem. Phys. 93, 8658 (1990).

[13]   S. Abrash, S. T. Repinec, and R. M. Hochstrasser, J. Chem.
         Phys. 93, 1041 (1990); S. T. Repinec, R. J. Sension,
         A. Z. Szarka, and R. M. Hochstrasser, J. Phys. Chem. 95,
         10380 (1991).
[14]   H. Petek, K. Yoshihara, Y. Fujiwara, Z. Lin, J. H. Penn, and
         J. H. Frederick, J. Phys. Chem. 94, 7539 (1990).
[15]   B. Bagchi and G. R. Fleming, J. Phys. Chem. 94, 9 (1990).
[16]   J. M. Rodier, X. Ci, and A. B.Myers, Chem. Phys. Lett. 183,
         55 (1991); J. M. Rodier and A. B. Myers, J. Amer. Chem. Soc.,
         in press.
[17]   J. Syage, W. R. Lambert, P. M. Felker, A. H. Zewail, and
         R. M. Hochstrasser, Chem. Phys. Lett. 88, 266 (1982); A. Amirav and
         J. Jortner, Chem. Phys. Lett. 95, 295 (1983).
[18]   J. Troe, Chem. Phys. Lett. 114, 241 (1985).
[19]   G. Rothenberger, D. K. Negus, and R. M. Hochstrasser,
         J. Chem. Phys. 79, 5360 (1983).
[20]   J. Schroeder, J. Troe, and P. Vöhringer, to be published.
[21]   M. W. Balk and G. R. Fleming, J. Phys. Chem. 90, 3975 (1986).
[22]   J. Schroeder, D. Schwarzer, J. Troe, and F. Voß, J. Chem. Phys. 93,
         2393 (1990).
[23]   J. Schroeder, J. Troe, and P. Vöhringer, Chem. Phys. 203,
         255 (1993).
[24]   Ch. Gehrke, J. Schroeder, D. Schwarzer, J. Troe, and F. Voß,
         J. Chem. Phys. 92, 4805 (1990).

THE REORIENTATIONAL DYNAMICS IN LIQUID METHANOL

T. Bultmann*, K. Kemeter*, Ch. Rusbüldt**, Ph. A. Bopp**
and N. P. Ernsting*
* Max-Planck-Institut für Biophysikalische Chemie, Abt. Laserphysik,
Postfach 2841, D-3400 Göttingen-Nikolausberg
** Institut für Physikalische Chemie,
Rheinisch-Westfälische Technische Hochschule, Templergraben 59,
D-5100 Aachen

Abstract

The results from nuclear magnetic resonance (NMR) relaxation time measurements, time resolved ultrafast spectroscopy and classical molecular dynamics (MD) computer simulations are combined to yield a detailed understanding of the reorientational dynamics in liquid methanol (CH_3OH) at room temperature. It is found that the reorientational motions of single molecules are anisotropic, the correlation times for molecule-fixed vectors perpendicular to the H-bond is about half that for the O-H vector. Furthermore, the dielectric relaxation of methanol after strong perturbation was measured to be temperature dependent.

Introduction

The structure of liquid alcohols is often described as consisting of "chains" of molecules interconnected by hydrogen bonds. This predominantly one-dimensional arrangement facilitates the description of the microscopic structure and dynamics when compared for instance to water with its three dimensional hydrogen-bond network. Consequently, these liquids and most dominantly methanol, being the simplest one, have been the center of theoretical (1-12) and experimental efforts (13-17, see also the references in (6) and (13)). We focus here on the reorientational motions of single molecules in pure liquid methanol at room temperature. If the solvent is not too strongly perturbed by the solute, these dynamics can also be relevant for reactions involving either encounters between solute molecules or their dissociations. On the other hand, it is also possible to investigate the dynamics of a solvent perturbed by a polar solute upon modification (e.g. charge jump) by computer simulations. This is exactly analogous to typical time-resolved spectroscopy experiments.

J. Jortner et al. (eds.), Reaction Dynamics in Clusters and Condensed Phases, 383–391.

Simulations and Experiments

Pure methanol under various conditions (1-5), methanolic mixtures (6-8), and ionic solutions in methanol (9,10) have been studied in a number of MD computer simulations. In most cases, the methanol molecules are represented by a "three-site" model, i.e. the CH_3-group is treated as one force center (called Me). This kind of model is also adopted for the present study. The intermolecular interaction potential consists of Coulomb-interactions between partial charges located at the atomic sites together with empirical potentials. Furthermore, the molecules are assumed to be flexible by including an intramolecular potential between oxygen, hydrogen, and Me. This potential contains harmonic and anharmonic contributions of intramolecular stretches and bends up to fourth order. The total potential is assumed to be the sum of these inter- and intramolecular terms, the details and the potential constants are given in (1). (Note that a slightly modified set of parameters for the Me-Me and Me-O interactions is given in (6)). These models have been explored previously (6-10).

The simulation results presented here are essentially obtained from an NVE-MD simulation of 200 model molecules at a density of 0.7866 g/cm^3 and a temperature of 294 K. The total simulation time is 50 ps (200000 integration steps of 2.5 10^{-16} seconds each). An Ewald procedure is used for the computation of the energies and forces resulting from the Coulomb interactions. As has been noted in (5), very long simulations are indeed needed to determine the reorientational motions with sufficient accuracy, especially at low temperature. Here, however, we shall compare results computed for room temperature with NMR-experimental results.

NMR relaxation time measurements on isotopically substituted samples are a sensitive way to probe the reorientational dynamics of single molecules (18). For liquid methanol, the most reliable value for the reorientational correlation time τ_2 of the intramolecular O-H bond was extracted from the additional proton relaxation rate induced by ^{17}O in $CD_3^{17}OH$ (13).

We note here that the average intramolecular O-H distance r_{OH} is required to evaluate the correlation time τ_2 from the measured relaxation time T_1. We have (13,19):

$$\frac{1}{T_1} \propto r_{OH}^{-6} \cdot \tau_2 \quad .$$

The simulations with the flexible model allow one to estimate the average elongation of this bond in the liquid with respect to the gas-phase value. We obtain a value of $\Delta r_{OH}=0.02$Å, which, assuming a gas-phase value of 0.96 Å, leads to $< r_{OH} >= 0.98$ Å for the liquid. $r_{OH} = 0.99$ Å was used in (13), in agreement with constraints originating from the deuterium quadrupole coupling constant. The dipolar coupling constant (the proportionality factor in the equation above) must be known to determine τ_2. A number of assumptions were made in the past (see (15,16)), for example the same value of τ_2 has been taken for methanol and for water. This leads to the scattering of the experimental values given below. We note here that quadrupole coupling constants have recently been independently determined from combined ab-initio and simulation work (20). The values obtained by this procedure for water were found to be in excellent agreement with the most recent experimental ones.

The orientational relaxation of polar solvent molecules can also be measured by observing the time-resolved emission or absorption of suitable solute molecules. Generally speaking, one pumps the solute with a first ultrafast laser pulse into an excited state which has a different charge distribution compared to the ground state. Thus, the polar solvent surrounding the solute undergoes a dielectric orientational relaxation. The absorption/emission spectrum of the solute is sensitive to this change of the solvent polarisation. After a certain time, one probes the spectrum with a second broadband laser pulse (pump and probe spectroscopy). The temporal development of the spectrum contains the information on the dynamics of the solution (see Figure 1). If $\nu(t)$ denotes the time dependent spectral position of the first moment of the absorption band, the normalized spectral relaxation function NSR(t) defined as

$$\mathrm{NSR}(t) = \frac{\nu(t) - \nu(\infty)}{\nu(0) - \nu(\infty)}$$

characterizes the relaxation. Both continuum and molecular models (21) have been used to interpret such data.

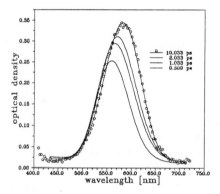

Figure 1. Time development of the absorption spectrum of BPADS
(bis-(p-aminophenyl)-disulfide) in methanol at 43° Celsius.

Nonequilibrium MD-simulations have been used in a few cases to study the underlying phenomena because they provide acces to the NSR(t), too (22). For a large number of starting conditions of the solute in solution, which are needed to sample the phase space, the electrical property of the solute is changed instantaneously. The mean value of the electrical energy of the solute in the solution as a function of time, E(t), provides the information about the simulated solvent relaxation. After normalisation, the following holds (23):

$$\mathrm{NSR}(t) = \frac{E(t) - E(\infty)}{E(0) - E(\infty)}$$

We are presently conducting preliminary simulation studies on the solvent dynamics of methanol following the dissociation of a model solute molecule with geometric characteristics and a charge

distribution reminiscent of bis-(p-aminophenyl)-disulfide (BPADS). This process is also studied
by photodissociating BPADS in a pump-probe experiment. Two identical radicals with dipole
moments of about 4.5 D, at an angle of about 105 degrees with respect to each other, are created
upon dissociation, which give a total dipole moment of about 5.5D. Compared to the dipole moment
of BPADS of 1.5D, this is a strong perturbation of the solvent. Hence, we may not be in the linear
response regime (24). The relaxation process, i.e. the rotation of the radicals relative to each
other and the orientational relaxation of the solvent molecules, is probed by taking an absorption
spectrum (25). The aim of the simulation is to get some insight into the molecular behaviour of
the solvent during relaxation, its dependency on the size and electrical properties of the solute and
on the energy relaxation rate. We note, however, that the solvent structure in the vicinity of the
solute will not be identical to the structure of the pure liquid.

Results and Discussion

The correlation function

$$C_2^{OH}(t) = \frac{1}{NM} \sum_i^N \sum_j^M \left(P_2\left(e_{OH}^i(t_j) \cdot e_{OH}^i(t_j + t)\right) \right)$$

where $e_{OH}^i(t_j)$ is the unit vector along the O-H-bond of the i-th molecule at time j and P_2 is the
second Legendre polynomial

$$P_2(e_1 \cdot e_2) = \frac{3}{2} \cos^2(\angle e_1 e_2) - \frac{1}{2} \quad ,$$

is computed from the simulation. N and M are the number of molecules and the number of time
origins t_j used. $C_2^{OH}(t)$ is shown in Figure 2 together with similar functions computed for other
molecule-fixed unit vectors. The non-exponential behaviour of all these functions at times $t < 0.5$ ps
is quite obvious.

Since the correlation functions $C_2(t)$ are non-exponential, some approximation must be used
to extract correlation times τ_2 which can be compared to the experimental one. We procede in the
following way:

$$\tau_2 = \int_0^\infty C_2(t)dt$$

$$= \int_0^{t_1} C_2(t)dt + \int_{t_1}^\infty \hat{C}_2(t)dt$$

The first integral is computed nummerically from the simulated correlation function. The function
$\hat{C}_2(t)$ is an exponential function fitted to $C_2(t)$ for times $t > t_1$. The time t_1 is varied between 4 ps
and 8 ps. Results for τ_2 from the simulation are compared in Table 1 with experimental ones.

The reorientation correlation time can be computed in a similar way for any other molecule-
fixed vector. Figure 2 shows $C_2(t)$ for the two vectors perpendicular to the O-H vector, one in the
plane of the (three-site) molecule, V_2, and one perpendicular to it, V_3, and for the C-O vector.
The correlation times τ_2 are also listed in Table 1.

Table 1. Reorientation correlation times τ_2 for various molecule-fixed
vectors from the simulation and from NMR experiments, in ps.

	ref.	T/K	O-H	V_2	V_3	C-O
MD	this work	294	5.2	2.7	2.1	2.7
NMR	13	298	5.1	-	-	1.8*
MD	4	300	2.6	a	a	1.1
MD	5	300	4.7	-	-	2.7
NMR	15	298	3.7 5.4 b			
			4.3 c			
			7.8 d			
NMR	16	298	3.6			

*: at 308 K, preliminary results (27)

a: values 2.4 ps and 1.5 ps are reported for the
molecule-fixed axes y and z, z being close to the O-C
vector, and y perpendicular to it. A value of 2.0 ps is given
for the dipole moment direction.

b: O-D, deuteron relaxation, evaluation with different deuteron
quadrupole coupling constants.

c: proton relaxation.   d: ^{17}O-relaxation.

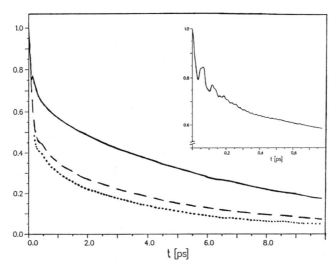

Figure 2. Correlation function $C_2(t)$ for the unit vectors in the molecule-
fixed frame of methanol: O-H (solid), V_2 (dashed), V_3.(dotted).
Insert: short time part of C_2^{OH}.

Both the experimental and the simulated τ_2 values for the O-H vector scatter by a factor of about two. As already mentioned, the evaluation of the experimental results involves the knowledge of the coupling constants and of the average intramolecular distances. Uncertainties in the results obtained from the simulations may originate either from insufficient lengths of the simulation runs or from insufficient energy conservation (in the NVE ensemble) in the long runs or from the mechanism used to maintain the temperature constant in other ensembles (26). The model parameters may also be inadequate. However, the agreement between the results from the present simulations and from the most recent experiments (13) is very satisfactory, the two values for τ_2 are indeed well within their mutual error bars.

This agreement motivated us to study the reorientational motions in more detail. In agreement with a previous simulation (5), it was found that the reorientational motions are anisotropic. The correlation times for the two vectors perpendicular to the O-H vector are about 50 percent of that of the O-H vector itself (Table 1). The C-O vector is close to $\mathbf{V}_2$ and its correlation time can be determined experimentally.

This direct experimental verification of the anisotropy of the reorientation, using ^{13}C-NMR in combination with ^{17}O enrichment to determine the correlation time τ_2 of the C-O vector, is in progress (27). Preliminary results indicate that this correlation time is indeed about half the value for the O-H vector, see Table 1.

A large fraction of the molecules belongs to hydrogen-bonded chains in liquid methanol at room temperature (1). It has been attempted to determine from computer simulations whether the reorientation time of a molecule is dependent upon its position in the chain (4), see also ref. (28). It was found that molecules with less than two H-bonds, i.e. molecules at the ends of a chain, or monomers, reorient faster than the ones inside the chain. A problem arises, however, from the fact that the lifetime of a bond itself is of the same order of magnitude as the reorientation time. It has been speculated (29) that the presence of a fourth neighbour in the vicinity of a molecule may enhance its reorientational process. This is analogous to a mechanism discussed for water (30).

We have also studied the self-diffusion process. From the simulation, the self-diffusion coefficient is computed independently from the integral over the velocity autocorrelation function (Green-Kubo relation) and from the Einstein relation. We obtain values of 1.84 cm^{2}s^{-1} and 1.95 cm^{2}s^{-1}, respectively, for a simulation temperature of 294 K. Extensive measurements of the self-diffusion in methanol have been reported in (31). A fit formula to the experimental values yields $D = 2.3\pm0.4$ cm^{2}s^{-1} for CH$_3$OD at 294 K.

Figure 3 shows recent measurements of the normalized spectral relaxation function, NSR(t), obtained for BPADS in methanol at different temperatures (32). The pump-probe cross correlation time is approximately 200 fs. Two correlation times, τ_1 and τ_L, are extracted with a biexponential fit. The larger of the two obtained values is the longitudinal relaxation time τ_L. The numerical values are shown in Table 2. For comparison, we measured the relaxation times at 25° Celsius in other alcohols (32) and obtained in iso-propanol 0.6/13.9 ps, in ethanol 0.4/8.0 ps, and in methanol 0.6/4.6 ps.

Table 2. Reorientational polarisation relaxation times τ_1 and τ_L, in ps,
for BPADS in methanol at different temperatures, in Celsius.

	10	20	25	43
τ_1	0.5	0.5	0.6	0.4
τ_L	10.1	5.0	4.6	2.4

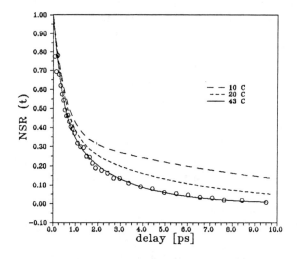

Figure 3. NSR(t) of BPADS in methanol measured at different temperatures

Summary and Conclusions

The reorientational dynamics of single methanol molecules has been studied by molecular dynamics computer simulations. The reorientational process was characterized by a correlation time τ_2. This quantity was also accessible from NMR relaxation time measurements. The orientational relaxation of methanol solutions after a strong perturbation of the electrical properties of the solute was investigated by ultrafast pump-probe spectroscopy. Nonequilibrium MD-simulations may be performed to get a detailed microscopic understanding of the underlying dynamics.

Acknowledgement

This work was largely financed by the grants Er 154/3-1,2 and Bo 744/5-1,2 under the auspices of the "Schwerpunktprogramm schnelle intermolekulare Prozesse in Flüssigkeiten" of the Deutsche Forschungsgemeinschaft (DFG). Grants of computer time on the "Landesvektorrechner NRW (SNI S600/20)" at the computer center of the RWTH and on the AIX-cluster at the Gesellschaft für wissenschaftliche Datenverarbeitung Göttingen (GWDG) are also gratefully acknowledged. P.B.

also acknowledges helpful discussions with R.Ludwig and M.D.Zeidler, and the award of the Heisenberg Fellowship from the DFG.

References

(1) Pálinkás, G., Hawlicka, E., and Heinzinger, K. (1987) "A Molecular Dynamics Study of Liquid Methanol with a Flexible Three-site Model", J.Phys.Chem. **91**, 4334-4341.

(2) Haughney, M., Ferrario, M., and McDonald, I.R., (1987) "Molecular Dynamics Simulation of Liquid Methanol", J.Phys.Chem. **91**, 4934-4940

(3) Pálinkás, G., Tamura, Y., Spohr, E., and Heinzinger, K. (1988) "Molecular Polarity and the Structure of Liquid Methanol", Z.Naturforsch. **43a**, 43-46.

(4) Matsumoto, M., and Gubbins, K.E.,(1990) "Hydrogen bonding in liquid methanol" J.Chem.Phys. **93**, 1981-1994

(5) Sindzingre, Ph., and Klein, M.L., (1992) "A molecular dynamics study of methanol near the liquid-glass transition", J.Chem.Phys. **96** 4681-4692

(6) Pálinkás, G., Hawlicka, E., and Heinzinger, K. (1991) "Molecular dynamics simulations of water-methanol mixtures", Chem.Phys. **158**, 65-76.

(7) Pálinkás, G., and Bakó, I. (1991) "Excess Properties of Water-Methanol Mixtures as Studied by MD Simulations" Z.Naturforsch. **46a**, 95-99 (1991)

(8) Pálinkás, G., Bakó, I., Heinzinger, K., and Bopp, Ph. (1991) "Molecular dynamics investigation of the inter- and intramolecular motions in liquid methanol and methanol-water mixtures", Mol.Phys. **73**, 897-915.

(9) Tamura, Y., Spohr, E., Heinzinger, K., Pálinkás, G., and Bakó, I. (1992) " A Molecular Dynamics and X-ray Diffraction Study of $MgCl_2$ in Methanol", Ber.Bunsen.Phys.Chem. **96**, 147-158

(10) Heinzinger, K., and Pálinkás, G. (1993) "On the Solvation of Ions in Methanol", preprint.

(11) Jorgensen, W.L., (1981) "Transferable Intermolecular Potential Functions. Application to Liquid Methanol Including Internal Rotation", J.Am.Chem.Soc. **103**, 341-345

(12) Kolafa, J., and Nezbeda, I., (1987) "Monte Carlo simulations on primitive models of water and methanol", Mol.Phys. **61** 161-175

(13) Ludwig, R., Gill, D.S., and Zeidler, M.D., (1991) "Molecular Reorientation in Liquid Methanol", Z.Naturforsch. **46a**, 89-94

(14) Mashimo, S., Kuwabara, S., Yagahira, S., and Higasi, K., (1989) "The dielectric relaxation of mixtures of water and primary alcohols", J.Chem.Phys., **90**, 3292-3294

(15) Versmold, H., and Yoon, C., (1972) "Oxygen-17 NMR studies of Methanol and Ethanol", Ber.Bunsen.Phys.Chem. **76**, 1164-1168

(16) Versmold, H., (1980) "NMR Studies of the Reorientational Motions in Methanol and Methanol-Water Mixtures", Ber.Bunsen.Phys.Chem. **84**,168-173

(17) Fung, B.M., and McGaughy, T.W., (1976) "Molecular motions in liquid. I. Rotation of water and small alcohols studied by deuteron relaxation", J.Chem.Phys. **65**, 2970-2976

(18) Lankhorst, D., Schriever, J., and Leyte, J.C., (1982) "Determination of the Rotational Correlation Time of Water by Proton NMR Relaxation in $H_2{}^{17}O$ and Some Related Results" Ber.Bunsen.Phys.Chem., **86** 215-221

van der Maarel, J.R.C., (1989) "A H, D, and ^{17}O Nuclear Magnetic Relaxation Study on the Structure and Dynamics of Water in Concentrated $ZnCl_2$" J.Magn.Res. **81**, 92-103; and references therein

(19) Abragam, A., (1961) "The Principles of Nuclear Magnetism", Clarendon, London

(20) Eggenberger, R., Gerber, S., Huber, H., Searles, D., and Welker, M., (1992) " Ab-initio calculation of the deuterium quadrupole coupling in liquid water", J.Chem.Phys. **97**, 5898-5904

(21) Barbara, P.F., and Jarzeba, W. (1990) "Ultrafast Photochemical Intramolecular Charge Transfer and Excited State Solvation" Adv.Photochem. **15**, 1-68

(22) Maroncelli, M., (1991) "Computer Simulations of Solvation Dynamics in Acetonitrile" , J.Chem.Phys. **94**, 2084-2103
Carter, E.A., and Hynes, J.T. (1991) "Solvation Dynamics for an Ion in a Polar Solvent: Time-Dependent Fluorescence and Photochemical Charge Transfer", J.Chem.Phys. **94**, 5961-5979

(23) Maroncelli, M., and Fleming, G.R. (1988) "Computer Simulations of the Dynamics of Aqueous Solvation", J.Chem.Phys. **89**, 5044-5069

(24) Fonseca, T., and Ladanyi, B.M. (1991) "Breakdown of Linear Response for Solvation Dynamics in Methanol", J.Phys.Chem. **95**, 2116-2119

(25) Ernsting, N.P., and Kaschke, M. (1991) "A Reliable Pump-Probe, broadband spectrometer for subpicosecond transient absorption", Rev.Sci.Instrum. **62**, 600-608

(26) Toxwaerd, S., and Olsen, O.H. (1990) " Canonical Molecular Dynamics of Molecules with Internal Degrees of Freedom", Ber.Bunsen.Phys.Chem. **94**, 274-278

(27) Ludwig, R., and Zeidler, M.D., to be published

(28) Bermejo, F.H., Batallan, F., Howells, W.S., Carlile, C.J., Enciso, E., Garcia-Hernandez, M., Alvarez, M., and Alonso, J., (1990) "A quantitative assessment of the effect of hydrogen bonding on microscopic dynamics by neutron quasielastic scattering". J.Phys.: Condens.Matter **2**, 5005-5013

(29) Pottel, R., (1993) personal communication

(30) Geiger, A., Mausbach J., Schnitker, J., Blumberg, R.L., and Stanley, H.E. (1984) "Structure and Dynamics of the Hydrogen Bond Network in Water by Computer Simulation", J. de Physique, coll C7 suppl. **45**, C7-13 - C7-30

(31) O'Reilly, D.E., and Peterson, E.M., (1971) "Self-Diffusion Coefficients and Rotational Correlation Times in Polar Liquids II", J.Chem.Phys. **55**, 2155-2163

(32) Bultmann, T. and Ernsting, N.P., to be published

MOLECULAR THEORY OF VIBRATIONAL
ENERGY RELAXATION IN GASES AND LIQUIDS

Steven A. Adelman
Department of Chemistry
Purdue University
West Lafayette, Indiana 47907–1393

I. Introduction

Vibrational energy relaxation (VER) in liquids (1), in part because of its fundamental significance for solution chemical reaction dynamics (2), is a process, which despite its venerable history (3), continues to be of interest (4).

Elsewhere (5) we have developed a molecular theory of liquid phase VER, based on our general approach to problems of chemical reaction dynamics in liquid solution (6). The main results of the new theory are expressions for the vibrational-translational-rotational and vibrational-vibrational contributions to the relaxation time T_1. These expressions permit evaluation of both contributions to T_1 from molecular properties, interaction potentials, and equilibrium site-site pair correlation functions. The expressions thus yield a first-principles evaluation of the VER rate constant which permits its numerical study (7) as a function of system and thermodynamic state. In addition to providing computational algorithms for the rate constant, the theory provides qualitative insights into its temperature, density, and isotope dependencies. For example the theory provides a molecular basis (7) for interpreting the observed correlations (8) between gas and liquid phase rates. This basis includes molecular level criteria (7) for the emergence of an isolated binary collision (IBC) model (8) type factorization of the rate constant and formulas (7) for its isothermal density dependence which are more realistic than those available from existing (8) IBC models. [These developments are possible since, as shown in Ref. 7, the theory predicts the full density dependence of the rate constant, i.e. it is not restricted to the dense fluid regime.]

In this paper, we present a synopsis of the theory and describe some of its first applications focusing our discussion, for simplicity, on the prototype problem of vibrational-translational (VT) energy transfer from a diatomic solute to a monatomic solvent.

We begin by noting that our work is based on the formulation of VER in fluids described in Ref. 5a. Within this formulation, vibrational energy transfer arises from dissipation processes due to the high frequency friction exerted by the

393

J. Jortner et al. (eds.), Reaction Dynamics in Clusters and Condensed Phases, 393–409.
© 1994 Kluwer Academic Publishers. Printed in the Netherlands.

solvent on the relaxing solute normal mode. Specifically, one obtains the following result (5a,9) for T_1,

$$T_1 = \beta^{-1}(\omega_\ell) .\qquad\qquad 1$$

In Eq. (1), $\beta(\omega)$ is the friction kernel of the relaxing solute normal mode, which we evaluate using Eq. (2), and ω_ℓ is the liquid phase frequency of this mode given in Eq. (8). The expressions for T_1 mentioned above are obtained from Eq. (1) using a new statistical mechanical methodology, developed in Ref. 5, for evaluating friction at the molecular (as opposed to, say, the continuum dynamics (10)) level. This molecular friction methodology will be outlined below.

Before proceeding further, we discuss in two points the significance of the present work and place it in context in the literature.

POINT 1.   The present theory, whether implemented analytically or by molecular dynamics simulation (4,5a), provides the first practical method for the rigorous ab initio evaluation of VER rates in liquids. Unlike the widely used IBC model (8), the theory does not require as input the gas phase rate constant but rather requires only the elementary input mentioned earlier.

POINT 2.   Our new molecular level procedures (5) provide a significantly more realistic evaluation of friction than that provided by the hydrodynamic (10a) and dielectric friction (10b) continuum model treatments available prior to our work. They, for example, provide expressions for the friction which depend upon quantities which reflect the microscopic solute-solvent interactions rather than upon properties like viscosity and dielectric constant (10) which determine the macroscopic solvent response.

Moreover, a molecular level treatment of friction is needed to realistically describe VER in fluids. This is indicated, for example, by the results of the IBC model (8) which show the equilibrium solvent structure microscopically near the solute, a factor ignored by the continuum dynamics treatments (10), is a critical determinant of the density dependence of T_1.

We next turn to the physical basis of our lifetime formula Eq. (1). This physical basis follows from the principles of our theory (6) of the solvent effect on solute reaction dynamics. This theory explains the solvent effect in terms of thermodynamic restoring forces, which play an enhanced role in reactions.

This enhancement derives from the asymmetric nature of the interactions in the solution, the asymmetry arising because only the solute molecules experience the special forces which result from typical [e.g. 1 ev activation barrier] gas phase potential energy surfaces. Because of these special forces, the nature of the solute motions which occur in reactions is such that the solvent following of these motions is highly imperfect relative to its following of thermal solute motions. That is the solvent cannot even approximately maintain equilibrium with the solute molecules as these molecules change their positions. Rather at any instant, the arrangement of solvent molecules around the solute is likely to be non-representative of the canonical ensemble distribution of arrangements. Thus forces, acting on both the solute and the solvent molecules, which attempt to convert these non-representative solute-solvent configurations into configurations more representative of thermodynamic equilibrium play a critical role. These restoring forces dominate the solvent dynamics and determine the solvent effect on the solute dynamics. The reaction dynamics is governed by the total force acting on the solute, which is the sum of the solute restoring force [which at any time depends on the solvent arrangement at that time and hence on the forces which have acted on the solvent] and the force arising from the solute gas phase potential energy surface.

These principles also apply to many non-reactive liquid solution phenomena. For VER, for example, imperfect following holds if (5a) $\omega_\ell >> \beta(\omega_\ell)$. Since $\beta(\omega)$ is large only for $\omega \sim$ the typical translational frequencies of the solvent $[\sim <\omega^2>^{1/2}$ defined in Eq. (4)], the criterion $\omega_\ell >> \beta(\omega_\ell)$ is normally satisfied for high frequency solute normal modes.

When imperfect following holds for VER, it manifests itself as motion of the relaxing solute normal mode in an environment which responds only weakly to this motion. This picture permits development of a simple classical model for VER (which ignores vibrational dephasing). When analyzed quantitatively (5a), this model yields Eq. (1) for T_1. It may be summarized qualitatively as follows. In zeroth order, the relaxing solute normal mode executes underdamped harmonic vibrations of frequency ω_ℓ in a hypothetical non-responding solution which moves as if the normal mode coordinate was fixed at its equilibrium value. In first order, these zeroth order motions "probe" the solution at frequency ω_ℓ inducing a weak response, which increases with the magnitude of the solution's frequency spectrum at $\omega_\ell, \sim \beta(\omega_\ell)$, and an associated energy transfer which gives rise to VER with T_1 given by Eq. (1).

We next discuss our procedures for evaluating $\beta(\omega)$ appearing in Eq. (1). We perform this evaluation within our partial clamping model (11). This model requires the assumption [reasonable, as discussed in Ref. 7, for VER processes occurring at the low temperatures required for liquid phase studies] that the displacements of the relaxing solute normal mode be restricted to small amplitudes. This restriction permits determination of $\beta(\omega)$ from the linear response of the solvent to these normal mode displacements, yielding a result for $\beta(\omega)$ of the familiar linear response theory type. Namely, $\beta(\omega)$ depends on the structure and dynamics of the liquid solution in the absence of the displacements, i.e., for the solute mode coordinate fixed at equilibrium. Specifically one obtains

$$\beta(\omega) = [k_B T]^{-1} \int_0^\infty <\tilde{\mathscr{F}}(t)\tilde{\mathscr{F}}>_0 \cos\omega t \, dt . \qquad 2$$

In Eq. (2), $<\tilde{\mathscr{F}}(t)\tilde{\mathscr{F}}>_0$ is the autocorrelation function of the fluctuating generalized force $\tilde{\mathscr{F}}[Q] \equiv \mathscr{F}[Q] - <\mathscr{F}>_0$ exerted by the solvent on the coordinate of the solute mode conditional that this coordinate is fixed. This prescription for $\beta(\omega)$ may be readily implemented. For example, $\beta(\omega)$ may be evaluated[4,5a] from Eq. (2) in a constrained molecular dynamics simulation[4,5a] in which the coordinate of the relaxing solute mode is kept at its equilibrium value while the remaining degrees of freedom of the solution (both solute and solvent) are allowed to more freely subject to this single restriction.

We finally note that the fictitious constrained dynamics of the partial clamping model (11) is the analogue within the present theory [Ref. 6] of the fictitious projected dynamics of the Mori theory. It is the relative ease of implementation of constrained dynamics relative to projected dynamics which gives Eq. (2) its advantage over the Mori formula for the friction kernel.

We next develop Eq. (6) for T_1. This development is based on the following simplification. For a relaxing solute normal mode of sufficiently high frequency ω_ℓ, T_1 may be evaluated from Eq. (1) using an approximate form for $\beta(\omega)$ which realistically describes only its asymptotic wings. These wings of $\beta(\omega)$, however, depend only on the <u>short-time</u> part of $<\tilde{\mathscr{F}}(t)\tilde{\mathscr{F}}>_0$ since the slowly decaying "tail" of this autocorrelation function [which is difficult to evaluate accurately] contributes negligibly to the wings. The simplest approximation to the short-time part is, however, provided by the following Gaussian model

$$<\tilde{\mathscr{F}}(t)\tilde{\mathscr{F}}>_0 = <\tilde{\mathscr{F}}^2>_0 \exp\left[-\tfrac{1}{2}<\omega^2>t^2\right] \qquad 3$$

where

$$<\omega^2> = \frac{<\ddot{\tilde{\mathscr{F}}}^2>_0}{<\tilde{\mathscr{F}}^2>_0} . \qquad 4$$

Comparison of Eqs. (2)–(4) yields the following asymptotic result for $\beta(\omega)$

$$\beta(\omega) = \frac{<\tilde{\mathscr{F}}^2>_0}{2k_BT}\left[\frac{2\pi<\tilde{\mathscr{F}}^2>_0}{<\ddot{\tilde{\mathscr{F}}}^2>_0}\right]^{1/2} \exp\left[-\frac{1}{2}\frac{<\tilde{\mathscr{F}}^2>_0}{<\ddot{\tilde{\mathscr{F}}}^2>}\omega^2\right] . \qquad 5$$

Use of the Gaussian approximation Eq. (5) for the wings of $\beta(\omega)$ has a partial theoretical foundation and also has support (5a,12) for Lennard–Jones liquids from the results of molecular dynamics simulations. Comparison of Eqs (1) and (5) then yields the required result for T_1,

$$T_1 = \left[\frac{2k_BT}{<\tilde{\mathscr{F}}^2>_0}\right]\left[\frac{<\ddot{\tilde{\mathscr{F}}}^2>_0}{2\pi<\tilde{\mathscr{F}}^2>}\right]^{1/2}\exp\left[\frac{1}{2}\frac{<\tilde{\mathscr{F}}^2>_0}{<\ddot{\tilde{\mathscr{F}}}^2>_0}\omega_\ell^2\right] . \qquad 6$$

To evaluate T_1 from Eq. (5), we require expressions for ω_ℓ, $<\ddot{\tilde{\mathscr{F}}}^2>_0$, and $<\tilde{\mathscr{F}}^2>_0$. We begin the development of these expressions by discussing the origin of the results for ω_ℓ given in Eqs. (7)–(9). We emphasize the origin of Eq. (7) for ω_e the liquid phase contribution to ω_ℓ. This origin follows from our definition of ω_ℓ as the (ensemble average) frequency of the relaxing solute mode in the non—responding solution, which is equivalent to the constrained solution discussed earlier in conjunction with the partial clamping model. The frequency ω_e, for example, may be evaluated from $\mathcal{F}(y) \equiv <\mathcal{F}[Q;y]>_0$, where $\mathcal{F}[Q;y]$ is the liquid phase generalized force acting on the relaxing mode coordinate, when it has the value y , and where $<...>_0$ denotes a constrained solution canonical ensemble average. $\mathcal{F}(y)$ is a thermodynamic restoring force which acts to return the normal mode coordinate to its equilibrium value y_0 . This follows since $\mathcal{F}[Q;y]$ is the force on the mode coordinate when it is displaced to a value y, while $<...>_0$, in contrast, is an average over a distribution of solution phase space coordinates Q which is the equilibrium distribution only if the mode coordinate is kept at y_0 . The frequency ω_e determines $\mathcal{F}(y)$ to the lowest non—trivial (linear)(5) order in $y - y_0$. It is defined as follows

$$\omega_e^2 \equiv - \left[\frac{\partial \mathcal{F}(y)}{\partial y} \right]_{y=y_0} = - \left\langle \left[\frac{\partial \mathcal{F}[Q;y]}{\partial y} \right]_{y=y_0} \right\rangle_0$$

or equivalently

$$\omega_e^2 = - \left\langle \frac{\partial \mathcal{F}}{\partial y} \right\rangle_0 \qquad\qquad 7$$

where $\frac{\partial \mathcal{F}}{\partial y} \equiv \left[\frac{\partial \mathcal{F}[Q;y]}{\partial y} \right]_{y=y_0}$. The frequency ω_ℓ may be analogously evaluated

from the <u>total</u> generalized force $F[Q;y]$ via the relation $\omega_\ell^2 = - \left\langle \left[\frac{\partial F[Q;y]}{\partial y} \right]_{y=y_0} \right\rangle_0$. This yields (5) the following result for ω_ℓ^2 ,

$$\omega_\ell^2 = \omega_g^2 + \omega_{cf}^2 + \omega_e^2 \qquad\qquad 8$$

where ω_g and ω_{cf} arise from the intramolecular potential and centripetal force components of the <u>gas phase</u> contribution to $F[Q;y]$. The explicit forms for ω_g and ω_{cf} , developed elsewhere (5), are

$$\omega_g^2 = \begin{array}{l} \text{square of the fundamental} \\ \text{gas phase frequency of the} \\ \text{diatomic solute} \end{array} \qquad\qquad 9a$$

and

$$\omega_{cf}^2 = \frac{6k_BT}{I} , \qquad\qquad 9b$$

where I is the moment of inertia of the diatomic solute molecule.

We next outline the derivation of our results for ω_ℓ , $<\dot{\mathcal{F}}^2>_0$, and $<\ddot{\mathcal{F}}^2>_0$ for the diatomic solute—monatomic solvent prototype. [The full development of our molecular friction methodology, valid for molecular solvents, (which is lengthy) is given in Ref. 5b]. The central theme underlying the derivations is the expression of the generalized force $\mathcal{F}[Q] \equiv \mathcal{F}$ in terms of solute–solvent pair potentials u_{ik} and the reduction of the constrained solution equilibrium phase space average $<...>_0$ to integrals over ensemble averaged densities of solvent atomic sites.

In our derivations, we specify the phase point of the N_s solvent molecules by their Cartesian coordinates $q = (\vec{q}_1 \; \vec{q}_2 \cdots \vec{q}_{N_s})$ and the corresponding Cartesian momenta $p_q = (\vec{p}_{q_1} \; \vec{p}_{q_2} \cdots \vec{p}_{q_{N_s}})$. In contrast, we specify the phase point of the solute in generalized coordinates, since VER involves energy transfer from the solute normal mode coordinate y which is a generalized coordinate. Specifically, we describe the solute molecule by its center of mass translational and bond–axis rotational coordinates $z \equiv (X, Y, Z, \theta, \phi)$, along with the conjugate momenta p_z, and by its normal mode coordinate y, along with the conjugate momentum p_y. The solute-solvent potential energy function $K_{vu}[yzq]$ is assumed to be pairwise additive, i.e. [n = 2 for a diatomic solute]

$$K_{vu} = \sum_{i=1}^{n} \sum_{\lambda=1}^{N_s} u_{ik}[\vec{r}_i, \vec{q}_\lambda] \qquad\qquad 10$$

where $u_{ik}[\vec{r}_i, \vec{q}_\lambda]$ is the pair potential linking solute atom i at point $\vec{r}_i$ with solvent atom λ at point $\vec{q}_\lambda$. The generalized force $\mathcal{F}$ acting on the solute normal mode coordinate y is given by

$$\mathcal{F} = -\frac{\partial K_{vu}[yzq]}{\partial y}. \qquad\qquad 11$$

To proceed further, we express $\frac{\partial}{\partial y}$ in terms of the gradients, $\frac{\partial}{\partial \vec{r}_i}$, with respect to the solute atomic Cartesian coordinates $\vec{r}_i$. We use the expansion

$$\frac{\partial}{\partial y} = \sum_{i=1}^{n} \vec{\mu}^{i1}(yz) \cdot \frac{\partial}{\partial \vec{r}_i} = \sum_{i=1}^{n} \vec{\mu}^{i1}(z) \cdot \frac{\partial}{\partial \vec{r}_i} \qquad\qquad 12$$

where the transformation coefficients $\vec{\mu}^{i1}(yz) \equiv \frac{\partial \vec{r}_i}{\partial y}$ and where the final equality in Eq. (12) is proven in Ref. 5a. Eq. (12) follows immediately from the fact (5) that y is determined from the solute atomic Cartesian coordinates by a point canonical transformation, i.e. $y = y(\vec{r}_1, \vec{r}_2)$. Comparing Eqs. (10)–(12) then yields the expansion

$$\mathscr{F} = -\sum_{i=1}^{n} \sum_{\lambda=1}^{N_s} \vec{\mu}^{i1}(z) \cdot \frac{\partial u_{ik}[\vec{r}_i, \vec{q}_\lambda]}{\partial \vec{r}_i}. \qquad 13$$

To illustrate our general methods of derivation, we first consider the development of Eq. (21) for $<\mathscr{F}>_0^2$. This quantity is required to evaluate $<\tilde{\mathscr{F}}^2>_0$ from

$$<\tilde{\mathscr{F}}^2>_0 = <\mathscr{F}^2>_0 - <\mathscr{F}>_0^2. \qquad 14$$

We then briefly outline how this simplest evaluation may be extended.

We begin by noting that $<\mathscr{F}>_0$ is a constrained solution equilibrium average, i.e. an average over the phase point (z, p_z, q, p_q) conditional that the relaxing solute normal mode coordinate is fixed at its equilibrium value y_0. For isotropic liquids, this average may be simplified to yield

$$<\mathscr{F}>_0 = <\mathscr{F}>_{y_0 z} \qquad 15$$

where $<...>_{y_0 z}$ denotes a canonical ensemble average over only the solvent phase point (q, p_q) conditional that the solute is fixed in the liquid with configuration point $y_0 z$. Eq. (15) follows by first averaging $\mathscr{F}$ over q, p_q with the solute fixed to obtain $<\mathscr{F}>_{y_0 z}$ and then by averaging $<\mathscr{F}>_{y_0 z}$ over z, p_z to obtain $<\mathscr{F}>_0$. However for isotropic liquids $<\mathscr{F}>_{y_0 z}$ is independent of z (the location and orientation of the solute in the liquid) yielding Eq. (15). Comparing Eqs. (13) and (15) then yields

$$<\mathscr{F}>_0 = -\sum_{i=1}^{n} \sum_{\lambda=1}^{N_s} \vec{\mu}^{i1}(z) \cdot \left\langle \frac{\partial u_{ik}[\vec{r}_i, \vec{q}_\lambda]}{\partial \vec{r}_i} \right\rangle_{y_0 z}. \qquad 16$$

Eq. (16) may be straightforwardly expressed in terms of solvent atomic site densities as

$$<\mathscr{F}>_0 = -\sum_{i=1}^{n} \vec{\mu}^{i1}(z) \cdot \int \rho^{(n+1)}[b_g; \vec{q}] \frac{\partial u_{ik}[\vec{r}_i, \vec{q}]}{\partial \vec{r}_i} d\vec{q} \qquad 17$$

where

$$\rho^{(n+1)}[b_g; \vec{q}] \equiv \sum_{\lambda=1}^{N_s} \left\langle \delta(\vec{q} - \vec{q}_\lambda) \right\rangle_{y_0 z} \qquad 18$$

is the ensemble averaged density of solvent atoms at point $\vec{q}$ conditional that the solute is fixed (with its equilibrium internuclear separation b_g.)

The site density integrals in Eq. (17) are most easily evaluated in the solute body-fixed frame (5) whose origin is located at the center of mass of the solute molecule and whose z–axis unit vector $\vec{e}_z$ points from solute atom 1 to solute atom 2, i.e. $\vec{e}_z = \dfrac{\vec{r}_2 - \vec{r}_1}{|\vec{r}_2 - \vec{r}_1|}$. To make this body–fixed evaluation, we require the transformation coefficient $\vec{\mu}^{i1}(z)$ evaluated in the solute body–fixed frame. The required result, which is derived in Ref. 5 from the explicit form of the transformation relation $y = y[\vec{r}_1, \vec{r}_2]$, is

$$\vec{\mu}^{i1}(z) = (-)^i \frac{\mu^{1/2}}{m_i} \vec{e}_z , \, i = 1 \text{ or } 2 , \tag{19}$$

where μ and m_i are, respectively, the solute reduced mass and the mass of solute atom i. To complete the derivation, we further assume central field site–site potentials, i.e. $u_{ik}[\vec{r}_i, \vec{q}] = u_{ik}[y_i]$ where $y_i = |\vec{y}_i|$ with $\vec{y}_i = \vec{q} - \vec{r}_i$. Comparison of Eqs. (17) and (19) shows that we require

$$\vec{e}_z \cdot \frac{\partial}{\partial \vec{r}_i} u_{ik}[\vec{r}_i, \vec{q}] = -[\vec{e}_z \cdot \vec{\iota}] \frac{du_{ik}[y_i]}{dy_i} \tag{20}$$

where the unit vector $\vec{\iota} = y_i^{-1} \vec{y}_i$. Comparing Eqs. (17), (19), and (20) then yields the following result for $<\mathcal{F}>_0^2$,

$$<\mathcal{F}>_0^2 = 3M^{-1} \sum_{i,j=1}^n T^{ij} [\vec{e}_z \cdot I_{ik}][\vec{e}_z \cdot I_{jk}] , \tag{21}$$

where $M = m_1 + m_2$ is the total mass of the solute molecule, where [i or j = 1 or 2]

$$T^{ij} \equiv (-)^{i+j} \frac{1}{3} \left[\frac{m_1 m_2}{m_i m_j} \right] \tag{22}$$

and where the site density integrals $\vec{e}_z \cdot I_{ik}$ and $\vec{e}_z \cdot I_{j\ell}$ are given by

$$\vec{e}_z \cdot I_{ik} = \int \rho^{(n+1)} [b_g; \vec{q}][\vec{e}_z \cdot \vec{\iota}] \frac{du_{ik}[y_i]}{dy_i} d\vec{q} \tag{23a}$$

and

$$\vec{e}_z \cdot I_{jk} = \int \rho^{(n+1)} [b_g; \vec{q}\,'][\vec{e}_z \cdot \vec{\jmath}\,'] \frac{du_{jk}[y_{j'}]}{dy_{j'}} d\vec{q}\,' \tag{23b}$$

where $y_{j'} \equiv |\vec{y}_{j'}|$ with $\vec{y}_{j'} = \vec{q}\,' - \vec{r}_j$ and where $\vec{\jmath}\,' = y_{j'}^{-1} \vec{y}_{j'}$.

The liquid phase frequency ω_e may be evaluated similarly. Comparing Eqs. (12) and (13) yields

$$\frac{\partial \mathscr{F}}{\partial y} = - \sum_{i=1}^{n} \sum_{\lambda=1}^{N_s} \vec{\mu}^{i1}(z)\, \vec{\mu}^{i1}(z) : \frac{\partial^2 u_{ik}\,[\,\vec{r}_i, \vec{q}\,]}{\partial \vec{r}_i^{\,2}}. \qquad 24$$

A site density integral expression for w_e^2 may be developed from Eq. (7) and (24) following steps analogous to those which lead to Eq. (21). One obtains

$$w_e^2 = 3M^{-1} \sum_{i=1}^{n} T^{ii}\, \vec{e}_z \cdot K_{ik} \cdot \vec{e}_z\,, \qquad 25$$

where the site density integral $\vec{e}_z \cdot K_{ik} \cdot \vec{e}_z$ is given by

$$\vec{e}_z \cdot K_{ik} \cdot \vec{e}_z = \int \rho^{(n+1)}[b_g; \vec{q}]\, B_{zz}^{ik}\,[\vec{y}_i] d\vec{q}\,, \qquad 26$$

where for arbitrary unit vectors $\vec{e}_\alpha$ and $\vec{e}_\beta$

$$B_{\alpha\beta}^{ik}\,[\vec{y}_i] \equiv [\vec{e}_\alpha \cdot \vec{i}][\vec{e}_\beta \cdot \vec{i}]\, U_{ik}[y_i] + [\vec{e}_\alpha \cdot \vec{e}_\beta]y_i^{-1}\frac{du_{ik}[y_i]}{dy_i}, \qquad 27$$

with

$$U_{ik}[y_i] \equiv \frac{d^2 u_{ik}[y_i]}{dy_i^{\,2}} - y_i^{-1}\frac{du_{ik}[y_i]}{dy_i}. \qquad 28$$

Thus $B_{zz}^{ik}[\vec{y}_i]$ in Eq. (26) is given by $B_{ik}^{zz}[\vec{y}_i] = [\vec{e}_z \cdot \vec{i}]^2\, U_{ik}[y_i] + y_i^{-1}\frac{du_{ik}[y_i]}{dy_i}$.

To complete the evaluation of $<\tilde{\mathscr{F}}^2>_0$ from Eq. (14), one requires site density integral expressions for $<\mathscr{F}^2>_0$. The required expression may be readily obtained from Eq. (13) by following steps similar to those used to derive Eqs. (21) and (25). The result is

$$<\mathscr{F}^2>_0 = 3M^{-1} \sum_{i,j=1}^{n} T^{ij}\,[\vec{e}_z\vec{e}_z : I_{ijkk} + \vec{e}_z\vec{e}_z : \hat{I}_{ijkk}]. \qquad 29$$

The site density integrals in Eqs. (29) are given by [cf. Eqs. (23)]

$$\vec{e}_z\vec{e}_z : I_{ijkk} = \int \rho^{(n+1)}[b_g; \vec{q}][\vec{e}_z \cdot \vec{i}][\vec{e}_z \cdot \vec{j}]\frac{du_{ik}[y_i]}{dy_i}\frac{du_{jk}[y_j]}{dy_j}d\vec{q} \qquad 30a$$

and

$$\vec{e}_z\vec{e}_z : \hat{I}_{ijkk} = \int \rho^{(n+2)}[b_g; \vec{q}\,\vec{q}\,'][\vec{e}_z \cdot \vec{i}][\vec{e}_z \cdot \vec{j}\,']\frac{du_{ik}[y_i]}{dy_i}\frac{du_{jk}[y_j\,']}{dy_{j'}}d\vec{q}\, d\vec{q}\,' \qquad 30b$$

where y_i, y_j', $\vec{\textbf{i}}$ and $\vec{\textbf{j}}'$ have been defined earlier and where we analogously

define $y_j = |\vec{\textbf{y}}_j|$ with $\vec{\textbf{y}}_j = \vec{\textbf{q}} - \vec{\textbf{r}}_j$ and $\vec{\textbf{j}} = y_j^{-1} \vec{\textbf{y}}_j$. Also in Eq. (30a),

$\rho^{(n+1)}[b_g; \vec{\textbf{q}}]$ is the single solvent atom density defined in Eq. (18) and

$\rho^{(n+2)}[b_g; \vec{\textbf{q}}\,\vec{\textbf{q}}\,']$ is an analogous equilibrium ensemble averaged solvent atomic

pair density. This pair density is proportional to the probability that a pair of

solvent atoms are simultaneously located at points $\vec{\textbf{q}}$ and $\vec{\textbf{q}}\,'$ conditional that

the solute is fixed in the liquid (with its equilibrium internuclear separation b_g)

and is defined by an expression analogous to Eq. (18).

To complete the evaluation of T_1 from Eq. (6), we require site density

integral expressions for $<\dot{\tilde{\mathscr{F}}}^2>_0$. We next briefly outline how these expressions

are obtained. We began with $\dot{\tilde{\mathscr{F}}} = iL_0 \mathscr{F}$ where L_0 is the Liouville operator of

the constrained solution. $\dot{\tilde{\mathscr{F}}}$ may be decomposed as

$$\dot{\tilde{\mathscr{F}}} = \dot{\tilde{\mathscr{F}}}_{d,t} + \dot{\tilde{\mathscr{F}}}_{i,t} + \dot{\tilde{\mathscr{F}}}_{i,r} \qquad 31$$

where $\dot{\tilde{\mathscr{F}}}_{d,t}$ and $\dot{\tilde{\mathscr{F}}}_{i,t(r)}$ arise, respectively, from the solvent translational and

solute translational (rotational) contributions to $\dot{\tilde{\mathscr{F}}}$. The explicit form of $\dot{\tilde{\mathscr{F}}}_{d,t}$,

for example, is

$$\dot{\tilde{\mathscr{F}}}_{d,t} = \sum_\alpha \frac{P_{z\alpha}}{M} \frac{\partial \mathscr{F}}{\partial z_\alpha} \qquad 32$$

where, in Eq. (32), z_α and $p_{z\alpha}$ are, respectively, the solute center of mass

translational coordinates and associated conjugate momenta. Defining

$<\dot{\tilde{\mathscr{F}}}^2>_{od,t} = <\dot{\tilde{\mathscr{F}}}^2_{d,t}>_0$, etc. and using Eq. (31) to evaluate $<\dot{\tilde{\mathscr{F}}}^2>_0$ yields

(because cross–terms vanish since, e.g., $<p_z p_q>_0 = 0$)

$$<\dot{\tilde{\mathscr{F}}}^2>_0 = <\dot{\tilde{\mathscr{F}}}^2>_{od,t} + <\dot{\tilde{\mathscr{F}}}^2>_{oi,t} + <\dot{\tilde{\mathscr{F}}}^2>_{oi,r} , \qquad 33$$

where, e.g.,

$$<\dot{\tilde{\mathscr{F}}}^2>_{od,t} = \frac{k_B T}{M} \sum_\alpha \left\langle \frac{\partial \mathscr{F}}{\partial z_\alpha} \frac{\partial \mathscr{F}}{\partial z_\alpha} \right\rangle_0 . \qquad 34$$

Eq. (34) holds since $\left\langle p_{z_\alpha} p_{z_\beta} \right\rangle = M k_B T \, \delta_{\alpha\beta}$. Site density integral expressions

for $<\dot{\mathscr{F}}^2>_{od,t}$ may be obtained from Eqs. (13) and (34) using an expansion of

$\dfrac{\partial}{\partial z_{e\alpha}}$ analogous to Eq. (12). Site density integral expressions for $<\dot{\mathscr{F}}^2>_{oi,t(r)}$

may be obtained analogously, permitting the evaluation of $<\dot{\mathscr{F}}^2>_o$ in terms of site density integrals using Eq. (33). The derivations are similar to those used to evaluate $<\mathscr{F}^2>_o$ and, moreover, the results are similar in structure to Eqs. (29) and (30) for $<\mathscr{F}^2>_o$. Both the evaluations and the results (omitted here for brevity) are given in Refs. 5.

We next give an approximate reduction of the site density integrals to numerically straightforward quadratures over readily calculable equilibrium pair correlation functions of the liquid solution. This reduction converts our molecular friction procedures into a practical tool for the numerical evaluation of T_1 for specific systems.

This reduction is based on the use of Kirkwood superposition approximations (SA's) for the site densities. The SA for the single–atom density $\rho^{(n+1)}[b_g; \vec{q}]$ defined in Eq. (18) is

$$\rho^{(n+1)}[b_g; \vec{q}] = \rho_0 \prod_{i=1}^{n} g_{ik}[y_i] \qquad\qquad 35a$$

where g_{ik} is the equilibrium pair correlation function linking solute atom $i = 1$ or 2 with a solvent atom, the solute atom being present at infinite dilution in the solvent. The SA for the two–atom density is correspondingly

$$\rho^{(n+2)}[b_g; \vec{q}\,\vec{q}\,'] = \rho_0^{-1} \, \rho^{(n+1)}[b_g; \vec{q}] \, \rho^{(n+1)}[b_g; \vec{q}\,'] g_{kk}[|\vec{q} - \vec{q}\,'|] ,$$

$$35b$$

where we evaluate the single–atom densities using Eq. (35a), and where g_{kk} is the equilibrium pair correlation function linking a pair of solvent atoms in the pure solvent.

We next note that the site density integrals $\vec{e}_z \cdot I_{ik(jl)}$ [Eqs. (23)],

$\vec{e}_z \cdot K_{ik} \cdot \vec{e}_z$ [Eq. (25)], and $\vec{e}_z \vec{e}_z : I_{ijkk}$ [Eq. (30a)] may be reduced to easily performed two-dimensional numerical quadratures. This is accomplished by using the spherical polar coordinates $(q, \theta, \phi) = \vec{q}$ and performing the ϕ–integrals analytically (possible for diatomic solutes because of the axial symmetry of the integrands in the body–fixed frame). The site density integral $\vec{e}_z \vec{e}_z : \hat{I}_{ijkk}$ [Eq. (30b)], however, requires a six–dimensional numerical quadrature.

In our first applications, we have made a simplified evaluation of $\vec{e}_z \vec{e}_z :$ $\hat{I}_{ijkk}$ [and of analogous integrals which appear in the expression for $<\ddot{\tilde{\mathscr{F}}}^2>_0$.] This evaluation is made by setting $g_{kk}[|\vec{q} - \vec{q}'|] = 1$ in the SA Eq. (35b). With this approximation, comparison of Eqs. (23) and (30b) and (35b) yields the approximate factorization

$$\vec{e}_z \vec{e}_z : \hat{I}_{ijkk} \doteq [\vec{e}_z \cdot I_{ik}][\vec{e}_z \cdot I_{jk}] . \qquad 36$$

Comparing Eqs. (14), (21), (29), and (36) then yields the following approximate expression for $<\ddot{\tilde{\mathscr{F}}}^2>_0$,

$$<\ddot{\tilde{\mathscr{F}}}^2>_0 = 3M^{-1} \sum_{i,j=1}^{n} T^{ij} \vec{e}_z \vec{e}_z : I_{ijkk} . \qquad 37$$

Thus within the approximation of Eq. (36), $<\ddot{\tilde{\mathscr{F}}}^2>_0$ may be evaluated from straightforward two–dimensional numerical quadratures over equilibrium pair correlation functions. Using analogous approximations for the site density integrals appearing in the expressions (5) for $<\ddot{\tilde{\mathscr{F}}}^2>_0$, T_1 may be evaluated from Eq. (6) from two-dimensional quadratures. We next present some early applications of our new procedures.

Detailed applications of the present theory to experimental systems (1,8) which permit interpretation of the observed temperature, density, and isotope dependencies of their VER rate constants are currently underway.

Here we briefly summarize a simpler application (7) to the diatomic solute–monatomic solvent prototype system. For this prototype system, the required equilibrium pair correlation functions are atomic solute–monatomic solvent radial distribution functions which we compute as solutions to the Percus–Yevick integral equation.

The applications were performed for model Lennard–Jones solutions whose potential energy parameters are given in Table I. These model solutions are designed to simulate molecular iodine dissolved in liquid xenon at temperature T = 298 K and molecular bromine dissolved in liquid argon at temperatures T = 295 K and T = 1500 K. For brevity we will refer to these three model solutions as I298, BR295, and BR1500.

TABLE I

Lennard–Jones Potential Parameters Describing Solvent-Solvent(VV) and Solute-Solvent (UV) Interactions for the Model Solutions.

System	σ_{vv}(Å)	ϵ_{vv}(K)	σ_{uv}(Å)	ϵ_{uv}(K)
I_2/X_e	4.10	229	3.94	324
Br_2/Ar	3.42	120	3.51	143

Evaluation of the isothermal density dependencies of the VER rates in the I298, BR295 and BR1500 model solutions based on Eqs (1), (8), (9), (25), (33), (37), and the site density integral expressions for $<\dot{\mathscr{F}}^2>_{od,t}$ and $<\dot{\mathscr{F}}^2>_{oi,t(r)}$ [given in Ref. 5b] are given in Tables II–V. Specifically we give results for ω_ℓ, $<\dot{\mathscr{F}}^2>_o$, and $<\dot{\mathscr{F}}^2>_o$ in, respectively, Tables II–V for the I298, BR295, and BR1500 solutions over a range of packing fractions [packing fraction $= PF = \frac{1}{6}\pi\rho_o \sigma_{vv}^3$, where ρ_o is the solvent number density and where σ_{vv} is the solvent Lennard–Jones diameter given in Table I]. Notice the liquid–phase frequencies ω_ℓ are nearly independent of density (i.e. $\omega_\ell \doteq \omega_g$) for all three solutions.

TABLE II

Instantaneous Frequency ω_e, Gaussian Model Parameters, and VER Time T_1 for the I298 Liquid Solution as a Function of Packing Fraction (PF)[a]

PF	ω_e(cm^{-1})	$<\dot{\mathscr{F}}^2>_o$	$<\dot{\mathscr{F}}^2>_{od}$	$<\dot{\mathscr{F}}^2>_{oi,t}$	$<\dot{\mathscr{F}}^2>_{oi,r}$	ω_ℓ(cm^{-1})	T_1(nsec)
0.1	11.57	6.28	266.91	138.08	92.37	215.10	35.84
0.2	13.70	9.15	389.91	201.71	132.14	215.22	25.09
0.3	15.66	12.48	536.37	277.48	174.71	215.36	18.50
0.4	19.26	19.86	867.42	448.73	256.34	215.65	11.99
0.5	24.93	34.92	1553.80	803.78	376.78	216.23	8.05

[a] ω_e, $<\dot{\mathscr{F}}^2>_o$ and the components of $<\dot{\mathscr{F}}^2>_o$ are computed by the procedures described above. $<\dot{\mathscr{F}}^2>_o$ and the components of $<\dot{\mathscr{F}}^2>_o$ are multiplied by a

factor of $(k_BT)^{-1}$ yielding $<\tilde{\mathscr{F}}^2>_o$ in units of psec^{-2} and $<\dot{\tilde{\mathscr{F}}}^2>_{od}$, and $<\dot{\tilde{\mathscr{F}}}^2>_{oi,t(r)}$ in units of psec^{-4}. The solute liquid phase frequency ω_ℓ is computed from Eq. (8) using $\omega_g = 214.6$ cm^{-1} and $\omega_{cf} = 9.64$ cm^{-1} (computed from Eq. (9b)).

TABLE III

Same as Table II Except for the BR295 Liquid Solution[a]

PF	ω_e(cm^{-1})	$<\dot{\tilde{\mathscr{F}}}^2>_o$	$<\dot{\tilde{\mathscr{F}}}^2>_{od}$	$<\dot{\tilde{\mathscr{F}}}^2>_{oi,t}$	$<\dot{\tilde{\mathscr{F}}}^2>_{oi,r}$	ω_ℓ(cm^{-1})	T_1(nsec)
0.01	3.25	0.45	69.66	17.41	12.91	323.52	115.23
0.05	7.22	2.26	349.24	87.30	64.76	323.59	23.07
0.10	10.19	4.59	708.10	177.01	131.37	323.67	11.42
0.20	14.67	10.00	1543.1	385.74	285.86	323.84	5.28
0.40	23.55	29.40	4574.5	1143.5	795.90	324.36	1.84

[a]For this system, $\omega_g = 323.2$ cm^{-1} and $\omega_{cf} = 14.11$ cm^{-1}.

TABLE IV

Same as Table II Except for the BR1500 Liquid Solution[a]

PF	ω_e(cm^{-1})	$<\dot{\tilde{\mathscr{F}}}^2>_o$	$<\dot{\tilde{\mathscr{F}}}^2>_{od}$	$<\dot{\tilde{\mathscr{F}}}^2>_{oi,t}$	$<\dot{\tilde{\mathscr{F}}}^2>_{oi,r}$	ω_ℓ(cm^{-1})	T_1(psec)
0.02	6.91	1.89	1396.9	349.19	264.19	324.84	80.05
0.05	11.15	5.00	3688.5	922.03	698.58	324.95	30.35
0.10	16.34	10.98	8109.5	2027.2	1539.9	325.17	13.83
0.20	24.85	26.74	19777.0	4943.7	3768.8	325.71	5.70
0.30	32.65	49.04	36325.0	9080.3	6911.0	326.40	3.13

[a]For this system $\omega_g = 323.2$ cm^{-1} and $\omega_{cf} = 31.82$ cm^{-1}.

The results of Tables II–IV are in accord with the experimentally inferred (8) factorization of the rate constant into density dependent and density–independent contributions. This may be seen from Table V where we give results for the decay time τ of the Gaussian model fluctuating force autocorrelation function. This decay time is found to be nearly independent of density for all three model solutions from the ideal to the dense fluid regimes.

TABLE V

Gaussian Relaxation Time $\tau = 10^3[<\tilde{\mathscr{F}}^2>_0/<\dot{\tilde{\mathscr{F}}}^2>_0]^{1/2}$ for the I298, BR295, and BR1500 Liquid Solutions as a Function of PF.

Liquid Solution	PF	τ(fsec)
I298	0.0001	112.34
	0.1	112.37
	0.2	112.41
	0.3	112.37
	0.4	112.39
	0.5	113.00
BR295	0.0001	67.20
	0.01	67.20
	0.05	67.20
	0.10	67.20
	0.20	67.19
	0.40	67.18
BR1500	0.0001	30.69
	0.02	30.68
	0.05	30.68
	0.10	30.66
	0.20	30.64
	0.30	30.62

The fact that τ is nearly independent of density leads to the factorization. This may be seen if, following convention [8], we define the liquid–phase VER rate constant as

$$k_{liq}(T) = T_1^{-1}$$
38

and the corresponding gas–phase rate constant per unit density as

$$k_{gas}(T) = \lim_{\rho_0 \to 0} [\rho_0^{-1} k_{liq}(T)] .$$
39

Using the fact that τ (and also ω_ℓ) is nearly independent of density and comparing Eqs. (6) and (38)–(39) yield the following result for $k_{liq}(T)$:

$$k_{liq}(T) = <\tilde{\mathscr{F}}^2>_0 \lim_{\rho_0 \to 0} \left[\frac{<\tilde{\mathscr{F}}^2>_0}{\rho_0} \right]^{-1} k_{gas}(T) .$$
40

Notice that since $<\tilde{\mathscr{F}}^2>_0$ is proportional to the density ρ_0 as $\rho_0 \to 0$, Eq. (40) shows that $k_{liq}(T)$ factorizes into a dynamical contribution, namely $\lim\limits_{\rho_0 \to 0} [\rho_0^{-1} <\tilde{\mathscr{F}}^2>_0]^{-1} k_{gas}(T)$, which is a pure function of temperature and a liquid–phase structural contribution, namely $<\tilde{\mathscr{F}}^2>_0$, which is a function of temperature and density. The IBC formula [8] for $k_{liq}(T)$, namely

$$k_{liq}(T) = \rho_0 \frac{g_{liq}(R^*)}{g_{gas}(R^*)} k_{gas}(T) , \qquad\qquad 41$$

has an analogous factorized form. However, in our Eq. (40) the liquid–phase structural information is carried by the mean–square fluctuating force $<\tilde{\mathscr{F}}^2>_0$, while in the IBC formula [Eq. (41)] it is carried by the contact pair correlation function $g_{liq}(R^*)$. This difference is the basis of some of the advantages of the present formulation over the IBC model.

Acknowledgement

Support of this work by the National Science Foundation under Grant Number CHE–8803938 is gratefully acknowledged.

References

1.   There is an extensive experimental and theoretical literature on vibrational energy relaxation (VER) in liquids. For a recent review with many references to earlier work see Harris, C. B., Smith, D. E., and Russell, D. J. (1990) "Vibrational Relaxation of Diatomic Molecules in Liquids" Chem. Rev. 90, 481.

2.   See, for example, Xu, X., Yu, S. C., Lingle, R. Jr., Zhu, H., and Hopkins, J. B., (1991) "Ultrafast Transient Raman Investigations of Geminate Recombination and Vibrational Energy Relaxation in Iodine: The Role of VV Relaxation Pathways J. Chem. Phys. 95, 2445.

3.   Herzeld, K. F., and Litovitz, T. A. (1959) "Absorption and Dispersion of Ultrasonic Waves" Academic Press, New York.

4.   Whitnell, R. M., Wilson, K. R., and Hynes, J. T. (1992) "Vibrational Relaxation of a Dipolar Molecule in Water" J. Chem. Phys. 96, 5354.

5.   (a) Adelman, S. A. and Stote, R. H. (1988) "Theory of Vibrational Energy Relaxation in Liquids: Construction of the Generalized Langevin Equation for Solute Vibrational Motion in Monatomic Solvents" J. Chem. Phys. 88, 4397. Stote, R. H. and Adelman, S. A. (1988) "Theory of Vibrational Energy Relaxation in Liquids: Diatomic Solutes in Monatomic Solvents" J. Chem. Phys. 88, 4415.

(b)   Adelman, S. A. Ravi, R., Muralidhar, R., and Stote, R. H. (1993) "Molecular Theory of Liquid Phase Vibrational Energy Relaxation" Adv. Chem. Phys. **84**, 73.

6.   Adelman, S. A. (1983) "Chemical Reaction Dynamics in Liquid Solution" Adv. Chem. Phys. **53** 61. For a paper with many references to our more recent work, see Adelman, S. A. and Muralidhar, R. (1991) "Theory of Liquid Phase Activated Barrier Crossing: The Instantaneous Potential and the Parabolic Model" J. Chem. Phys. **95**, 2752.

7.   Adelman, S. A., Muralidhar, R., and Stote, R. H. (1991) "Time Correlation Function Approach to Vibrational Energy Relaxation in Liquids: Revised Results for Monatomic Solvents and a Comparison with the Isolated Binary Collision Model" J. Chem. Phys. **95**, 2738.

8.   For a comprehensive review of the isolated binary collision (IBC) model with many references, which deals with both its theoretical foundation and its application to experimental systems, see Chesnoy, J. and Gale, G. M. (1984) "Vibrational Energy Relaxation in Liquids" Ann. Phys. Fr. **9**, 893.

9.   See Eq. (2.18) of Oxtoby, D. W. (1981) "Vibrational Population Relaxation in Liquids" Adv. Chem. Phys. **47**, 487.

10.   (a)   Zwanzig, R. and Bixon, M. (1970) "Hydrodynamic Theory of the Velocity Correlation Function" Phys. Rev. A2, 2005.

        (b)   Zwanzig, R. (1970) "Dielectric Friction on a Moving Ion. II. Revised Theory" J. Chem. Phys. **52**, 3625.

11.   Adelman, S. A., (1984) "The Method of Partial Clamping and Formulation of the Equation of Motion in Generalized Coordinates, J. Chem. Phys. **81**, 2776.   Adelman, S. A. (1987) "Generalized Langevin Theory for Many–Body Problems in Chemical Dynamics: Explicit Coordinate Dynamics in Molecular Solvents" Int. J. Quantum Chem. Symp. **21**, 199.

12.   Lee, L. L., Li, Y. S., and Wilson, K. R. (1991) "Reaction Dynamics from Liquid Structure" J. Chem. Phys. **95**, 2458, Fig. 1 and Table 1.

QUANTUM MECHANICAL CALCULATIONS OF TUNNELING RATES IN CONDENSED PHASE SYSTEMS

John Lobaugh and Gregory A. Voth
Department of Chemistry, University of Pennsylvania,
Philadelphia, Pennsylvania 19104-6323, USA

1. Introduction

Tunneling processes in condensed phase systems represent a significant computational challenge to the theoretical chemist. In the present article, the focus will be on studies of proton transfer processes in polar solvents, but much of the methodology discussed herein is applicable to other condensed phase tunneling problems. Proton transfer reactions in polar solvents have generated a considerable degree of experimental and computational interest. Computationally, efforts have focused on the calculation of the solvent activation free energy [1-3], studies of intramolecular effects arising from the modulation of the proton tunneling by molecular vibrations [1,4-6], and molecular dynamics simulations of adiabatic proton transfer in strongly hydrogen-bonded complexes [7]. Some studies [1-5] have been carried out in the nonadiabatic, or weak coupling, limit of proton transfer, while others have been in the adiabatic, or strong coupling, limit [7]. A central conclusion reached in all of these studies is that proton transfer in polar solvents involves a complex interplay between solvent activation dynamics, intramolecular mode coupling, and quantum tunneling effects.

The Feynman path integral quantum transition state theory (QTST) [8] approach has been used previously by us to study proton transfer reactions. This approach is capable of calculating the solvent and intramolecular contributions to the quantum mechanical activation free energy for realistic liquid state systems. Another strength of the method is that it is applicable in both the adiabatic and nonadiabatic limits of proton transfer, as well as to situations where the proton itself becomes classically activated. Within the path integral theory, the quantum tunneling of the proton is explicitly described on an equal footing with the solvent activation process [6], and highly nonlinear intramolecular couplings between the proton transfer coordinate and other molecular modes in the system can be included in the calculation.

In the present article, the path integral QTST approach is used to calculate the total quantum activation free energy for a model intramolecular proton transfer reaction in a polar liquid. This computational study is concerned with a determination of the relevant features of proton transfer, including the competition

411

J. Jortner et al. (eds.), Reaction Dynamics in Clusters and Condensed Phases, 411–422.
© 1994 *Kluwer Academic Publishers. Printed in the Netherlands.*

between proton tunneling and solvent activation, the influence from intramolecular vibrational modulation of the proton transfer coordinate, and the role of polarizability in both the solute and the solvent. Changes in the total activation free energy, and hence the reaction probability, due to the different effects are calculated. These calculations "set the stage" for future analytic work on the more complex features of the proton transfer problem.

The sections of this article are organized as follows: In Sec. 2, a summary of the path integral QTST is presented with a specific emphasis on its computational implementation to study proton transfer. Then, the different proton transfer models used in the calculations are reviewed in Sec. 3 and the results of the calculations presented in Sec. 4. Concluding remarks are given in Sec. 5.

2. Computational Method

As shown previously [6], the fundamental quantum rate constant for an activated proton transfer process can be written as

$$k_{PT} = \kappa \frac{k_B T}{h Q_R} exp(-\beta F_c^*) \quad , \tag{1}$$

where κ is a prefactor of order unity, Q_R is the partition function of the entire solute-solvent system with the proton in its reactant configuration, and F_c^* is a quantum mechanical excess free energy, given by

$$F_c^* = -k_B T \, ln \left[Q_c(q^*)/(\mu/2\pi\hbar^2\beta)^{1/2} \right] \quad . \tag{2}$$

The quantity $Q_c(q^*)$ in eq. (2) is the path integral centroid density [8], given by

$$Q_c(q^*) = \int \cdots \int \mathcal{D}\mathbf{x}(\tau) \, \mathcal{D}\mathbf{X}(\tau) \, \delta(q^* - \tilde{q}_0) \, exp\{-S[\mathbf{x}(\tau), \mathbf{X}(\tau)]/\hbar\} \quad , \tag{3}$$

where the coordinates $\mathbf{x}$ are the Cartesian coordinates of the proton and the coordinates $\mathbf{X}$ are all other general coordinates of the system, including solvent, intramolecular, and polarization modes. The coordinate q is the proton transfer reaction coordinate having a reduced mass μ. An example of this coordinate is the asymmetric stretch mode of an A–H–A proton transfer complex involving the proton and two heavy atoms to which it is bound in the reactant and product configurations, respectively. The centroid variable of the reaction coordinate path $q(\tau)$ is defined as [8]

$$\tilde{q}_0 = \frac{1}{\hbar\beta} \int_0^{\hbar\beta} d\tau \, q(\tau) \tag{4}$$

and is constrained to be at the dividing surface between reactant and product states of the proton [6]. The action $S[\mathbf{x}(\tau), \mathbf{X}(\tau)]$ in eq. (3) is the imaginary time path integral action functional [9], which depends on both the paths of the proton coordinates $\mathbf{x}(\tau)$ and those of the remaining "bath" coordinates $\mathbf{X}(\tau)$. The specific expression for the action functional in eq. (3) is given by

$$S[\mathbf{x}(\tau), \mathbf{X}(\tau)] = \int_0^{\hbar\beta} d\tau \left\{ \frac{m}{2} \dot{\mathbf{x}}(\tau)^2 + \sum_{i=1}^{N} \frac{M_i}{2} \dot{X}_i(\tau)^2 + V[\mathbf{x}(\tau), \mathbf{X}(\tau)] \right\} , \quad (5)$$

where $V[\mathbf{x}(\tau), \mathbf{X}(\tau)]$ is the general many-body potential. All possible quantum paths of the coordinates are integrated over in the functional integration in eq. (3). The classical approximation for any of the variables $\mathbf{X}$ in eq. (3) replaces their quantum paths $\mathbf{X}(\tau)$ with their classical Boltzmann configurations in the path integration [9]. Such an approximation is not necessary, but is it likely to be a good one for orientational modes in most solvents and certainly reduces the computational demands of the calculations.

Equation (1) can be written in a form particularly well suited for computer simulation as

$$k_{PT} = \kappa \frac{\omega_c}{2\pi} exp(-\beta \Delta F_c^*) , \quad (6)$$

where $\mu\omega_c^2$ is the curvature of the reaction coordinate centroid free energy near the reactant configuration (i.e., for $\tilde{q}_0$ near q_r), and the difference in centroid free energies between the reactant and transition state is given by

$$\begin{aligned} \Delta F_c^* &= - k_B T \, ln[Q_c(q^*)/Q_c(q_r)] \\ &= - k_B T \, ln[P_c(q_r \to q^*)] . \end{aligned} \quad (7)$$

The probability $P_c(q_r \to q^*)$ to move the reaction coordinate centroid variable from the reactant configuration to the transition state is readily calculated by path integral Monte Carlo techniques (for reviews, see ref. [10]) combined with umbrella sampling [11]. To increase convergence properties of the free energy calculations the method outlined in ref. [6] was modified to use an improved sampling algorithm outlined by Owicki [12].

The quantum activation free energy (ΔF_c^*) in the present studies was calculated as a function of the path centroid [cf eq. (4)] of the asymmetric stretch coordinate, defined as

$$q_{as} = 2^{-1/2}(r_1 - r_2) , \quad (8)$$

where r_1 and r_2 are the distances between the proton and the heavy solute atoms A and A', respectively, to which it is bound the reactant and product states. Path integral Monte Carlo was used in the calculations with the so-called "primitive discretization" [10]. The value of discretization parameter P for the quantum limit was determined by calculating the free energy curves in the absence of solvent and determining at which value of P the quantum free energy curve became independent of P.

3. Model

3.1 EMPIRICAL VALENCE BOND MODEL

The simulations reported in the present article are for model proton transfer systems in a polar fluid. However, the method used to model the potential energy interactions was the realistic empirical valence bond (EVB) method used extensively by Warshel and co-workers [13].

3.1.1 *Nonadiabatic proton transfer model*

The intramolecular proton transfer complex used in the present calculations of proton transfer in the non-adiabatic limit is a three-body solute of the form A−H−A. The solvent model used was a Stockmayer fluid [14] (i.e., a Lennard-Jones fluid with permanent dipoles). Two diabatic EVB states were chosen to represent the symmetric product and reactant configurations of the complex. The diabatic state EVB matrix elements were given by

$$V_{ii}^{total} = V_{ii} + V_{LJ} + V_{solv} \qquad (i = 1, 2) \qquad (9)$$

where V_{LJ} are the short range Lennard-Jones interactions between the solute and solvent, V_{solv} are the solvent-solvent interactions in the Stockmayer fluid, and the remaining terms are given by

$$V_{11} = \frac{1}{2}m\omega^2(|\mathbf{x}(\tau) - \mathbf{R}_A| - x_0)^2 - \sum_{i=1}^{N} \frac{e_A \mathbf{m}_i \mathbf{R}_{iA}}{R_{iA}^3} - \sum_{i=1}^{N} \frac{e_H \mathbf{m}_i \mathbf{R}_{iH}(\tau)}{R_{iH}^3(\tau)} \qquad (10)$$

$$V_{22} = \frac{1}{2}m\omega^2(|\mathbf{x}(\tau) - \mathbf{R}_{A'}| - x_0)^2 - \sum_{i=1}^{N} \frac{e'_A \mathbf{m}_i \mathbf{R}_{iA'}}{R_{iA'}^3} - \sum_{i=1}^{N} \frac{e_H \mathbf{m}_i \mathbf{R}_{iH}(\tau)}{R_{iH}^3(\tau)}. \qquad (11)$$

Here, $\mathbf{x}$ is the position of the proton and $\mathbf{R}_A$ ($\mathbf{R}_{A'}$) is the position of heavy atom A (A'). Each diabatic state in eqs. (9)−(11) consists of a harmonic bond between the proton and one of the A atoms with frequency $\omega = 1693$ cm^{-1} and minima $x_0 = 1.0$Å. The electronic charge distribution of the complex with consists of partial charges of $e_{A,A'} = -0.5$ for the respective A and A' atom in each diabatic state and a charge of $e_H = +0.5$ on the proton in both diabatic state. The second term in the diabatic states is the interaction of the solvent with the electronic distribution of the solute compex where $\mathbf{R}_{iA,A',H}$ is the vector connecting the appropriate solute atom to a solvent particle with dipole moment $\mathbf{m}_i$. The entire complex was constrained to be linear, though this constraint is not essential.

The proton moves on the adiabatic energy surface resulting from the diagonalization of the 2×2 EVB matrix. The off-diagonal EVB matrix element V_{12} was chosen to be 6.92 kcal/mole, resulting in a 40 k_BT barrier in the absence of any coupling of the solute complex to the solvent. In calculations, the A and A′ atoms were given a mass of 40 amu each. In some calculations, the A−A′ distance was held rigid, while in others the A and A′ atoms were given a harmonic interaction with a frequency of 250 cm^{-1} and an equilibrium bond length of 3.0 Å.

The dipole moment of the Stockmayer particles was taken to be that of methanol (2.20 Debye) and the Lennard-Jones (LJ) parameters were chosen to be $\epsilon = 0.1532$ kcal/mole and $\sigma = 3.166$ Å. Solvent-solvent, as well as the solute-solvent, interactions were evaluated using a sharp spherical cutoff of half the simulation box length. The charge-dipole interactions between the solute and the solvent were supplemented with LJ interactions between the A atoms and solvent particles and were present in both diabatic states. The parameters of the LJ interactions were chosen to be the same as those between solvent particles. The simulation box size was chosen to give a solvent density of 0.0335 particles/Å. The simulations were run at 298 K with 216 particles in total, one solvent particle having been replaced by the solute.

3.1.2 Adiabatic proton transfer model

The same linear triatomic as in Sec. 3.1.1 was used in the adiabatic model except the harmonic diabatic terms were replaced with Morse oscillators to give a 10 k_BT barrier along the proton asymmetric stretch coordinate. The Morse oscillator parameters were chosen to be $D = 19.342$ kcal/mole, $a = 3.789$ Å^{-1}, and $r_e = 0.95$ Å. The A-A′ distance was constrained to be rigid in the adiabatic system with a seperation of 2.5 Å.

3.2 NON-POLARIZABLE SOLUTE MODEL

In order to investigate the issue of solute polarizability, the diabatic states representation of the model was modified so that the EVB diabatic states were written in terms of the solute properties alone (i.e., not in solution). The charges on the A and A′ atoms were thus considered to be functions of the solute EVB adiabatic energy surface. The electric field produced by the solute at solvent particle i was then set to

$$\mathbf{E}_i(\tau) = \langle \Psi(\tau) | \left[|1\rangle \frac{e_A \mathbf{R}_{iA}}{R_{iA}^3} \langle 1| + |2\rangle \frac{e_{A'} \mathbf{R}_{iA'}}{R_{iA'}^3} \langle 2| + \frac{e_H \mathbf{R}_{iH}(\tau)}{R_{iH}^3(\tau)} \hat{\imath} \right] | \Psi(\tau) \rangle ,$$

(12)

where the EVB eigenfunction $| \Psi(\tau) \rangle$ of the solute is given by

$$| \Psi(\tau) \rangle = c_1(\tau) | 1 \rangle + c_2(\tau) | 2 \rangle .$$

(13)

The coefficients for this wavefunction are determined by a diagonalization of the EVB matrix at each imaginary time slice of the proton path $\mathbf{x}(\tau)$.

3.3 POLARIZABLE SOLVENT MODEL

In order to model the effects of a polarizable solvent on the quantum proton transfer activation free energy, a polarizable Stockmayer fluid model was developed using quantum Drude oscillators [15]. To avoid the numerical overhead of explicitly treating the quantum degrees of freedom of the solvent, the quantum Drude oscillators were integrated out. To this end, the total path integral of the system can be subdivided. Within the context of a nonpolarizable solute model, the quantum Drude oscillator portion of the total path integral in eq. (5) is given by

$$
Z_D = \int \mathcal{D}\mathbf{p}(\tau) \exp \left\{ -\frac{1}{\hbar} \int_0^{\beta\hbar} d\tau \left[\sum_{i=1}^N \frac{\dot{\mathbf{p}}_i^2}{2\alpha\omega_0^2} + \sum_{i=1}^N \frac{\mathbf{p}_i^2}{2\alpha} \right. \right.
$$
$$
\left. \left. - \sum_{i<j}^N (\mathbf{m}_i + \mathbf{p}_i(\tau)) \mathcal{T}_{ij} (\mathbf{m}_j + \mathbf{p}_j(\tau)) - \sum_{i=1}^N (\mathbf{m}_i + \mathbf{p}_i(\tau)) \mathbf{E}_i(\tau) \right] \right\} \quad (14)
$$

where $\mathcal{T}_{ij}$ is the $3N \times 3N$ dipole-dipole interaction tensor, given by

$$
\mathcal{T}_{ij} = \frac{1}{R_{ij}^3} \left(1 - \frac{3\mathbf{R}_{ij}\mathbf{R}_{ij}}{R_{ij}^2} \right) \quad (i \neq j) \ ,
$$
$$
\mathcal{T}_{ij} = 0 \quad\quad\quad\quad\quad\quad\quad (i = j) \ . \quad (15)
$$

The Stockmayer particles permanent dipoles $\mathbf{m}$ and Drude dipoles $\mathbf{p}(\tau)$ couple to the electric field $\mathbf{E}(\tau)$ produced by the solute complex. The Drude oscillator portion contains all the solvent-solvent interactions as well as the coupling of the solvent to the electric field produced by the solute complex. The electric field $\mathbf{E}(\tau)$ is given by eq. (12).

For convenience, the permanent dipoles are also treated as quantum variables and the quantum paths of both the permanent dipoles and Drude oscillator are expanded in Fourier series as

$$
\mathbf{p}_i(\tau) = \sum_{n=-\infty}^{\infty} \tilde{\mathbf{p}}_{i,n} e^{-i2\pi n/\beta\hbar}, \quad (16)
$$

$$
\mathbf{m}_i(\tau) = \sum_{n=-\infty}^{\infty} \tilde{\mathbf{m}}_{i,n} e^{-i2\pi n/\beta\hbar}. \quad (17)
$$

These expansions diagonalize the kinetic energy and quadratic Drude terms, permiting an integration over the Drude oscillator modes. One then obtains the following expression for Z_D,

$$Z_D = \left\{ \prod_{n=-\infty}^{\infty} \left[\frac{2\pi\alpha_n}{\beta} \right]^{3N} \frac{1}{\det A_n} \right\}^{1/2}$$

$$\exp\left\{ \beta \sum_{n=-\infty}^{\infty} \left[\frac{1}{2\alpha_n} \left(\sum_{i,j}^{N} (\alpha_n \tilde{\mathbf{E}}_{i,n}^* + \tilde{\mathbf{m}}_{i,n}^*)[A_n]_{ij}^{-1} (\alpha_n \tilde{\mathbf{E}}_{j,n} + \tilde{\mathbf{m}}_{j,n}) \right. \right. \right.$$

$$\left. \left. \left. - \sum_{i=1}^{N} \tilde{\mathbf{m}}_{i,n}^* \tilde{\mathbf{m}}_{i,n} \right) \right] \right\} \quad (18)$$

where $\tilde{\mathbf{E}}_{i,n}$ is the nth Fourier component of the electric field due to the solute complex at the ith particle, α_n is the nth Fourier component of the Drude polarizability given by

$$\frac{1}{\alpha_n} = \frac{1}{\alpha} \left\{ \left[\frac{2\pi n}{\beta \hbar \omega_0} \right]^2 + 1 \right\} \quad , \quad (19)$$

and the matrix A_n is given by

$$A_n = 1 - \alpha_n T \quad . \quad (20)$$

The prefactor in eq. (18) contains the quantum many-body dispersion interactions of the Drude oscillators [15]. In the extreme quantum limit (i.e. $\beta\hbar\omega_0 \to \infty$) this interaction reduces to the pairwise additive $1/R^6$ term of the Lennard-Jones potential. Since this term is already present in the Stockmayer potential, the contribution from the prefactor is neglected.

Upon expanding A_n^{-1} in a Taylor expansion, i.e.,

$$A_n^{-1} = \sum_{m=0}^{\infty} (\alpha_n T)^m \quad , \quad (21)$$

and collecting terms up to the order of induced dipole−permanent dipole interactions, the Drude oscillator quantum partition function becomes

$$Z_D \approx \exp\left\{ -\beta \left[-\sum_{i>j}^{N} \mathbf{m}_i T_{ij} \mathbf{m}_j \right] - \frac{1}{\beta\hbar} \int_0^{\beta\hbar} d\tau \left[-\sum_{i=1}^{N} \mathbf{m}_i \mathbf{E}_i(\tau) \right. \right.$$

$$\left. \left. -\alpha \sum_{i=1}^{N} \sum_{j=1}^{N} \mathbf{m}_i T_{ij} \mathbf{E}_j(\tau) \right] + \sum_{i=1}^{N} \int_0^{\beta\hbar} \frac{d\tau}{\beta\hbar} \int_0^{\beta\hbar} \frac{d\tau'}{\beta\hbar} \alpha(\tau' - \tau) \mathbf{E}_i^*(\tau') \mathbf{E}_i(\tau) \right\}. \quad (22)$$

The model developed here differs from the nonpolarizable solvent model by only the last two terms in eq. (8). The next to last term represents an induced dipole—permanent dipole interaction, while the last term represents the inter-action of the electric field of the solute complex with the electric field induced at each particle.

In the quantum limit $\beta\hbar\omega_0 \to \infty$ the kernel of the influence functional $\alpha(\tau'-\tau)$ is proportional to a delta function. The last term in the exponential in eq. (8) then becomes

$$\lim_{\beta\hbar\omega_0\to\infty} \sum_{i=1}^{N} \int_0^{\beta\hbar} \frac{d\tau}{\beta\hbar} \int_0^{\beta\hbar} \frac{d\tau'}{\beta\hbar} \alpha(\tau'-\tau)\mathbf{E}_i^*(\tau')\mathbf{E}_i(\tau) \;=\; \alpha \sum_{i=1}^{N} \frac{1}{\beta\hbar} \int_0^{\beta\hbar} |\mathbf{E}_i(\tau)|^2 d\tau \;.$$

$$(23)$$

The quantum activation energy for the polarizable Stockmayer model was calculated using eqs. (6) and (7) for a polarizability of 22.3 bohr^3 and with the nonpolarizable solute, adiabatic proton transfer model outlined in section 3.1.2.

4. Results

4.1 EFFECT OF SOLVENT ACTIVATION

The total quantum activation free energy curves as a function of the proton asymmetric stretch coordinate are depicted in Fig. 1 for several models. Shown are the activation curves for the solute with rigid A−A′ distance in isolation (dotted curve) and in the nonpolarizable Stockmayer fluid (dot-dashed curve). Also shown are results for the case in which the A−A′ distance can fluctuate both in isolation (long-dashed curve) and in solution (short-dashed curve). The solid curve is the classical free energy for the rigid solute isolation. The results shown here are for the nonadiabatic proton transfer model described in Sec. 3.1.1. As can be seen from Fig. 1, the solvent provides an important contribution to the quantum activation free energy in all cases. In addition, the influence from the proton tunneling is evident from comparing any of the quantum curves with the classical limit.

4.2 EFFECT OF INTRAMOLECULAR MODE COUPLING

As can be seen from Fig. 1, the inclusion of the intramolecular A−A′ vibration causes a decrease in the total quantum activation free energies in both the presence and absence of solvent. This mode fluctuation effect thus enhances the proton transfer rate in this model as predicted in other work [4-6]. This effect is not surprising since inward fluctuations of the A−A′ distance lead to a lowering of the classical barrier along the poton reaction coordinate. However, it is interesting to note that the solvent contribution to the activation free energy is *smaller* in the flexible case than the rigid case (1.5 versus 2.9 kcal/mole). This nonlinear effect is due to the fact that the solute complex can "present" a smaller dipole moment change to the solvent during the proton tranfer during an inward fluctuation of the A−A′ distance. This effect might be called "solvent activation avoidance" and should be present to some degree in all proton transfer reactions.

4.3 EFFECT OF SOLUTE POLARIZABILITY

The solution of the full EVB adiabatic energy function from Sec. 3.1.1 can give rise to a situation where the solvent continues to solvate one diabatic state even though the proton is at the transition state along its asymmetric stretch reaction coordinate. In fact, this is an energetically preferred situation due to the fact that the solvent prefers to solvate the large dipole associated with a given diabatic state with the proton at the transition state. In order to investigate this EVB solute "polarization" effect on the quantum activation free energy, the nonadiabatic proton transfer model was modified according to the non-polarizable solute prescription in Sec. 3.2. This modification results in the solute complex always having a zero dipole moment when the proton asymmetric stretch is at its transition state value. The solvent contribution to the activation free energy for the flexible polarizable (dot-dashed curve) and non-polarizable (dashed curve) solute models are shown in Fig. 2. Interestingly, the flexible solute has a very similar solvent activation free energy for the polarizable and nonpolarizable models. One would have expected that the the non-polarizable solute model would have a larger contribution from the solvent in the activation free energy since the solvent dipoles are "forced" into the transition state in that model. The added feature of solute flexibility apparently allows for a somewhat different interaction between the solute and the solvent, leading to a diminished effect from solute polarizability.

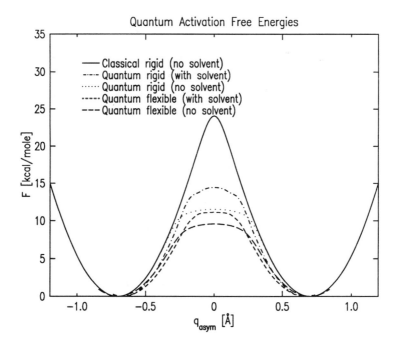

Figure 1

4.4   EFFECT OF SOLVENT POLARIZABILITY

The effect of solvent polarizability on the proton transfer quantum activation process was studied for the adiabatic model outlined in Sec. 3.1.2 with the Drude model from Sec. 3.3. In Fig. 3, the different activation free energy curves are shown. The effect of the solvent polarizability (long-dashed curve) is to raise the solvent contribution to the activation free energy by approximately 0.5 kcal/mole over the non-polarizable solvent limit (short-dashed curve). Future research will focus on the origins of this effect.

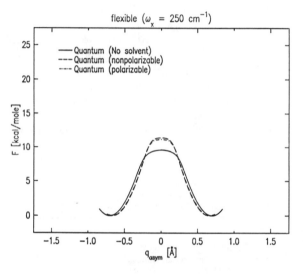

Figure 2

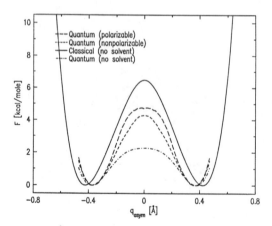

Figure 3

5. Concluding Remarks

The quantum activation free energy for proton transfer reactions in polar solvents has been studied using path integral QTST methods. Several features of the problem have been investigated, include the effect of solvent activation, the role of intramolecular or intra-complex mode fluctuations, and the influence of both solute and solvent polarizability. The most general conclusion one can draw from these studies is that proton transfer reactions are fundamentally complex, probably more so than electron transfer reactions. In order to *quantitatively* understand the rate of proton transfer reactions, one must deal with a number of complex, nonlinear interactions. Examples of such interactions include the nonlinear dependence of the solute dipole on the position of the proton, the coupling of the solute dipole to *both* the proton coordinate and to other vibrational modes, and the intrinsically nonlinear interactions arising from both solute and solvent polarizability effects. Analytic studies to understand these effects in more detail, and to predict their magnitude for any given proton transfer reaction, will be the subject of future research.

Acknowledgements

This research was supported by a grant from the United States National Science Foundation. GAV is a recpient of a National Science Foundation Presidential Young Investigator Award, a David and Lucile Packard Fellowship in Science and Engineering, an Alfred P. Sloan Foundation Research Fellowship, and a Dreyfus Foundation New Faculty Award. JL is a recipient of a Graduate Fellowship from the Natural Sciences and Engineering Research Council (Canada).

References

[1] Warshel, A. (1982) "Dynamics of reactions in polar solvents. Semiclassical trajectory studies of electron-transfer and proton-transfer reactions", J. Phys. Chem. **86**, 2218-2224; Warshel, A., and Chu, Z. T. (1990) "Quantum corrections for rate constants of diabatic and adiabatic reactions in solutions", J. Chem. Phys. **93**, 4003-4015.

[2] Borgis, D., and Hynes, J. T. (1991) "Molecular-dynamics simulation for a model nonadiabatic proton transfer reaction in solution", J. Chem. Phys. **94**, 3619-3628.

[3] Cukier, R. I. and Morillo, M. J. (1989) "Solvent effects on proton transfer reactions", Chem. Phys. **91**, 857-863.

[4] Borgis, D., Lee, S. and Hynes, J. T. (1989) "A dynamical theory of nonadiabatic proton and hydrogen atom transfer reaction rates in solution", Chem. Phys. Lett. **162**, 19-26; Borgis, D., and Hynes, J. T. (1993) "Dynamical theory of proton tunneling transfer rates in solution: General formulation", Chem. Phys. **170**, 315-346.

[5] Cukier, R. I. and Morillo, M. (1990) "On the effects of solvent and intermolec-

ular fluctuations in proton transfer reactions", J. Chem. Phys. **92**, 4833-4838; Suárez, A. and Silbey, R. (1991) "Hydrogen tunneling in condensed media" J. Chem. Phys., **94** 4809-4816.

[6] Lobaugh, J. and Voth, G. A. (1992) "Calculation of quantum activation free energies for proton transfer reactions in polar solvents", Chem. Phys. Lett. **198**, 311-315; Li, D. H. and Voth, G. A. (1991) "Feynman path integral approach for studying intramolecular effects in proton-transfer reactions" J. Phys. Chem. **95**, 10425-10431.

[7] Borgis, D., Tarjus G., and Azzouz, H. (1992) "Solvent-induced proton transfer in strongly H-bonded complexes: An adiabatic dynamical simulation study", J. Phys. Chem. **96**, 3188-3191; Azzouz, H. and Borgis, D. (1993) "A quantum molecular-dynamics study of proton-transfer reactions along asymmetrical H bonds in solution", J. Chem. Phys. **98**, 7361-7374; Borgis, D., Tarjus G., and Azzouz, H. (1992) "An adiabatic dynamical simulation study of the Zundel polarization of strongly H-bonded complexes in solution", J. Phys. Chem. **97**, 1390-1400; Laria, D., Ciccotti, G., Ferrario, M. and Kapral, R. (1992) "Molecular-dynamics study of adiabatic proton transfer reactions in solution", J. Chem. Phys. **97**, 378-388.

[8] Voth, G. A., Chandler, D. and Miller, W. H. (1989) "Rigorous formulation of quantum transition state theory and its dynamical corrections", J. Chem. Phys. **91**, 7749-7760; Voth, G. A. (1990) "Analytic expression for the transmission coefficient in quantum mechanical transition state theory", Chem. Phys. Lett. **170**, 289-296; Voth, G. A. (1993) "Feynman path integral formulation of quantum mechanical transition state theory", J. Phys. Chem. in press.

[9] Feynman, R. P. (1972) Statistical Mechanics, AddisonPWesley, Reading, MA.

[10] Berne, B. J. and Thirumalai, D. (1987) "On the simulation of quantum systems: Path integral methods", Annu. Rev. Phys. Chem. **37**, 401-424.

[11] Valleau, J. P. and Torrie, G. M. (1977) "A guide to Monte Carlo for statistical mechanics: 2. Byways", in B. J. Berne (ed.), Statistical Mechanics, part A, Plenum Press pp. 169-191.

[12] Owicki, J. C. (1978) "Optimization of sampling algorithms in Monte Carlo calculations on fluids", ACS Symp. Ser. **86**, 159-171.

[13] Warshel, A. and Weiss, R. M. (1980) "An emperical valence bond approach for comparing reactions in solutions and in enzymes", J. Am. Chem. Soc. **102**, 6218-6226.

[14] Pollock, E. L. and Alder, B. J. (1980) "Static dielectric properties of Stockmayer fluids", Physica **102A**, 1-21.

[15] Cao, J. and Berne, B. J. (1992) "Many-body dispersion forces of polarizable clusters and liquids", J. Chem. Phys. **97**, 8628-8636.

HOW DO THE PROPERTIES OF WATER IN CONFINED VOLUMES DIFFER FROM THOSE IN THE NORMAL LIQUID?

G. Wilse Robinson and S.-B. Zhu*
Subpicosecond and Quantum Radiation Laboratory
Departments of Chemistry and Physics
Texas Tech University
Lubbock TX, USA   79409-1061

1. Introduction

What we are trying to do in our laboratory is to combine pico- and femtosecond experiments with theory and computational molecular dynamics for the study of condensed phase chemical reactions, primarily, but not entirely, in the water solvent.   More recently, we have become interested in these types of reactions near surfaces.   The paper will mainly concern this latter aspect.

2. Description of Liquid Water

Anyone interested in water as a solvent quickly becomes aware of certain basic features that distinguish it from all other solvents.   These features arise from structural changes taking place in the liquid with changing thermodynamic state.   One effect of these changes is a cooperative "softening" of the structure of the liquid with increasing temperature or pressure, and a concomitant flattening of the intermolecular potential surfaces. Another effect is the creation of significantly more compact intermolecular bonding components as the temperature, or of course also the pressure, is raised.

2.1. Structural softening. The structural softening phenomenon enters the liquid water problem in a similar way that it does for rotations in solids.   When the molecules begin to rotate, the activation barrier for the rotation of other molecules in the neighborhood becomes lower.   This idea was elegantly described by R. H. Fowler in his book on statistical mechanics [1] and can be quoted in essence. "The directional terms in the field to which any one molecule is subject are of course mainly due to the lack of spherical symmetry in the combined fields of the

423

J. Jortner et al. (eds.), Reaction Dynamics in Clusters and Condensed Phases, 423–440.
© 1994 *Kluwer Academic Publishers. Printed in the Netherlands.*

surrounding molecules. This lack of symmetry will be greatly weakened and might in fact be almost destroyed by a sufficient degree of rotation among the surrounding molecules. Thus, the orientational activation energy itself cannot be a constant but must depend on the degree of rotation already present among the molecules." This scenario seems to occur, not only for orientations but also for translations, when liquid water is heated or pressurized. This leads to lower activation energies for various dynamical processes. In addition, the flattening of intermolecular potential surfaces causes a lowering of librational frequencies as the temperature or pressure is raised. In turn, thermodynamic properties are affected (see below). That activation energies for various transport and relaxation phenomena decrease with increasing temperature and pressure is a well established experimental fact for liquid water [2-5]. It is also known that librational frequencies in liquid water strongly decrease with increasing temperature [6,7]. See Figure 1.

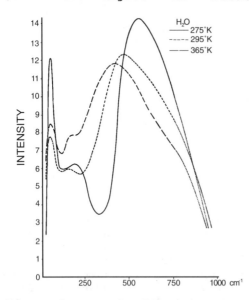

Fig. 1. Inelastic neutron scattering spectra of A_2 librational mode showing the temperature dependence. After Larsson and Dahlborg [6].

2.2. *Heat capacity anomaly.* State-dependent librational frequencies $\nu(p,T)$ clearly give rise to an anomaly in the heat capacity, which can be roughly expressed as,

$$C_p(T) = U^0(d\ln\nu/dT) + C_p^0[1-(T/\nu)(d\nu/dT)] , \qquad (1)$$

where U^0 and C_p^0 are the normal internal energy and heat capacity for an ensemble of harmonic oscillators in the absence of $d\nu/dT$ terms, but calculated with the appropriate temperature-dependent frequency at each temperature in question. The dominant second term on the RHS of Equation (1) substantially raises $C_p(T)$ over what it would be for temperature independent librational frequencies because $d\nu/dT$ is strongly negative. For example, the A_2

orientational librational mode shown in Figure 1 shifts from about 580 cm^{-1} at 275K to about 435 cm^{-1} at 365K. This effect gives a large contribution to $C_p(T)$ from this and the other five local librational modes, which are also temperature dependent. In fact, the second term in Equation (1) is more than twice as large as it would be if $d\nu/dT$ were zero. This is quantitatively in keeping with the observed heat capacity anomaly of the liquid. Thus, the thermodynamic properties of liquid water do not merely depend on the thermal population of a fixed set of energy levels. Rather, the energy levels themselves change with p and T because of the potential surface changes. A similar, but less specific, way of understanding the anomalous heat capacity of liquid water has been to attribute it to "configurational contributions" [5] or to "the unusually large temperature dependence of the structure" [8].

2.3. Denser intermolecular bonding. Besides these softening effects, the structural changes taking place in liquid water also cause a transition from locally open to locally dense intramolecular bonding as T or p is raised. This is best appreciated by realizing that small water clusters can have a wide range of configurations with very different structures but nearly equivalent energies [9].

The multiplicity of intermolecular structures in water clusters carries over to the various forms [5] of ice, creating subtle complications, and of course leads to even greater complications in the liquid, where entropy encourages the existence of a multiplicity of structural forms. Since the local open tetrahedral bonding among water molecules is in a sense a limit, new bonding forms have little choice but to show an increase in the density. Actually, experimental evidence, to be described below, does indicate that more compact bonding components in the liquid structure, locally resembling various dense forms of ice, grow in as the temperature of the liquid is raised.

The arrangement of hydrogen bonds in ice-II resembles that of normal ice-Ih, except that half the hexagonal tunnels are filled with more densely packed molecules, and the hydrogen atom positions are ordered, so the Pauling entropy is absent. For a good pictorial comparison, see Figures 41 and 42 of the book by Pauling and Hayward [10]. It is interesting that, without this ordering, ice-II, with a density of ~1.18 g cm^{-3}, would have roughly equal stability as ice-Ih, whose density is about 0.92 g cm^{-3}. This would, of course, have had an immeasurable impact on life forms on Earth. It is also of interest that this ordering is apparently stable under electric field perturbations, since it is responsible for a very low dielectric constant (T = 240K) in ice-II, ~4 *vs.* ~100 for normal ice, where the hydrogen atoms, under field perturbations, are free in ice-Ih to tunnel and change the macroscopic electrical properties of this substance [5].

In both ice-III and ice-V the hydrogen bonds are disordered, but strains in these structures caused by the 5- and 4-member rings, respectively, makes them less stable enthalpically than ice-Ih or ice-II. In all these forms, ice-Ih, -II, -III and -V, it is most important to note that, while the nearest-neighbor hydrogen-bonded $O \cdots O$ distances are nearly the same, ranging from about 2.74 Å to 2.87 Å, non-hydrogen-bonded $O \cdots O$ neighbors have very different spacings. They are shortened from about 4.5 Å in ice-Ih to 3.24-3.47 Å in the other forms. This is the main reason for the huge density difference in these ice structures: around 0.92 g cm^{-3} in ice-Ih and 1.15-1.26 g cm^{-3} in ice-II, -III and -V.

How might these known structural properties of the ice polymorphs play a role in bonding in the liquid state? Figure 2, from the work of Bosio, Chen and Teixeira [11], shows that, as the temperature of the liquid increases, the emerging structural form contains a preponderance of $O \cdots O$ intermolecular distances near 3.3 Å. This is an $O \cdots O$ distance that is totally absent in ice-Ih! Such a next-nearest-neighbor $O \cdots O$ separation then acts as a fingerprint for the emergence with increasing temperature in liquid water of higher density bonding forms, and furthermore that these structures, at least with respect to their $O \cdots O$ distances, resemble those of ice-II, -III and -V. In the liquid, with the curtailment of the Pauling

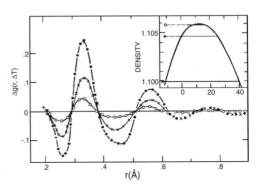

Fig. 2. Isochoric differential x-ray scattering from liquid water after Bosio, et al. [11] showing the growth of the ~3.3 Å $O \cdots O$ peak with temperature.

entropy, the main factor stabilizing ice-Ih with respect to ice-II, a number of these energetically similar local structures must play a role.

2.4. The density anomaly. In light of the above discussion, it has in fact been found possible [4,12] to fit the specific volumes of liquid water from T = 238-423K within at least 4-decimal point accuracy using the following simple equation,

$$V(p,T) = f(p,T) \, V_I(p,T) + [1-f(p,T)] \, V_{II}(p,T), \qquad (2)$$

where $V_I(p,T)$ and $V_{II}(p,T)$ are the p,T-dependent specific volumes of ice-Ih and ice-II (or -III or -V). The factor $f(p,T)$, which can be independently assessed [4] from the activation energies of dynamic processes, is the proportion of the liquid that is composed of open ice-Ih-type structures.

A volume minimum (density maximum) occurs in Equation (2) in spite of the fact that both $V_I(p,T)$ and $V_{II}(p,T)$ increase with T. This is because the proportion of the dense form rapidly increases with increasing T, causing a sharply decreasing $f(p,T)$. This behavior is indicated by the relaxation data shown in Figure 1 of Reference [4].

The isotope dependence of the density maximum is caused [4] by the fact that the proportions of I-type and II-type structures at any given temperature depend on $f(p,T)$, which is sensitive to frequencies and zero-point effects in the temperature-dependent intermolecular potentials, particularly those related to molecular rotational librations, which have the largest H, D, T isotope effects. This is why the density maximum in D_2O occurs at a temperature over 7°C higher than in H_2O.

2.5. Transport properties. The pressure behavior of the shear viscosity provides additional insights concerning the behavior of $f(p,T)$. Ordinarily, the viscosity of a liquid increases with increasing pressure, since the molecules become squeezed together, inhibiting rotational and translational freedom. For liquid water at sufficiently low temperatures, the opposite effect occurs [5]. This observation confirms that the denser II-type bonding component in the liquid, whose proportion increases with pressure, corresponds to flatter intermolecular potential energy surfaces and more facile transport properties. Such a conclusion may go against certain intuitions, but it is in agreement with all the experimental facts.

Raising the temperature gets rid of the anomalous pressure effect on the viscosity, and the anomalous pressure effect on other properties as well. This is because increasing the temperature also leads to structural transformations that raise the proportionality of the dense component, so at higher temperatures "there is nothing anomalous left for the pressure to do". The isotope effect on the librational motions, which are related to the activation energies for dynamical processes, is responsible for the large isotope effects on the viscosity and various relaxation times in liquid water.

2.6. Isothermal compressibility. In addition to density considerations, by differentiating Equation (2) with respect to the pressure at fixed temperatures T, a minimum in the isothermal compressibility is found at the same temperature at which it occurs in the real liquid [4,12]. Again, the minimum is not primarily caused by the derivatives of the volumes $V_I(p,T)$ and $V_{II}(p,T)$ with p, but rather is a result of the rapid disappearance of the pressure dependance of $f(p,T)$ with increasing T, a phenomenon that was described above and is also easily seen from experimental relaxation times as a function of pressure [2,3]. See also Figure 5 of Reference [4].

2.7. A take-home lesson. The most important take-home lesson from these considerations is that the "breakdown of hydrogen-bonding" in liquid water must be viewed as an atypical breakdown, since it has to do, not with nearest neighbors, but with next nearest neighbors. This fact could possibly explain the difficulty that chemists, with their intuitive knowledge of molecular bonding, have had trying to understand the bonding in liquid water. In fact, many previous ideas about the temperature effects in liquid water have concentrated on nearest-neighbor changes, for which there is little supporting evidence. The proportion of ~2.78 Å nearest-neighbor O···O structure does decrease in liquid water as the temperature is raised (Figure 2), but this is only a minor fraction of the total: compare the ~-0.15 change [11] from -12°C to +40°C with the total amplitude, ~3.1, of the first peak of $g_{OO}(r)$ [13].

3. Interfacial Water

In the present paper, a glance at some published and some heretofore unpublished computational results which bear on the interfacial water problem will be presented. Experimental data on interfacial water are expected to accumulate rapidly over the next few years: using second-harmonic-generation (SHG) techniques to measure molecular orientations [14,15]; using infrared-visible sum-frequency generation (SFG) to measure intramolecular vibrational frequencies at the surface [16]; and using fluorescence probe studies [17], where direct ultrafast dynamic competition between probe diffusion and probe photodynamics near an interface can be monitored. These types of experimental data combined with much improved computational results, as computer power continues to grow, should help greatly in achieving a much fuller understanding of the interfacial water problem by the close of the millenium.

Water in very small volumes plays a dominant role as the medium that controls structure, function, dynamics and thermodynamics near biological membranes or in other confined regions of space. Strong structural perturbations at a surface are incongruent with the ordinary directional binding between water molecules in the bulk liquid. The resulting surface-induced perturbative disruptions therefore modify water's normal properties. Questions still abound as to the distance these perturbations extend from the surface, their influence on diffusion and other physical properties, their effect on chemical reactions, such as acid-base equilibria, and on phenomena taking place in biological microsystems [18]. These questions are of great interest because of the fundamental role that interfacial water plays in many physical, geophysical, chemical, biological and industrial processes.

Since the structures within liquid water are so fragile with respect to temperature and pressure changes, it would be expected that effects similar to those in the bulk liquid might be even more prominent in water perturbed by surfaces. If the surface is chemically and physically inert, the inability of the water molecules to extend their hydrogen bond structure into such a surface could have the same effect as a thermal increase. If the surface interacts with water, the inability of those molecules, forcibly oriented near the surface, to maintain normal hydrogen bonding with their neighbors creates the same type of perturbation. The open I-type tetrahedral structure would have to give way to a denser bonding: more tightly packed hexagonal channels combined with 4- and 5-membered ring structures. The viscosity would be lower, as would the librational frequencies. Transport and relaxation phenomena would be enhanced. How well do the current computational studies bear out these expectations?

4. Some Computational Results

Any water model that is designed to sort out the various issues concerning bulk or interfacial water should not be burdened with inadequacies that could call into question revelations it is hoping to promote. Unfortunately, because of computer expense, all existing water models contain undesirable simplifications. Rapidly responsive electronic polarization is often missing in these models, intramolecular bonds are usually taken to be rigid, the electric quadrupole tensor is poorly represented and the sample size is not sufficiently large. Realistically, forces on the water molecules spatially vary on time scales that correspond to molecular translations and orientational motions. The absence of an appropriate

molecular response to these forces could cast doubt on any
result obtained. Interfacial water presents a particularly
difficult problem in these respects.

4.1. FP models. In an effort to remove some of the
simplifications in water modeling, our laboratory has
developed three flexible-polarizable (FP) models [19-21].
The most advanced of these [21] is a modified five-site ST2
model [22], which is
abbreviated MST-FP. An
interesting, but
controversial, example of
interfacial MST-FP water
is an ultra thin (~22 -
44 Å) film, forming a
double-sided liquid-vapor
interface [23]. In
earlier work [24] with
a lesser model (SPC-FP)
[20], interfacial water in
a thin film between
Lennard-Jones walls was
considered. In that
study, comparisons were
made when the wall was
neutral and when it was
electrically charged.

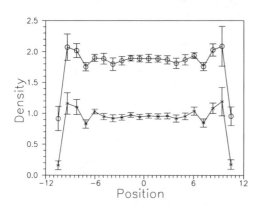

Fig. 3. MST-FP relative
density profiles for the 256
molecule film: H-atom (top),
O-atom (bottom).

Described in the present
paper are some results
from both these studies
concerning profiles as a
function of distance (Å) from the film center: of the
O- and H-atom mass densities, the O-H bond lengths and the
H-O-H bond angles, the molecular orientations and, briefly,
some of the transport properties.

The MST-FP model for water [21], unlike the
original ST2 model [22], starts with the correct gas phase
structure and dipole moment, together with intramolecular
anharmonic coupling that gives the correct gas phase
vibrational frequencies. Employed was a simplified method
for including rapidly responsive polarization. In addition,
an orientationally stiffening Morse interaction for the
hydrogen bonds was included to counteract the softening
influences [20] of the flexibility, and the four point-
charges in ST2 were replaced by distibuted charges.

When a "bulk liquid" sample having the correct
density is made up from 256 MST-FP water molecules,
quantiative agreement with the liquid state dipole moment
and molecular structure is automatically obtained; and
good agreement with the liquid state librational and
intramolecular frequencies also arises: the intramolecular

stretching frequencies become lower and the bending frequency becomes higher in the liquid, as is the case in real water. Thus, in spite of certain reservations to the contrary [25], there is no need to include quantum mechanics to obtain correct gas-liquid intramolecular frequency shifts.

On the other hand, as in many other water models, the gas-phase quadrupole tensor elements in MST-FP water are about 40% too small, though they have the correct signs [26,27]. Also, the dielectric constant is about 25% too low, and transport and relaxation processes are too slow, possibly because of an overly stiff intermolecular Morse-type orientational binding in this model. Of course, the 256-molecule ensemble is not nearly large enough to allow realistic long range cooperative effects. Even so, rather reasonable agreement with the collective dipole moment relaxation time was found. In spite of these deficiencies, the MST-FP model is current state-of-the-art. Moreover, it is sufficiently inexpensive computationally to allow a variety of interfacial water studies on currently available supercomputers.

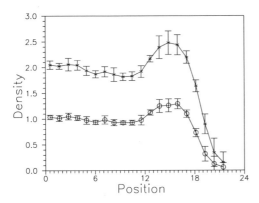

Fig. 4. Same Profiles as Fig. 3 for the 512 molecule film, but here the film sides are averaged.

4.2. Density profiles. Figure 3 shows the mass density profiles for the MST-FP liquid-vapor interface in a liquid film about 22 Å thick, containing 256 molecules. Periodic boundary conditions in the transverse directions were employed. A somewhat similar density profile (Figure 4), though not as sharply structured, was obtained for 512 molecules when the film thickness is doubled. Clearly, there is a measurable effect, arising from reflections from the opposite surface, which could possibly eliminate the oscillations in thick films. This is a very clear indication of the importance of ensemble size restrictions on current water modeling. Adopted in these studies was a gravity-like external force proportional to mass in order to hinder evaporation from the film. The O- and H-atoms thereby gain identical additional accelerations back into the film, as long as their distances from the surface are the same. This external potential must be designed to give a near zero torque on the water molecules in order to minimize perturbations on the film. Details will be given elsewhere [23]. See also the Figure 5 caption.

From the greater than twofold ratio of the H-atom density compared with the O-atom density very close to the vapor side, it is seen that the water molecules must be oriented with at least one of the protons pointing preferably outwards towards the vapor. In addition, the profiles for the oxygen and hydrogen atoms are structured. Neither of these results agrees very well with earlier computational studies [28,29] using rigid water models, but the computed orientational properties of MST-FP water may be more in line with recent experimental data [16]. The vector from the O-atom to the H-H bisector of MST-FP water was found on the average to point 20-30° out of the liquid-vapor surface plane towards the vapor, instead of lying in the surface plane [28]. See Figure 5.

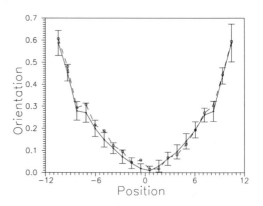

Fig. 5. Orientational profile of H-O-H bisector vector: averaged cosine of the angle between this vector and the surface normal vs. distance from film center in Å. For a random distribution the average is zero. The dashed line is for a mass independent restraint potential. See text.

It was Frenkel [30] who first suggested that the quadrupole moment of the water molecule is responsible for preferred molecular orientations at the liquid-vapor interface. His analysis was based on the concept that molecules in the surface layer tend to orient so as to immerse their electric field as much as possible in the high dielectric region of the fluid, thereby minimizing the free energy. Extending this argument, it could well be that the more polarizable O-atom would prefer to immerse itself in the dielectric, thus enhancing the molecule's dipole moment, whatever the signs and magnitudes of the quadrupole moments may be. Nonpolarizable water models would not be able to pick up inevitable variations of the induced dipole, quadrupole, etc. moments near a surface.

In any case, the orientational properties of water molecules at such a surface must depend strongly on the water model used: where the hydrogen and lone-pair charges are placed and whether or not rapidly responsive polarization is included. It should also be remarked that studies using the thicker film indicate that the range of perturbative effects from the surface depend on the type of property being investigated. Molecular orientational perturbations were found to extend much farther from the surface than those for the density.

The structure in Figures 3 and 4 could be caused by the perhaps overly stiff orientationally dependent intermolecular interaction incorporated into the MST-FP model. While undoubtably a real effect for these films and this particular water model, such a density oscillation may not occur or be difficult to detect experimentally in real water, though supercooled temperatures could help. In any event, the discovery of such well defined density maxima at a liquid-vapor interface is unprecedented, all previous studies of Lennard-Jones fluids [31] or various water models [28,29] having shown no definitive evidence for density oscillations near such an interface. Of course, from the discussion in Sections 2 and 3, a density maximum

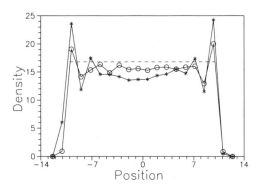

Fig. 6. O-atom mass density profile for SPC-FP water between L-J walls: neutral (o), charged (*), SPC-FP bulk liquid density ---.

for interfacial water is expected to occur near any surface because of the expected transformation from open I-type to dense II-type local structuring. This is certainly the case for liquid-solid interfaces, as everyone would now agree. See Figures 6 and 7.

It should be remarked here that density oscillations near a liquid-vapor interface reported earlier by us for SPC-F (nonpolarizable) water were much less distinct and more far ranging than for MST-FP water. Through a crude error analysis, it was believed those oscillations to be real, but there was certainly room for doubt. This doubt is absent in the present study as evidenced by the agreement between the independently assessed density profiles on each side of the 256 molecule

film. Also, the error analysis here is more reasonable:
following equilibration, the equivalent of 350 ps in real
time was employed to accumulate data; then this long study,
consuming about 108 CPU hours on the Pittsburgh YM-P/832
CRAY, was grouped into 14 pieces, each lasting 25 ps. The
error bars in Figures 3-5, also 8 and 9, are twice the
standard deviations from this set of measurements.

      4.3. Intramolecular structure. While the density
profiles and molecular orientations at the liquid-vapor
interface of water are still controversial issues, the
effect of this type of interface on the intramolecular
structure is more straightforward. Figures 8 and 9 depict
the profiles at the MST-FP liquid-vapor interface for
the average O-H bond
length and H-O-H bond
angle. As expected, the
molecular geometry trends
towards that of the
isolated monomer as
molecules on the liquid
side approach the low
density region near the
vapor. These structural
changes mark the
transition from relatively
strong hydrogen bonding in
the bulk liquid to zero
hydrogen bonding in the
vapor.

      It is most
interesting that the bond
length reaches a maximum
and the bond angle reaches
a minimum at the same
position, just inside the

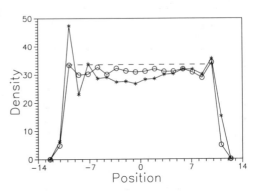

**Fig. 7. H-atom mass density
profile for same system as
Fig. 6.**

interface, where the density profile shows a minimum. This
indicates a greater degree of hydrogen bonding in this
narrow region, i.e. a structure that is more ice-Ih-like.
This result certainly supports the validity of the density
profiles obtained for this water model. Shifts found for
the intramolecular vibrational frequencies in this region
are also consistent with this picture. Thus, all these
MST-FP liquid-vapor profiles indicate a transition from the
vapor side, first to dense II-type structure, where it is
always to be remembered that 4- and 5-membered rings may
play a role, then to a I-type structure before the bulk
liquid properties are attained.   Again, though this
picture seems reasonable from the discussion in Sections 2
and 3, it may not occur or be extremely difficult to detect
in real water.

4.4. Electrically charged surfaces. Reproduced in the above Figures 6 and 7 are some results [24] using the simpler water model, SPC-FP [20], confined between interactive Lennard-Jones (L-J) walls. Clear density oscillations, about which there is little dispute for this type of interface, are evident. Furthermore, the oscillations become more prominent on the application of a field -- in the present case when the left wall is positively charged and the right wall is negatively charged. Compare with the results obtained by Lee *et al.* [32] for Stockmayer particles, where similar results were obtained.

It is interesting that the H-atom density maximum near the positive wall on the left is much higher than it is on the right near the negative wall. This seems backwards, but is merely caused by the H-atoms being attached intramolecularly to O-atoms, so they have no other choice but to follow the O-atoms to the positively charged surface. Note that the H-atom density very near the negative surface does increase about threefold, while the O-atom density near the positive surface increases about sixfold over that when the walls are uncharged.

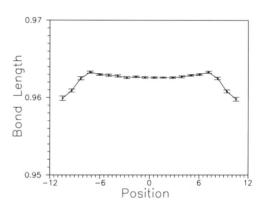

As expected, the bond lengths were found to decrease and the bond angles were found to increase near either the neutral or the charged surfaces, in fact, very similar to the case for the liquid-vapor interface. Interestingly,

Fig. 8. Profile of O-H bond length at MST-FP liquid-vapor interface.

both these intramolecular structural features become more gas-like near the negatively charged surface containing a preponderance of H-atoms, but less gas-like near the positively charged surface where O-atoms abound. This must be a direct result of the effects of polarizability, which resides primarily on the O-atom.

4.5. Relaxation Properties. In order to gain knowledge about the change of dynamic properties in the MST-FP water film, a number of time-dependent autocorrelation functions were investigated. The center-of-mass self-diffusion coefficient transverse to the surface plane is larger than in the bulk. This can be attributed to the softening near the surface of the hydrogen bonding in the region where dense II-type structure is prominent. On the other hand, the longitudinal component of the center-of-mass velocity autocorrelation function shows a faster initial decay followed by a fairly deep well, indicating reflections from the surface, as implicated earlier from the MST-FP density profiles as a function of film thickness.

While translational diffusion is definitely affected by the interfacial perturbations, and in the way expected from the density profiles and the discussion in Sections 2 and 3, the angular velocity autocorrelation functions of the H-H vector and the H-O-H bisector were found not to change much compared with those in the bulk phase. In other words, perturbations from the liquid-vapor interface do not change the rotational velocity properties of the water molecules significantly. This could be caused by the more subtle, longer range effects of interfacial perturbations on orientational properties. On the other hand, reorientations of the H-O-H bisector in the interfacial region, as measured by orientational autocorrelation functions, are slower than the process taking place in the bulk state.

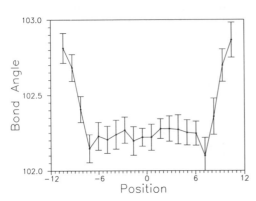

Fig. 9 Profile of H-O-H bond angle at MST-FP liquid-vapor interface.

Dipole correlation functions are also of interest. No significant change was observed for single molecule dipole moment relaxations, whether along or normal to the surface. On the other hand, the longitudinal component of the collective molecular dipole moment relaxation process was found to be accelerated in the interfacial environment, while the transverse components are slowed down. Again, however, one must be aware of possible corruptions of any of the long range orientational effects because of the small ensemble size.

In summary then, some of the relaxation results from current computational studies seem a bit more complicated than expected. Of course, any of these results, either static or dynamic, would be expected to depend on the nature of the water model used, whether polarizable or flexible, and where and how the charges are placed; on the ensemble size, which affects results at long range, particularly orientational data; and on the strength and orientational properties employed for the surface-water interactions. To date, very little systematic variation of any of these properties has been attempted.

5. Concluding Remarks

How close do current computational models of water come to giving realistic descriptions of the static and dynamic effects of water near an interface? This and other questions will continue to be gnawing until they are finally settled by future generations of effort, both in computational and experimental fields of endeavor. Meanwhile, however, information cracks in the "water fortress" are beginning to open, and this can only improve future intuition, suggest future paths to follow, and, most importantly, act as a guide for future interpretations of experimental data.

Acknowledgments

Financial support for this work has been shared by the National Science Foundation (CHE-9112002), the State of Texas Advanced Research Program (1306) and the Robert A. Welch Foundation (D-0005 and D-1094).

References

[1] Fowler, R. H. (1966) Statistical Mechanics, 2nd ed., Cambridge University Press, Cambridge. See page 811.

[2] Lang, E., and Lüdemann, H.-D. (1980) "Pressure and Temperature Dependence of the Longitudinal Deuterium Relaxation Times in Supercooled Heavy Water to 300 MPa and 188 K", Ber. Bunsen-Ges. Phys. Chem. **84**, 462-470.

[3] E. W. Lang, E. W., and Lüdemann, H.-D. (1981) "High Pressure O-17 Longitudinal Relaxation Time Studies in Supercooled H_2O and D_2O", Ber. Bunsen-Ges. Phys. Chem. **85**, 603-611.

[4] Bassez, M.-P., Lee, J., and Robinson, G. W. (1987) "Is Liquid Water Really Anomalous?", J. Phys. Chem. **91**, 5818-5825.

[5] Eisenberg, D., and Kauzmann, W. (1969) The Structure and Properties of Water, The University Press, Oxford.

[6] Larsson, K. E., and Dahlborg, U. (1962) "Some Vibrational Properties of Solid and Liquid H_2O and D_2O Derived from Differential Cross-Section Measurements", J. Nuc. Energy **B16**, 81-89. The authors thank Pergamon Press for permission to reproduce Figure 1.

[7] Walrafen, G. E., Fisher, M. R., Hokmabadi, M. S., and Yang, W.-H. (1986) "Temperature Dependence of the Low- and High-Frequency Raman Scattering from Liquid Water", J. Chem. Phys. **85**, 6970-6982.

[8] Chandler, D. (1987) Introduction to Modern Statistical Mechanics, Oxford University Press, New York.

[9] Reimers, J. R., and Watts, R. O. (1984) "The Structure and Vibrational Spectra of Small Clusters of Water Molecules", Chem. Phys. **85**, 83-112.

[10] Pauling, L. and Hayward, R. (1964) The Architecture of Molecules, W. H. Freeman, San Francisco.

[11] Bosio, L., Chen, S.-H., and Teixeira, J. (1983) "Isochoric Temperature Differential of the X-ray Structure Factor and Structural Rearrangements in Low-Temperature Heavy Water", Phys. Rev. **A27**, 1468-1475. The authors thank Professor S.-H. Chen and the American Institute of Physics for permission to reproduce Figure 2.

[12] Vedamuthu, M., Singh, Surjit, and Robinson, G. W., "Comments on the Properties of Liquid Water. The Density Maximum and the Isothermal Compressibility Minimum", to be published.

[13] Soper, A. K., and Phillips, M. G. (1986) "A New
     Determination of the Structure of Water at 25°C",
     Chem. Phys. 107, 47-60.
[14] Goh, M. C., Hicks, J. M., Kemnitz, K., Pinto, G.
     R., Bhattacharyya, K., Eisenthal, K. B., and
     Heinz, T. F. (1988) "Absolute Orientation of Water
     Molecules at the Neat Water Surface", J. Phys.
     Chem., 92, 5074-5075.
[15] Yang, B., Sullivan, D. E., Tjipto-Margo, B., and
     Gray, C. G. (1991) "Molecular Orientational
     Structure of the Water Liquid/Vapor Interface", J.
     Phys.: Condens. Matter 3, F109-F125.
[16] Du, Q., Superfine, R., Freysz, E., and Shen, Y. R.
     (1993) "Vibrational Spectroscopy of Water at the
     Vapor/Water Interface", Phys. Rev. Lett. 70, 2313-
     2316.
[17] Fillingim, T. G., Zhu, S.-B., Yao, S., Lee, J., and
     Robinson, G. W. (1989) "Chemically Stiff Water:
     Ions, Surfaces, Pores, Bubbles and Biology", Chem.
     Phys. Lett. 161, 444-448.
[18] Water and Ions in Biomolecular Systems (1990)
     D. Vasilescu, J. Jaz, L. Packer, and B. Pullman
     (eds.), Birkhauser Verlag, Basel.
[19] Zhu, S.-B., and Robinson, G. W. (1989) "Molecular
     Dynamics Simulation on Liquid Water with Non-Pair
     Additive Interactions", in L. P. Kartashev and
     S. I. Kartashev (eds.), Proc. 4th Internat. Conf.
     Supercomp., Vol. II, International Supercomputing
     Institute, St. Petersburg FL, pp. 189-197.
[20] Zhu, S.-B., Yao, S., Zhu, J.-B., Singh, Surjit, and
     Robinson, G. W. (1991) "A Flexible/Polarizable
     Simple Point Charge Water Model", J. Phys. Chem.
     95, 6211-6217.
[21] Zhu, S.-B., Singh, Surjit, and Robinson, G. W. (1991)
     "A New Flexible/Polarizable Water Model", J. Chem.
     Phys. 95, 2791-2799.
[22] Stillinger, F. H., and Rahman, A. (1974) "Improved
     Simulation of Liquid Water by Molecular Dynamics",
     J. Chem. Phys. 60, 1545-1557.
[23] Zhu, S.-B., and Robinson, G. W., "Molecular Dynamics
     Study of an UltraThin Water Film", to be published.
[24] Zhu, S.-B., and Robinson, G. W., (1991) "Structure and
     Dynamics of Liquid Water between Plates", J. Chem.
     Phys. 94, 1403-1410.
[25] Ojamäe, L., Hermansson, K., and Probst, M. (1992) "The
     OH Stretching Frequency in Liquid Water
     Simulations: the Classical Error", Chem. Phys.
     Lett. 191, 500-506.

[26] Verhoeven, J., and Dymanus, A. (1970) "Magnetic
     Properties and Molecular Quadrupole Tensor of the
     Water Molecule by Beam-Maser Zeeman Spectroscopy",
     J. Chem. Phys. 52, 3222-3233.
[27] Zhu, S.-B., Singh, Surjit, and Robinson, G. W. (1993)
     "Field Perturbed Water", Adv. Chem. Phys. LXXXV(3),
     in press.
[28] Townsend, R. M., and Rice, S. A. (1991) "Molecular
     Dynamics Studies of the Liquid-Vapor Interface of
     Water", J. Chem. Phys. 94, 2207-2218.
[29] Wilson, M. A., Pohorille, A., and Pratt, L. R. (1987)
     "Molecular Dynamics of the Water Liquid-Vapor
     Interface", J. Phys. Chem. 91, 4873-4878.
[30] Frenkel, J. (1955) Kinetic Theory of Liquids, Dover,
     New York.
[31] Nijmeijer, M. J. P., Bruin, C., van Woerkom, A. B.,
     Bakker, A. F., and van Leeuwen, J. M. J. (1992)
     "Molecular Dynamics of the Surface Tension of a
     Drop", J. Chem. Phys. 96, 565-576.
[32] Lee, S. H., Rasaiah, J. C., and Hubbard, J. B. (1986)
     "Molecular Dynamics Study of a Dipolar Fluid
     between Charged Plates", J. Chem.Phys. 85, 5232-
     5237.

---

* Dr. Sheng-Bai Zhu's current address is, IBM Almaden
  Research Center, 650 Harry Road, San Jose CA,
  95120-6099, USA.

SOLVENT MEAN FORCE PERTURBATIONS OF MOLECULAR VIBRATION, ISOMERIZATION AND DISSOCIATION

DOR BEN-AMOTZ and LUÍS E. S. DE SOUZA
Purdue University, Department of Chemistry
West Lafayette, IN 47907-1393

1. Introduction

The mean force potential represents the effect of solvent-solute interactions on solvation thermodynamics, and thus on solute chemical potentials and equilibrium constants. Chemical reaction dynamics, on the other hand, may involve additional non-equilibrium contributions to the solvation of short-lived intermediates. Nevertheless, the solvent mean force potential places significant constraints on reaction dynamics as well as thermodynamics by defining the equilibrium structure of the entire reactive potential surface. Perturbed hard sphere fluid theories [1-3], which make optimal use of analytical statistical mechanical expressions for the thermodynamic properties of hard sphere fluids in predicting the properties of real liquids, offer an appealing formalism for modeling such effects [4-6].

In this work we illustrate the capabilities of a particularly simple perturbed "hard fluid" theory [7] in describing solvent effects on chemical processes. Recent studies of pressure and temperature induced vibrational frequency shifts of diatomic [7,8] and polyatomic [9] solutes dissolved in wide variety of solvents suggest the predictive utility of such an approach. Here we extend this to treat solvent effects on diatomic dissociation and polyatomic isomerization processes, and compare the results with available computer simulations and experimental measurements.

2. Perturbed Hard Fluid Theory

The perturbed hard fluid theory separates the solvent contributions to molecular vibration, dissociation and isomerization into repulsive (hard sphere reference system) and attractive (perturbation) contributions. Excess chemical potential changes, $\Delta\mu$, (relative to those in an ideal gas solvent at the same density and temperature) are thus separated into repulsive, $\Delta\mu_0$, and attractive, $\Delta\mu_a$, contributions.

$$\Delta\mu = \Delta\mu_0 + \Delta\mu_a = \Delta G \qquad (1)$$

The last equality which relates the total chemical potential to the corresponding excess partial molar Gibb's free energy change, ΔG, holds for any process carried out at constant pressure P and temperature T.

J. Jortner et al. (eds.), Reaction Dynamics in Clusters and Condensed Phases, 441–460.
© 1994 Kluwer Academic Publishers. Printed in the Netherlands.

The repulsive excess chemical potential change, $\Delta\mu_0$, is calculated by modeling the solvent as a hard sphere fluid and the reactant and product solute species as appropriately chosen hard spheres or hard diatomics. The corresponding chemical potential change is related to the hard sphere two-cavity distribution function, $y(r)$, as a function of cavity separation, r, solvent density, composition and temperature. The logarithm of this distribution function is proportional to the chemical potential change associated with bringing two hard spheres together from infinite separation to a separation r, in a hard sphere fluid (where k is Boltzmann's constant).

$$\Delta\mu_0^{assoc} = - kT \ln y(r) = \Delta G_0^{assoc} \tag{2}$$

The hard fluid model offers an analytical expression for the $y(r)$, whose form is derived from exact results in the continuum and dilute gas solvent limits [7,9].

$$\ln\{y(r)\} = A + B\,r + C\,r^3 + D\,(1/r) \tag{3}$$

The coefficients, A, B, C, D, in this expression are explicit functions of the solvent component diameters and partial densities (see appendix 5.2. for details). The results are exact at low density and have been shown to very accurately represent the mean force potential in dense hard sphere fluids [6,7].

Repulsive contributions to molecular vibration, isomerization and dissociation reactions are approximated using the above cavity distribution function by associating the process of interest with that involving appropriately chosen diatomics dissolved in a hard sphere fluid. In particular, solvent effects on vibrational potentials are determined from derivatives of $y(r)$ with respect to cavity separation (bond length) while isomerization reactions are modeled as a larger amplitude bond length displacement of an appropriately chosen pseudo-diatomic, and dissociation reactions are represented by allowing the product bond length to go to infinity (see below 2.1.-2.3. for details).

Long range attractive contributions to the above processes are modeled using the van der Waals mean field approximation. This well know approximation [3,10] implies a linear dependence of the attractive excess chemical potential change on solvent density.

$$\Delta\mu_a = C_a\,\rho = \Delta G_a \tag{4}$$

where ρ is the density of the solvent and C_a is a coefficient representing the change in attractive solvation energy in the process of interest. This approximation amounts to neglecting any entropy changes associated with the long range attractive force interactions and assuming that the distribution of the solvent about the solute is density independent. The first assumption may be justified by noting that at high density the structure, and therefore the entropy, of liquids is dominated by repulsive excluded volume interactions. The second assumption is exact for sufficiently long range solute-solvent interactions since these probe the average solvent density on a length scale that is long compared to structural correlations in the liquid. Equation 4 thus follows from the approximate proportionality between the local solvent density around a solute and the bulk solvent density.

Equation 1 can be used to derive other thermodynamic functions such as the excess partial molar entropy, ΔS, enthalpy, ΔH, and volume, ΔV [6,21], changes for a given chemical process using the following standard thermodynamic relations.

$$\Delta S = - \left(\frac{\partial \Delta G}{\partial T}\right)_P \qquad (5)$$

$$\Delta H = \Delta G + T \Delta S \qquad (6)$$

$$\Delta V = \left(\frac{\partial \Delta G}{\partial P}\right)_T \qquad (7)$$

In order to make practical use of the above expressions, however, the parameters representing repulsive and attractive solvent-solute interactions must be determined. The key repulsive force parameters are the diameters of the solute atoms (or pseudo-atoms) and the effective hard sphere diameters of the solvent molecules. Of these, the latter is more critical than the former, as a few percent change in solvent diameter typically has about the same effect on the thermodynamics of reaction as a ten percent change in solute diameter. Previous studies indicate that reliable effective solvent hard sphere diameters can be derived from the compressibility of polyatomic liquids [11,12]. Solute pseudo-atom diameters, on the other hand, may either be inferred from the corresponding molecular hard sphere diameters or derived from tables of atomic and molecular "van der Waals" volume increments [13].

In this work the attractive mean field coefficient, C_a, is treated as the one independently adjustable parameter in the theory. This parameter is determined using a single experimental measurement of one of the thermodynamic functions in equations 1 and 5-7. Furthermore, since C_a is expected to be roughly temperature independent, it may also be used in variable temperature calculations (see below 2.1.). In this case, however, the slight, but typically not insignificant, temperature dependence of the solvent and solute effective hard sphere diameters must be taken into account (see appendix 5.1. for details).

2.1. VIBRATIONAL FREQUENCY SHIFTS

The influence of the solvent potential of mean force on molecular vibrational frequencies is calculated by making use of the following quantum mechanical perturbation theory result, first derived by Buckingham, for the gas to liquid frequency shift of an anharmonic oscillator [14]

$$\frac{\Delta \nu}{\nu_e} = - \frac{3}{2}\frac{g}{f^2} F + \frac{1}{f} G \qquad (8)$$

where ν_e is the harmonic frequency of the isolated solute, the coefficients f and g are the harmonic and anharmonic force constants of the isolated solute with a vibrational potential V and equilibrium bond length r_e,

$$V(r - r_e) = \frac{1}{2} f (r - r_e)^2 + \frac{1}{2} g (r - r_e)^3 + \dots \qquad (9)$$

while F and G are averages over all solvent configurations of first and second derivatives of the total solute excess internal energy, U.

$$F = \left\langle \frac{\partial U}{\partial r} \right\rangle \tag{10}$$

$$G = \frac{1}{2} \left\langle \frac{\partial^2 U}{\partial r^2} \right\rangle \tag{11}$$

An expression similar to equation 8 has been obtained for a Morse oscillator [15]. This gives essentially identical results except for very anharmonic (e.g. hydrogen) vibrations [16]. The above frequency shift expression does not include the effects of centrifugal forces, which contribute an additional shift to the average vibrational frequency of a motionally narrowed Raman or IR band in solution. The centrifugal contribution to the observed shift may readily be estimated from the parameters of the isolated solute [8,9].

In the perturbed hard sphere fluid theory the total frequency shift (excluding centrifugal effects) is separated into repulsive and attractive contributions

$$\Delta \nu = \Delta \nu_0 + \Delta \nu_a \tag{12}$$

where

$$\frac{\Delta \nu_0}{\nu_e} = -\frac{3}{2} \frac{g}{f^2} F_0 + \frac{1}{f} G_0 \tag{13}$$

The hard fluid cavity distribution function is used to calculate repulsive force coefficients, F_0 and G_0 (see appendix 5.2. and 5.3.1. for details) and the attractive frequency shift is taken to have a linear density dependence.

$$\Delta \nu_a = C_a^{vib} \rho \tag{14}$$

The attractive force coefficient, C_a^{vib}, thus represents the combined influence of the attractive mean field on F_a and G_a.

Figure 1 illustrates a comparison of the predictions of this theory with experimental results for the vibrational frequency shift of pure fluid nitrogen (N_2) as a function of density at three different temperatures. These experimental results were obtained using Raman scattering at variable pressure and temperature [8,17]. The attractive mean field coefficient, C_a^{vib}, used in generating the theoretical curves is determined by fitting the calculated total frequency shifts to the lowest density experimental frequency shift at 295 K (after centrifugal correction of the experimental shifts). The higher density shift predictions at 295 K, as well as those at 77 K and 935 K are calculated with no further parameter adjustment. In other words, C_a^{vib} is assumed to be temperature and density independent, while the diameter of N_2 is taken to have a temperature dependence consistent with its compressibility [11].

The above analysis assumes that both the diameter and centrifugal potential of N_2 are density independent. Previous studies of N_2 indicate that at higher densities than those in figure 1 this is no longer be a reasonable assumption [8]. On the other hand, the good agreement between the experimental and theoretical frequency shifts at 295 K and 935 K suggests that such density dependent effects may reasonably be neglected even at relatively high liquid densities (note that the 77 K experimental point corresponds to normal liquid N_2). The discrepancy between the theoretical and experimental shift at 77 K is perhaps not surprising since at such low temperatures the system is expected to deviate more significantly from one composed of hard spheres with mean field attractive interactions. The results in figure 1 suggest that at high temperatures *the attractive mean field coefficient may be taken to be an approximately density and temperature independent constant.* This has significant practical consequences since it allows the calculation of the temperature dependence of solvent induced frequency shifts (and other thermodynamic properties) using an attractive mean field coefficient determined from experimental data at a single temperature.

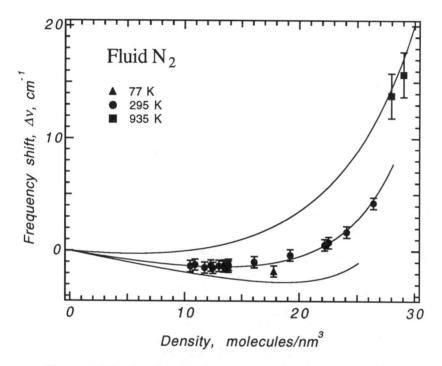

Figure 1. Vibrational frequency shifts in pure supercritical fluid nitrogen. The experimental shifts relative to the Q-branch origin (rotation free) in the dilute vapor phase are corrected for centrifugal stretching forces at the experimental temperatures [8,17].

Figure 2 reveals another general relationship of potential practical importance. This is illustrated by the linear correlation between the attractive mean field coefficients obtained from the frequency shifts of N_2 dissolved in a wide variety of solvents and the polarizability, α_s, of the the corresponding solvents. Such a linear correlation is consistent with the assumption that attractive solvation forces induced principally by London dispersion interactions [8]. A similar predominance of dispersion interactions has been found in polyatomic systems containing dipolar solvent and/or solute molecules [9]. These results suggest that *solvent dependent changes in attractive mean field solvation energies may be estimated using experimental data in a single solvent.*

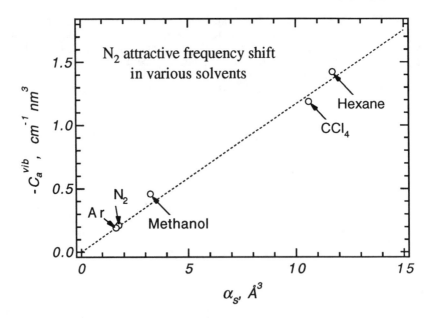

Figure 2. Linear correlation between the mean field attractive frequency shift coefficient, C_a^{vib}, and the polarizability of the solvent, α_s.

2.2. REACTION FREE ENERGY AND VOLUME CHANGES

Diatomic dissociation reactions in spherical solvents offer the simplest tests of solvent effects on chemical reactions. Although we have not been able to find any direct experimental measurements of the excess (solvent contribution) to diatomic dissociation thermodynamics, reported computer simulation measurements may be used to extract such information [18,19,20]. These can in fact yield information not only on the dependence of dissociation equilibrium constants on density and temperature but also on the magnitude of solute-solvent interactions in the bound and dissociated states.

In particular, Lennard-Jones (LJ) fluid simulation measurements have been performed to obtain the change in excess chemical potential (or partial molar Gibbs free energy) for the dissociation of a diatomic solute composed of atoms whose interactions with the solvent are characterized by the same LJ parameters, σ_{LJ} and ε_{LJ}, as those of the pure solvent [18,19,20]. Furthermore, simulation results have also been obtained for the chemical potential of LJ spheres dissolved in LJ fluids, in which the solute-solvent and solvent-solvent interaction parameters are not the same [19]. Thus, for example, results for spheres of the same size as the solvent but with a deeper interaction well depth (larger ε_{LJ}) can be used to evaluate the effects of increasing solute-solvent coupling in the dissociated state on the excess free energy of dissociation.

Repulsive contributions to the thermodynamics of dissociation are modeled using a system consisting of a homonuclear diatomic solute (cavity pair) whose bond length goes from a finite value such as, $r_0 = 0.5 \, \sigma_{LJ}$ (σ_{LJ} is the solvent LJ diameter), to infinity upon dissociation. Thus the repulsive excess Gibb's free energy change is simply related to the value of the cavity distribution function at $r = r_0$.

$$\Delta G_0^{diss} = kT \ln y(r_0) \qquad (15)$$

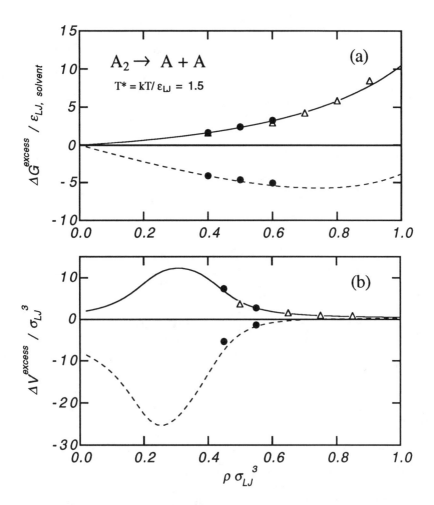

Figure 3. Comparison of theoretical and computer simulation [18,19,20] results for the excess free energy (a) and reaction volume (b) change for two diatomic dissociation reactions. The solid curves pertain to a reaction in which no change in solvent-solute coupling, ε_{LJ}, occurs upon dissociation, while the dashed curves pertain to a reaction with a 50% increase in ε_{LJ} in the dissociated state. In both cases the diatomic bond length is $r_0 = 0.5 \, \sigma_{LJ}$.

Attractive contributions to the dissociation free energy are, in accordance with the mean field approximation, taken to be proportional to the solvent density (see appendix 5.3.2 for details).

$$\Delta G_a^{diss} = C_a^{diss} \, \rho \qquad\qquad (16)$$

Figure 3a shows simulation (points) and theoretical (curves) results for the density dependence of the excess free energy of dissociation. The solid curve and associated simulation points [18,19,20] represent results for a system in which the solute atom-solvent atom coupling is the same in the bound and dissociated states. The dashed curve and associated simulation points [19,20] represent results for a system in which the dissociated atoms have a 50% larger attractive solute atom-solvent atom well depth, ε_{LJ}, than the bound atoms. In other words the solvent is more strongly attracted to the product species. A change in attractive coupling of this order is comparable to that expected for the dissociation of simple diatomics such as Br_2 dissolved in argon [21]. Clearly such a change in solvent-solute coupling is sufficient to produce significant changes in the thermodynamics of reaction.

In each of the above dissociation reactions the only adjustable parameter used in generating the perturbed hard fluid curves is again the attractive mean field coefficient, C_a^{diss}. In this case C_a^{diss} is determined by constraining the theoretical ΔG curves to go through the lowest density simulation point. The good agreement between the predicted and simulated ΔG values at higher densities thus represents a significant confirmation of the perturbed hard fluid theory predictions.

The striking difference between the excess free energies for the two dissociation reactions reflects the delicate balance of repulsive and attractive solvation forces. For the reaction with no change in the solvent-solute coupling upon dissociation the excess free energy is positive at all densities. This reflects the fact that work must be performed by the dissociating solute, against the predominantly repulsive solvent forces. On the other hand, in the reaction with a 50 % increase in solvent-solute coupling upon dissociation the excess free energy is negative at all densities. In this case the increased attractive solvation energy in the dissociated state is sufficient to overcome the repulsive work required to separate the solute atoms.

Figure 3b illustrates even more strikingly the sensitivity of solvation thermodynamics to the delicate balance of repulsive and attractive forces. This figure displays the changes in excess reaction volume as a function of solvent density [6,21] for the same two dissociation reactions. The theoretical curves are in this case obtained using the same attractive mean field coefficients (and other parameters) used in the ΔG calculations. The most striking features of these results are the positive and negative maxima in ΔV for the two reactions. Similar large excursions in the partial molar volumes and reaction volumes of solutes have been noted in previous experimental and theoretical studies [22-26]. In particular, Debbenedetti and coworkers [25,26] have extensively investigated and discussed such effects in terms of the cluster breaking (repulsive) or cluster forming (attractive) behaviors of near critical fluid systems.

Qualitatively the change in sign of ΔV reflects the fact that the total volume of the system may either increase or decrease in the course of the dissociation reaction, depending on the balance of repulsive and attractive forces. For the reaction with no increase in ε_{LJ} the volume of the system increases upon dissociation, reflecting the larger volume and increased repulsion in the dissociated state. On the other hand, for the reaction in which ε_{LJ} increases by 50% upon dissociation the solvent tends to collapse around the product atoms, leading to a decrease in the system volume upon dissociation.

The extrema in ΔV occur when the solvent is most compressible, that is near its critical point, $\rho_c{}^* \approx 0.30$ and $T_c{}^* \approx 1.32$. In this region the solvent is most readily able to accommodate small changes in solvent-solute repulsion or attraction by adjusting its packing configuration around the solute. At higher density the solvent is less compressible and therefore less able to accommodate its structure to the solute. In fact at liquid densities the solvent becomes so incompressible that its structure is dominated by repulsive packing forces, thus obviating any possibility of significant contraction around the dissociated solute even when the solvent-solute coupling increases significantly upon dissociation. Thus at liquid densities the reaction volume is always positive, and similar in magnitude to the small intrinsic volume change of the solute upon dissociation [6]. The detailed agreement between the predicted curves and the simulation results for ΔV again confirms the predictive utility of the perturbed hard fluid theory, particularly in view of the fact that no reaction volume data has been used to constrain the theoretical results.

2.3. ISOMERIZATION ENTHALPY AND VOLUME CHANGES

Isomerization can be viewed as a model chemical reaction in which the size, shape and polarity (or polarizability) of the solute changes upon reaction. These changes may be expected to produce corresponding changes in both repulsive and attractive solvent-solute interaction energies. Repulsive interactions are again modeled using an appropriate hard sphere reference fluid containing solutes representing the reactant and product species and attractive interactions are represented by a linearly density dependent attractive mean field.

In particular, we model the gauche-trans isomerization of 1,2-disubstituted ethanes as pseudo-diatomic processes in which the separation between the pseudo-atoms representing the two ends of the solute increases in the transformation from the more compact gauche (syn) to the more extended trans (anti) structure (see appendix for details). A similar model has previously been applied by Pratt, Hsu and Chandler [4] to the gauche-trans isomerization of n-butane, although in that work attractive solvation force changes were entirely neglected. The predicted reaction volume, $\Delta V \approx 2$ cm³/mole, was found to be a factor of two larger than the experimentally measured value, $\Delta V \approx 1$ cm³/molc [27] whereas the predicted excess enthalpy change, $\Delta H \approx 0.4$ kJ/mole was significantly lower than most experimentally reported values, 0.5 kJ/mole $\leq \Delta H_{expt} \leq 2.2$ kJ/mole [27-33]. However, the significant variation in experimental enthalpies (particularly the isolated molecule ΔH reference value) suggests the difficulty of accurately measuring such small enthalpy changes.

In this work we illustrate the application of the perturbed hard fluid model to the gauche-trans isomerization of 1,2-dichloroethane and 1,2-dibromoethane in the pure liquid state. In each of these systems the solute and solvent molecules are the same (although the latter are treated as spheres while the former are treated as pseudo-diatomics). The repulsive excess Gibb's free energy change for the transformation from the gauche to the trans isomers (carried out at constant temperature and pressure) is,

$$\Delta G_0^{isom} = kT \ln\left[\frac{y(r_{gauche})}{y(r_{trans})}\right] \qquad (17)$$

In order to evaluate this expression the diameter of the solvent molecules is estimated using hard sphere volume increments, which are in turn derived from the analysis of the compressibility of a wide variety of similar compounds [12]. The solute pseudo-atom diameters are derived from van der Waals volume increment tables and the "bond lengths" of the pseudo-diatomics in the gauche and trans states are determined geometrically (see appendix 5.3.3. for details) [13]. In order to calculate the attractive isomerization free energy the attractive mean field coefficients, C_a^{isom}, are determined using experimental values for ΔH at ambient temperature and pressure.

$$\Delta H_a^{isom} = \Delta G_a^{isom} + T\Delta S_a^{isom} = C_a^{isom}\left[\rho - T\left(\frac{\partial\rho}{\partial T}\right)_P\right] \tag{18}$$

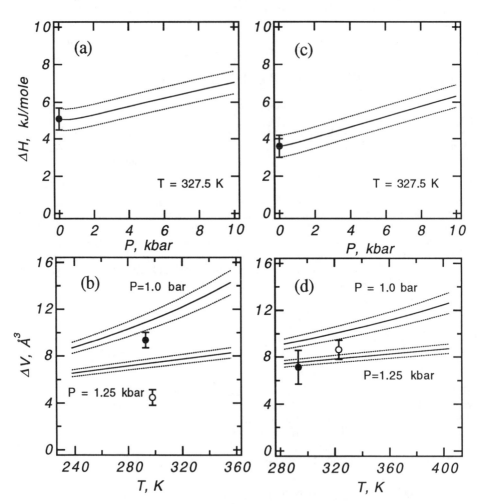

Figure 4. Theoretical and experimental [34-36] results for the enthalpy (a and c) and volume (b and d) change for the gauche-trans isomerization of 1,2-dichloroethane (a and b) and 1,2-dibromoethane (b and c) as a function of pressure and temperature.

Figure 4a shows the predicted excess enthalpy of reaction as a function of pressure for 1,2–dichloroethane. Note that although the theoretical curve is forced to go through the one experimental data point [34] (by fixing C_a^{isom}), the high pressure behavior of ΔH offers a significant, although untested, prediction of the theory. The dashed lines in the figure reflect the uncertainty derived from the reported experimental precision of the 1 bar ΔH value. The theoretically predicted increase in ΔH with increasing pressure arises from the additional repulsive work which the solute must perform in order to increase the system volume upon isomerization. On the other hand, the absolute magnitude of ΔH at 1 bar contains nearly equal attractive and repulsive contributions.

Figure 4b shows the predicted reaction volume as a function of temperature and pressure for 1,2-dichloroethane. Notice that in this case no further parameter adjustment has been performed and that the quantitative agreement between the predicted and experimental [34,35] reaction volumes, although not perfect, is significantly better than it was in the earlier n-butane calculations [4].

The predicted increase in ΔV with increasing temperature arises from the thermal expansion of the system, and thus the larger partial molar volumes of the gauche and trans isomers at higher temperature. The predicted decrease in the reaction volume with increasing pressure is more interesting since such a decrease has often been observed in experimental reaction and activation volume measurements [37]. Although such pressure dependent volume changes have often been presented as quite mysterious, they are again simply a manifestation of the reduced partial molar volumes of reactants and products at high pressure [6].

The experimental points and predicted curves in figures 4c and 4d represent the corresponding results for 1,2-dibromoethane. The agreement between the experimental and predicted ΔV values is again not prefect, although the discrepancies are in this case essentially within experimental error [34,36]. The general similarity of the ΔH and ΔV predictions for the two compounds when plotted as as a function of pressure and temperature belie the significant molecular differences between these two compounds. Attempts to offer a simple interpretation of the small differences between the two compounds is frustrated by the fact that both the solute and solvent species are changing in going from one system to the other, as well as by the fact that both the solvent and solute may exist in two conformational isomers. Additional experimental studies of these two solutes dissolved in simpler spherical (e.g. rare gas) solvents would offer more significant further tests of the perturbed hard fluid theory, and a deeper insight into the effects of solvation on molecular potential energy surfaces and chemical reactivities.

3. Conclusions

Several general conclusions may be drawn from the results of these studies. Most importantly, the perturbed hard fluid theory, which treats the effects of solvation on diatomic (and pseudo-diatomic) processes using a hard sphere cavity distribution function plus a van der Waals attractive mean field, has proved to be remarkably successful. This suggests that complex repulsive packing forces in dense fluid systems may reasonably be approximated using appropriately chosen hard sphere reference fluids. Furthermore, the diameter of the solvent reference hard sphere fluid, which is required in order to carry out any practical calculations, may be conveniently determined from the compressibility of the pure solvent (see appendix 5.).

The magnitude of the attractive mean field coefficient, C_a, for a given chemical process has been treated as the only independently adjustable parameter in this theory. This coefficient is determined using a single experimental data point. The magnitude of C_a for different processes determines the degree of competition between attractive and repulsive interactions in solution. In general, repulsive forces tend to dominate at high densities and temperatures. This arises naturally in the perturbed hard fluid theory, since repulsive solvation free energies increase strongly (non-linearly) with density and nearly linearly with temperature, while attractive solvation free energies are linear in density and nearly temperature independent.

The magnitude of C_a for various processes reveals further interesting molecular details of solvation. In general, attractive solvation forces can only compete effectively with repulsion at low density, and then only if there is a significant change in solvent-solute attraction in the process of interest. This is illustrated clearly by the dissociation results in section 2.2., as well as by frequency shift studies in model fluid systems [38,39]. In both cases a significant change in polarizability (LJ well depth, ε_{LJ}) is required in order to produce attractive force dominated solvation behavior ($\Delta G < 0$ or red frequency shift). For isomerization reactions, on the other hand, both attractive and repulsive solvation forces make a positive contribution to the excess free energy of solvation at all densities. The sign of the attractive contribution is consistent with an enhanced attractive solvation energy for the more dipolar gauche state, while that of the repulsive contribution derives from the larger volume excluded to the solvent by the more extended trans confromation.

The proportionally smaller contribution which attractive solvation makes to isomerization as opposed to vibration or dissociation processes conforms with the expectation that conformational changes do not perturb the electronic structure of a solute as significantly as bond stretching or breaking process. In other words, attractive solvation forces may be expected to play a greater role in bond stretching and dissociation because electrons gain degrees of freedom (polarizability) in these processes, while conformational processes nominally do not involve any change in solute atom covalent bond distances and therefore little or no change in polarizability, although the overall dipole (or multipole) moment of a solute may depend on its conformational state.

On the other hand, dipolar interactions are often found to contribute relatively weakly to solvation thermodynamics, even in very polar systems. This is exemplified both by recent frequency shift studies of acetonitrile, CH_3CN [9], and by the isomerization results for 1,2–dicholoroethane and 1,2–dibromoethane (which have a significant dipole moments in the gauche state and virtually no dipole moment in the trans state). Even for 1,2–dichloroethane, which has the largest C_a^{isom}, the calculated attractive contribution to the free energy of solvation at low density is less than twice that of repulsion (at 327.5 K). This should be contrasted with the frequency shift of N_2, for example, for which attraction contributes about 4 times more than repulsion to the low density frequency shift (at 295 K).

4. Acknowledgements

Support for this work from the United States Office of Naval Research (N00014-92-1559), the National Science Foundation (CHE-9157535) and the Exxon Education Foundation are gratefully acknowledged. We would also like to thank Fred LaPlant, George Devendorf, Nick Miklusak and Willie Nichols for assisting in the collection of some of the N_2 frequency shift data reported in this work.

5. Appendix

5.1. PERTURBED HARD SPHERE EQUATION OF STATE

The perturbed hard sphere Carnahan-Starling-van der Waals (CS-vdW) equation of state [11] is used to describe the PVT properties of atomic and molecular single component fluids. The values of the hard sphere diameter, σ, and attractive mean field coefficient, τ, in this equation have previously been determined from the compressibility of a wide variety of compounds [11] and more recently extended using a sub-group increment analysis, to include countless other compounds [12].

The density and temperature derivatives of the predicted pressures may be expressed analytically in terms of the packing fraction $\eta = \pi/6\,\rho\,\sigma^3$, density, ρ, and temperature, T, of the fluid.

$$\left(\frac{\partial P}{\partial \rho}\right)_T = kT\left[\frac{1 + 4\eta + 4\eta^2 - 4\eta^3 + \eta^4}{(1-\eta)^4} - 8\eta\,\frac{\tau}{T}\right] \tag{A.1}$$

$$\left(\frac{\partial P}{\partial T}\right)_\rho = k\rho\left\{\frac{2\eta(2-\eta)}{(1-\eta)^3} - 4\eta\frac{\partial\tau}{\partial T} + \frac{6\eta}{\sigma}\left(\frac{\partial\sigma}{\partial T}\right)\left[\frac{(2+2\eta-\eta^2)T}{(1-\eta)^4} - 2\tau\right] + 1\right\} \tag{A. 2}$$

The ratio of these derivatives determines the derivative of the density with respect to temperature at constant pressure.

$$\left(\frac{\partial \rho}{\partial T}\right)_P = -\left(\frac{\partial P}{\partial T}\right)_\rho \bigg/ \left(\frac{\partial P}{\partial \rho}\right)_T \tag{A.3}$$

The molecular hard sphere diameter, σ, and attractive force coefficient, τ, as well as their temperature derivatives, are determined from the analysis of compressibility data for the corresponding fluid [11,12]. For the special case a spherical LJ fluid with diameter, σ_{LJ}, and well depth, ε_{LJ}, the temperature dependence of the hard sphere diameter can be expressed as [11],

$$\sigma = 1.1532\,\sigma_{LJ}\left[1 + 1.3775\,(kT/\varepsilon_{LJ})^{1/2}\right]^{-1/6} \tag{A.4}$$

and the equation of state of a LJ fluid is obtained from smoothed computer simulation data [40].

5.2. REPULSIVE SOLVENT MEAN FORCE CALCULATIONS

General expressions for the hard sphere two-cavity distribution function coefficients A, B, C and D (in equation 3) for a system with arbitrary solute cavity diameters, solvent diameters, composition and density are presented in previous studies [7,9]. In this work all of the repulsive solvation mean force calculations pertain to the special case of a solute modeled as a homonuclear cavity pair, with a cavity diameter of σ_0, dissolved in a single component hard sphere solvent of diameter σ and number density ρ (corresponding to a packing fraction of $\eta = \pi/6\,\rho\,\sigma^3$, and a diameter ratio $d = \sigma_0/\sigma$). In this case the cavity distribution function parameters are [8],

$$A = \ln[y(0)] \tag{A.5}$$

$$B = \left[\frac{\partial \ln y(r)}{\partial r}\right]_{r=0} \tag{A.6}$$

$$C = \frac{1}{\sigma_0^3}\{\ln[y(\sigma_0)] - \ln[y(0)] - B\sigma_0\} \tag{A.7}$$

$$D = 0 \tag{A.8}$$

Where the value of y(r) and its derivative at zero cavity separation, r = 0, are,

$$\ln y(0) = d\,\eta\left[\frac{(-3d^2+3\,d+3)\,\eta^2 + (6\,d^2-9d-6)\eta - d^2+6\,d+3}{(1-\eta)^3}\right] + (2\,d^3-3d^2+1)\,\ln\left[\frac{1}{(1-\eta)}\right] \tag{A.9}$$

$$\left[\frac{\partial \ln y(r)}{\partial r}\right]_{r=0} = -\frac{3\eta}{2\sigma}\frac{\left[1+2d+d^2+(-2-d+d^2)\eta+(1-d)\eta^2\right]}{(1-\eta)^3} \tag{A.10}$$

and the value of the the distribution function at contact separation, r = σ_0 is,

$$y(\sigma_0) = \frac{1 + \left(\frac{3}{2}d-2\right)\eta + \left(\frac{1}{2}d^2-\frac{3}{2}d+1\right)\eta^2}{(1-\eta)^3} \tag{A.11}$$

The expressions above may be used to calculate the excess ΔG_0 for vibration, dissociation and isomerization (as described in section 2.). The other repulsive thermodynamic functions are obtained using

$$\Delta S_0 = -\left(\frac{\partial \Delta G_0}{\partial T}\right)_P = -\left\{k\ln\left[\frac{y(r_{reactant})}{y(r_{product})}\right] + kT\left(\frac{\partial\{\ln[y(r_{reactant})/y(r_{product})]\}}{\partial T}\right)_P\right\} \tag{A.12}$$

and

$$\Delta V_0 = \left(\frac{\partial \Delta G_0}{\partial P}\right)_T = -kT\left(\frac{\partial\{\ln[y(r_{reactant})/y(r_{product})]\}}{\partial\rho}\right)_T\left(\frac{\partial\rho}{\partial P}\right)_T \tag{A.13}$$

Notice that the repulsive y(r) depends on temperature even at constant density since σ depends on T [11].

Excess attractive free energy changes, ΔG_a, are calculated from equation 4, while other attractive thermodynamic properties are obtained using

$$\Delta S_a = -\left(\frac{\partial \Delta G_a}{\partial T}\right)_P = -C_a^{rxn}\left(\frac{\partial \rho}{\partial T}\right)_P \qquad (A.14)$$

and

$$\Delta V_a = \left(\frac{\partial \Delta G_a}{\partial P}\right)_T = C_a^{rxn}\left(\frac{\partial \rho}{\partial P}\right)_T \qquad (A.15)$$

5.3. APPLICATIONS

5.3.1. *Frequency shift calculations*

Calculation of the repulsive contributions to vibrational frequency shifts requires evaluation of the hard sphere solvent configuration averaged first and second solute bond length derivatives of the solvent-solute interaction energy, F_0 and G_0, respectively.

$$F_0 = -kT\left(\frac{\partial \ln y(r)}{\partial r}\right)_{r_e} = -kT\left[B + 3C\,r_e^2 - D\left(\frac{1}{r_e^2}\right)\right] \qquad (A.16)$$

$$G_0 \approx -\frac{kT}{2}\left(\frac{\partial^2 \ln y(r)}{\partial r^2}\right)_{r_e} = -3\,kT\,C\,r_e \qquad (A.17)$$

The above expression for G_0 is not exact but represents a lower bound to the true G_0 [7]. The linear dependence of Δv_a on density follows from the mean field approximation, since within this approximation the attractive excess chemical potential of the solute is equal to its attractive excess internal energy and this is in turn is assumed to be proportional to the solvent density, both F_a and G_a and therefore Δv_a must also be proportional to density.

Table 1. Vibrational attractive mean field coefficients.

Solute	Solvent	C_a^{vib} (cm^{-1} ·nm^3)
Nitrogen	Nitrogen	-0.21
Nitrogen	Argon	-0.19
Nitrogen	Methanol	-0.46
Nitrogen	Carbontetrachloride	-1.18
Nitrogen	n-hexane	-1.42

The solvent and solute hard sphere diameters used in the frequency shift calculations are derived from the analysis of compressibility data [11], and the nitrogen atom diameters are fixed by equating the volume of the diatomic with that of the effective hard sphere representing N_2. The internal (and centrifugal) parameters of N_2 are the same as those used in previous studies [8]. The attractive mean field coefficients derived from the experimental gas to liquid (or dense fluid) frequency shifts of N_2 dissolved in various solvents are given in table 1.

5.3.2. *Dissociation calculations*

In the calculation of $y(r_0)$, the solvent hard sphere diameter σ in a LJ fluid is given by equation A.4 [11]. The solute hard sphere diameter is assumed to be equal to that of the solvent, regardless of the value of the solute-solvent coupling strength, ε_{LJ}. The attractive mean field coefficients derived from the low density simulation measurements of ΔG are given in table 2.

Table 2. Dissociation attractive mean field coefficients.

Reaction	$\Delta\varepsilon_{LJ}$	$C_a^{diss} / (\varepsilon_{LJ}\,\sigma^3_{LJ})$
$A_2 \rightarrow A + A$	0%	-1.1
$A_2 \rightarrow A + A$	50%	-15.4

5.3.3. *Isomerization calculations*

The separation between the pseudo-atoms representing the two -CH_2A groups (where A = Cl or Br) on either end of the molecule is taken as that between the "center of volume" of each pseudo-atom, defined as the center of mass for a group of atoms, where each atom is assumed to have a mass proportional to its van der Waals volume. The atomic van der Waals volumes and bond lengths used in the calculations are listed in table 3. We assumed that all bond angles are equal to 109.5 ° .

Table 3. Atomic volumes and bond lengths

van der Waals volume [13]	bond length [41]
$H = 7.24$ Å^3	C-H = 1.101 Å
$C = 15.00$ Å^3	C-C = 1.541 Å
$Cl = 22.45$ Å^3	C-Cl = 1.767 Å
$Br = 26.52$ Å^3	C-Br = 1.937 Å

The resulting pseudo-atom cavity separations in the gauche, r_{gauche}, and trans, r_{trans}, states, along with the estimated pseudo-atom diameters at 20 °C, $\sigma_{0,20}$ [12] are given in table 4.

Table 4. Solute pseudo-atoms data

Solute	r_{gauche}	r_{trans}	$\sigma_{0,20}$
1,2-dichloroethane	2.328 Å	2.553 Å	4.107 Å
1,2-dibromoethane	2.455 Å	2.762 Å	4.275 Å

The hard sphere diameter of the solvents and their temperature derivatives at 20°C (σ_{20} and $[T(\partial\sigma/\partial T)]_{20}$) are listed in table 5 [12], along with other parameters necessary to obtain the equations of state of the corresponding liquid [11,12]. The pseudo-atom diameters are assumed to be proportional (in their temperature dependence) to the solvent hard sphere diameter.

Table 5. Equation of state parameters [11,12].

Molecule	σ_{20}	$[T(\partial\sigma/\partial T)]_{20}$	τ_{20}	$T(\partial\tau/\partial T)]_{20}$
1,2-dichloroethane	5.056 Å	-0.15 Å	1976 K	-354 K
1,2-dibromoethane	5.304 Å	-0.15 Å	2462 K	-550 K

The attractive mean field coefficients derived from the experimental ΔH values shown in figure 4 are given in table 6.

Table 6. Isomerization attractive mean field coefficients.

Molecule	C_a^{isom} (kJ·mole^{-1}·nm^3)
1,2-dichloroethane	+0.290
1,2-dibromoethane	+0.022

References

1. Chandler, D., Weeks, J. D., and Andersen, H. C. (1983) "Van der Waals Picture of Liquids, Solids, and Phase Transformations", Science 220, 787-794

2. Barker, J. A., and Henderson, D. (1976) "What is 'liquid'? Understanding the states of matter", Rev. Mod. Phys. 48, 587-671

3. Hansen, J. P., and McDonald, I. R. (1986) Theory of Simple Liquids, Academic Press, London

4. Pratt, L. R., Hsu, C. S., and Chandler, D. (1978) "Statistical mechanics of small chain molecules in liquids. I. Effects of liquid packing on conformational structures", J. Chem. Phys. 68, 4202-4212

5. Schweizer, K. S., and Chandler, D. (1982) "Vibrational dephasing and frequency shifts of polyatomic molecules in solution", J. Chem. Phys. 76, 2296-2314

6. Ben-Amotz, D. (1993) "Chemical Reaction Volumes in Model Fluid Systems. 1. Hard-Sphere Solvation and Diatomic Dissociation Processes", J. Phys. Chem. 97, 2314-2319

7. Ben-Amotz, D., and Herschbach, D. R. (1993) "Hard Fluid Model for Solvent-Induced Shifts in Molecular Vibrational Frequencies", J. Phys. Chem. 97, 2295-2306

8. Devendorf, G. S., and Ben-Amotz, D. (1993) "Vibrational Frequency Shifts of Fluid Nitrogen up to Ultrahigh Temperatures and Pressures", J. Phys. Chem. 97, 2307-2313

9.   Ben-Amotz, D., Lee, M.-R., Cho, S. Y., and List, D. J. (1992) "Solvent and pressure-induced perturbations of the vibrational potential surface of acetonitrile", J. Chem. Phys. **96**, 8781-8792

10. Longuet-Higgins, H. C., and Widom, B. (1964) "A rigid sphere model for the melting of argon", Mol. Phys. **8**, 549-556

11. Ben-Amotz, D., and Herschbach, D. R. (1990) "Estimation of Effective Diameters for Molecular Fluids", J. Phys. Chem. **94**, 1038-1047

12. Ben-Amotz, D., and Willis, K. G. (1993) "Molecular Hard Sphere Volume Increments", J. Phys. Chem., in press.

13. Bondi, A. (1964) "van der Waals Volumes and Radii", J. Phys. Chem. **68**, 441-451

14. Buckingham, A. D. (1958) "Solvent effects in infra-red spectroscopy", Proc. Roy. Soc. A **248**, 169-182; Buckingham, A. D. (1960) "Solvent effects in vibrational spectroscopy", Trans. Farad. Soc. **56**, 753-760; Buckingham, A. D. (1960) "A theory of frequency, intensity and band-width changes due to solvents in infra-red spectroscopy", Proc. Roy. Soc. A **255**, 32-39

15. Dijkman, F. G., and van der Maas, J. H. (1977) "Inhomogeneous broadening of Morse oscillators in liquids", J. Chem. Phys. **66**, 3871-3878

16. Zakin, M. R., and Herschbach, D. R. (1988) "Density dependence of attractive forces for hydrogen stretching vibrations of molecules in compressed liquids", J. Chem. Phys. **89**, 2380-2387

17. Etters, R. D., Belak, J., and LeSar, R. (1986) "Thermodynamic character of the vibron frequencies and equation of state in dense, high-temperature, fluid N_2" Phys. Rev. B **34**, 4221-4223

18. Llano-Restrepo, M., and Chapman, W. G. (1992) "Bridge function and cavity correlation function for the Lennard-Jones fluid from simulation", J. Chem. Phys. **97**, 2046-2054

19. Shing, K. S., Gubbins, K. E., and Lucas, K. (1988) "Henry constants in non-ideal fluid mixtures. Computer simulation and theory", Mol. Phys. **65**, 1235-1252

20. Ghonasgi, D., Llano-Restrepo, M., and Chapman, W. G. (1993) "Henry's law constant for diatomic and polyatomic Lennard-Jones molecules", J. Chem. Phys. **98**, 5662-5667

21. Ravi, R., de Souza, L. E. S., and Ben-Amotz, D. (1993) "Reaction Volumes in Model Fluid Systems II. Diatomic Dissociation in Lennard-Jones Solvents", in preparation.

22. Shim, J.-J., and Johnston, K. P. (1991) "Phase Equilibria, Partial Molar Enthalpies, and Partial Molar Volumes Determined by Supercritical Fluid Chromatography", J. Phys. Chem. **95**, 353-360

23. Peck, D. G., Mehta, A. J., Johnston, K. P. (1989) "Pressure Tuning of Chemical

Reaction Equilibria in Supercritical Fluids", <u>J. Phys. Chem.</u> **93**, 4297-4304

24. Kimura, Y, Yoshimura, Y, and Nakahara, M. (1989) "Chemical reaction in medium density fluid. Solvent density effects on the dimerization equilibrium of 2-methyl-2-nitrosopropane in carbon dioxide", <u>J. Chem. Phys.</u> **90**, 5679-5686

25. Debenedetti, P. G. (1987) "Clustering in Dilute, Binary Supercritical Mixtures: A Fluctuation Analysis", <u>Chem. Eng. Sci.</u> **42**, 2203-2212

26. Debenedetti, P. G., and Mohamed, R. S. (1989) "Attractive, weakly attractive, and repulsive near-critical systems", <u>J. Chem. Phys.</u> **90**, 4528-4536

27. Devaure, J., and Lascombe, J. (1979) "Étude par spectrométrie Raman de l'effet de la pression et de la température sur les conformations du butane, du méthyl-2-butane et du diméthyl-2-3-butane a l'état liquide pur", <u>Nouv. J. Chim.</u> **3**, 579-581

28. Kint, S., Scherer, J. R., and Snyder, R. G. (1980) "Raman spectra of liquid *n*-alkanes. III. Energy difference between *trans* and *gauche n*-butane", <u>J. Chem. Phys.</u> **73**, 2599-2602

29. Verma, A. L., Murphy, W. F., and Bernstein, H. J. (1974) "Rotational isomerism. XI. Raman spectra of *n*-butane, 2-methylbutane, and 2, 3-dimethylbutane", <u>J. Chem. Phys.</u> **60**, 1540-1544

30. Durig, J. R., Wang, A., Beshir, W., and Little, T. S. (1991) "Barrier to Asymmetric Internal Rotation, Conformational Stability, Vibrational Spectra and Assignments, and *Ab Initio* Calculations of *n*-Butane-d_0, d_5 and d_{10}", <u>J. Raman Spectrosc.</u> **22**, 683-704

31. Murphy, W. F., Fernández-Sánchez, J. M., and Raghavachari, K. (1991) "Harmonic Force Field and Raman Scattering Intensity Parameters of *n*-Butane", <u>J. Phys. Chem.</u> **95**, 1124-1139

32. Gassler, G., and Hüttner, W. (1990) "The gauche-trans Energy Difference of n-Butane from a Doppler Limited Investigation of the 740 cm^{-1} CH_2-Rocking Region", <u>Z. Naturforsch.</u> **45a**, 113-125

33. Hoyland, J. R. (1968) "Internal rotation in butane", <u>J. Chem. Phys.</u> **49**, 2563-2566

34. Nomura, H., Murasawa, K, Ito, N., Iida, F., and Udagawa, Y. (1984) "Pressure Effect on Conformational Equilibria of 1,2-Dichloroethane and 1,2-Dibromoethane by Means of Raman Spectroscopy", <u>Bull. Chem. Soc. Jpn.</u> **57**, 3321-3322

35. Seki, W., Choi, P.-K., and Takagi, K. (1983) "Ultrasonic relaxation and the volume difference between the rotational isomers in 1,2-dichloroethane", <u>Chem. Phys. Lett.</u> **98**, 518-521

36. Takagi, K., Choi, P.-K., and Seki, W. (1983) "Rotational isomerism and ultrasonic relaxation in 1,2-dibromoethane", <u>J. Chem. Phys.</u> **79**, 964-968

37. Asano, T, and le Noble, W. J. (1978) "Activation and Reaction Volumes in Solution", Chem. Rev. **78**, 407-489

38. Pratt, L. R., and Chandler, D. (1980) "Effective intermolecular potentials for molecular bromine in argon. Comparison of theory with simulation", J. Chem. Phys. **72**, 4045-4048

39. Herman, M. F., and Berne, B. J. (1983) "Monte Carlo simulation of solvent effects on vibrational and electronic spectra", J. Chem. Phys. **78**, 4103-4117

40. Nicolas, J. J., Gubbins, K. E., Street, W. B., and Tildesley, D. J. (1979) "Equation of state for the Lennard-Jones fluid", Mol. Phys. **37**, 1429-1454

41. Weast, R. C., Astle, M. J., and Beyers, W. H. (eds.) (1986) CRC Handbook of Chemistry and Physics, CRC Press, Boca Raton, Florida.

MOLECULAR THEORY OF OPTICAL ABSORPTION LINESHAPES OF DILUTE SOLUTES IN LIQUIDS: A PROBE OF SOLVENT DYNAMICS

J. G. Saven and J. L. Skinner
Theoretical Chemistry Institute and Department of Chemistry
University of Wisconsin
Madison, WI 53706-1322
USA

Introduction

Spectroscopy remains a powerful tool for obtaining information about structure and dynamics in liquids. One particularly useful approach uses a distinct transition of a dilute species ("chromophore" or "solute") to probe local properties of the solvent. If the solvent is static on a time scale defined by the inverse of the chromophore absorption linewidth (in Hz) then the (inhomogeneous) lineshape provides some information about the static distribution of local environments, that is, the solute-solvent structure. If, on the other hand, the solvent is moving appreciably on this same time scale, then "motional narrowing" occurs, and the lineshape provides information about solvent dynamics as well as structure.

Some time ago a beautiful and intuitive (but phenomenological) model of the lineshape was developed that encompasses the two regimes described above, and which has found application to many different branches of spectroscopy. This theory was reviewed in the seminal paper of Kubo's entitled "A Stochastic Theory of Line Shape" [1]. It is assumed that the chromophore transition frequency is a time-dependent Gaussian random variable described by only two parameters: its root-mean-square fluctuation Δ and correlation time τ. The limit $\Delta\tau \gg 1$ produces an inhomogeneous Gaussian lineshape, and the limit $\Delta\tau \ll 1$ leads to a (homogeneous) motionally-narrowed Lorentzian lineshape. While the Kubo theory has been useful in the analysis of many experiments [2, 3, 4, 5], it would be more valuable still if one could establish a microscopic foundation for the model. That is, from a statistical mechanical perspective one would like to understand under what conditions the model is valid, and if it is valid, to provide a molecular interpretation of the parameters Δ and τ. This is one of the aims of the present paper. To this end our efforts are directed toward the problem of the electronic spectra of dilute nonpolar chromophores in nonpolar solvents, but some of the ideas and techniques discussed may be more generally applicable to other systems and types of spectroscopy.

Electronic spectra in liquids are, in fact, typically inhomogeneously broadened. To understand why this is so, we first note that the difference in ground and excited state solute-solvent interaction pair potentials sets the frequency scale for the

461

J. Jortner et al. (eds.), Reaction Dynamics in Clusters and Condensed Phases, 461–469.
© 1994 *Kluwer Academic Publishers. Printed in the Netherlands.*

linewidth. If this difference is large, then the linewidth is large, and the solvent will be static on the time scale of the inverse linewidth. A nice example involves the $^1S_0 \rightarrow {}^3P_1$ transition of Xe in sub- and super-critical Ar [6]. The large change in the Xe-Ar pair potential upon Xe excitation produces a FWHM absorption linewidth, $\Delta\tilde{\nu}$, on the order of 1000 cm^{-1}, leading to an associated time scale $(\pi c\Delta\tilde{\nu})^{-1}$ of 10 fs, on which Ar is essentially static. On the other hand, for $\pi \rightarrow \pi^*$ transitions of organic chromophores the ground and excited state solute-solvent potential difference will be much smaller, leading to a narrower linewidth. Thus for example, the 6^1_0 vibronic feature of the $^1A_{1g} \rightarrow {}^1B_{2u}$ transition of benzene in Ar fluid [7] and clusters [8] gives linewidths on the order of 10-200 cm^{-1}, leading to a time scale of 50 fs to 1 ps. In this case the Ar dynamics may well be important. Indeed, it is not easy to explain the observed temperature and density dependence of the bulk fluid experiments invoking only inhomogeneous broadening [9], and there are indications that dynamics is important for the cluster experiments as well [10].

While it may be true that experimentally the motionally-narrowed limit will never be observed in electronic spectroscopy because of the large changes in the solute-solvent potential with solute excitation as discussed above, nonetheless it seems important to have a theory that is valid into the motionally-narrowed regime, especially because in other branches of spectroscopy (e.g. vibrational [11, 12, 13]) this regime will be easier to access. In addition, a dynamic theory valid for long times should not be difficult to generalize in order to describe exciting new time-domain experiments involving photon echoes [3], transient hole-burning [14], or the time-dependent Stokes shift [15].

Much progress has been made in developing molecular theories of inhomogeneous broadening in liquids [6, 16, 17, 18]. These theories lead to a Gaussian lineshape whose parameters can be expressed in terms of the solvent density, the solute-solvent and solvent-solvent radial distribution functions, and solute-solvent interaction potentials for the ground and excited state solute.

Most theoretical studies investigating the effect of solvent dynamics on spectroscopy [2, 19, 20] have involved computer simulation, Kubo's stochastic theory, or the related phenomenological multimode Brownian oscillator model of Yan and Mukamel [21]. An exception to the above is a molecular theory of the photon echo for nonpolar fluids [22], but here only the shortest time ballistic motion is considered.

In this paper, we outline a microscopic theory of the electronic absorption lineshape that, like the Kubo theory, smoothly bridges the limits of homogeneous and inhomogeneous broadening. In the static solvent limit our theory recovers the inhomogeneous broadening results mentioned above [17, 18].

Molecular Theory

We begin by considering a particular vibronic transition of a single solute atom or molecule in an atomic or molecular solvent. Throughout, however, we will suppress all intramolecular nuclear degrees of freedom for solvent and solute, and will take all solute-solvent potentials to be isotropic. Therefore it is probably best to think of an electronic transition of an atomic solute in an atomic solvent. Within the two-state, Born-Oppenheimer, and semiclassical approximations we can write the absorption lineshape as [23]

$$I(\omega) = \frac{1}{2\pi} \int_{-\infty}^{\infty} dt e^{-i\omega t} \, \phi(t),$$ (1)

$$\phi(t) = \left\langle \exp\left(\frac{i}{\hbar} \int_0^t dt' U(t')\right) \right\rangle,$$ (2)

where the angular brackets denotes a classical equilibrium statistical mechanical average over initial coordinates and momenta of all the particles (i.e. a phase space integral weighted by the Boltzmann factor for the ground state Hamiltonian), and $U(t)$ is the time-dependent difference of the excited and ground state Born-Oppenheimer potential surfaces evaluated as the particles undergo classical dynamics on the ground state surface.

To develop the molecular theory of the lineshape for simplicity we assume pairwise additive isotropic potentials, so that the ground and excited state potential surfaces are given respectively by

$$V_0 = \sum_i v_0(r_i) + \sum_{i<j} v_s(r_{ij}),$$ (3)

$$V_1 = E_0 + \sum_i v_1(r_i) + \sum_{i<j} v_s(r_{ij}).$$ (4)

In the above, the summation indices run from 1 to N (the number of solvent atoms), r_i is the distance of the ith solvent atom from the solute, $v_0(r)$ is the ground state solute-solvent pair potential, r_{ij} is the distance between solvent atoms i and j, $v_s(r)$ is the solvent-solvent pair potential, $v_1(r)$ is the excited state solute-solvent potential, and E_0 is the reference (gas phase) transition energy of the solute. The classical time-dependent potential difference $U(t)$ can now be written as

$$U(t) = E_0 + \sum_i [v_1(r_i(t)) - v_0(r_i(t))],$$ (5)

where, as discussed above, the time-dependent distances arise from classical dynamics on the ground state surface.

To proceed we perform a cumulant expansion of the semiclassical expression for the relaxation function $\phi(t)$, truncating at second order to obtain [23, 24, 25, 26]:

$$\phi(t) = \exp\{i\omega_0 t - \int_0^t dt'(t - t')C(t')\},$$ (6)

where

$$\omega_0 = E_0/\hbar + \langle \sum_i v(r_i) \rangle,$$ (7)

$$v(r) = [v_1(r) - v_0(r)]/\hbar,$$ (8)

$$C(t) = \langle \Omega(t)\Omega(0) \rangle,$$ (9)

$$\Omega(t) = \sum_i v(r_i(t)) - \langle \sum_i v(r_i) \rangle. \tag{10}$$

We can easily obtain a microscopic expression for ω_0, which is [23]

$$\omega_0 = E_0/\hbar + \rho \int dr v(r) g(r), \tag{11}$$

where ρ is the solvent density and $g(r)$ is the solute-solvent radial distribution function. Similarly, we can obtain an approximate expression for the time-correlation function:

$$\begin{aligned} C(t) &= \rho \int d\mathbf{r}_1' d\mathbf{r}_1 v(r_1') v(r_1) G(\mathbf{r}_1', t|\mathbf{r}_1, 0) g(r_1) \\ &+ \rho^2 \int d\mathbf{r}_1' d\mathbf{r}_1 d\mathbf{r}_2 v(r_1') v(r_2) G(\mathbf{r}_1', t|\mathbf{r}_1, 0) g(r_1) g(r_2) [g_s(r_{12}) - 1]. \end{aligned} \tag{12}$$

In the above $g_s(r)$ is the solvent-solvent radial distribution function, and $G(\mathbf{r}_1', t|\mathbf{r}_1, 0)$ is the conditional probability that if at time 0 solvent molecule 1 has position $\mathbf{r}_1$ (relative to the solute), then it has position $\mathbf{r}_1'$ at time t. An approximate theory for G that interpolates between inertial motion at short times and diffusive motion at long times has been suggested by Haan [27]. This theory, which involves a Smoluchowski equation with the potential of mean force and a nonlinear time, is adapted for the present study [23]. To implement this theory, we need the distribution function $g(r)$ and the diffusion constants for solvent and solute motion.

In analogy with Kubo's stochastic theory, we can define the root-mean-square transition frequency fluctuation by $\Delta^2 = C(0)$ and the correlation time by $\tau = \int_0^\infty dt C(t)/C(0)$. Like the Kubo theory, the limits $\Delta\tau \gg 1$ and $\Delta\tau \ll 1$ correspond to inhomogeneous and homogeneous broadening respectively [23].

Model

For explicit calculations we consider a simple model of a single solute in a fluid of N solvent particles, which interact with each other with the Lennard-Jones potential

$$v_s(r) = 4\epsilon \left[\left(\frac{\sigma}{r}\right)^{12} - \left(\frac{\sigma}{r}\right)^6 \right], \tag{13}$$

where ϵ is the well-depth of the potential, and σ is the effective diameter of the solvent atom or molecule. The solvent particles interact with the solute in its ground and excited states respectively with the Lennard-Jones potentials

$$v_0(r) = 4\epsilon_0 \left[\left(\frac{\sigma_0}{r}\right)^{12} - \left(\frac{\sigma_0}{r}\right)^6 \right], \tag{14}$$

$$v_1(r) = 4\epsilon_1 \left[\left(\frac{\sigma_1}{r}\right)^{12} - \left(\frac{\sigma_1}{r}\right)^6 \right]. \tag{15}$$

In fact, in what follows the diameter of the solute will remain unchanged upon excitation, so that $\sigma_1 = \sigma_0$, and the excited state potential will have a larger well-depth than the ground state potential, so that $\epsilon_1 > \epsilon_0$. This corresponds to an excited state that is more polarizable than the ground state, which is quite generally the case. Furthermore the ground state solute-solvent potential is taken to be identical to the solvent-solvent potential, so that $\sigma_0 = \sigma$ and $\epsilon_0 = \epsilon$. This last simplification, though unrealistic, leads to a considerable improvement of the statistics of the computer simulation (to be described below), since now we can consider every particle in the fluid to be a solute. It also means that $g_s(r) = g(r)$ and that the diffusion constants and masses of the solute and solvent are identical.

We define the dimensionless coupling parameter

$$\lambda = \frac{\epsilon_1 - \epsilon}{\epsilon}, \tag{16}$$

which characterizes the difference in the ground and excited state solute-solvent potentials. Therefore we can write (from Eq. 8)

$$v(r) = \lambda \frac{4\epsilon}{\hbar} \left[\left(\frac{\sigma}{r}\right)^{12} - \left(\frac{\sigma}{r}\right)^6 \right]. \tag{17}$$

From Eq. 12 we can see that λ directly determines the magnitude of the root-mean-square fluctuation Δ ($\Delta \propto \lambda$). On the other hand, τ, the correlation time of the fluctuations, is independent of λ. Therefore varying λ changes the product $\Delta\tau$, and so we can "tune" λ to access the complete range from inhomogeneous to homogeneous broadening.

We will present and discuss our results in terms of the dimensionless time $t^* = t(\epsilon/m\sigma^2)^{1/2}$. The dimensionless density, temperature, and mass are defined by $\rho^* = \rho\sigma^3$, $T^* = kT/\epsilon$, and $m^* = m\sigma^2\epsilon/\hbar^2$. Taking the parameters for Ar of [28] $\epsilon/k = 119.8K$, $\sigma = 3.405$ Å, and $m = 6.634 \times 10^{-23}g$, gives $m^* = 1144$. We choose thermodynamic variables to give a supercritical fluid: $\rho^* = 0.5, T^* = 1.41$ ($\rho_c^* = 0.35, T_c^* = 1.35$ [29]). Note that with these parameters $t^* = 1$ corresponds to 2.16 ps. All frequencies are presented in dimensionless units of $\omega^* = \omega(m\sigma^2/\epsilon)^{1/2}$. $\omega^* = 1$ corresponds to 4.63×10^{11} rad/s or 2.46 cm^{-1}. Finally, the diffusion constant will be presented in the dimensionless form $D^* = D(m/\sigma^2\epsilon)^{1/2}$.

The molecular dynamics simulation used a velocity-Verlet algorithm to propagate a fluid of 864 particles in a cubic box of edge length 12σ. We used periodic boundary conditions and the minimum image convention, and the pair potential was truncated at half the box length. The system was integrated using a time step of $\delta t^* = 0.010$.

Results and Discussion

From the simulation the radial distribution function $g(r)$ was calculated, and then the potential of mean force was obtained using $w(r) = -kT \ln g(r)$. From the single particle mean-square displacement we calculated the diffusion constant, finding $D^* = 0.314 \pm 0.001$. (In this and what follows, all error bars are two standard deviations.) With the above we can now calculate $C(t)$; our theoretical result [23] for $C(t)/C(0)$, which is independent of λ, is shown in Fig. 1.

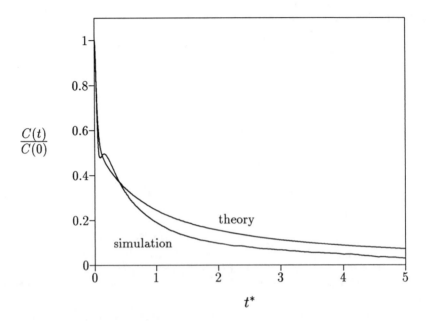

Fig. 1: The normalized transition frequency autocorrelation function, $C(t)/C(0)$ vs t^*, as calculated from the microscopic theory and from the molecular dynamics simulation.

As seen, it is distinctly nonexponential. The short time decay is due to inertial motion, while the long time decay results from the thermally activated escape of a solvent molecule from the first solvent shell over the barrier in the potential of mean force. The time integral of $C(t)/C(0)$ was determined numerically, yielding $\tau^* = 1.70$. The simulation result for $C(t)/C(0)$ is also shown in Fig. 1. It shows the same rapid decay due to inertial motion, followed by a small increase that is presumably due to the librational motion of atoms in the first solvent shell. The correlation function then settles into a slow decay, presumably due to the exchange of solvent molecules between first and second solvation shells. One sees that at long times the amplitude of the simulation correlation function is somewhat less than that for theory. Numerical integration of $C(t)/C(0)$ from simulation gives $\tau^* = 0.73$. Our theoretical calculation of the zero-time value of the correlation function gives $\Delta^*/\lambda = 53.03$, which is in reasonable agreement with $\Delta^*/\lambda = 52.25 \pm 0.28$ from the simulation. For the solvent-induced frequency shift (setting $E_0 = 0$) Eq. 11 gives $\omega_0^*/\lambda = -225.97$, in good agreement with the simulation result $\omega_0^*/\lambda = -225.61 \pm 0.14$.

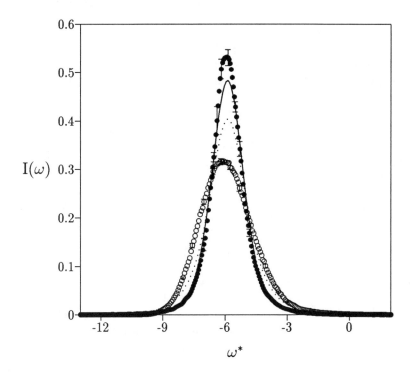

Fig. 2: The absorption spectrum, $I(\omega)$ vs ω^*, for $\lambda = 0.026$. Presented are the theoretical spectrum (—), and the simulation results for both the spectrum (•) and the distribution of transition frequencies (∘). The error bars for the simulation results are two standard deviations. Also shown is the lineshape as calculated using the Kubo model (⋯).

We are now in a position to calculate the theoretical lineshape using Eqs. 1 and 6. As an example, we choose $\lambda = 0.026$. With the simulation results for Δ and τ above, this produces a product of $\Delta\tau = 1.0$, which is in the regime intermediate between homogeneous and inhomogeneous broadening. The calculated lineshape is shown in Fig. 2. This lineshape can be compared with the numerically exact (within the model) result from Eqs. 1 and 2 and the molecular dynamics simulation. These results are also shown in Fig. 2. For comparison we have also determined from the simulation the distribution of transition frequencies, which would be identical to the lineshape if the the latter were completely inhomogeneous, and have plotted this in Fig. 2. We see that the simulation lineshape and the distribution do not agree, showing the pronounced effect of dynamics on the lineshape. The theory is in reasonble agreement with simulation. Also shown in Fig. 2 is the Kubo model result, where the parameters Δ, τ, and ω_0 were determined from simulation. As shown, this model is not as accurate as our theory, because of its assumption that the correlation function decays exponentially.

To summarize, in this paper we have endeavored to go beyond Kubo's stochastic model in developing a molecular theory of the lineshape. Like the stochastic model, where the second cumulant truncation is exact, we also make a second cumulant truncation. Unlike the stochastic model the theory produces a correlation function of the fluctuating transition frequency that is distinctly nonexponential. In one sense our theory provides support for the Kubo model in that it verifies its general features and provides microscopic expressions for its phenomenological parameters. On the other hand we show that the Kubo model cannot produce accurate lineshapes in the regime intermediate between inhomogeneous and homogeneous broadening because of the exponential form of its transition frequency correlation function. It is our hope and expectation that a molecular theory such as the one described herein will enable time- and frequency-domain spectroscopy to provide a detailed probe of solvent dynamics.

Acknowledgements

The text and figures of this paper are adapted from a manuscript that has been submitted to *The Journal of Chemical Physics*. The authors thank Dr. Brian Laird for helpful discussions at an early stage of this research and acknowledge NSF for support from grants CHE90-96272 and CHE92-19474 and from a Graduate Fellowship to JGS.

References

[1] R. Kubo, Adv. Chem. Phys., **15**, 101 (1969).

[2] S. Mukamel, Adv. Chem. Phys., **70**, 165 (1988).

[3] E.T.J. Nibbering, K. Duppen, and D.A. Wiersma, J. Photochem. Photobiol. A, **62**, 347 (1992).

[4] J.-Y. Bigot, M.T. Portella, R.W. Schoenlein, C.J. Bardeen, A. Migus, and C.V. Shank, Phys. Rev. Lett., **66**, 1138 (1991).

[5] A.B. Myers and B. Li, J. Chem. Phys., **92**, 3310 (1990).

[6] I. Messing, B. Raz, and J. Jortner, J. Chem. Phys., **66**, 2239 (1977).

[7] R. Nowak and E.R. Bernstein, J. Chem. Phys., **87**, 2457 (1987).

[8] X. Li, M.Y. Hahn, M.S. El-Shall, and R.L. Whetten, J. Phys. Chem., **95**, 8524 (1991).

[9] J.G. Saven and J.L. Skinner. (unpublished).

[10] L.E. Fried and S. Mukamel, J. Chem. Phys., **96**, 116 (1992).

[11] D. Oxtoby, Adv. Chem. Phys., **40**, 1 (1979).

[12] D. Oxtoby, Adv. Chem. Phys., **47 (part 2)**, 487 (1981).

[13] J. Chesnoy and G.M. Gale, Adv. Chem. Phys., **70 (part 2)**, 297 (1988).

[14] J. Yu, T.J. Kang, and M. Berg, J. Chem. Phys., **94**, 5787 (1991).

[15] P.F. Barbara and W. Jarzeba, Adv. Photochem., **15**, 1 (1990).

[16] S.H. Simon, V. Dobrosavljević, and R.M. Stratt, J. Chem. Phys., **93**, 2640 (1990).

[17] H.M. Sevian and J.L. Skinner, Theoretica Chimica Acta, **82**, 29 (1992).

[18] H.M. Sevian and J.L. Skinner, J. Chem. Phys., **97**, 8 (1992).

[19] S. Mukamel, Ann. Rev. Phys. Chem., **41**, 647 (1990).

[20] L.E. Fried and S. Mukamel, Adv. Chem. Phys. (in press).

[21] Y.J. Yan and S. Mukamel, J. Chem. Phys., **94**, 179 (1991).

[22] A.M. Walsh and R.F. Loring, Chem. Phys. Lett., **186**, 77 (1991).

[23] J.G. Saven and J.L. Skinner, submitted to J. Chem. Phys. (1993).

[24] R. Kubo, *in: Fluctuation, Relaxation, and Resonance in Magnetic Systems, Ed. D. TerHaar*, (Oliver and Boyd, Edinburgh, 1962).

[25] R. Kubo, J. Phys. Soc. Jpn, **17**, 1100 (1962).

[26] R. Kubo, J. Math. Phys., **4**, 174 (1963).

[27] S.W. Haan, Phys. Rev. A, **20**, 2516 (1979).

[28] J.O. Hirschfelder, C.F. Curtiss, and R.B. Bird, *Molecular Theory of Gases and Liquids*, (Wiley, New York, 1954).

[29] J.J. Nicolas, K.E. Gubbins, W.B. Streett, and D.J. Tildesley, Mol. Phys., **37**, 1429 (1979).

Multiple Time Scales in Molecular Dynamics: Applications to Vibrational Relaxation

B. J. Berne

Department of Chemistry and Center for Biomolecular Simulations
Columbia University, New York, NY 10027

1 Introduction

Many physical systems of interest in chemistry, physics and biology are characterized by multiple time scales. Such systems pose problems for numerical simulations where small integration time steps are required for stable integration of the fast degrees of freedom and a large number of time steps must therefore be used to simulate the relaxation of the slow degrees of freedom. The presence of multiple time scales has thus been a road block to the direct simulation of many physical systems. A simple example is that of a very stiff diatomic molecule dissolved in a monatomic system. Vibrational relaxation in this system is so slow that it has not been possible to determine vibrational energy relaxation times and dephasing times directly by simulation. Instead, approximate theories such as the Kubo relaxation theory[1, 2] are invoked giving the dephasing time T_2 as,

$$\frac{1}{T_2} = \int_0^\infty d\tau \langle \delta\omega(0)\delta\omega(\tau) \rangle \tag{1}$$

where $\delta\omega(t) \equiv \omega(t) - \bar{\omega}$ is the fluctuation in the vibrational frequency from its average value caused by the interaction of the oscillator with the solvent atoms. From the interaction potential the instantaneous frequency fluctuation can be expressed as a function of the nuclear positions of the system Molecular dynamics is then used to evaluate the autocorrelation function of the frequency fluctuation in a system in which the bond length of the diatomic molecule is fixed[3, 4]. Thus the fast dynamics is eliminated from the system being simulated at the cost of using an approximate theory.

In a similar vein the energy relaxation time, T_1, which is very slow in this system must be expressed in terms of a correlation function. This is a bit more difficult since

J. Jortner et al. (eds.), Reaction Dynamics in Clusters and Condensed Phases, 471–494.
© 1994 *Kluwer Academic Publishers. Printed in the Netherlands.*

the theory for anharmonic oscilators requires many approximations. If the molecule is harmonic its energy relaxation time can be calculated using the generalized Langevin equation (GLE)[5]

$$\frac{1}{T_1} = \frac{\zeta'(\tilde{\omega})}{\mu} \tag{2}$$

where μ is the reduced mass of the oscillator, $\tilde{\omega}$ is the renormalized vibrational frequency due to the solvent shift and $\zeta(\tilde{\omega}) = Re\zeta(\tilde{\omega})$ is the real part of the frequency dependent friction coefficient at the renormalized vibrational frequency. This latter property can be calculated in molecular dynamics by fixing the bond length[6, 7, 8] and computing the frequency transform of the dynamic friction

$$\zeta(t) = \beta\langle\delta f(0)\delta f(t)\rangle \tag{3}$$

where f is the force on the bond, $\bar{f}$ is the average force on the bond and $\delta f(t) \equiv f(t) - \bar{f}$. Thus one studies the dynamics of a rigid rotor dissolved in the liquid and infers from it the vibrational energy relaxation time. In the same vein one can derive a similar expression for the dephasing time as

$$\frac{1}{T_2} = \frac{\zeta'(\tilde{\omega})}{2\mu} = \frac{1}{2T_1} \tag{4}$$

The trouble with this approach is that both of these analytical expressions are approximate and are not easily generalized to anharmonic systems. In this talk we shall report on a systematic study comparing full molecular dynamics, using our multiple time scale algorithm, with theoretical predictions[5].

Stochastic molecular dynamics offers another vehicle for simulating systems with multiple time scales. An example is the generalized Langevin equation. For the above system one can write[9]

$$\mu\ddot{x}(t) = -\frac{\partial W(x)}{\partial x} - \int_0^t d\tau\zeta(t-\tau)\dot{x}(\tau) + R(t) \tag{5}$$

where again molecular dynamics is used to determine the potential of mean force, $W(x)$ and the dynamic friction coefficient $\zeta(t)$ by evaluating the autocorrelation function defined in Eq. (3). In the molecular dynamics simulations it has been shown that the fast molecular time scale can be omitted by performing simulations on the diatomic molecule with fixed or constrained bond length[8]. The GLE is then solved numerically for the reaction coordinate[10] (or vibrational coordinate) $x(t)$ and the corresponding velocity $v(t)$ and the energy and dephasing times are determined from these trajectories[5]. Unfortunately the GLE is itself an approximation to the true dynamics so that it behooves the investigator to demonstrate that the GLE is accurate.

Recently we have introduced new molecular dynamics algorithms, the reversible RESPA and NAPA algorithms, that make possible the direct simulation of vibrational dephasing and energy relaxation of stiff oscillators[5, 10, 11, 12, 13, 14]. It is now possible to determine the vibrational dephasing and energy relaxation of molecules in the liquid state without making any of the approximations inherent in the Kubo theory.

In this talk I will show how one can now simulate the full dynamical system with both the fast and slow time scales without having to invoke an approximate theory by using new multiple time scale algorithms, RESPA and NAPA, recently devised by us at Columbia[11, 12, 13, 14, 10, 5]. These methods are derived from the Liouville formalism of classical mechanics and have now been used to study many other systems with multiple time scales. The results of the full molecular dynamics are compared with GLE stochastic dynamics and with various theories of vibrational relaxation for harmonic and anharmonic molecules. It is found that the GLE gives very good agreement with full molecular dynamics.

2 Theory

In a recent paper[5] we have derived several important theoretical results from the GLE which are summarized in this section. The results are for different forms of the potential of mean force $W(x)$. We first consider the case of a harmonic potential of mean force and then consider the case of a cubic anharmonic potential of mean force where a cubic anharmonicity is added to the quadratic term.

2.1 Harmonic Potential of Mean Force

The harmonic potential of mean force is

$$W(x) = \frac{1}{2}\mu\tilde{\omega}^2 x^2 \tag{6}$$

where $\tilde{\omega}$ is the renormalized harmonic frequency defined by

$$\tilde{\omega}^2 = \frac{kT}{\mu\langle q^2\rangle} \tag{7}$$

where $q \equiv x - \bar{x}$ is the vibrational displacement from the average bond length $\bar{x}$.

We are most interested in the case of a very stiff oscillator where $\tilde{\omega} \gg \gamma(t = 0)$ where $\gamma(t) \equiv \zeta(t)/\mu$ is the dynamic damping coefficient. Then it can be shown, by solving for the roots of the dispersion of the GLE in the high frequency limit, that the normalized velocity correlation function is[5]

$$C_{vv}(t) = e^{-\gamma'(\tilde{\omega})t/2}\left[\cos(\Omega t) - \frac{\gamma'(\tilde{\omega})}{2\tilde{\omega}}\sin(\Omega t)\right] \tag{8}$$

where $\Omega = \tilde{\omega} + \gamma''(\tilde{\omega})/2$ and where $\gamma''(\tilde{\omega})$ is the imaginary part of the frequency dependent damping coefficient. Likewise the normalized displacement autocorrelation function is[5]

$$C_{xx}(t) = e^{-\gamma'(\tilde{\omega})t/2} \left[\cos(\Omega t) + \frac{\gamma'(\tilde{\omega})}{2\tilde{\omega}} \sin(\Omega t) \right] \tag{9}$$

From this it is clear that these correlation functions decay with an exponential envelope. The vibrational dephasing time for stiff harmonic oscillator embedded in a "slow" fluid is thus seen to be

$$\frac{1}{T_2} = \frac{1}{2}\gamma'(\tilde{\omega}). \tag{10}$$

We have also been able to derive an expression for the energy relaxation[5]

$$\langle \delta\epsilon(t) \rangle_{x(0),v(0)} = \delta\epsilon_K(0)C_{vv}^2(t) + \delta\epsilon_P(0)C_{xx}^2(t) + \delta\epsilon(0)\frac{1}{\tilde{\omega}^2}\dot{C}_{xx}^2(t) \tag{11}$$

where the the left hand side of the equation gives the average deviation of the energy from the thermal average kT for an ensemble of oscillators all starting in the same initial state $\{x(0), v(0)\}$ and the quantities $\delta\epsilon_K(0)$, $\delta\epsilon_P(0)$, and $\delta\epsilon(0)$ are respectively the initial deviations of the harmonic kinetic, potential and total full energy from the corresponding thermal averages $kT/2, kT/2$ and kT respectively. Substitution of Eqs. (8) and (9) shows that the envelope of the energy relaxation is exponential with an energy relaxation time T_1 given by

$$\frac{1}{T_1} = \gamma'(\tilde{\omega}). \tag{12}$$

and thus by Eq. (4).

For completeness one can show that the normalized autocorrelation function of energy fluctuation is

$$C_{\epsilon\epsilon}(t) = \frac{1}{2}C_{vv}^2(t) + \frac{1}{2}C_{xx}^2(t) + \frac{1}{\tilde{\omega}^2}\dot{C}_{xx}^2(t) \to e^{-\gamma'(\tilde{\omega})t} \tag{13}$$

where the right arrow indicates the high frequency limit. The derivation of this is given in ref [8].

For this potenial of mean force the vibrational energy relaxation time T_1 and the vibrational dephasing time T_2 can be found in the limit where the vibrational frequency is large.

2.2   Cubic Anharmonic Potential of Mean Force

The harmonic oscillator with cubic anharmonicity is given by

$$W(x) = \frac{1}{2}\mu\tilde{\omega}^2 x^2 + \frac{1}{6}\tilde{f}^3. \tag{14}$$

Using the Kubo theory, Oxtoby[3, 4] and later Pollak and Levine[15] showed that if the spectral density of the bath is very small at the vibrational frequency the dephasing time is dominated by the anharmonicity; furthermore if the anharmonicity is a small perturbation to the harmonic force, then they show that[3, 4, 15]

$$\frac{1}{T_2} = \frac{f^2}{4\mu\omega_0^6\beta}\tilde{\gamma}(s = 0), \tag{15}$$

where f and ω_0 are the bare anharmonicity and vibrational frequency and $\tilde{\gamma}(s = 0)$ is the static damping rate which is expected to be much larger than $\gamma'(\tilde{\omega})$ for stiff oscillators with the consequence that the dephasing time for a stiff anharmonic oscillator will be shorter than the dephasing time in its corresponding harmonic system.

The, derivation of Eq. (15) was based on the Kubo theory[1, 2]. Recently we have shown how one can derive the dephasing time directly from the GLE[5] for a stiff oscillator with the result that,

$$\frac{1}{T_2} = \frac{1}{2}\gamma'(\tilde{\omega}) + \frac{\tilde{f}^2}{4\mu\tilde{\omega}^6\beta}\tilde{\gamma}(s = 0) + O(\tilde{f}^4). \tag{16}$$

Comparison of Eqs.(15) and (16) shows that the former does not contain the energy relaxation where the latter does. This expression is similar to the usual assumption in NMR that

$$\frac{1}{T_2} = \frac{1}{2T_1} + \frac{1}{T_2^*}. \tag{17}$$

One cannot derive an exact expression for the energy autocorrelation function for an anharmonic oscillator as was done for the harmonic oscillator in Eq. (13).

All of the results for the anharmonic oscillator have been derived using dynamic perturbation theory. It would be of interest to derive the corresponding theory for the Morse oscillator. For purposes of comparison we introduce a Morse oscillator that is chosen to fit the properties of the cubic oscillator.

$$W(x) = D_0 \left(1 - e^{-\alpha x}\right)^2 \tag{18}$$

where the parameters D_0 and α are chosen so that harmonic and cubic terms in the Taylor expansion of this potential match the cubic potential of Eq.(14), i.e.,

$$D_0 = \frac{9\mu^3\tilde{\omega}^6}{2\tilde{f}^2}$$

$$\alpha = \frac{\tilde{f}}{3\mu\tilde{\omega}^2} \tag{19}$$

3   Multiple Time Scale Molecular Dynamics

In order to test the GLE, it is necessary to consider extremely high frequency oscillators in relatively low frequency baths. In the systems considered here, a frequency ratio between oscillator and bath as high as 15 was considered. Even for moderate frequencies, there can be a severe separation of time scales between the oscillator and the surrounding solvent atoms. Recently, we have developed algorithms (NAPA and RESPA) to handle the separation of time scales in both the Molecular Dynamics and the Stochastic Dynamics simulations[11, 12, 13, 14, 10, 5]. In the case of Molecular Dynamics, the reversible version of NAPA and RESPA presented in in ref.[14] lead to significant improvement over earlier forms of these integrators and even though the GLE simulations are not reversible, the use of "reversible-like" methods have improved the performance in the GLE as well.

The reversible NAPA and RESPA methods are based on the Trotter expansion of the classical Liouville propagator $\exp(iLt)$, where the Liouville operator is $iL = \{..., H\}$. H is the system Hamiltonian. Let x represent the fast relative coordinate and let y represent all of the solvent degrees of freedom and the center of mass motion of the diatomic. The Liouville operator can then be written in the form

$$iL = \left[\frac{p_x}{\mu}\frac{\partial}{\partial x} + F_r(x)\frac{\partial}{\partial p_x}\right] + f(x,y)\frac{\partial}{\partial p_x} + \frac{p_y}{m_y}\frac{\partial}{\partial y} + F_y(x,y)\frac{\partial}{\partial p_y} \qquad (20)$$

where $f(x,y)$ is the force on x due to the solvent. We choose a reference system for which the Liouville operator is $iL_r = (p_x/\mu)\partial/\partial x + F_r(x)\partial/\partial p_x$. Then $iL = iL_r + iL_y$, where iL_y contains all other terms in Eq. (20). The classical propagator is factorized according to the Trotter expansion

$$e^{iL\Delta t} = e^{iL_y\Delta t/2}e^{iL_r\Delta t}e^{iL_y\Delta t/2} + O(\Delta t^3). \qquad (21)$$

This expansion is good to order Δt^3 and is manifestly reversible in time.

The derivatives in Eqs. (20) and (21) act on the current state. The operator $\exp(iL_y\Delta t/2)$ on the left is further factorized into such that

$$\exp\left(iL_y\Delta/2\right) \approx \exp\left(\frac{\Delta t}{2}F_y(x,y)\frac{\partial}{\partial p_y}\right)\exp\left(\frac{\Delta t}{2}f(x,y)\frac{\partial}{\partial p_y}\right)\exp\left(\frac{\Delta}{2}\dot{y}\frac{\partial}{\partial p_y}\right) \qquad (22)$$

and the transpose of this is used for the operator on the right of the reference system. The resulting propagator is still reversible. When these factorizations are substituted into Eq. (21) and the resulting operator is applied to the initial state of the system one finds the integration scheme:

$$x(\Delta t) = x_r\left[\Delta t; x(0), \dot{x}(0) + \frac{\Delta t}{2\mu}f(0)\right]$$

$$\dot{x}(\Delta t) \;=\; \dot{x}_r \left[\Delta t; x(0), \dot{x}(0) + \frac{\Delta t}{2\mu} f(0) \right] + \frac{\Delta t}{2\mu} f(\Delta t)$$

$$y(\Delta t) \;=\; y(0) + \Delta t \dot{y}(0) + \frac{\Delta t^2}{2m_y} F_y(0)$$

$$\dot{y}(\Delta t) \;=\; \dot{y}(0) + \frac{\Delta t}{2m_y} \left[F_y(0) + F_y(\Delta t) \right] \tag{23}$$

where, again, x_r and $\dot{x}_r$ represent the evolution of the position and velocity under the action of $\exp(iL_r \Delta t)$. It should be noted that when $F_r(x)$ is a Hooke's Law type force $F_r(x) = -\mu\omega^2 x$, the analytical solution is readily obtained

$$e^{iL_r \Delta t} x = x \cos \omega \Delta t + \frac{\dot{x}}{\omega} \sin \omega \Delta t$$

$$e^{iL_r \Delta t} \dot{x} = \dot{x} \cos \omega \Delta t - \omega x \sin \omega \Delta t \tag{24}$$

This presentation is terse. The interested reader should consult the original papers where we combine the Trotter factorization presented here with a factorization of the solvent-solvent and solvent-solute forces into long and short range components[13]. The combination of these two Trotter factorizations leads to a combined reduction in cpu time equal to the product of the reductions for these two procedures taken separately. We have found for this system that we can reduce the the cpu time by as much as a factor of 50 over straightforward use of the velocity Verlet algorithm.

The application of the reversible NAPA and RESPA to the GLE was presented in ref.[5]. The reference system in this algorithm can be evaluated analytically or numerically as before. In both the MD and GLE simulations, Δt may be chosen according to the time scale of the solvent motion (the slow motion) without loss of accuracy. In the GLE simulations, we prefer the use of analytical solutions for both the harmonic and cubic reference systems simply because a numerical reference system requires a greater fraction of the cpu time required per step. However, the use of a numerical reference is perfectly acceptable.

4   Results

The system studied consisted of 64 Lennard-Jones particles at reduced temperature $\hat{T} = 2.5$ and reduced density $\rho\sigma^3 = 1.05$ in which is imbedded a single diatomic with either a harmonic or cubic bond potential. Some of the cases considered here were run using 500 particles to check system size dependence, and these were found to agree with the small system results. The diatomic is kept at a fixed spatial orientation along a body diagonal of the cubic cell so that rotational anharmonicities are not present. Using the integrator of Eq.(23), simulations were run for 2×10^6 or more steps using a big time step of 2×10^{-3} in all cases. In all simulations, the time step used gave energy conservation $\Delta\hat{E} \sim 10^{-3}$ measured in this manner independent of frequency or cubic coupling.

Most molecules are anharmonic. How well does the GLE describe the energy decay of these molecules? Tuckerman and Berne[5] through a combination of full molecular dynamics simulations, and stochastic simulations were able to show that even in dense realistic fluids with strong short-range forces the GLE does an excellent job. The calculations proceeded as follows:

4.1   Harmonic diatomic imbedded in Lennard-Jones fluid

The first set of studies were carried on a harmonic diatomic $U(x) = \mu\omega^2 x^2/2$, with frequency choices of $\omega = 60$, 90, 120, and 150. For the harmonic potential, we use an analytic reference system solution with the reversible NAPA scheme in Eq.(23). The peak of the spectral density of the solvent is around $\omega = 20$, so that $\omega = 60$ and 90 for the diatomic solute are well within the significant part of the spectral density, whereas 120 and 150 are not. Thus there is a sharp separation of time scales between the solvent and the vibrational motion of the diatomic at the two higher frequencies. The manifestation of this time scale separation can be seen in Fig. 1 in which the decay envelopes of the velocity autocorrelation functions of the oscillator at the four frequencies are plotted. There is a drastic increase in the decay time between $\omega = 90$ and $\omega = 120$. This is also manifest in the decay time of the energy autocorrelation functions plotted in Fig. 2.

To test the predictions based on the GLE, it is necessary to carry out simulations in which the GLE is integrated numerically. The inputs to the GLE are a friction kernel $\zeta(t)$, and a random force $R(t)$ such that Eq.(3) is satisfied. One way to obtain the friction kernel from Molecular Dynamics is the method of Straub and Berne[6]. If we multiply the GLE for the harmonic oscillator on both sides by $\dot{x}$ and average over a canonical ensemble, we obtain a memory function equation of the Volterra type for the velocity autocorrelation function

$$\dot{C}_{vv}(t) = - \int_0^t d\tau\, K(t-\tau) C_{vv}(\tau) \tag{1}$$

where the kernel $K(t) \equiv \tilde{\omega}^2 + \gamma(t)$. Using the velocity autocorrelation function from the Molecular Dynamics simulations, it is a simple matter to invert the memory function equation to obtain the friction kernel. It should be noted, however, that the inversion process is stable only at low frequency, and in this case works well only for the cases $\omega = 60$ and 90. To find the friction on an extremely high frequency bond, the friction on the bond may be approximated by the friction on a rigid bond[8, 16]. It is shown in ref.[8] that in the infinite frequency limit, the true friction and the friction on the rigid bond are equal. Alternatively, an autoregression technique can be used to determine the friction[17, 18]. In Fig. 3 (a), we show the friction kernels as a functions of time for the case $\omega = 60$ and for the rigid bond, and in Fig. 3 (b), we show the corresponding Fourier transforms. The random force can be obtained in a simple way if it is assumed that $R(t)$ is a Gaussian random process. Then the method of Rice can be used as described in ref. [19, 20].

GLE simulations were carried out on the diatomic molecule for the four frequencies discussed above using an analytic reference system. The GLE requires not the bare frequency ω but the renormalized frequency $\tilde{\omega}$ from the potential of mean force. The potential of mean force surface for this problem has been fit by Straub et al.[21], and we use the renormalized frequencies from this fit which for bare frequency values $\omega = 60, 90, 120$, and 150 are $\tilde{\omega} = 59.447, 89.632, 119.724$, and 149.780, respectively. The decay envelopes of the velocity autocorrelation functions calculated from the GLE agree with Molecular Dynamics with the consequence that the predictions of $1/T_2$ from Eq. (16) are in excellent agreement with the MD simulations. In Table 1, we show the values of $1/T_2$ for each value of ω as predicted by MD, by the GLE, and by perturbation theory on the GLE (cf. Eq.(4)). We see that as the frequency increases,

ω	$\left(\frac{1}{T_2}\right)_{MD}$	$\left(\frac{1}{T_2}\right)_{GLE}$	$\frac{\gamma'(\tilde{\omega})}{2}$
60	2.71	2.70	2.97
90	0.74	0.73	0.78
120	0.19	0.18	0.18
150	0.08	0.08	0.08

Table 1: Vibrational relaxation rates for harmonic oscillator from MD, GLE and Eq.(2.19).

the GLE simulation results and the GLE perturbation theory results come into closer agreement as expected, since the perturbation theory assumed high frequency. The MD and GLE agree well at all frequencies. In Fig. 2, we show the energy autocorrelation functions from the GLE together with the MD. The GLE energy autocorrelation functions can be computed directly from the simulation or using the formula Eq.(13). These results serve as a test of the assumption of a Gaussian random force. It is interesting to note that the GLE autocorrelation functions consistently decay faster than those from the MD although the discrepancy is small. In Table 2, we show the values of $1/T_1$ for each value of ω for the MD, GLE, and GLE perturbation theory.

Again, we see a consistently small discrepancy between the GLE and MD results. In Fig. 4, we show that the solvent force is not well described by Gaussian statistics by plotting the autocorrelation function of the square of the random force computed in the MD simulations using the formula

$$R(t) = \mu\ddot{x} + \mu\tilde{\omega}^2 x^2 + \int_0^t d\tau \zeta(t-\tau)\dot{x}(\tau) \qquad (2)$$

ω	$\left(\frac{1}{T_1}\right)_{MD}$	$\left(\frac{1}{T_1}\right)_{GLE}$	$\gamma'(\tilde{\omega})$
60	5.10	5.56	5.94
90	1.36	1.45	1.56
120	0.34	0.36	0.36
150	0.14	0.15	0.15

Table 2: Energy relaxation rates for harmonic oscillator from MD, GLE and Eq.(2.20).

If $R(t)$ is a Gaussian random process, then, since $\langle R(t)\rangle = 0$, the autocorrelation function of the square of the random force should be given by

$$\langle R^2(0)R^2(t)\rangle = \langle R^2(0)\rangle^2 + 2\langle R(0)R(t)\rangle^2 \tag{3}$$

The solid line in Fig. 4 shows the autocorrelation function of the square of $R(t)$ and the dashed line shows the plot of the factorization based on Eq.(3). The top curve is for a diatomic with frequency $\omega = 60$ and the bottom curve is for a frequency of $\omega = 150$. We see that the higher the frequency, the more dramatic the departure of the random force statistics from that those of a Gaussian random process. It is also interesting to note that the initial value $\langle R^4(0)\rangle$ is consistently larger than the prediction $3\langle R^2(0)\rangle^2$ of Eq.(3). This fact suggests that for short times, the statistics of the solvent force (which will be determined by strong collisions coming from the short range part of the force) are non-Gaussian and we conjecture that these should be treated as a Poisson process.

A possible explanation for the small discrepancy between the GLE and molecular dynamics is the following. In the real fluid the vibrational displacement suffers infrequent strong collisions and frequent soft collisions. Strong collisions occur when energetic solvent atoms approach either of the atomic sites on the molecule very closely. These are binary collisions and should be described by statistics that are more like Poisson statistics. The soft collisions are due to the superposition of longer range forces from many solvent atoms and thus, according to the Central Limit Theorem, should be described by the statistics of Gaussian random variables. We find, consistently, that Gaussian statistics gives rise to faster energy relaxation than what is observed from the MD simulations.

It should be possible to devlope a model which includes both Gaussian and Poissonian collisions in the GLE in such a way that Eq.(3) is satisfied. This would describe the situation envisioned here where the molecule suffers frequent soft collisions punctuated infrequently by strong collisions. This has the flavor of the old Rice-Allnatt theory of liquids[22, 23], and similar effects were discused by Berne et al.[24]. It should be noted that such models could be important when strong infrequent pair

interactions contribute to the energy relaxation. We expect that the stiffer the oscillator, and the lower the solvent density, the more important such effects might
be.

4.2   Anharmonic   Diatomic   imbedded   in   Lennard-Jones Fluid

The second set of studies were carried out on an anharmonic diatomic molecule with
bond potential $U(x) = \mu\omega^2 x^2/2 + f x^3/6$ imbedded in the same Lennard-Jones fluid.
Each simulation requires a value for ω and f. The parameters chosen were $f = 10000$
for $\omega = 60$, $f = 30000$ for $\omega = 90$, $f = 90000$ for $\omega = 120$, $f = 1.8\text{x}10^5$ for $\omega = 150$,
and $f = 1.0\text{x}10^6$ for $\omega = 300$. It is worth examining in what sense, these choices for
f are perturbations on the harmonic potential. The cubic potential has its maximum
at $x_c = -2\mu\omega^2/f$, and the height of the maximum is

$$V_c = \frac{2\mu^3\omega^6}{3f^2} \qquad (4)$$

The heights of the maximum are $V_c = 15.5kT$, $19.68kT$, $12.29kT$, $11.72kT$, and
$24.3kT$ corresponding to $\omega = 60$, 90, 120, 150, and 300, respectively with $kT = 2.5$. Therefore, for energies around kT, the effect of the anharmonicity is small. For
$\omega = 120$ and 150 two other f values 40000 and 60000, respectively were also chosen.
These f values give $62.2kT$ and $105kT$ for V_c, respectively. That the anharmonicity
is small can be seen as well from the forces. Since the particle is near the bottom of
a deep well, we expect that

$$\mu\omega^2 x \gg \frac{1}{2}f x^2 \qquad (5)$$

which is verified in the MD simulations.

In Fig. 5, we plot the decay envelopes of the velocity autocorrelation functions
computed from the Molecular Dynamics simulations for the five frequencies. It is
interesting to note that even though the cubic anharmonicity is a small perturbation
on the harmonic potential, the high frequency dephasing times are dramatically different from the harmonic results. The GLE results are also plotted in this figure. We
see that even for the cubic diatomic, the GLE results are in good agreement with
MD, at least in the prediction of $1/T_2$. This is a particularly interesting result, since
the GLE cannot be generally derived for the case that both the bath and the bond
potential are anharmonic. In the case of an anharmonic bond potential, the GLE
can be derived from the Zwanzig Hamiltonian[9]. The agreement of the GLE with
MD for the cubic potential suggests that there may be an underlying effective harmonic bath (i.e., a particular choice of the g_α's) which reproduces the effects of the
Lennard-Jones potential at the chosen temperature and density. The GLE derived in
this manner is valid even for processes far from equilibrium. For a general bath, the

ω	f	$\left(\frac{1}{T_2}\right)_{MD}$	$\left(\frac{1}{T_2}\right)_{GLE}$	Eq.(2.42)	Kubo	mod. Kubo
60	10^4	3.03	3.05	3.41	0.85	3.04
90	$3 \cdot 10^4$	1.42	1.41	1.30	0.70	1.33
120	$4 \cdot 10^4$	0.53	0.52	0.44	0.31	0.52
120	$9 \cdot 10^4$	1.71	1.64	0.91	0.84	1.02
150	$6 \cdot 10^4$	0.28	0.27	0.22	0.21	0.25
150	$1.8 \cdot 10^5$	2.00	2.01	0.85	0.79	0.86
300	10^6	0.36	0.36	0.35	0.35	0.35

Table 3: Dephasing rates for cubic oscillator from MD, GLE, Eq.(2.42), Kubo theory, and modified Kubo theory

GLE can be derived in linear response theory, in which case, it will appear to have an effective harmonic potential. In the next section, we shall study a system in which the GLE remains valid for the anharmonic oscillator even for highly non-equilibrium states, a fact which strongly suggests the notion of an underlying effective harmonic bath.

The predictions of $1/T_2$ for given ω and f are summarized in Table 3, where we show the data from MD, GLE, Eq.(16), Kubo theory and a modified Kubo theory.

The Kubo theory prediction is obtained by computing the fluctuating frequency and then evaluating $1/T_2^*$ from Eq.(1). For a derivation of the fluctuating frequency, see Appendix D. We see that the agreement is extremely poor except for the case of $\omega = 300$. In light of this finding, we propose that the Kubo theory be modified to include energy relaxation according to Eq.(17), that is

$$\frac{1}{T_2} = \frac{1}{2T_1} + \int_0^\infty d\tau \langle \delta\omega(0)\delta\omega(\tau) \rangle. \tag{6}$$

This modification is *ad hoc* and cannot be derived with the framework of classical Kubo theory. However, we see Table 3 that including energy relaxation in the Kubo theory gives agreement with the Eq.(16) as expected. These results suggest that the contribution of $F_2(t)$ in

$$\omega(t) = \sqrt{\omega^2 + \frac{2F_2(t)}{\mu} - \frac{fF_1(t)}{\mu^2\omega^2}} \approx \omega + \frac{F_2(t)}{\mu\omega} - \frac{fF_1(t)}{2\mu^2\omega^3} \tag{7}$$

is relatively unimportant (see Appendix D in ref. [5] for the derivation and for the definition of the configuration dependent terms $F_1(t)$ and $F_2(t)$). The failure of Kubo theory at low frequency suggests that energy relaxation is an important contribution to the decay of the velocity autocorrelation function, whereas at extremely high frequency, there is almost no energy relaxation over the time needed for this function to decay so that the decay is due entirely to dephasing. Table 3 also indicates that Eq.(16) fails for $\omega = 120$ and 150 at the higher f values. This must be due to the fact that the cubic perturbation is too strong to be treated by perturbation theory, since for the lower values of f Eq.(16) predicts the relaxation rate accurately. The Kubo theory with energy relaxation is worked out to the same order in f, hence it will fail where the GLE perturbation theory fails.

4.3   Non-Equilibrium Energy Decay

The GLE accurately predicts vibrational dephasing times for anharmonic diatomics in the liquid in the linear response regime. A stringent test of the GLE, however, is its prediction of energy relaxation from non-equilibrium initial states. If an initial state is chosen not too far from equilibrium, then the energy relaxation should be well described by the Onsager relation (linear response theory)

$$\frac{\bar{\epsilon}(t) - \bar{\epsilon}(\infty)}{\bar{\epsilon}(0) - \bar{\epsilon}(\infty)} = C_{\epsilon\epsilon}(t) \tag{8}$$

where $\bar{\epsilon}(t)$ is the average over a non-equilibrium ensemble, and $C_{\epsilon\epsilon}(t)$ is the equilibrium energy autocorrelation function. Clearly, $\bar{\epsilon}(\infty) = \langle \epsilon \rangle = kT$, which provides a criterion for knowing when the energy has relaxed. As long as the Onsager relation is valid, it is expected that the GLE will agree with MD simulations of energy relaxation from the non-equilibrium state, since linear response will hold. Whitnell and coworkers have tested the validity of Eq.(8) in studies of fast diatomics in polar solvents[16] However, it is not clear that the GLE will be able to predict energy relaxation from initial states which are not in the linear response regime. This question is examined by comparing the relaxation predicted by MD and the GLE for a diatomic with an internal Morse potential in the Lennard-Jones liquid. The Morse potential is used here instead of the cubic potential as it allows the oscillator to start at much higher energies without "going over the edge" as would happen with the cubic potential.

The Morse potential is of the form given by Eq. (18) For $\omega = 120$ and $f = 90000$, the parameters are $D_0 = 207.36$ and $\alpha = 4.167$. Initial conditions for the oscillator are chosen such that $x(0) = 0$ so that all the energy initially is kinetic. The initial velocity is chosen so as to cause the relative separation of the atoms to increase. For the MD simulations, an ensemble of initial conditions must be prepared. This is done by placing a rigid diatomic with the bond length fixed at the equilibrium separation and allowing the fluid to equilibrate around it. In order to insure that the initial conditions of the bath are distributed canonically the system is run with the Nosé-Hoover chain dynamics[25]. Once the fluid has equilibrated around the rigid diatomic, configurations are written out every 300 steps until 100 total initial states are accumulated. The 300-step time interval is a long enough to avoid correlations in time between initial states. This procedure should also correspond directly to the canonical sampling of realizations of the random force which is done for the GLE simulations. Finally, to simulate the energy relaxation, the Nosé-Hoover chains are removed, and trajectories at constant total energy are run starting from each of the 100 sampled initial fluid states. The initial conditions of the diatomic are the same for all trajectories, as described above. For very high energy initial states, the giving off of energy by the diatomic to the fluid can cause the fluid to heat up, thus changing the spectral density. To minimize this effect, the fluid must contain a large number of atoms, a fact which makes the computations extremely intensive. In order to increase the efficiency of the simulations, the double RESPA scheme of ref.[13] with long and short range force breakup is used. This gives an overall factor of 32 in the time step over velocity Verlet, and an overall saving in CPU time of 16 over velocity Verlet.

MD simulations using 500 fluid atoms and a single diatomic were run for initial oscillator energies of $20kT$, and $40kT$ as well as GLE simulations. GLE simulations were also run for a variety of low energies to determine how high the energy of the diatomic must be in order to be outside the linear response regime. It was found that the linear response theory Eq.(8) is valid up to energies around $10kT$ for this system so that energies of 20 and $40kT$ are well outside this regime. In Fig. 6, we show the comparison of the MD and GLE simulations of the non-equilibrium energy decay (i.e., the left side of Eq.(8)). Also shown in the figure is the equilibrium energy autocorrelation function for the Morse potential determined from the full GLE simulations of Sec.4.2. We see that even up to energies of $20kT$ and $40kT$, the GLE predicts very accurately, the non-equilibrium relaxation rate, even though these are not well described by linear response theory. In addition, MD and GLE simulations were also run for a harmonic diatomic far from equilibrium for initial bond energies up to $90kT$, and all were found to be well described by linear response theory.

In all cases studied, the nonlinear GLE predicts that the energy relaxes faster than or is bounded from above by what would be predicted by the GLE in the linear response regime. This is expected since, for the nonlinear oscillator, the vibrational period is an increasing function of the vibrational energy. (The vibrational frequency is therefore a decreasing function of energy.) Hence, a highly excited oscillator will have a vibrational frequency lying closer to the accepting modes of the bath and will thus exchange energy with the bath more readily. As the oscillator relaxes, its vibrational frequency gets larger and detunes away from the bath modes with the result that the vibrational energy transfer slows down and eventually overlaps what is observed in the linear response regime. This is similar to what was observed in the relaxation of I_2 and Br_2 in argon[26, 27]

To understand these results, consider that when the bond potential and the bath are both anharmonic, the GLE cannot be shown to be a valid description of the system. However, if the bath can be well described by an equivalent effective harmonic bath, then the GLE should be valid for any bond potential even far from equilibrium. These results suggest, then, that there should an effective harmonic bath underlying the true Lennard-Jones fluid used for these studies. The notion of an underlying effective harmonic bath for this system is not altogether surprising when one considers that the reduced density $\rho\sigma^3 = 1.05$ is just below the crystal density for reduced temperature $\hat{T} = 2.5$. Although the fluid is a hot, dense fluid, it is not far from the solid phase. It is expected that if the density were lowered, the notion of an underlying effective harmonic bath would break down as strong impulsive collisions would become increasingly important. In this case, the GLE with a Gaussian random force would probably not be a good description of the non-equilibrium energy decay. However, as discussed in Sec. 4.1, if impulsive, Poissonian type collisions could be built into it, the GLE might serve as a good description of the harmonic bath is a good approximation for the general bath.

5   Conclusion

In this paper full molecular dynamics simulations of vibrational relaxation in simple fluids show that the Generalized Langevin Equation is an excellent quantitative model. We have derived expressions for the velocity, position, and energy autocorrelation functions from the GLE for a harmonic potential of mean force and extracted the relaxation time T_2 and T_1 from these. We have also used perturbation theory to derive the vibrational relaxation time T_2 from the GLE with cubic anharmonicity. These results have been compared to full molecular dynamics simulations of a single oscillator at high frequency and a range of cubic couplings in a Lennard-Jones fluid. In order to solve the multiple time scale problem, the reversible version of RESPA and a new multiple time scale algorithm for the GLE based on the reversible version of RESPA were used. The friction kernel for the GLE was calculated using the method of Straub and Berne[6] or that of Berne, et al [8] from the force autocorrelation function on a rigid diatomic. The GLE and MD were found to be in good agreement for both predictions of T_1 and T_2 for the harmonic diatomic. At high frequency, a small discrepancy was found between the GLE and MD for T_1. We attribute this to the fact that the GLE misses the rare impulsive collisions which occur in real systems by assuming the random force is a Gaussian random process. Therefore, this discrepancy is expected to grow as the frequency increases, and could become important at low frequencies for less dense systems. We derived an expression for the deviation in the energy autocorrelation function in the Gaussian random force approximation and the true energy autocorrelation function for short times and found the predicted discrepancy consistent with our observations from the GLE-MD comparison.

The predictions of vibrational relaxation times of an anharmonic diatomic with cubic anharmonicity from MD and the GLE were also compared. These were found to be in good agreement at all frequencies and cubic couplings. The perturbation theory result Eq.(16) was found to break down for $\omega = 120$ and 150, when f was too large. When the perturbation was made smaller, the agreement was better. The MD and GLE results were also compared to Kubo theory, which was found to be in poor agreement with the simulation results for all except the highest frequency

($\omega = 300$). We conjecture that the breakdown of Kubo theory is due to the fact that energy relaxation can contribute significantly to the vibrational relaxation of an anharmonic oscillator at low frequencies, and Kubo theory leaves this contribution out. By modifying Kubo theory to include the energy relaxation, the modified predictions were found to be in much better agreement with the simulation results. Since the GLE cannot be shown to be valid for an anharmonic diatomic interacting with a an anharmonic bath, except in linear response, the agreement of the GLE with MD for all frequencies and cubic couplings is somewhat surprising. The GLE can be derived from the Zwanzig Hamiltonian[9], however, a fact which suggests that the Lennard-Jones bath at $\hat{T} = 2.5$ and $\rho\sigma^3 = 1.05$ may be well described by an underlying effective harmonic bath.

To test the validity of the GLE for nonlinear oscillators and the notion that the bath may be described well by an underlying effective harmonic bath, the energy relaxation from non-equilibrium states for an diatomic with an internal Morse potential was studied. The parameters of the Morse potential were chosen to agree with the cubic potential for $\omega = 120$ and $f = 90000$ by expanding the Morse potential and equating the harmonic and cubic terms. It was found that, even when the non-equilibrium decay was poorly described by linear response theory, the simulation of the nonlinear GLE predicted the same relaxation rate as MD. As expected, the non-equilibrium relaxation of a harmonic diatomic, a linear system, was found to agree with linear response theory for all initial states up to $90kT$. The fact that the GLE predicts the same relaxation rate as MD even for highly non-equilibrium states supports the notion that the fluid can be well described by an effective harmonic bath. It is expected that for lower densities, in which impulsive collisions will be more frequent and more important, the GLE would begin to break down for highly non-equilibrium initial states. However, if the GLE could be supplemented to include the impulsive Poissonian-type collisions along with the Gaussian random force in a way which preserved the second fluctuations dissipation theorem, then the range of validity would, most likely, be increased to include low density, high temperature situations.

Acknowledgments

This research was funded by grants from the NSF and NIH. In addition we wish to thank IBM Thomas J. Watson Research Laboratories for generous allocation of computer time. We also gratefully acknowledge a grant of computer time from the NCSA at the University of Illinois. I would like to thank Dr. Mark Tuckerman for many discussions and for providing me with copies of the figures.

References

[1] R. Kubo. *Adv. Chem. Phys.*, 13:101, 1963. J. Math. Phys. **4**, 174 (1963).

[2] R. Kubo. In D. Ter Haar, editor, *Fluctuations, Relaxation, and Resonance in Magnetic Systems*. Plenum Press, New York, NY, 1962.

[3] D. Oxtoby. *J. Chem. Phys.*, 70:2605, 1979.

[4] D. Oxtoby, D. Levesque, and J. Weiss. *J. Chem. Phys.*, 68:5528, 1978.

[5] M. E. Tuckerman and B. J. Berne. *J. Chem. Physics*, 98:7301, 1993.

[6] J. E. Straub, M. Borkovec, and B. J. Berne. *J. Phys. Chem.*, 91:4995, 1987.

[7] J. Straub, B. J. Berne, and Benoit Roux. *J. Chem. Phys.*, 93:6804, 1990.

[8] B. J. Berne, M. E. Tuckerman, John E. Straub, and A. L. R. Bug. *J. Chem. Phys.*, 93:5084, 1990.

[9] R. Zwanzig. *J. Stat. Phys.*, 9:215, 1973.

[10] M. E. Tuckerman and B. J. Berne. *J. Chem. Physics*, 95:4389, 1991.

[11] M. Tuckerman, G. Martyna, and B. J. Berne. *J. Chem. Phys*, 93:1287, 1990.

[12] M. Tuckerman, G. Martyna, and B. J. Berne. *J. Chem. Phys*, 94:6811, 1991.

[13] M.E. Tuckerman and B. J. Berne. *J. Chem. Phys*, 95:8362, 1991.

[14] M.E. Tuckerman, G. J. Martyna, and B. J. Berne. *J. Chem. Phys*, 97:1990, 1992.

[15] A. M. Levine, M. Shapiro, and E. Pollak. *J. Chem. Phys.*, 88:1959, 1988.

[16] R. M. Whitnell, K. R. Wilson, and J. T. Hynes. *J. Phys. Chem.*, 94:8628, 1991.

[17] D. E. Smith and C. B. Harris. *J. Chem. Phys.*, 92:1304, 1990. Unfortunately this technique is impracticle for determining the long time and low frequency dependence of the friction, a regime necessary for determining the dephasing times in anharmonic systems.

[18] D. E. Smith and C. B. Harris. *J. Chem. Phys.*, 92:1312, 1990.

[19] S. O. Rice. *Bell Tel. J.*, 23:282, 1944.

[20] S. O. Rice. *Bell Tel J.*, 25:46, 1945.

[21] J. E. Straub, M. Borkovec, and B. J. Berne. *J. Chem. Phys.*, 89:4833, 1988.

[22] S. A. Rice and A. R. Allnatt. *J. Chem. Phys.*, 34:2144, 1961.

[23] A. R. Allnatt and S. A. Rice. *J. Chem. Phys.*, 34:2156, 1961.

[24] B. J. Berne, J. L. Skinner, and P. G. Wolynes. *J. Chem. Phys.*, 73:4314, 1980.

[25] G. J. Martyna, M. Klein, and M. Tuckerman. *J. Chem. Phys.*, 97:2635, 1992.

[26] J. T. Hynes, R. Kapral, and G. M. Torrie. *J. Chem. Phys.*, 72:177, 1980.

[27] F. G. Amar and B. J. Berne. *J. Phys. Chem.*, 88:6720, 1984.

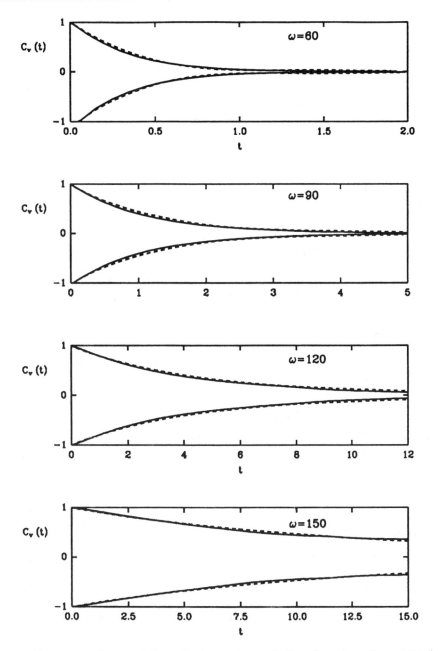

Fig. 1 Decay envelopes of the velocity autocorrelation functions from MD (solid line) and GLE (dashed line) simulations for a harmonic oscillator in Lennard-Jones fluid for $\omega = 60$, 90, 120, and 150.

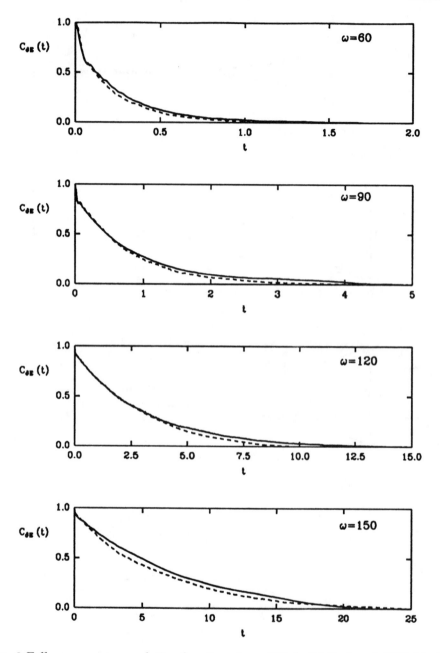

Fig. 2 Full energy autocorrelation functions from MD (solid line) and GLE (dashed line) simulations of a harmonic oscillator in Lennard-Jones fluid for $\omega = 60$, 90, 120, and 150.

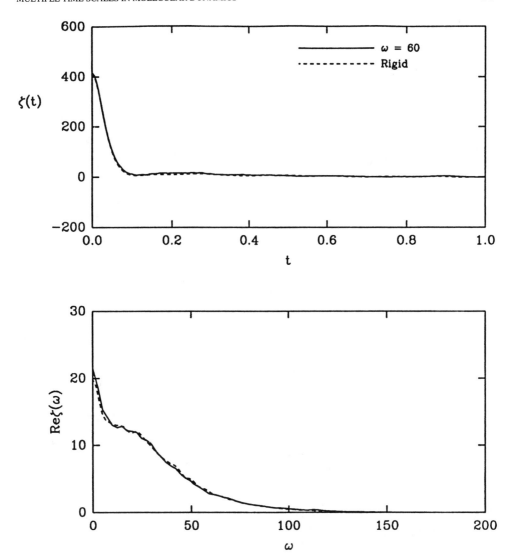

Fig. 3 The friction kernel used in the GLE simulations as computed using the method of Straub and Berne[6] based on Eq.(1) (solid line) and using the method of Berne *et al* from the velocity autocorrelation function on a rigid bond (dashed line). The upper curve shows the time dependent friction kernel, and the lower curve shows the frequency dependence from the Fourier cosine transform.

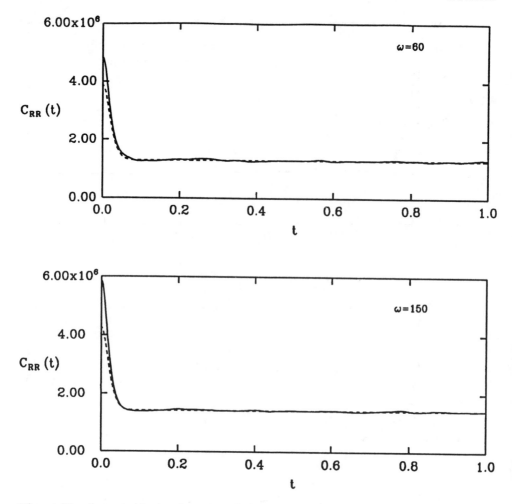

Fig. 4 The Autocorrelation function of the square of the random force as computed from MD using Eq.(2) (solid line) and the Gaussian approximation to this correlation function from Eq.(3) (dashed line) for $\omega = 60$ (upper curve) and $\omega = 150$ (lower curve)

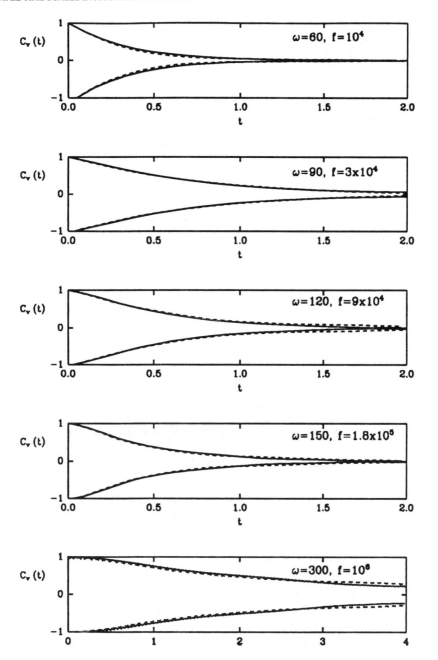

Fig. 5 Decay envelopes of the velocity autocorrelation functions from MD (solid line) and GLE (dashed line) simulations for a cubic oscillator (*cf* Eq.(14)) in Lennard-Jones fluid for $\omega = 60, 90, 120, 150$ and 300 with cubic couplings shown

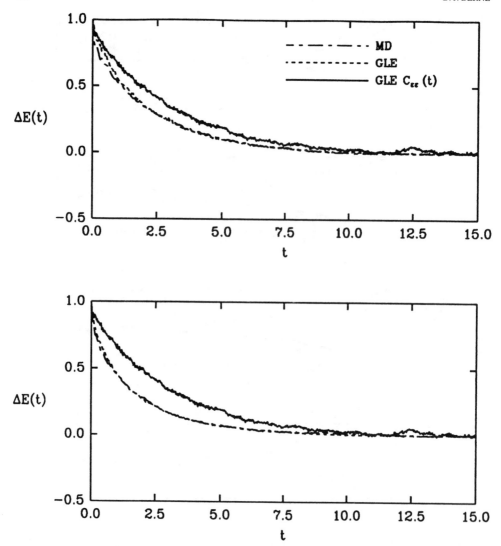

Fig. 6 Non-equilibrium energy decay from the left side of Eq.(8) for a Morse oscillator in a Lennard-Jones fluid as computed from the GLE (dashed line) and MD (double dashed line) using 500 particles for initial conditions corresponding to energies of $20kT$ (upper curve) and $40kT$ (lower curve). Also shown in each figure, with a solid line, is the equilibrium energy autocorrelation function (i.e., linear response theory) from full GLE simulations.

NEW APPROACHES TO SOLUTION REACTION DYNAMICS: QUANTUM CONTROL AND ULTRAFAST DIFFRACTION

Bern Kohler, Jeffrey L. Krause, Ferenc Raksi, Christoph Rose-Petruck,
Robert M. Whitnell, Kent R. Wilson, Vladislav V. Yakovlev and YiJing Yan
Department of Chemistry, University of California, San Diego
La Jolla, CA 92093-0339, USA

1. Introduction

Much progress has been made in recent years in the theoretical and computational treatments of solution reaction dynamics. On the computational side alone, several hundred papers have been published over the past 20 years as described in a recent review article.[1] Our own work on the molecular dynamics of solution reactions (much of which has been in collaboration with Casey Hynes and Raphy Levine) has covered: I_2 and ICN photodissociation in rare gas solution,[2-6] atom replacement in rare gas solution,[7-10] S_N2[11-15] and S_N1[16] reactions in water, the effect of rare gas solvent on unimolecular dissociation and IVR,[17] parallels to gas phase dynamics,[7, 18] stochastic models of solution reactions,[19] and the link between reaction dynamics and solution structure.[20]

While it is gratifying that to see these achievements on the theoretical and computational sides, there is still much work to be done on the third leg of the triad, that of experiment. Examples have been reported in the areas of photodissociation and solvation dynamics, but there are still few compellingly clear experimental tests of the many microscopic theoretical models that have already been developed. In this work, we discuss our vision of new experimental directions designed to probe more directly and with greater precision the dynamics of solution reactions. Such experiments will allow tests of the theoretical and computational ideas which have been developed by many workers over the past two decades.

In the following, we will discuss two new approaches to the study of molecular dynamics, quantum control and ultrafast diffraction, for discovering the microscopic reality of how reactions happen in solution. These ideas, however, are much more general, and are in principle equally applicable to gas phase, surface, cluster and condensed phase dynamics.

We will begin with quantum control. Our goal here is to develop methods to probe the motions of molecules by focusing quantum wavepackets in time, position, and momentum, thus observing the progress, by a subsequent interaction pulse, of the

J. Jortner et al. (eds.), Reaction Dynamics in Clusters and Condensed Phases, 495–507.

reaction in these variables. We discuss below the theoretical, computational and experimental implementations of these ideas.

Many current experimental tests of solution reaction dynamics, including the ones we are developing, rely on ultrafast spectroscopy. In favorable cases, ultrafast spectroscopy can provide snapshots of wavepacket dynamics and thereby give detailed knowledge of wavepacket evolution. However, especially in larger systems, the dependence of ultrafast spectra on quantities that are often not precisely known can present problems in interpreting the data. We therefore proposed a decade ago[3, 4] a more direct measure of dynamics, ultrafast diffraction, which can in principle produce a series of molecular structure snapshots that can then be strung together to form a movie. The experimental challenge of actually doing this with either x-rays or electrons is formidable, and neither we, nor to our knowledge anyone else at the time of this writing, has measured the actual dynamics of a chemical reaction by ultrafast diffraction, although the reactants and the products have been observed by time-resolved electron diffraction.[21, 22] We briefly discuss below our work designed to test the feasibility of observing molecular dynamics with ultrafast diffraction of x-rays and electrons.

2. Quantum Control

Ultrafast spectroscopy has in general been implemented by exciting a system with a short light pulse (for example, exciting a diatomic molecule onto a dissociative potential energy curve), and probing the resulting time-evolved wavepacket with short light pulses at various delay times with respect to the initial pulse. While this method has proven useful in many systems, it is a passive technique with respect to the dynamics on the excited state. One often can derive only rough conclusions about the excited state dynamics. If one is interested, for example, in using a dissociating molecule to probe solute-solvent structure and dynamics, then active control of excited state dynamics, in which the experimentalist purposely intervenes in the dynamics, can be more useful. One can achieve this type of control of the quantum evolution of the system with tailored light pulses.[23-25] These pulses allow the evolving matter packet (wavefunction) to be focused in time, position and momentum, in a predictable manner, as we shall demonstrate.[26] The dynamic behavior of the system (gas, liquid, cluster or surface) can be analyzed by a subsequent light pulse as a function of these variables and in principle, be controlled as well.

In collaboration with Shaul Mukamel, we have developed a density-matrix quantum control theory[26, 27] which allows just such an approach to solution reaction dynamics (as well as the dynamics of reactions in clusters or on surfaces). One can consider thermal distributions (mixed states),[27, 28] as well as stochastic dephasing and energy decay in both the nuclear and electronic degrees of freedom.[27] A hierarchy of rigor is possible, encompassing quantum, semiclassical and classical approaches or a mixture of these within the same calculation.

In the weak field limit, Yan et al.[26, 27] have derived an equation (hereafter referred to as the "Yan equation"), which allows the globally optimal light field, or the field that best drives an initial state to a selected final state, to be computed from an eigenequation, in which the eigenvalues are the yields and the eigenfunctions are the fields. The eigenvector associated with the largest eigenvalue is the globally optimal light field, or "divine lightwave," in the weak field limit. Specifically, given an initial density matrix $\rho(-\infty)$, we wish to calculate the electric field which best achieves a particular target operator A. For example, $\hat{A}$ can represent a projection onto a particular pure state $|\Phi\rangle$, and therefore take the form $\hat{A} = |\Phi\rangle\langle\Phi|$. For simplicity, we limit our consideration of control here to the case in which the initial density matrix is on a ground potential energy surface and the target operator is a projection operator onto a distribution in phase (postion/momentum) space on an excited potential energy surface. The yield of the process is defined by

$$A(t_f) \equiv \text{Tr}\left[\hat{A}\rho_e(t_f)\right], \tag{1}$$

where $\rho_e(t_f)$ is the density matrix on the excited state at the target time t_f. In the weak field limit, through second order in perturbation theory, the yield is given by[26, 27]

$$A(t_f) = \int_0^{t_f}\int_0^{t_f} d\tau' \, d\tau M_e^S(\tau,\tau')E^*(\tau)E(\tau'), \tag{2}$$

where $E(\tau)$ can be any arbitrary electric field and $M_e^S(\tau,\tau')$ is a (Hermitian) molecular response function that depends only on the initial density matrix, the dynamics of the system on the excited state in the absence of any field and the nature of the target. If one considers variations of $E(\tau)$ about the globally optimal field, an eigenequation (the "Yan equation") for the optimal field results:[26, 27]

$$\int_0^{t_f} d\tau' \, M_e^S(\tau,\tau')E(\tau') = \lambda E(\tau), \tag{3}$$

where λ is related to the yield corresponding to the field $E(\tau)$ through the relation

$$A(t_f) = \lambda \int_{t_0}^{t_f} d\tau |E(\tau)|^2. \tag{4}$$

Thus, the eigenfunction corresponding to the largest eigenvalue of Eq. (3) (or the equivalent matrix form found by discretizing time) is the globally optimal field, the "divine lightwave", which is the ideal solution in the weak field limit.

We have employed these techniques to demonstrate wavepacket focusing in time, position and momentum, using as illustrations an iodine "molecular cannon" and "molecular reflectron."[26] These focused wavepackets are an initial step to controlling chemical reactions by, for example, a subsequent probe interaction to a higher-lying state or back to the ground state. The molecular cannon produces an outgoing minimum uncertainty wavepacket in the dissociative region of the excited potential surface in which the separated atoms are focused in both position and momentum. Figure 1 shows the results of focusing of such an outgoing wavepacket on the iodine B state at an internuclear distance of 5.84 Å at a final time of 1100 fs. The initial state is the $v = 0$ level of the iodine X state; thus, this calculation is performed at 0 K.

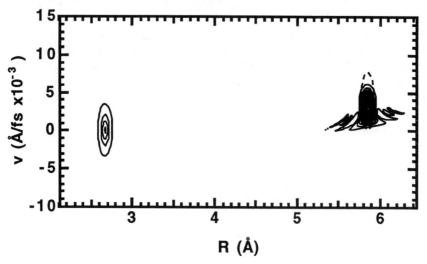

Figure 1: Iodine molecular cannon. Wigner representation in phase space of the focused wavepacket at $t_f = 1100$ fs. The solid contours on the right are the focused wavepacket; the dotted contours are the target. The solid contours on the left are the initial $v = 0$ state on the iodine X state. The internuclear velocity, rather than the momentum, is plotted on the vertical axis for ease of interpretation.

We define a measure $\alpha(t)$ of how well the field creates a wavepacket that overlaps the target. We have termed this measure the "achievement function" and defined it by

$$\alpha(t) = \left\{ \frac{\text{Tr}\left[\hat{A}\rho_e(t)\right]}{\text{Tr}\hat{A}\,\text{Tr}\rho_e(t_f)} \right\}^{1/2}, \tag{5}$$

in which $\alpha(t_f) = 1$ for perfect control. In the case where both the excited state density matrix and the target operator can be expressed in terms of single pure states, the achievement function reduces to the overlap of the wavefunction on the excited state, $\psi_e(t)$, with the target function, Φ. For the molecular cannon shown in Figure 1, Eq. (5) gives an achievement of 0.92.

Figure 2 shows the globally optimal weak field which produces the focused wavepacket shown in Figure 1. This field is displayed in a Wigner time-frequency representation, modified in the interest of clarity to suppress the self-interference of the field. We wish to make three points about this field. First, it is simple. The temporal envelope of the field is smooth, and the Wigner representation shows a monotonic increase in the average frequency over the course of the pulse. Second, the globally optimal field is robust. We have shown that fitting this field to a Gaussian function, incorporating linear and quadratic chirps, results in a field which gives an achievement, as calculated from Eq. (5), that is 94% that of the actual globally optimal field.[26] This result is obtained even though the temporal envelope of the globally optimal field is not Gaussian. However, the fit to the Gaussian function is able to reproduce the chirp quite well, and this appears to be the key to the success of the molecular control. We have also tested the sensitivity of the results with respect to variations in the parameters of the best fit field, and found that reasonable changes in the parameters do not dramatically reduce the control.[28] Finally, pulses such as these are experimentally conceivable. The frequency bandwidth, temporal duration and chirp are within the state-of-the-art.

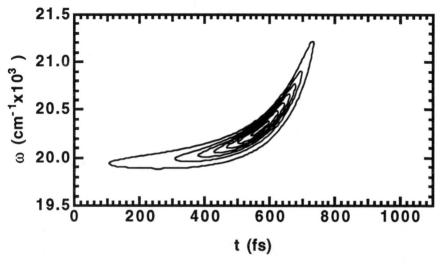

Figure 2: "Divine lightwave" for the iodine molecular cannon in a Wigner representation. The frequency on the vertical axis is the photon energy from the $v = 0$ initial state.

Our second illustrative "molecular machine" is an iodine "molecular reflectron" which produces an incoming minimum uncertainty wavepacket with a negative average momentum. In the example calculations, we choose to target the reflectron at 3.72 Å with an average energy small enough that the excited-state wavepacket is bound. Figure 3 shows the final wavepacket for the molecular reflectron, while Figure 4 displays the Wigner representation of the globally optimal field which leads to this focusing. The achievement for the reflectron is 0.97. The globally optimal field is again robust, with the fit to a Gaussian giving 97% of the achievement of the globally optimal field. Modifications to this Gaussian,[28] including changing the temporal width of the pulse and setting the quadratic chirp to zero have only small effects on the amount of control achieved. Changing the linear chirp by a substantial factor (by changing its sign, setting it equal to zero or changing it by a factor of 2), however, leads to a large decrease in the achievement.

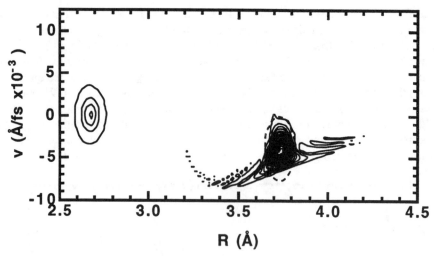

Figure 3: Iodine molecular reflectron at 0 K. Wigner representation in phase space of the focused wavepacket at t_f = 550 fs. Contours as in Figure 1.

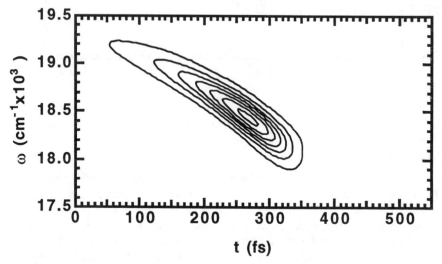

Figure 4: Wigner representation of the "divine lightwave" for the iodine molecular reflectron. The frequency on the vertical axis is the photon energy.

It is not easy to perform experiments at 0 K, especially in solution, and the temperature dependence of control is therefore quite important. Because our theory has been cast in a density matrix formalism, the extension to thermal mixed states is straightforward.[27, 28] Figure 5 shows the final state of the reflectron at 300 K. One can see from this figure that both the initial phase space distribution on the ground state and the final distribution on the excited state are broader than those for the 0 K case shown in Figure 3. However, the achievement, 0.91, is still quite good. A surprising result[28] is that the globally optimal fields for the 0 K and 300 K cases are almost identical, even though the I_2 ground state vibrational population at 300 K is only 0.64. Once again, the robustness of the optimal field is evident. Even a substantial increase in temperature does not significantly alter the amount of control achieved, or the form of the optimal field.

After the matter has been focused using quantum control, it can be detected by probing with a second short pulse of light. If the Franck-Condon factors are right, the focused packet will be excited either to another excited electronic state, producing fluorescence,[29, 30] or back to the ground state by stimulated emission pumping (SEP).[31] Thus, through quantum controlled focusing by an initial tailored pulse of light, followed by a second probe pulse, we can, in principle observe molecular dynamics as a function of time, position and momentum (within the limits of the uncertainty principle). The method to calculate such a response to two pulses has been discussed by us elsewhere.[5, 32]

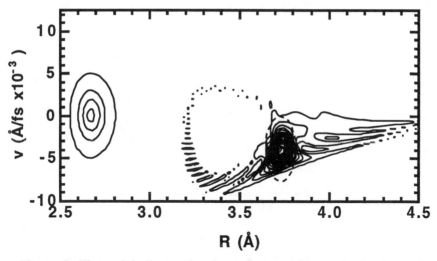

Figure 5: Thermal iodine molecular reflectron. Shown is the focused wavepacket at an initial temperature of 300 K at $t_f = 550$ fs. Solid and dotted contours on the right as in Figure 1. The solid contours on the left represent the equilibrium phase space representation on the ground electronic state at 300 K as computed using the Boltzmann weights of the vibrational levels.

These techniques can be extended to the solution phase by two methods. First, one can represent the effect of the solvent stochastically, as we have already illustrated,[27] for quantum control, in both the solute nuclear and electronic degrees of freedom. Second, we can go beyond the 3 or 4 atom computational limit for a full quantum treatment by turning to classical and semiclassical approximations.

We take as our example a classical version of the control theory in which we focus an outgoing iodine wavepacket in the gas phase and in Ar solution. The parameters of the target are the same as for the reflectron, but the sign of the average momentum is chosen to be outgoing rather than incoming. We discuss the construction of the M matrix of Eq. (3) for classical systems in detail elsewhere,[33] and note here that the excitation from the X to B state of iodine is treated in the classical Condon approximation, or static dephasing limit, and that it is necessary to calculate classical trajectories only on the B state in order to compute the M matrix for the gas phase system. The initial distribution is taken to be a 300 K classical distribution on the ground electronic state. Figure 6 shows a comparision of the Wigner representations of the globally optimal fields for focusing quantum and classical systems at 300 K. The classical calculations were performed in 1-D, ignoring rotations, The time-frequency structure of the two fields are very similar, indicating that classical mechanics gives a very good representation of the full quantum mechanical system in this case. The achievements are high in both cases, being 0.9 for the classical system and 0.95 for the quantum case.

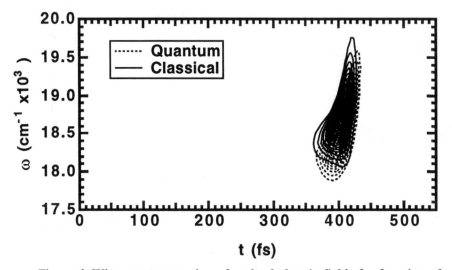

Figure 6. Wigner representation of optimal electric fields for focusing of outgoing I_2 wavepacket. Shown are the results from classical mechanics (solid line) and quantum mechanics (dotted line). Both calculations were performed at 300 K.

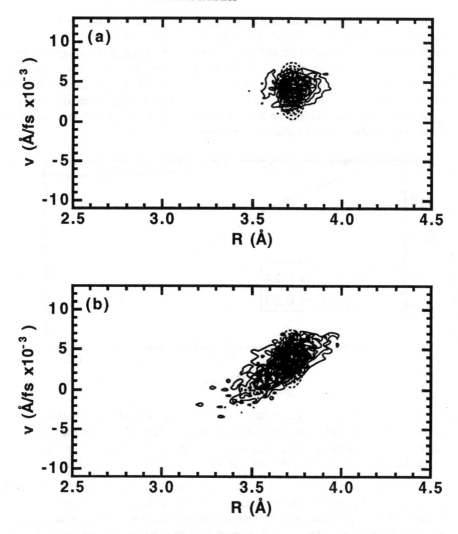

Figure 7: Classical control for a focused outgoing iodine wavepacket in Ar solution. Reduced density (a) $\rho^* = 0.24$; (b) $\rho^* = 0.83$. Solid contours are the classical phase space distribution on the B state under the influence of the globally optimal field at the target time t_f. Dotted contours are the target. These calculations are performed using 3000 trajectories in each case.

We have also performed calculations for I_2 in solution using two different reduced densities of Ar solvent, $\rho^* = 0.24$, corresponding to 100 bar Ar and $\rho^* = 0.83$, corresponding to approximately 3000 bar Ar. The former density is in effect a high pressure gas, while the latter density is more similar to liquid densities. The classical calculation of the globally optimal field can be performed exactly as before, but now the average over the ensemble of classical systems includes an average over solvent dynamics as well. For the target described here, the achievement at the lower density of Ar is 0.86 while in the higher density solvent the achievement is 0.68. The phase space distributions at t_f for these two cases are shown in Figure 7. From these distributions, one can see, even for the "divine lightwave," the strong randomizing effect of the higher density solvent, while the distribution at the lower density of Ar is much more localized and is still very similar to the gas phase distribution.

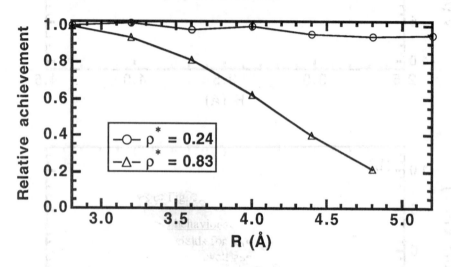

Figure 8: Effect of rare gas solution on control of molecular dynamics. Shown here is the achievement relative to the gas phase for the focusing of an outgoing wavepacket in rare gas solution at two different densities. Due to statistical uncertainties, differences of less than ~3% are not significant.

Solvent effects on the dynamics can readily be studied by attempting to focus matter packets at progressively larger distances. We have therefore performed classical control calculations on a series of minimum uncertainty phase space targets at a series of I_2 internuclear distances. The average momenta of the targets are chosen so that the energy of the center of the target, $\bar{p}^2 / 2m + V(\bar{q})$, is constant. This energy is chosen to be $4720\ cm^{-1}$ above the minimum of the B state which is approximately $350\ cm^{-1}$ above the dissociation limit. The achievement relative to gas phase control as a function of the position center of the target is shown in Figure 8. The low density solvent does not substantially affect the attainable degree of control until fairly large internuclear distances. However, at larger internuclear distances, the high density solvent progressively impedes the ability to focus the dynamics of the iodine molecule. At the largest internuclear distances studied, focusing in the high density solvent is essentially nonexistent. These results are *optimal* (within the weak field limit and the approximation of classical mechanics). Thus, the resulting fields are not only the very best for the task, but the resulting achievements are also the best one can do. The ability to observe very clearly how the solvent progressively affects the solute dynamics is evident in Figure 8, and provides an example of using wavepacket control to resolve solvent dynamics effects in space and time.

Using classical (and semiclassical) treatments, we can easily (albeit approximately) treat polyatomic, cluster, surface, and condensed phase systems. One must always, however, keep in mind the "radius of convergence" of any such approximation; for example classical treatments will break down when interference or tunnelling becomes important. Nonetheless, such approximate treatments, appropriately applied, can be very valuable in understanding the limits of experimental feasibility in systems too large for exact theoretical treatment.

The proof of the ideas expressed above will be in the experiments they engender. With this goal, we are building an instrument[34] designed to control quantum dynamics and to use this control to better understand molecular properties and dynamics, including solution reaction dynamics. It is designed to produce a wide palette of tailored light fields, which will be used to both excite and detect quantum control. Figure 9 shows a schematic of our currently operating light source,[35] which we are refining in order to approach the globally optimal fields predicted by our theoretical studies.

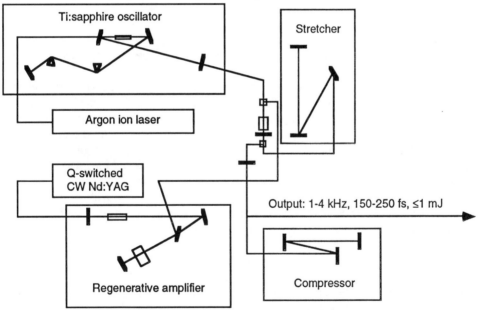

Figure 9: Schematic of Ti:sapphire regenerative amplifier laser system.

3. Ultrafast Diffraction

In principle, ultrafast diffraction (x-rays or electrons) can observe molecular dynamics directly, by measuring the molecular structure as a function of time.[3, 4, 22, 26] This idea has been a long-term theme in our group. In 1982, in a collaboration with the Mourou group, we turned ultrafast light pulses into short electron pulses and by accelerating these electron pulses into an anode, made short x-ray pulses.[4] In 1983 and 1986, we published simulations of ultrafast x-ray diffraction of I_2 photodissociation in solution.[3, 4] More recently, in the Mourou group, the production of short duration and wavelength x-ray (and probably γ-ray) pulses at 10 Hz, from plasmas driven by very intense subpicosecond light pulses, has been demonstrated.[36] We are simulating the performance and optimizing the design of both ultrafast x-ray and electron diffraction instruments, to discover what time resolution is practical, and what experiments will have a sufficient signal-to-noise ratio to be feasible. It is not yet clear what the limits of such techniques will be, and whether they can, with presently available technology, achieve sufficient time resolution to provide a temporal description of the reaction dynamics, but the possible rewards are large enough to make them worth investigating.

4. Conclusion

The control of molecular dynamics using tailored light fields is rapidly coming to experimental fruition. We have shown here theoretical and computational evidence that wavepacket focusing is possible not only in zero temperature, gas phase systems, but also in thermal systems and in solution. The ability to efficiently perform such wavepacket focusing makes possible a new set of theoretical and experimental methods for probing solution reaction dynamics as a function of time, position, and momentum. In turn, the ability to create controlled wavepackets calls for new ways of probing the dynamics of those wavepackets. While ultrafast spectroscopy should prove a very useful tool in this endeavor, we have also discussed methods for ultrafast x-ray and electron diffraction which might allow more direct probes of wavepacket dynamics. We believe that the combination of these theoretical and experimental techniques can provide new and better ways to observe, understand and control molecular dynamics in gases, liquids, clusters, and on surfaces.

References

[1] R. M. Whitnell and K. R. Wilson. in *Reviews in Computational Chemistry* , edited by K. B. Lipkowitz and D. B. Boyd. (VCH, New York, 1993).

[2] P. Bado, P. H. Berens and K. R. Wilson. Proc. Soc. Photo-Optic. Instrum. Engin. **322**, 230 (1982).

[3] P. Bado, P. H. Berens, J. P. Bergsma, M. H. Coladonato, C. G. Dupuy, P. M. Edelsten, J. D. Kahn, K. R. Wilson and D. R. Fredkin. Laser Chem. **3**, 231 (1983).

[4] J. P. Bergsma, M. H. Coladonato, P. M. Edelsten, J. D. Kahn, K. R. Wilson and D. R. Fredkin. J. Chem. Phys. **84**, 6151 (1986).

[5] Y. J. Yan, R. M. Whitnell, K. R. Wilson and A. H. Zewail. Chem. Phys. Lett. **193**, 402 (1992).

[6] I. Benjamin and K. R. Wilson. J. Chem. Phys. **90**, 4176 (1989).

[7] J. P. Bergsma, P. M. Edelsten, B. J. Gertner, K. R. Huber, J. R. Reimers, K. R. Wilson, S. M. Wu and J. T. Hynes. Chem. Phys. Lett. **123**, 394 (1986).

[8] J. P. Bergsma, J. R. Reimers, K. R. Wilson and J. T. Hynes. J. Chem. Phys. **85**, 5625 (1986).

[9] I. Benjamin, B. J. Gertner, N. J. Tang and K. R. Wilson. J. Am. Chem. Soc. **112**, 524 (1990).

[10] I. Benjamin, A. Liu, K. R. Wilson and R. D. Levine. J. Phys. Chem. **94**, 3937 (1990).

[11] J. P. Bergsma, B. J. Gertner, K. R. Wilson and J. T. Hynes. J. Chem. Phys. **86**, 1356 (1987).

[12] B. J. Gertner, J. P. Bergsma, K. R. Wilson, S. Lee and J. T. Hynes. J. Chem. Phys. **86**, 1377 (1987).

[13] B. J. Gertner, K. R. Wilson, D. A. Zichi, S. Lee and J. T. Hynes. Faraday Discuss. Chem. Soc. **85**, 297 (1988).

[14] B. J. Gertner, K. R. Wilson and J. T. Hynes. J. Chem. Phys. **90**, 3537 (1989).

[15] B. J. Gertner, R. M. Whitnell, K. R. Wilson and J. T. Hynes. J. Am. Chem. Soc. **113**, 74 (1991).

[16] W. P. Keirstead, K. R. Wilson and J. T. Hynes. J. Chem. Phys. **95**, 5256 (1991).

[17] Y. S. Li, R. M. Whitnell, K. R. Wilson and R. D. Levine. J. Phys. Chem. **97**, 3647 (1993).

[18] Y. S. Li and K. R. Wilson. J. Chem. Phys. **93**, 8821 (1990).

[19] I. Benjamin, L. L. Lee, Y. S. Li, A. Liu and K. R. Wilson. Chem. Phys. **152**, 1 (1991).

[20] L. L. Lee, Y. S. Li and K. R. Wilson. J. Chem. Phys. **95**, 2458 (1991).

[21] J. D. Ewbank, W. L. Faust, J. Y. Luo, J. T. English, D. L. Monts, D. W. Paul, Q. Dou and L. Schäfer. Rev. Sci. Instr. **63**, 3352 (1992).

[22] J. C. Williamson, M. Dantus, S. B. Kim and A. H. Zewail. Chem. Phys. Lett. **196**, 529 (1992).

[23] S. A. Rice. Science **258**, 412 (1992).

[24] P. Brumer and M. Shapiro. Annu. Rev. Phys. Chem. **43**, 257 (1992).

[25] W. S. Warren, H. Rabitz and M. Dahleh. Science **259**, 1581 (1993).

[26] J. L. Krause, R. M. Whitnell, K. R. Wilson, Y. J. Yan and S. Mukamel. J. Chem. Phys. (1993), submitted.

[27] Y. J. Yan, R. E. Gillilan, R. M. Whitnell, K. R. Wilson and S. Mukamel. J. Phys. Chem. **97**, 2320 (1993).

[28] B. Kohler, J. L. Krause, F. Raksi, C. Rose-Petruck, R. M. Whitnell, K. R. Wilson, V. V. Yakovlev and Y. J. Yan. J. Phys. Chem. (1993), submitted.

[29] R. M. Bowman, M. Dantus and A. H. Zewail. Chem. Phys. Lett. **161**, 297 (1989).

[30] A. H. Zewail, M. Dantus, R. M. Bowman and A. Mokthari. J. Photochem. Photobiol. A: Chem. **62**, 301 (1992).

[31] Y. Chen, L. Hunziker, P. Ludowise and M. Morgon. J. Chem. Phys. **97**, 2149 (1992).

[32] Y. J. Yan. Chem. Phys. Lett. **198**, 43 (1992).

[33] J. L. Krause, R. M. Whitnell, K. R. Wilson and Y. J. Yan. in *Ultrafast Reaction Dynamics and Solvent Effects*, edited by Y. Gauduel and P. Rossky. (1993). in press.

[34] Y. J. Yan, B. E. Kohler, R. E. Gillilan, R. M. Whitnell, K. R. Wilson and S. Mukamel. in *Ultrafast Phenomena VIII* (Springer, Berlin), in press.

[35] F. Salin, J. Squier, G. Mourou and G. Vaillancourt. Opt. Lett. **16**, 1964 (1991); available from Clark-MXR, Inc.

[36] J. C. Kieffer, G. Korn, G. Mourou, F. Raksi, J. Squier and D. Umstadter, private communication.

DYNAMICS OF THE CAGE EFFECT FOR MOLECULAR

PHOTODISSOCIATION IN SOLIDS

R.B. Gerber and A. Krylov
Department of Physical Chemistry
Institute for Advanced Studies
and Fritz Haber Research Center
The Hebrew University of Jerusalem
Jerusalem 91904
ISRAEL
and
Department of Chemistry
University of California, Irvine
CA 92717
U.S.A.

ABSTRACT

The photodissociation of small guest molecules in rare-gas crystals is studied by Molecular Dynamics simulations, and the results are compared to recent experiments. The findings throw light upon the physical mechanisms on the detailed atomic level whereby the cage effects occurs in solid-state reactions. Several types of cage effects are found and characterized in terms of the dynamical mechanisms involved, and their consequences on the yield of cage-exit by the product, and on the timescale of the mutual separation of the photofragments. The following main results are discussed:

(1) The photolysis of Cl_2 in Ar exhibits a delayed photofragment separation for low excitation energies (hv$\leq$ 8eV), with a pronounced change-over to a direct cage exit mechanism for higher energies.

(2) In the photolysis of F_2 in Kr, anomalously high yields are found at low temperature for the F atom cage exit. Long distance migration ($\lambda > 40$ A) is seen for some of the F atoms following photolysis.

(3) Cage exit of H atoms from photolysis of HCl in Ar occurs as two successive pulses in time - a pulsating cage effect.

(4) It is found that in photolysis of ICN in Ar, there is no cage exit of the CN photofragment, even at fairly high temperatures. This study suggests that the complete caging of molecular photogragments is rather general for reactions in solids, and contrasts with the corresponding behavior for liquids.

Detailed understanding emerges from the simulations as to the dynamical origins of these effects, and general insights are gained on the nature of the cage effect in solids.

509

J. Jortner et al. (eds.), Reaction Dynamics in Clusters and Condensed Phases, 509–520.
© 1994 *Kluwer Academic Publishers. Printed in the Netherlands.*

I.   Introduction

The cage effect is one of the central concepts in condensed-phase reaction dynamics. This term refers to a wide range of phenomena in which the immediate environment of solvent molecules that surrounds a reactive species can affect the chemical process. For instance, when a diatomic molecule undergoes photodissociation in chemically inert solvent, the surrounding "cage" of solvent molecules may in some cases completely prevent the separation of the photofragments, or may delay the mutual separation by a considerable span of time. The available knowledge on and understanding of the dynamical mechanisms whereby cage effects occur is, however, quite limited as yet. One of the specific areas in which considerable progress was made towards understanding the cage effect at the detailed molecular level is for photodissociation of small molecules in rare-gas matrices.[1-4]. The simplicity gained from having a fairly well-defined cage structure around the reagent molecule, and from the availability of reliable interaction potentials for these systems, has clearly been an important motivating force in pursuing reactions in a rare-gas solid as solvent.

One of the experimental points of focus has been the study of the yield for "permanent" separation of two photofragments produced upon photoabsorption[2,3]. Such yields have been successfully measured in recent years for several systems, in some cases as a function of the photoexcitation energy or as a function of temperature and pressure. Time-dependent studies of the cage effect in rare-gas solids, using femtosecond spectroscopy, have recently been reported, and a rapidly growing and promising direction.[5] Equally important has been the theoretical progress, based mostly on many-atom Molecular Dynamics simulations of the processes involved. The availability of potential functions for the rare-gas systems and their interactions with the guest molecules was among the main factors that made this possible. The theoretical studies began with investigations by Alimi et al[4] on photolysis of HI in Xe and of Cl_2 in Xe, first with the motivation of providing semiquantitative insights and some guidance for possible experiments. Cooperation between theory and experiment in this area has soon afterwards led to quantitative understanding of suitably chosen case studies, and to detailed evidence on the nature of the cage effects in the systems studied.[6,7] Very recently, Molecular Dynamics simulations were reported that include also explicitly the occurence of recombination between the photofragments.[8]

The purpose of the present paper is to survey recent theoretical studies on photodissociation in rare-gas solids, and discuss the nature of the cage effect and its dynamical mechanism in each case. As we hope to show, several distinct types of cage effects were found in these studies and can be characterized, and some general understanding of the cage effect for reactions in rare gas solids is beginning to emerge. Several new results will be reported, primarily on the important issue of the cage effect for a molecular (rather than atomic) product species.

It appears that this has not received precious theoretical attention and the difference between the cage effect for atoms and for molecules in solids is both striking and interesting.

In Sec.II we briefly comment on the Molecular Dynamics simulation methods used, and on the modelling of the systems involved. Sec. III discusses the various cage effects found and their dynamical interpretation. Concluding remarks are given in Sec. IV.

II.   The Molecular Dynamics Simulations

Most of the results described in this article were obtained by classical Molecular Dynamics (MD) simulations.[9] A description of these simulations for photodissociation in solids can be found in the earlier studies by Alimi et al on these processes.[4] Typically such calculations require a simulation box of at least several hundred (~400) atoms. If smaller simulations are used, then unphysical effects due to the limited size may influence the results, e.g. the heating up of the system due to the photodissociation energy released will significantly affect the results. Most of the simulations used a microcanonical system, in which, however, the temperature fluctuations in the initial state, prior to photolysis, were found to be quite small. The system consisting of the rare-gas atoms and the impurity molecule was propagated in time using the ground-state potentials, until equilibration was obtained at the temperature of interest. From the equilibrium ensemble so generated, initial configurations and momenta were sampled for the photodissociation process, by taking "snapshots" of the equilibrium simulation at random times, after equilibration has set in. Photoabsorption to the excited state was described by a vertical transition in the spirit of the classical Frank-Condon approach. That is, the ground state interactions were (suddenly) changed for the excited-state ones at the instant of photoabsorption, using the configuration the system had at that instant as the initial state for the propagation on the excited state. Typically, several hundred excited state trajectories were computed for each photodissociation event. Potentials from gas-phase data, mostly empirical ones, were used in all the calculations. The interactions used were based on pairwise potentials between the impurity species and the rare gas atoms. The evidence so far suggests that these potentials are adequate to describe the processes at a level compatible with the corresponding experimental data. To describe the dependence of the reactions on pressure and temperature, it is useful to employ the P,T ensemble, rather than the N,E ensemble, in the MD simulations of the initial state. This was done in a study of pressure effects on the photolysis of Cl_2 in solid Xe.[10]

Quantum effects in photolysis of hydrides were studied by a method that treats the light atom quantum-mechanically and the heavy atoms by classical dynamics.[11] Indeed, a recent study of photodissociation of HCl in the Ar...HCl cluster used a more refined method, in which the H atom was treated quantum-mechanically and the heavy atoms by semiclassical wavepackets.[12] The use of such hybride quantum/classical or quantum/semiclassical MD schemes is of considerable interest, since it appears that at least for H atoms there are important quantum effects in the context of cage phenomena.[18] We anticipate that such studies will be pursued much more extensively in the future.

Finally, one of the most severe limiations in conventional classical MD simulations is the neglect of non-adiabatic processes, of which recombination is an obvious important example. Studies which include nonadiabatic transitions for photochemical processes in matrices were recently reported by Gersonde and Gabriel,[8] and other calculations of non-adiabatic reactions in matrices and their effects on cage behavior are underway in our group. This important topic will, however, not be discussed here. Our estimate is that the timescales of the non-adiabatic processes are such that the results discussed here are not affected in a major way for the systems considered.

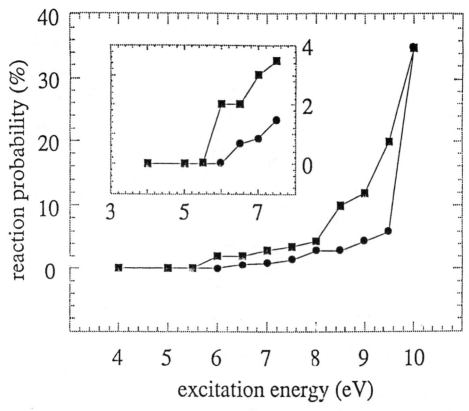

Fig. 1 Yield for photofragment permanent separation versus excitation energy, in the photolysis of Cl_2 in solid Ar. Theory (squares) and experiment (circles) are compared.

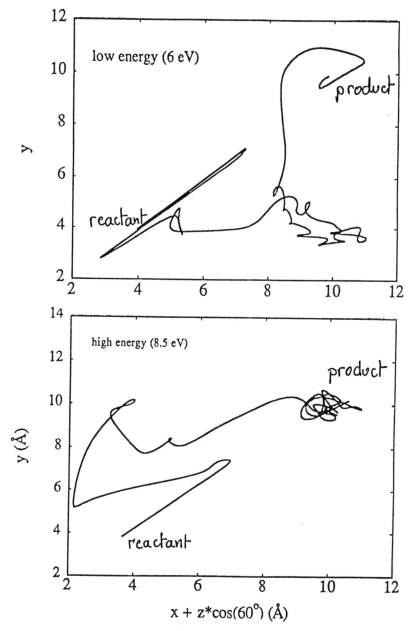

Fig. 2 Path of Cl photofragments in events leading to cage exit in photolysis of Cl_2 in Ar. One path shown is at low energies (delayed cage exit), the other for high energies. The Figure shows a projection of the path on a chosen plane through the crystal.

III.   Discussion of Cage Effects

I.   Direct vs. delayed cage exit. A combined experimental-theoretical study on the photolysis of Cl_2 in Ar has led to very interesting results on a transition between two types of cage behavior.[7] For low photoexcitation energies (~6eV-8.5eV) the photodissociation yield in this system is very low, 2-4% only. For these excitations, the Cl atoms have low excess energy, and are stopped by the cage atoms from exciting the cage in nearly all the trajectories. Indeed, the similar masses of Cl and Ar imply very efficient energy transfer in the impact of the Cl upon a cage atom, and the Cl deactivation is thus very rapid. Only for a few events do Cl atoms survive with sufficient energy to ultimately leave the cage, and this happens after a long delay in which the Cl atom has explored an extensive region untill almost statistically a "hole" in the surrounding potential wall was found. The situation is dramatically different for excitation energies $E > 8.5$ eV. The Cl atoms in many of the cases now have sufficient energy to knock a cage atom out of the way. The cage exit time in the direct mechanism is shorter than 1 psec, and the probability for cage exit events reaches 40% for excitation energies of about 10 eV. In conclusion, there is a change from a delayed, inefficient and "statistical" mechanism of cage exit at low energies, to an impulsive mechanism of high yields at high energies which involves a "knockout" of a cage atom by the Cl on its way. Fig. 1 shows the comparison between theory and experiment for the yields of photofragment separation.[7] Fig. 2 shows the paths (projected into a plane) of the Cl atom in two trajectories leading to cage exit: one at low energy, and one at high energy. The different mechanism is shown in the two cases of Fig. 2.

2.   The "porous cage" behavior for F atoms in Kr. Photolysis of F_2 in Kr or Ar leads to anomalously high yields for separation of the photofragments.[6] The yields reach almost 100% for energies well above threshold, at low temperatures (12°K). The MD simulations suggest that the origin of this behavior lies in the relatively very short range of the repulsive part of the F-Kr interaction. (This range is shorter than that of the repulsive part of the Ar-Kr potential).[6,14] As a result of the short range of these repulsive interactions, the nascent F atom formed in photolyis "sees" a very porous solid. It can exit this "porous" cage with high probability, and subsequent long-range migration of the F atom in the crystal is possible, since there are sufficiently large channels corresponding to attractive potential regions, and the repulsive potential ones do not suffice to block all possibilities for such motion. The agreement between the calculated cage-exit probabilities for F in Kr, and the corresponding experimental values measured by A.Apkarian and coworkers is excellent.[6]

3.   The "Pulsating Cage Effect". We consider now the cage exit behavior of a light particle in a host crystal of much heavier atoms. The calculations that will be discussed here were carried out by Blake et al [15] for the photolysis of HCl in Ar. These calculations are classical, and effects such as tunneling of the H atoms from the cage were ignored. The assumption used to justify this is that for high excess energies ($E > 2.5$ eV, say), the behavior of the H atom is expected to be predominantly classical. Certainly the energy available suffices for surmounting the barrier for cage exit, for the potential functions used. We note that for much lower excess energies of the H atom, classical cage exit becomes forbidden and the issue of cage exit by tunneling becomes important. Indeed, such an effect has been demonstrated in hybrid quantum classical MD calculations for the photolysis of HI in Xe.[13]

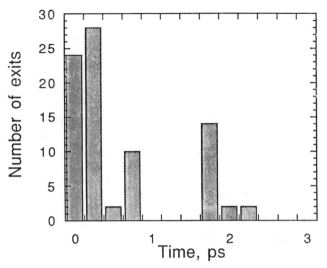

Fig.3 Probability of H atom to leave the cage versus time, in photolysis of HCl in solid Ar.

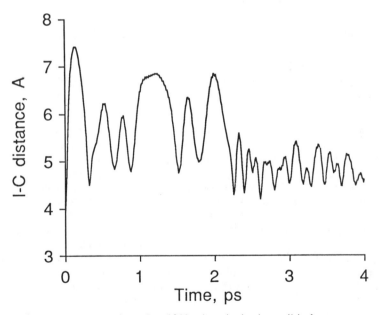

Fig.4 I-C distance versus time for ICN photolysis in solid Ar (typical trajectory).

Fig. 3 shows results from the calculations of Blake et al[15] on the cage exit probability of H in the photolysis of HCl in Ar, for excess energies (measured with regard to gas phase photolysis) of 2.7 eV. The quantity plotted at the figure is P(t), the probability of finding the H atom out the original reagent cage at time t after photolysis. The figure shows an interesting behavior consisting of two successive pulses: There is a first "pulse" of cage exit events for t $\sim$< 1 picosecond, then no noticable cage exit is found for about 0.7 ps, after which a second and weaker pulse of cage exit events is seen. The interpretation obtained from analysis of the simulations is as follows: The cage exit of the H atom is, in nearly all trajectories, at least somewhat delayed, as found also in the simulations of Alimi et al on the photolysis of HI in Xe.[4] The H-Ar repulsive potential is relatively long range, giving rise to relatively high cage-exit barriers. The "windows" corresponding to the transition states for cage exit, are not easily accessible. However, as the H atom hits the cage walls, the cage softens up and the windows for cage exit are substantially increased. Because of the large H/Ar mass discrepancy, the energy transfer is relatively inefficient, and the H atom still retains a substantial fraction of its initial energy, even after several collisions with the Ar "walls" of the cage. The first pulse is therefore mostly due to H atoms that left following at least some expansion of the cage. However, the 12 Ar atoms that surround the H, which constitute the cage, gain energy from the collisions of the H atom, and as a result a cage vibration, that can very crudely be thought of as being primarily a vibrational breathing of the near spherical cage, sets in. After 1 picosecond, the cage which first expanded, has contracted around the H in its vibration, closing the windows for cage exit. Cage exit is renewed in a second pulse only when the cage expands again in a second breathing oscillation. No additional pulses were seen by Blake et al, because the greater majority of H atoms already left, and also because relaxation effects involving more removed crystal atoms become effective at this stage, quenching the cage breathing vibration. The "pulsating cage" effect predicted by Blake et al is of considerable experimental interest, and should arise in many systems where a hybrid is photolyzed in an crystalline environment of heavy atoms.

4.   Caging of diatomic photofragments. So far we considered cage exit of atomic photoproducts. We shall now focus on several very important differences from the behavior discussed above when the photofragment is a diatomic species. The present authors have very recently studied the photolysis of ICN in solid Ar.[16] The light CN fragment carries initially the photodissociation energy, and the question arises whether the diatomic product can exit the cage and remain permanently dissociated from the I atom. MD simulations by Krylov and Gerber have shown that there is no cage exit of the CN in this system,[16] even at artificially high photodissociation energies (excess energies up to 4.0eV were used in some of the calculations), and at temperatures not far from the melting temperature. These results were confirmed in recent experimental studies by Fraenkel and Haas.[17] The behavior for a typical trajectory is shown in Fig. 4, where the I-C bond distance is plotted versus time, following photolysis. This distance grows initially almost to the dimensions of the cage, as the CN knocks against the Ar atoms that constitute the cage. However, the CN recoils back, and after several large amplitude oscillations in the cage is stabilized near the I atom (The calculations did not include treatment of recombination on the ground electronic state, which requires also the incorporation of non-adiabatic effects). The fact that cage exit does not occur at all in this case,

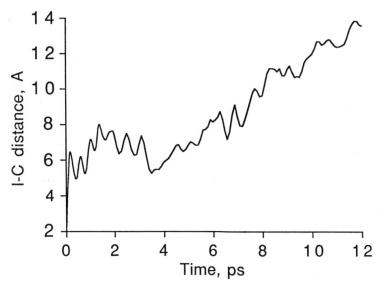

Fig.5 I-C distance versus time for ICN photolysis in liquid Ar.

unlike for the atomic species discussed above, is mostly due to the size and shape of the diatomic fragment. To exit the cage the CN fragment must reach the "window" for possible exit at special favorable orientations of the molecular axis, otherwise the species is too large for the window. However, this steric requirement is very restrictive, and not attained in the simulations. Another factor that operates against cage exit by the diatomic is that part of the photolysis energy goes into its rotational mode, reducing the energy available to the C.M. of the species for the cage exit. The fact that the CN fragment is not very elongated or large in size, suggests that the result found should be rather typical, if not general. Due to molecular size, steric factors and the participation of rotational modes, <u>no cage exit in solid is expected for diatomic (or larger) fragments.</u> The situation is very different for the liquid. Fig. 5 shows the result for a trajectory of ICN photolysis in liquid Ar, somewhat above the melting temperature. In this case, cage exit by the CN fragment occurs essentially directly, within less than a picosecond after photolysis. Diffusive motion of the CN is seen to set in after several more picoseconds. <u>The large thermal fluctuations in the structure of the liquid give rise to large possible "exit windows"</u>, so also molecular species can leave the cage.

The caging of the photoproducts in the solid gives rise to interesting isomerization dynamics in the $^3\Pi_0$ and $^1\Pi_1$ states involved.[16] We will not discuss this "cage induced" dynamics here. Of considerable interest are also the non-adiabatic processes in this system, both with regard to excited-state potential crossings and with regard to recombination induced by the cage atoms. These topics are now being pursued by the authors in extensive simulations.

5.   <u>Quantum effects in cage dynamics.</u> Gerber and Alimi[13] studied quantum effects in photolysis of HI in solid Xe, using a quantum wavepacket treatment of the H atom, and classical description of the heavy atoms. One of the interesting findings was an important difference between the behavior seen in these simulations, and the results of classical MD calculations, for the cooling in the cage of the hot H atoms initially produced. The difference, that builds up for t > 5 ps, say is due to the fact that in the quantum calculations the H atom relaxes to its ZPE in the cage, while in the classical case the relaxation is to the low thermal energy value. Tunneling effects for the H atom cage exit were also calculated. There have been no experimental tests of these predictions to date.

IV.   Concluding Remarks

Theoretical and experimental studies in recent years have thrown detailed light on important types of cage exit mechanisms in molecular photolysis in inert host solids. Molecular Dynamics simulations have played a very important role in providing atomic-scale interpretation for the dynamics of the processes involved. Furthermore, there have been several extremely encouraging examples of quantitative accord between MD simulations and experiments. Such agreement, and indeed rigorous testing of the simulations by experiments are critical for any confidence that the simulations, and the atomic-scale interpretation they provide, represent reality. The several successful interactions between theory and experiment in this area are above all due to the availability of relatively reliable interaction potentials for the systems studied, and these potentials were taken from gas-phase data. Thus, the major progress made in understanding the mechanisms of the cage effect in rare-gas solids is to a large extent due to the fact that for these systems, polarization effects are small, and gas phase potentials can be employed. The open important questions in this new field are many. One still little understood topic is that of non-adiabatic effects in solid-state (and other condensed-phase) reactions. These processes are clearly extremely important, and are essential, e.g. for exploring the path to recombination on the electronic ground state. Another extremely important aspect is the study of reactions in systems less ideal than rare gases, where polarization effects are important and unmodified gas-phase potential cannot be used. Progress in these areas will be a central goal in this field in the forthcoming years.

Acknowledgement:

The Fritz Haber Research Center is supported by the Minerva Gesellschaft fur die Forschung, Germany. This work was supported in part in the framework of the Saerree K. and Louis P. Feidler Chair in Chemistry at The Hebrew University, of which one of the authors is the incumbent. Part of this work was carried out on Silicon Graphics INDIGO workstations generously loaned by the Institute for Advanced Studies of The Hebrew University of Jerusalem by SG.

References

1.   Bondybey, V.E., and Fletcher, C.J. ,(1976), J. Chem. Phys. **64**, 3615 ; Bondybey, V.E., and Brus, L.E. (1975), J. Chem. Phys., **62**, 620 .

2.   Fajardo, M.E.,and Apkarian, V.A., (1986) J. Chem. Phys. **85**, 5660 ; Lawrence, W. , Okada, F., and Apkarian, V.A., (1988), Chem. Phys. Lett., **1 50**, 339; Fajardo, M.E., Whitnall, R., Feld J., Okada, F., Lawrence, W., Weideman, L. and Apkarian, V.A., Laser Chem. **9**, 1 (1988).

3.   Schriver, R., Chergiu, M. Kunz, H., Stepanenko, V., and Schriver, R., (1989) J. Chem. Phys. **91**, 4128 ; Kunz, H., McCaffrey, J.G., Schreiver, R.J., and Schwentner, N., (1990) J. Chem. Phys., **93**, 3245 .

4.   Alimi, R., Gerber, R.B., and Apkarian, V.A., (1988), J. Chem. Phys., **89**, 174 ; Alimi, R., Brokman , A., and Gerber, R.B., (1989), J. Chem. Phys. **91**, 1611 .

5.   Zadoyan, R., and Apkarian, V.A., Chem., Phys. Lett (In press).

6.   Alimi, R., Gerber, R.B., and Apkarian, V.A., (1991), Phys. Rev. Lett. **66**, 1295 .

7.   Alimi, R., Gerber, R.B., McCaffrey, J.G., Kunz, H., and Schwentner, N., (1992) Phys. Rev. Lett., **69**, 856 .

8.   Gersonde, I.M., and Gabriel, H., (1992), J. Chem. Phys.,**98**, 2094.

9.   Allen , M.P. and Tildesley, D.J., 1987 "Computer Simulations of Liquids" (Clarendon, Oxford,).

10.   Alimi, R., Apkarian, V.A., and Gerber, R.B., (1993) J. Chem. Phys. **98**, 331.

11.   Gerber, R.B., and Alimi, R., (1991) Isr. J. Chem. **81**, 383 .

12.   Garcia-Vela, A., and Gerber, R.B., (1993) J. Chem. Phys. **98**, 427 .

13.   Gerber, R.B., and Alimi, R., (1990) Chem. Phys. Lett, **173**, 393 .

14.   Alimi, R., Gerber, R.B., and Apkarian, V.A., (1990) J. Chem. Phys. **92**, 3551

15.   Blake, N.C., Apkarian, V.A., and Gerber, R.B., to be published.

16.   Krylov, A., and Gerber, R.B., to be published.

17.   Fraenkel R., and Haas, Y., to be published.

CAGE EFFECT AND MOLECULAR DYNAMICS OF Cl_2
IN RARE GAS SOLIDS

N. SCHWENTNER, M. CHERGUI, H. KUNZ and J. McCAFFREY*
Institute for Experimental Physics, FU Berlin, Arnimallee 14
D–1000 Berlin 33

1. Introduction

Over the past few years, a strong interest has developed for the study of elementary photochemistry of small molecules embedded in rare gas crystals. This stems from the fact that such systems are considered as models for condensed phase photochemistry and are more easily amenable to computer simulations. In addition, laser applications are foreseen.

The fascinating problem of cage exit of H atoms has been treated experimentally and theoretically for H_2O, D_2O, HCl and H_2S and an interesting discussion on the contribution of direct and delayed cage exit as well as on the role of nonadiabatic transitions has emerged. For F_2 in Ar and Kr matrices, exit across low barriers, preorientation and most stimulating long range F migration in terms of channeling has been reported. Photostimulated migration of atoms like Ag, Cu and O has been observed and treated theoretically. Systematic variations of the exit conditions by changing the matrix, the isotopic composition, the temperature and the external pressure as well as the first femtosecond investigations elucidate the cage exit in more and more detail. A general review of these different aspects was given previously /1/. This contribution will treat the dissociation of Cl_2 in rare gas matrices because both for experimental and theoretical reasons, the series of halogens represents an ideal testing ground for new models and provides comparisons with previous and actual studies in the liquid phase /2/. Finally, because of their large electron affinity, intermolecular interactions will contribute to photochemistry in addition to intramolecular excitations.

Photodissociation of molecules such as I_2, Br_2 and Cl_2 has been reported by several groups /1/ but Cl_2 is the halogen molecule that to date has most been studied as regards cage effects. Ever since the early demonstration of hindered dissociation of Cl_2 in Ar matrices by Bondybey and Fletcher /3/, a number of studies have been achieved and we will now focus on the recent experimental and theoretical investigations of Cl_2 fragmentation. This molecule and its fragments are in many

*Now at: Department of Chemistry, St. Patricks College, Maynooth, Co. Kildere, Ireland

J. Jortner et al. (eds.), Reaction Dynamics in Clusters and Condensed Phases, 521–537.

aspects complementary to the small and light F and H fragments already studied
/1/:

- the Cl atom–to–matrix atom mass ratios are not far from one (2:1 for Cl:Ne; 1:1 for Cl:Ar and 1:4 for Cl:Xe). The energy loss per collision will be large compared to the hydrogen case and the number of escape attempts with appreciable kinetic energy will be small for the Cl fragments
- the large size of the Cl atoms causes a strong repulsive interaction with the lattice in the dissociation process and the resulting barriers will be high
- the parent Cl_2 molecule and the Cl fragment are too large to fit respectively into the substitutional and interstitial sites of lighter rare gas matrices and substantial structural rearrangements are to be expected
- the tight fitting of the Cl_2 molecules in the host lattice leads in general to a preferential orientation and to librational motions which are limited by barriers with respect to rotation. Thus the access to escape cones by thermal activation can be important
- Cl_2 momomers and aggregates can be distinguished spectroscopically via the A′–X fluorescence and the influence of clustering on the dissociation efficiency can be probed.

The size dependent effects have been varied systematically from an extreme mismatch in the case of Ar matrices to the situation in a Xe matrix which provides a good fit of the Cl_2 and Cl species into substitutional and interstitial sites respectively, while the Kr case offers an intermediate situation.

2. Dissociation of Cl_2 in Ar matrices

2.1.   GENERAL REMARKS

In the following, examples of results from an extensive investigation of the spectroscopy and the dissociation processes in these 3 matrices will be presented which has been carried out under comparable preparation conditions and which included concentration and annealing studies /4/. The extended spectrum of photon energies at the synchrotron radiation facility BESSY in Berlin has been exploited in the photon energy range from 3 to about 15 eV to provide sufficient kinetic energy for the Cl fragments. The experiments have been complemented in the case of Xe matrices by excitation at 308 nm with a XeCl excimer laser. The presentation will start with an illustration of very large barrier heights in the case of Ar, then demonstrate the much lower ones in the looser Xe lattice and then show the consistency of the results in the superposition of the different effects observed for the intermediate case of Kr matrices.

2.2.   BARRIER HEIGHT, PROMPT AND DELAYED EXIT IN AR

The ground state Cl_2 molecule has a binding energy of 2.5 eV, and has the low lying bound excited A′, A and $B^3\Pi_u$ states. It can be excited in the 3 to 9 eV range to the repulsive $C^1\Pi_u$ and $^3\Sigma_u^+$ states /5/. Around 9 to 11 eV, the bound $1^1\Sigma_u^+$ and an $Ar^+Cl_2^-$ charge transfer state are reached in Ar matrices (see Fig. 14). No dissociation is observed up to a photon energy of 6.1 eV and the efficiency has to be below 10^{-6} in this low energy range. A rise in efficiency occurs above 6.1 eV but the efficiency remains small with values around 1 to 2% up to 9 eV and a prominent threshold shows up around 9.5 eV with an efficiency growing above 30% (Fig. 1). Subtracting the Cl_2 binding energy yields barrier heights for the two

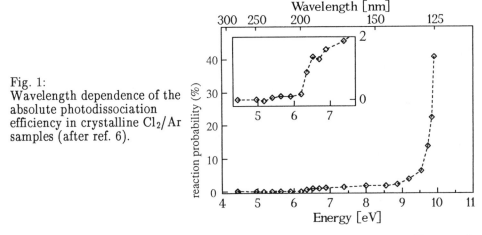

Fig. 1:
Wavelength dependence of the absolute photodissociation efficiency in crystalline Cl$_2$/Ar samples (after ref. 6).

onsets of 3.6 and 7 eV /6/. These large values infer that the cage effect can be extremely strong in a tight lattice and that a high kinetic energy is required for the fragments to leave the cage in a permanent dissociation.

A first guess about barriers can be derived from the Cl–Ar pair potential and the Ar fcc lattice structures /4,6/. The lowest barrier with an energy of 3.6 eV to leave a substitutional lattice site corresponds to a passage through the center of a square of Ar atoms, the next higher one with 7.7 eV is represented by the triangular base of a tetrahedron and in order to cross the line connecting two lattice atoms requires 15.2 eV in a stiff unrelaxed geometry.

A real understanding of the cage exit requires knowledge about the local geometry around the Cl$_2$ guest molecules and a modelling by molecular dynamics calculations. The experiments yield information about the local structure of the Cl$_2$ molecule and the Cl fragments as well. The A′–X emission (see Fig. 9) of Cl$_2$/Ar shows a well resolved isotope splitting indicating a weak interaction with the matrix. The Cl$_2$ molecule, on the other hand, is too large for a single substitutional place according to the pair potentials and implementation in a very tight cage should result in a broadened A′–X emission. Therefore a larger, especially a double vacancy site, is most likely. The observed distribution of Cl sites after dissociation confirms a double vacancy site for Cl$_2$ and shows that the Cl atoms finally occupy single substitutional sites. Cl atoms generated in disordered matrix show a superposition of lines in the Cl atomic absorption band due to the different local structures caused by the disorder /7/. Only one dominant site is observed for dissociation of Cl$_2$ in a cristalline matrix indicating that all Cl atoms have the same surrounding. This is only possible for the given Cl and Ar sizes if the Cl$_2$ molecule originally occupies a double vacancy site and each of the two produced Cl atoms takes one single substitutional site. In the dissociation process the lattice has to rearrange to guarantee separation of both Cl atoms by at least one matrix atom preventing recombination /7/.

A molecular dynamics calculation has been carried for Cl$_2$ immersed in a solid slab of 400 Ar atoms and the results are fully consistent with all the experimental findings /8/. The modelling will be treated more extensively in the contribution of

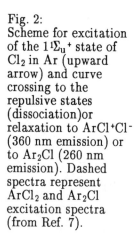

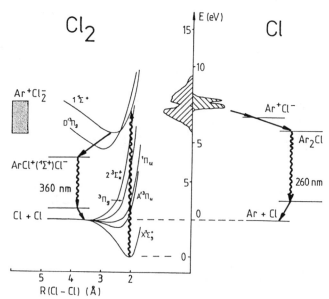

Fig. 2:
Scheme for excitation of the $1^1\Sigma_u^+$ state of Cl_2 in Ar (upward arrow) and curve crossing to the repulsive states (dissociation)or relaxation to $ArCl^+Cl^-$ (360 nm emission) or to Ar_2Cl (260 nm emission). Dashed spectra represent $ArCl_2$ and Ar_2Cl excitation spectra (from Ref. 7).

B. Gerber in this issue and therefore the results are only briefly summarized. The calculation yields both thresholds, around 6 and 9 eV, as well as the relative intensities which are reproduced. The double vacancy site for Cl_2 has a lower energy than the single substitutional site, while the efficiencies for Cl_2 at a single substitutional site would differ from the experimental findings. Furthermore, it was shown experimentally that the dissociation efficiency in the region of the prominent threshold does not depend on temperature between 5 to 23 K, which agrees with the molecular dynamics calculation. It is fascinating that the model explains the low energy onset by delayed cage exit taking place on a time scale of several picoseconds, whereas the threshold around 9 eV corresponds to prompt exit on a subpicosecond time scale.

2.3. PREDISSOCIATION AND CHARGE TRANSFER STATES

The region of efficient dissociation between 9 eV and 11 eV is dominated by excitation of the higher lying Cl_2 bound states and the contribution by direct excitation of the repulsive states is negligible according to the absorption spectra in Fig. 14. Therefore above the prominent threshold, we are dealing with predissociation. The molecular dynamics calculations are classical and do not include the predissociation step but start immediately on the repulsive surfaces. In the gas phase, the Cl_2 potential surfaces are quite complicated due to strong mixing of Rydberg and intramolecular charge transfer states /5/. The level scheme is significantly simplified in Ar (see section 5) due to a shift of the Rydberg states to higher energies and only the repulsive surfaces, the intramolecular $1^1\Sigma^+$ ionic state and an intermolecular $Ar^+Cl_2^-$ charge transfer state are relevant for the discussion as shown in Fig. 15 and sketched in Fig. 2. A comparison of the dissociation efficiency with the excitation efficiency of the Cl_2 A'–X emission in Fig. 3 and

Fig. 3:
Efficiency for Cl_2 in Ar of
a) Cl_2^* radiative decay $(ArCl_2;$
360 nm), b) overall excitation
$(Cl_2; A'-X; 867$ nm) c) Ar^+Cl^-
radiative decay $(Ar_2Cl; 260$ nm)
and, d) dissociation, versus
photon energy.

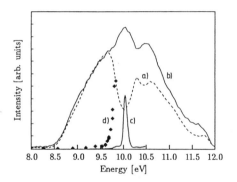

with the deconvolution into the $1^1\Sigma_u^+$ and $Ar^+Cl_2^-$ contribution (Fig. 15) shows that the threshold of dissociation is close to the onset of the $Ar^+Cl_2^-$ band. This finding is quite interesting since it seems to suggest at first sight that a harpooning reaction might be an important channel for dissociation in this case.

Dissociation via harpooning has been discussed recently /9,10/ and is based on the fact that the Cl_2^- potential surface is repulsive in the center of the Cl_2 ground state Franck–Condon region leading to a kinetic energy in the Cl fragment of about 0.3 eV. Thus a sequence

$$Ar + Cl_2 + h\nu \rightarrow Ar^+Cl_2^- \rightarrow Ar^+Cl^- + Cl(0.3 \text{ eV}) \qquad (1)$$

could lead to dissociation but the small kinetic energy of 0.3 eV makes this process rather unlikely in view of the large barriers in Ar. A byproduct would be the formation of Ar^+Cl^- which can be excited also in the Ar matrix at $h\nu \approx 7-8$ eV and 10.1 eV (see Fig. 3), leading to a strong emission at 260 nm. This emission is not observed in the threshold region for efficient dissociation between 9.5 and 9.9 eV and therefore we conclude that in the Ar case, harpooning is not important for dissociation. The 10.1 eV excitation band of the 260 nm band belongs to a Cl atomic transition which accidentally coincides with the Cl_2 absorption band but is much narrower and can be well separated from the dissociation threshold. Thus the results indicate that Cl_2 is excited in a superposition of the $^1\Sigma_u^+$ and $Ar^+Cl_2^-$ charge transfer states and that it predissociates to the repulsive surfaces aquiring kinetic energy and dissociating according to the modelling in the molecular dynamics calculations /8/. Alternatively, it relaxes to the bottom of the $^1\Sigma_u^+$ state forms an $ArCl^+Cl^-$ state and emits at 360 nm (Fig. 2).

A scheme which is consistent with observed intensities of the $Cl_2(A'-X)$ emission, the $ArCl^+Cl^-$ emission (360 nm) and the dissociation efficiency versus excitation energy is shown in Fig. 4. Below $h\nu=8.5$ eV, the repulsive surfaces are populated directly but the kinetic energy is too small for efficient dissociation. Between 8.5 and 9.5 eV excitation of $1^1\Sigma_u^+$ and $Ar^+Cl_2^-$ leads to relaxation to the $ArCl^+Cl^-$ state and with comparable probability 360 nm emission or slow (time scale of the 360 nm radiative decay time) curve crossing to the repulsive surfaces takes place. Due to the relaxation, the kinetic energy is still too small, dissociation is very inefficient, $A'-X$ emission occurs and the ratio of 360 nm and $A'-X$ is fixed. Above the prominent threshold (9.5–10 eV), about 50% of the excited molecules still relaxe and yield 360 nm and $A'-X$ emission. The other 50% cross resonantly to the

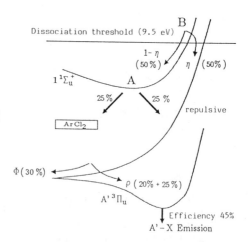

Fig. 4:
Scheme of branching ratios for Cl_2
in Ar for resonant curve crossing (η)
and relaxation in $1\,{}^1\Sigma_u{}^+$ $(1-\eta)$, for
radiative decay $(ArCl_2)$ and curve
crossing in the relaxed state, as
well as for dissociation (ϕ) and
recombination in $A'(\rho)$

repulsive surfaces, the kinetic energy is now higher and above threshold, and 30%
dissociate while 20% recombine and give $A'-X$. This picture is obvious from the
decrease of the $ArCl^+Cl^-$ emission with respect to $A'-X$ emission which coincides
with the increase in dissociation efficiency (Fig.3) because 360 nm emission and
dissociation are competing whereas $A'-X$ is stabilized by two recombination
contributions.

3. Dissociation of Cl_2 in Xe

As seen in the case of Cl_2 in Ar matrices, the pair potentials are extremely relevant
for the dissociation processes and, since in Xe, the Cl_2 and Cl atoms fit more easily
into the lattice, the barrier heights should be considerably lower. In addition the
intermolecular charge transfer state $Xe^+Cl_2^-$ is situated at lower energies (Fig. 14)
due to the lower Xe ionisation energy. Thus the extension to Xe represents an
important test of the consistency of the interpretations.

3.1. SMALL BARRIERS, UNSTABLE DISSOCIATION

Cl atoms produced by Cl_2 dissociation in Xe occupy two different sites in the

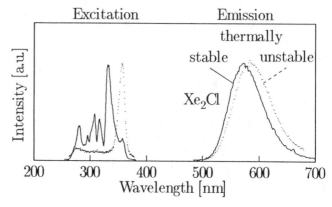

Fig. 5:
Excitation and
emission spectra of
Cl_2 in Xe for thermally
stable (solid line)
and unstable sites
(dotted line) from
Ref. 11.

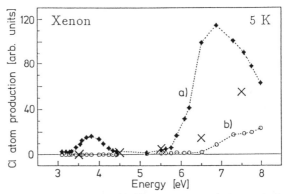

Fig. 6:
Efficiency for disso—
ciation of, a) Cl_2 in
Xe to unstable sites
(at 5 K) and, b) to
stable sites (annealing
at 40 K) from Ref. 4.
Calculation (crosses)
from Ref. 12.

lattice which can be distinguished by the differences in their emission and absorption (excitation) spectra (Fig. 5). Therefore the content of Cl atoms in the two sites can be determined by the intensities of the emission bands for an appropriate choice of excitation energies /11/. It turns out that the lower energetic site (points in Fig. 5) is irreversibly quenched by annealing the sample above 40 K. It will be called "metastable" or "unstable site" and will be treated in this section. The high energetic site (solid line) is not affected by annealing, is called "stable site" and will be treated in the following section.

The squares in Fig. 6a represent the dissociation efficiency with respect to the metastable site versus the dissociating photon energy. Annealing above 40 K would lead to a disappearance of this site for all dissociation energies. The dissociation efficiency in Fig. 6a is not corrected for the absorption probability to yield quantum efficiencies. A comparison of the low energy range of Fig. 6a between 3.5 and 4.5 eV with the $C^1\Pi_u$ absorption band in Fig. 7 shows that the dissociation efficiency follows it smoothly. Obviously the dissociation efficiency to the metastable site is already quite large for small photon energies of about 4 eV, the quantum efficiency is independent of photon energies in this range and a barrier for dissociation has to be smaller than 1 eV, if it exists at all. Qualitatively, this observation fits perfectly in the trend expected in going from Ar to Xe according to lattice structure and pair potentials. It has to be remembered that the Cl atoms in this site disappear by annealing, obviously due to thermally activated recombination to Cl_2. This means that cage exit is not complete in this site and the two atoms remain close to each others and are not even separated by a matrix atom.

Fortunately, an extensive molecular dynamics investigation of the Cl_2/Xe system has been carried out recently, clarifying the processes involved in the metastable site /12/. Potential surfaces are generated by the diatomics—in—molecules (DIM) method, the trajectories on these surfaces are calculated classically and nonadiabatic transitions inculding relaxation between excited states as well as recombination to the ground state are explicitly taken into account in the equation of motion. The Cl_2 molecule is stabilized in a single substitutional site, oriented along the [1, 1, 0] direction and the motion is restricted to small librations at 5 K in agreement with previous calculations /13/. For the Cl atoms, four potential minima are found. Besides the expected interstitial sites with octahedral and tetrahedral symmetry a special site which is rather sensitive to the pair potentials

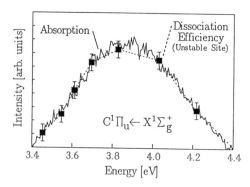

Fig. 7:
Comparison of absorption
profile and dissociation
efficiency in the $C^1\Pi_u$
state of Cl_2 in Xe (from
Ref. 11).

and a site with D_2 symmetry are obtained. The D_2 site corresponds to a Cl atom placed between two Xe atoms and will be relevant for the dissociation to the metastable site. The new results concerning dissociation differ significantly from a previous calculation /13/. The calculated dissociation efficiency (Fig. 6 crosses) rises smoothly above an excess energy of about 2 eV. The exit processes are mainly delayed by 1 to 2 ps and the initial molecular orientation is not crucial   The threshold energy below 1 eV from the experiment is not quantitatively reproduced but qualitative agreement is obtained that at 5 K, small excess energies are sufficient for dissociation. The dominant configuration after dissociation for small excess energies consists of a Cl atom trapped at the D_2 interstitial site next to the other Cl atom which remained in the substitutional site. Raising the temperature to 40 K in the calculation leads to a recrossing of the Cl atom in the D_2 site above the barrier back to the substitutional site and recombination. This mimics excellently the behaviour of the metastable site which dominates in the experiments at low excess energies. For excess energies above 3 eV, Cl atoms in D_2 sites further away from the original site show up in the calculations and are not detrapped up to 80 K. Also permanent displacement of host atoms as well as exit of both fragments happens. All these correspond to stable sites and lead us to the experimental results concerning them.

3.2.   STABLE DISSOCIATION AND HARPOONING

The dissociation efficiency to permanent sites (Fig. 6b) is always lower than that to metastable sites. It remains very low up to about 6.5 eV and shows a prominent increase above. The efficiency in the range of the repulsive $C^1\Pi_u$ state around 4 eV is not zero but very low and only about 10^{-3} that of the metastable site. This again is in qualitative agreement with the above mentioned theoretical predictions for stable sites. The rise around 6.5 eV in Fig. 6.b suggests that a new type of processes starts at that energy. This brings us to contributions by harpooning which are not included in the theoretical model.

The absorption spectrum (represented by the $Cl_2(A'-X)$ excitation spectrum in Fig. 8a) consists of a superposition of the intramolecular $^3\Sigma_u^+$ band of Cl_2 and the intermolecular $Xe^+Cl_2^-$ charge transfer state (see §5). The $Xe^+Cl_2^-$ state dominates according to the strength of the band (see Fig. 14). Thus dissociation by a harpooning reaction as described by equ. 1 becomes possible in this energy range /11/. For Xe, barriers are absent or small and therefore the small kinetic energy of 0.3 eV might be sufficient and the process is much more likely. Even a "negative

Fig. 8:
Comparison for Cl_2 in Xe of,
a) absorption efficiency,
b) excitation efficiency of
charge transfer state Xe^+Cl^-
(Xe_2Cl band at 570 nm) and,
c) dissociation efficiency
to stable sites (after Ref. 11).

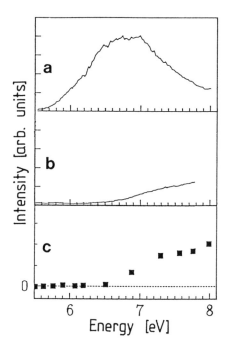

cage effect" might exist /10/. In this process an excited Xe^+Cl^- pair has to be formed and indeed, the corresponding emission from an excited Xe_2Cl center is observed for excitation in this range. The dependence of this emission band on excitation energy (Fig. 8b) follows exactly the efficiency curve for dissociation (Fig. 8c) in this range. It proves that the harpooning reaction is the dominating dissociation process in this range. Even the energy shift between the center of the absorption and the onset can be understood in this way. The center of the Franck–Condon region of the $Cl_2 \rightarrow Cl_2^-$ transition corresponds to a kinetic energy of 0.3 eV for the decay of Cl_2^- to Cl_*Cl^- /9/. The transition is centered at 6.8 eV in Fig. 8a and below 6.5 eV only bound Cl_2^- states are populated. The fragmentation can start energywise in this process only above 6.5 eV as observed in both channels of Fig. 8b and 8c.

4. Dissociation of Cl_2 in Kr

In Kr the size conditions for Cl_2, Cl in the matrix are intermediate to those of Ar and Xe. This leads, as will be demonstrated, to a superposition of the effects from both matrices. Therefore, this matrix is quite interesting because the different processes can be studied in one matrix which can serve as a rather rigorous test for the predicted trends. On the other hand, it is most challenging because it is hard to disentangle the mixed processes either experimentally or theoretically. At present only the consistency of predictions can be demonstrated qualitatively due to the complexity of the system.

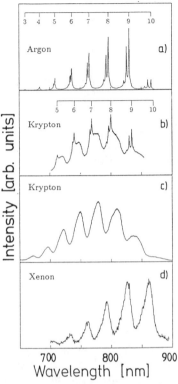

Fig. 9:
A′–X emission spectra for Cl₂ in,
a) Ar (0.01% Cl₂, deposited at 18 K),
b) Kr (0.001% Cl₂, deposited at 7 K),
c) Kr (0.001% Cl₂, deposited at 22K),
and, d) Xe (0.001% Cl₂, deposited at 40K)

4.1. SUPERPOSITION OF SINGLE AND DOUBLE SUBSTITUTIONAL SITES FOR Cl₂

The lineshape of the isotopically split vibrational members of the A′–X transition of Cl₂ is sensitive to the local structure given by, for example, single substitutional or double vacancy sites or by dimers or larger clusters. Aggregation leads to a broadening, a shift and a shortening of the decay time as has been exemplified for Ar /7/. Aggregation is a severe problem and it changes the dissociation processes significantly /7/. For all three matrices, concentration and condensation temperatures have been varied /4/ and here, only results at sufficiently low concentrations will be presented, ensuring that the monomers dominate. The comparison shows the well resolved isotope splitting of the A′–X emission in Ar (Fig. 9a) due to a small linewidth of the transition which is characteristic for the double vacancy site as has been discussed before. The bands are strongly broadened and red shifted in Xe (Fig. 9d) which is characteristic for the single substitutional site in this matrix. In a Kr matrix deposited a low temperature (Fig. 9b), a superposition of sharp bands with a background of broadened and red shifted bands is observed and according to the analogy, the bands are attributed to double vacancy and single substitutional sites, respectively. An investigation of the lifetimes, illustrated in Fig. 10 by gated spectra, shows that these two contributions (I, II) have the same lifetime of 49 ± 1 ms and that there is in between a third contribution (III) with a shorter lifetime of 13 ± 1 ms which we attribute to a weak contribution of clusters according to the Ar results /7/ and the

Fig. 10:
A′–X emission spectra of
0.001% Cl_2 in Kr deposited
at 5 K. a) prompt emission,
b) emission after 120 ms
delay. I is attributed to
double vacancy site, II to
single substitutional site
and III to clusters.

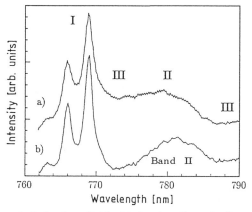

concentration studies. Hoffmann and Apkarian /14/ obtained rather similar spectra
and offered a somewhat different interpretation but our picture yields a consistent
assignment for all three matrices. The distribution of sites depends on the
condensation temperature and the single substitutional site can be enhanced by a
higher condensation temperature (Fig. 9c). The Cl atoms are able to occupy three
sites in the Kr matrix and the distribution is sensitive to the condensation,
annealing and dissociation conditions. The sites can be distinguished in the Kr_2Cl
($4^2\Gamma$) emission centered around 3.4 eV and its excitation spectrum around the B–X
and D–X transitions between 5.5 and 7 eV and the Rydberg series of charge
transfer states /15/ around 5 to 7 eV. The sites show a characteristic annealing
behaviour. The "stable" sites (excitation 5.58 eV, emission 3.49 eV) resist
annealing to 30 K, the "stimulated" sites, (excitation 5.34 eV, emission 3.35 eV)
increase by annealing and the "unstable" sites (excitation 5.13 eV, emission 3.17
eV) are weakened by annealing for certain dissociation conditions. A detailed
presentation of this interesting spectroscopic material goes beyond the scope of this
paper.

4.2. SUPERPOSITION OF STABLE, UNSTABLE AND STIMULATED DISSOCIATION

The complex interconnection of the processes will be exemplified by a set of
dissociation efficiencies taken for irradiation at 5 K after deposition at 22 K (Fig.
11) and a set of dissociation efficiencies for a similar sample for irradiation at 25 K
(Fig.12). The intensity of each site in Fig. 11 was determined after each irradiation
at 5 K (dots) and, following an annealing cycle to 25 K (circles), again at 5 K. No
Cl production is observed for irradiation with photon energies below 5 eV neither
at 5 K nor at 25 K. Thus as expected a barrier to dissociation occurs for Kr and it
is somewhat lower than that for the weak onset in the case of Ar (Fig. 1 and §2).
The efficiency curves differ drastically for the various sites and depend also on the
sample temperature. The stable site is generated above 6 eV (Fig. 11a) and its
efficiency increases up to 9 eV at 5 K. Annealing reduces somewhat the stable Cl
content in the low and high energy range and it stays constant for a production
around 8 eV. Generation of the same site at 25 K (Fig. 12a) shows a smooth and
pronounced increase in efficiency from 6 to 8 eV and a drop for higher energies.

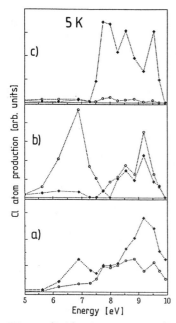 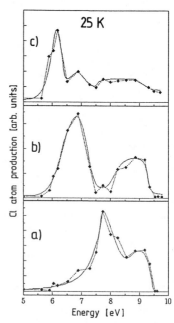

Fig. 11 (left): Dissociation efficiency of Cl_2 in Kr at 5K versus photon energy to, a) stable site, b) stimulated site and, c) unstable site before (dots) and after (circles) annealing to 25 K.
Fig. 12 (right): Analog of Fig. 11 but with dissociation at 25 K.

Production of the stimulated site starts also around 5 eV and the efficiency remains rather low at 5 K (Fig. 11b) until it rises more strongly above 8 eV. Annealing leads to a remarkable behaviour. The content rises strongly if photons between 6 to 8 eV have been used and it rises a little bit for the higher photon energies. Obviously, this site is generated by annealing and the Cl atoms come from the losses in the stable site (Fig. 11a). This thermal generation of sites is always active for dissociation at 25 K and the two maxima around 7 and 9 eV observed after annealing (Fig. 11b) are therefore also found (Fig. 12b).

Production of the unstable site starts for a 5 K sample (Fig. 11c) at much larger energies above 7.3 eV and the efficiency shows some oscillations on a large plateau. Annealing quenches essentially also the content of unstable sites. The results for 25 K (Fig. 12c) are quite surprising. A significant content of this "unstable" site can be generated and conserved at this high temperature and in addition it can be generated already with photon energies above 5.6 eV with a maximum at 6.2 eV.

The processes which generate the different sites seem to compete in the high temperature sample (Fig. 12). The unstable site starts at the lowest energy, reaches a maximum in efficiency and the efficiency decreases when the stimulated site can be generated. The efficiency of the stimulated site decreases also when the stable site reaches its maximum.

This complicated behaviour reflects the changes in the local Cl_2 geometry with annealing temperature (Fig. 9b, c), the rearrangements in the geometry around the Cl atoms with temperature and the dependence of the various dissociation channels on photon energy and temperature. Thus a modelling is required to shed more light into this complex variety of processes.

5. Guest–host electronic interactions

In the previous section dissociation on repulsive surfaces as well as harpoon reactions and predissociation of bound states by curve crossings have been discussed using the distribution of electronic states of the Cl_2 molecule in a matrix as a given fact. It will now be shown that electronic states in general and Rydberg, ionic and charge transfer states especially are significantly modified by the matrix inducing the severe changes in the dissociation dynamics demonstrated before. Here the physical origins of the general trends will be treated which are observed in the interaction of the electronic states of the guest molecule with the matrix.

The general picture is that valence states are little perturbed by the matrix as evidenced by small red gas–to–matrix shifts while Rydberg states are strongly perturbed and exhibit large blue shifts (~0.3 to ~1.0 eV depending on the matrix) /16/. Ionic (intramolecular charge transfer) states show relatively strong red shifts due to dipole solvation /17/. More important in the case of halogens (and due to their large electron affinity) is, the occurance of intermolecular charge transfer (C.T.) states between dopand and matrix species. Finally, the forbidden character of some transitions may be partly or fully lifted due to crystal field effects /18/ or heavy–atom effects /19/. Those largely different behaviours can be exploited to modify reaction pathways and create new ones in a controlled manner. In the following, we will present some examples concerning Cl_2 where, in addition to the geometrical and physical effects discussed above, the electronic interaction between host and guest strongly modifies the dissociation dynamics of the latter.

5.1. RYDBERG – IONIC STATES COUPLINGS

In the gas phase, the VUV excitations which give rise to the Cl_2 $D'\,^3\Pi_g$–$A'\,^3\Pi_{2u}$ laser emission at 4.7 eV, correspond to transitions from the ground state $X\,^1\Sigma_g^+$ to the $1\,^1\Sigma_u^+$ double–well adiabatic state resulting from an avoided crossing between a $1\,^1\Sigma_u^+$ Rydberg inner state and an ionic state of same symmetry dissociating to $Cl^+(^1D)+Cl^-(^1S)$. As a result, the intensity distributions of the excitation bands are erratic as can be seen in Fig. 13a /20/. The absorption at threshold is due to excitations into the ionic part of the double–well and the sharp intensity fall at ~9.2 is due to excitations above the barrier, induced by the avoided crossing.

In comparison, the overall excitation spectrum in a Ne matrix is very different (Fig. 13b), while the threshold region is identical to that in the gas phase, suggesting no matrix effect on the ionic part of the double–well. The broad observed excitation band between 9 and 10 eV, corresponds to the deperturbed ionic state as the Rydberg state has been shifted to higher energies. Indeed, as mentioned above, Rydberg states of matrix–isolated molecules are strongly blue shifted with respect to their gas phase positions /16/, while valence or ionic states are less affected. This tends to lift the configuration mixing between Rydberg and valence (or ionic) states and leave the latter unperturbed. In Fig. 13b, due to its

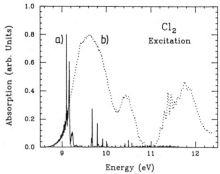

Fig. 13:
Comparison of Cl$_2$ laser
emission intensity versus
excitation energy in gas
phase (a) and in Ne matrix
(b), from Ref. 20.

small vibrational quantum and strong electron–phonon coupling, the vibrational
structure of the ionic state is not observed. The bands at energies above 10.5 eV
have not been assigned but may be due to blue shifted Rydberg states.

It should be mentioned that while matrices tend to lift perturbation of valence
(ionic) states existing in the gas phase, new perturbations are created in the regions
of blue shifted Rydberg states /21/. By an appropriate choice of matrix and by
density effects, the energy range of deperturbation and reperturbations can be
tuned in a controlled way /21/, thus opening new predissociation channels. In this
context it is also worthwhile mentioning that matrix effects on bound states are
retained in the case of purely repulsive Rydberg or valence states /22/.

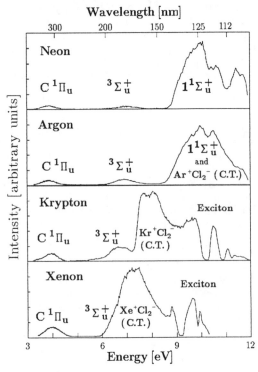

Fig. 14:
Excitation efficiency of
A′–X emission of 0.001%
Cl$_2$ in Ne, Ar Kr and Xe
matrices at 5 K from
Ref. 11.

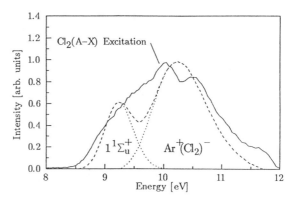

Fig. 15:
Simulation of the A–X'
excitation spectrum of
Cl₂ in Ar by the $1\,^1\Sigma_u^+$
valence and the $Ar^+(Cl_2)^-$
charge transfer state
from Ref. 7.

5.2 COVALENT AND CHARGE–TRANSFER STATES

In the above example (Fig. 13), it was seen that the ionic state was hardly shifted in Ne matrices with respect to its gas phase energy. However, when going to more polarisable matrices, one has to consider dipole solvation and, in the case of halogens, due to their large electron affinities, creation of intermolecular Charge Transfer (C.T.). states. A summary of the spectroscopy of molecular chlorine in Ne, Ar, Kr and Xe matrices is shown in Fig. 14, as monitored on the Cl₂ $A'\,^3\Pi_u–X\,^1\Sigma^+_{0g}$ emission in the red/near–infrared region. Assignment of the excitation features in Ne clearly indicates the presence of absorption to the purely repulsive covalent $C\,^1\Pi_{1u}$ and $^3\Sigma^+_u$ states, as well as the bound $1\,^1\Sigma^+_u$ ionic state discussed above. The dependence of the peak positions and intensities of these transitions on the host can easily be seen by comparing the panels in Fig. 14 /11/. Thus, the peak positions of the purely repulsive covalent states show little or no dependence on the host, while an increase of the relative intensity of the $^3\Sigma^+_u$ to the $C\,^1\Pi_{1u}$ state is noticable in going to heavier matrices, as a result of the heavy–atom effect on spin–forbidden transitions /19/.

Finally, an additional broad band (C.T.) appears which shifts red systematically from Ar to Xe. In Ne matrices it is missing and this band is due to intermolecular charge–transfer $(Cl_2)^-–Rg^+$ excitations which have been investigated in detail by Apkarian and co–workers /9,10/. The C.T. bands are superimposed on intramolecular excitation bands. In Ar, the $Ar^+(Cl_2)^-$ C.T. excitations are embedded in the manifold of ionic Cl^+Cl^- states around 10 eV, as also shown by a simulation /11/ (Fig. 15). In Kr, the $Kr^+(Cl_2)^-$ state lies in a region between the $^3\Sigma^+_u$ covalent state and the ionic states beyond 9 eV. In Xe, the $Xe^+(Cl_2)^-$ state is superimposed on the covalent $^3\Sigma^+_u$ state (Fig. 14). This energy coincidence will give rise to competing dissociative or predissociative reaction channels by intramolecular excitations or "harpooning" mechanisms.

5.3. INTRAMOLECULAR .VS. "HARPOONING" REACTION
MECHANISMS

The harpooning mechanism of the Cl₂/Rg systems is well known in the gas phase and has been extensively studied by Apkarian and co–workers for matrix–isolated halogens /9,10/ who postulated a "negative" cage effect for the production of atomic fragments via intermolecular charge transfer excitations, (see equ. 1). In the light of our spectra (Figs. 14,15) the question is raised as to which extent the harpooning mechanism dominates the dissociative reaction. As discussed above, and also in references 7 and 11, the answer depends on the matrix. We have seen that in Ar matrices, a "harpooning" mechanism can clearly be ruled out even though a comparison of the Cl yield curve (Fig. 3) with the simulated $Ar^+(Cl_2)^-$ absorption band (Fig. 15) shows that the C.T. state is the precursor. In fact, as discussed in ref. 7, dissociation occurs by an impulsive mechanism following curve crossing from the $Ar^+(Cl_2)^-$ state to a purely repulsive Cl_2 potential correlating with ground state Cl atoms. In case of Xe matrices, the situation is reversed as shown in Fig. 8, where it can be seen that the Xe_2Cl excitation spectrum coincides with the Cl atom yield curve and that both show the same threshold and wavelength dependence as the $Xe^+(Cl_2)^-$ C.T. excitation curve (ref. 11 and § 3.2). These conclusions are in agreement with the studies of Apkarian and co—workers /9,10/.

These examples were meant to show that by an appropriate choice of matrix and dopand, one can shift potential surfaces of the latter or create new intermolecular states which can fruitfully be exploited to carry out controlled photochemistry in the condensed phase.

References

/1/   Schwentner, N., Chergui, M., (1992) "Photochemistry, Charge Transfer States and Laser Applications of Small Molecules in Rare Gas Crystals" in Optical Properties of Excited States of Solids, ed.: B. Di Bartolo (Plenum Press, New York), 499–524 and references therein
      Chergui, M., Schwentner, N. (1993) "Cage effect on the photodissociation of small molecules in van der Waals clusters and crystals" in Research Trends: Chemical Physics, ed.: J. Menon (Trivandum, India) and references therein

/2/   Harris, A.L., Brown, J.K., and Harris, C.B. (1988) "The Nature of simple photodissociation reactions in Liquids on ultrafast time scales" Ann. Rev. Phys. Chem. 39, 341–366 and references therein

/3/   Bondybey, V.E., Fletcher, C. (1976) "Photophysics of low lying Electronic states of Cl_2 in rare gas solids", J. Chem. Phys. 64, 3615–3620

/4/   Kunz, H. (1991) "Photodissoziation von Cl_2 in Edelgasmatrizen und spektroskopische Untersuchung der Dissoziationsfragmente", thesis, FU Berlin

/5/   Peyerimhoff, S.D. and Buenker, R.J. (1991) "Electronically excited and ionized states of the Chlorine Molecule", Chem. Phys. 57, 279–296

/6/   Kunz, H., McCaffrey, J.G., Schriever, R., and Schwentner, N. (1991) "Spectroscopy and photodissociation of molecular chlorine in argon matrices", J. Chem. Phys. 94, 1039–1045

/7/   McCaffrey, J.G., Kunz, H., and Schwentner, N. (1992) "Spectroscopy and photodissociation of chlorine monomers and clusters in argon matrices", J. Chem. Phys. 96, 155–164

/8/   Alimi, R., Gerber, R.B., McCaffrey, J.G., Kunz, H., and Schwentner, N. (1992) "Delayed and direct cage exit in photodissociation of Cl_2 in solid Ar", Phys. Rev. Lett. 69, 856–859

/9/   Fajardo, M.E. Withnall, R. Feld, J., Okada, F., Lawrence, W., Wiedemann, L., and Apkarian, V.A. (1988) "Condensed phase laser induced harpoon reactions", Laser Chemistry, 9, 1–12

/10/ Fajardo, M.E. and Apkarian, V.A. (1988) "Charge Transfer Photodynamics in Halogen Doped Xenon Matrices II: Photoinduced Harpooning and the Delocalized Charge Transfer States of Solid Xenon Halides (F, Cl, Br)", J. Chem. Phys. 89, 4102–4123

/11/ McCaffrey, J.G., Kunz, H., and Schwentner, N. (1992) "Photodissociation of molecular chlorine in Xe matrices", J. Chem. Phys. 96, 2825–2833

/12/ Gersonde, I.H., and Gabriel, H. (1993) "Molecular dynamics of photodissociation in matrices including nonadiabatic processes", J. Chem. Phys. 98, 2094–2106

/13/ Alimi, R., Brokman, A. and Gerber, R.B. (1989) "Molecular dynamics simulations of reactions in solids: Photodissociation of Cl₂ in crystalline Xe", J. Chem. Phys. 91, 1611–1617

/14/ Hoffmann, J. and Apkarian, V.A. (1991) "Photodynamics of (Cl₂)ₙ clusters trapped in Solid Krypton", J. Phys. Chem. 95, 5372–5374

/15/ Kunz, H., McCaffrey, J.G., Chergui, M., Schriever, R., Ünal, Ö., Stepanenko, V., and Schwentner, N. (1991) "Rydberg series of charge transfer excitations: Cl and H in rare gas crystals", J. Chem. Phys. 95, 1466–1472

/16/ Chergui, M., Schwentner, N., Böhmer, W. (1986) "Rydberg states of NO in rare gas matrices", J. Chem. Phys. 85, 2472–2482

/17/ Bondybey, V.E., Brus, L.E. (1980) "Nonradiative processes in small molecules in low–temperature solids", Adv. Chem. Phys. 41, 260–320

/18/ Roncin, J.Y. (1968) "Electronic Transitions of CO, N₂ and NO Molecules trapped in Solid Rare gas Matrices: Qualitative Discussion", J. Mol. Spect. 26, 105–110

/19/ Robinson, G.W. (1967) "Intensity Enhancement of Forbidden Electronic Transitions by Weak Intermolecular Interactions", J. Chem. Phys. 46, 572–585

/20/ Gürtler, P., Kunz, H., Le Calve, J. (1989) "Vacuum ultraviolet spectroscopy of the Cl₂ molecule trapped in pure neon, pure argon or mixed neon–argon matrices", J. Chem. Phys. 91, 6020–6028

/21/ Chergui, M. Schwentner, N. (1992) "Rydberg~valence perturbations in matrix–isolated NO", J. Chem. Phys. 97, 2881–2890

/22/ Chergui, M., Gödderz, K.H., Schriever, R., Schwentner, N., Stepanenko, V. "Absorption by dissociative continua of matrix–isolated molecules: H₂O and HCl", J. Chem. Phys. (submitted)

ELECTRONIC ENERGY TRANSFER IN THE NAPHTHALENE – ANTHRACENE BICHROMOPHORIC MOLECULAR CLUSTER

Yoram KARNI and Shammai SPEISER
Department of Chemistry
Technion – Israel Institute of Technology
Haifa 32000, Israel

1. Introduction

In an ideal bichromophoric molecule the electronic absorption spectrum can be described by a simple superposition of the absorption spectra of the two chromophores. The bridge serves as a molecular spacer that does not influence the basic electronic structure of the two chromophores while preventing intrachromophore interaction in their ground state. However, excitation of either chromophore may lead to such interactions and to the observation of phenomena such as intramolecular complex formation [1, 2], intramolecular electron transfer [3], or intramolecular electronic energy transfer (intra–EET) [4–12].

In recent years we have investigated several aspects of intra–EET in solution. These studies involved the synthesis and characterization of specially designed bichromophoric molecules. Study of these molecules enabled us to elucidate the mechanism of intra–EET for symmetrically linked bichromophoric molecules [6–13].

Electronic energy transfer (EET) processes involve non-radiative transfer of electronic excitation from an excited donor molecule D* to an acceptor molecule A. The transfer may be an intermolecular process, that can be described in terms of a bimolecular kinetic process [14]

$$D^* + A \xrightarrow{k_Q} D + A^* \tag{1}$$

where the bimolecular quenching process is related to an intermolecular energy transfer rate by

$$k_{ET}^{inter} = k_Q[A] \tag{2}$$

Theoretically k_{ET} is attributed to two possible contributions. The long range coulombic contribution formulated by Förster [15] in terms of dipole-dipole interaction is well documented. It is particularly suitable for describing electronic energy transfer in solution whenever conditions for favourable spectroscopic overlap conditions between the emission of D* and the absorption of A are met.

The second contribution to EET can be realized whenever these conditions are not fulfilled. A short range exchange interaction, as formulated by Dexter, can then facilitate EET [16].

J. Jortner et al. (eds.), Reaction Dynamics in Clusters and Condensed Phases, 539–556.

Intra–EET processes in bichromophoric molecules are usually described in terms of the process:

$$D^* - (\ ) - A \xrightarrow{k_{ET}^{intra}} D - (\ ) - A^* \tag{3}$$

where the excitation energy is transferred from an excited donor D^* chromophore moiety to a ground state acceptor moiety A, resulting in quenching of D^* and sensitization of A and $-(\ )-$ denotes a molecular spacer bridge connecting the two chromophores.

Our previous results on intra–EET indicated that the transfer efficiency is strongly structure dependent, suggesting that the Dexter type exchange interaction [7] is responsible for singlet-singlet intra–EET between the chromophores in these bichromophoric molecules. It was shown that Dexter's approximation for the k_{ET}^{intra} [16] can account for the observations:

$$k_{ET}^{intra} = k_{ET}^{ex} = (2\pi/\hbar) \, K \, J \exp\left(-2R/L\right) \tag{4}$$

where L is the average orbital radius involved in the initial and final states, K is a constant not related to experimental parameters and J is the overlap integral:

$$J = \int_0^\infty \bar{F}_D(\bar{\nu}) \bar{\epsilon}_A(\bar{\nu}) d\bar{\nu} \tag{5}$$

where $\bar{F}_D(\bar{\nu})$ and $\bar{\epsilon}_A(\bar{\nu})$ are the normalized emission spectrum of D and the normalized extinction coefficient of A respectively:

$$\int_0^\infty \bar{F}_D(\bar{\nu}) d\bar{\nu} = 1; \qquad \int_0^\infty \bar{\epsilon}_A(\bar{\nu}) d\bar{\nu} = 1 \tag{6}$$

Bichromophoric molecules can be designed so as to make J small enough to eliminate contributions to Intra–EET from long range dipole-dipole interaction. The typical distance dependence for dipole-dipole long-range EET is given by the Förster relation [14]:

$$k_{ET}^{d-d} = \tau_D^{-1} (R_0/R)^6 \tag{7}$$

where τ_D is the fluorescence decay time of D^*, R is the distance between D and A, and R_0 is the critical transfer radius given by

$$R_0^6 = \frac{9000(\ln 10)\kappa^2 \phi_D}{128\pi^5 n^4 \tau_D} \int \frac{\bar{F}_D(\bar{\nu})\epsilon_A(\bar{\nu})d\bar{\nu}}{\bar{\nu}^4} \tag{8}$$

where κ^2 is an orientation factor for the relative orientations of D and A dipoles, n is the refractive index of the medium and ϕ_D is the fluorescence quantum yield of the donor [5].

Since J involves <u>normalized</u> $\bar{\epsilon}_A(\bar{\nu})$ as opposed to $\epsilon_A(\bar{\nu})$, that appears in eq. 8, energy transfer via forbidden transitions in A are allowed by the exchange interactions while they are much less probable in the Förster mechanism. Thus, we may conclude that unless $R_0 < 10\,\text{Å}$, energy will be transferred mainly by the Förster mechanism.

The exponential distance dependence of k_{ET}^{ex} eq. (4) reflects the spatial overlap of the molecular orbitals involved in the exchange interaction. In Dexter's approximation hydrogen atom orbitals were assumed thus excluding contributions to k_{ET}^{ex} from the relative orbital orientation. Recent calculations indicate [10] that such a contribution should generally be considered.

All the above discussion pertains to bichromophoric molecules in solution where solvent effects cannot be ruled out and vibrational relaxation of D^* is extremely fast and occurs prior to intra–EET process. In a supersonic jet expansion we have to consider intra–EET from a specific excited vibronic state of D^*. A recent study by Levy and coworkers [17] of the anisole-dimethylaniline bichromophoric system showed that the general features of intra–EET for this vibrationally unrelaxed system are not consistent with either simple Förster or Dexter formalism showing in addition a marked dependence on the particular donor-acceptor conformation.

Intra–EET between two chromophores in a bichromophoric molecular complex, formed in a supersonic jet expansion, is even more complicated. As in the case of a covalently bound bichromophoric molecule intra–EET in the jet involves specific vibronic states. In addition, the limited chemical stability of the complex introduces dissociative channels that might interfere with observation of the intra–EET process. However, study of intra–EET in bichromophoric clusters is interesting as a source of information on the relative importance of these competing relaxation processes. One question that may be addressed is whether or not an intra–EET can be observed as in solution studies of bichromophoric molecules, by dual fluorescence spectrum [6–12] typical of a bichromophoric D–A molecule or cluster.

The other system that we have chosen to study was the naphthalene-anthracene (NAP-ANT) D–A bichromophoric molecular complex formed in a supersonic jet expansion. The pioneering work of Schnep and Levy [4] on the NAP-ANT bichromophoric molecules indicated that intra–EET in this system, in solution, is complete upon excitation of the naphthalene moiety. This is not surprising as the $(CH_2)n$ bridge is too flexible and in addition the Förster critical transfer radius for this system is $R_0 = 22Å$ [18] resulting in very efficient intra–EET at the much shorter interchromophore distance.

The overlap integral (eq. (5)) under supersonic jet conditions should now be calculated for fluorescence of D^* from vibrationally unrelaxed states overlapping the corresponding absorption spectrum of A. Such an overlap integral will exhibit sharp resonances, that depending on the excitation energy, may result in a very low value of R_0, so that the dominating mechanism intra–EET in the D–A naphthalene anthracene complex will be the short-range Dexter type exchange interaction. Such an intra–EET process is another nonradiative channel open for the excited naphthalene. Figure 1 shows the energy levels of the naphthalene-anthracene system and the photophysical processes associated with them. It is the purpose of the present study to provide a quantitative evaluation of the mechanism of intra–EET in molecular complexes in supersonic jet expansion.

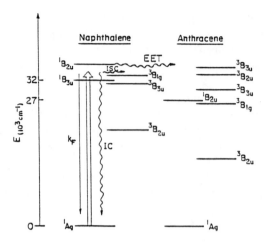

Figure 1: Naphthalene-anthracene energy level diagram.

2. Experimental

Naphthalene-anthracene gas mixtures were prepared by passing He at 4–5 atm over naphthalene held at $-50°$C in the gas entrance of a high-temperature solenoid valve. The valve,
built according to the design of Li and Lubman [19], serves as the hot oven and contains
the added anthracene. The hot oven in the valve is separated from the solenoid by a water
jacket which serves to control the naphthalene temperature and to protect the solenoid
and the spring.

The gas mixture was expanded through a 1 mm orifice into a stainless steel vacuum
chamber. The chamber was pumped by an untrapped CVC Goldine 6" diffusion pump
backed by a Welch model 1397 rotary pump. Typical background pressure was about 10^{-4}
torr.

The samples were excited by a Quanta-Ray PDL-1 Dye laser, pumped by the second
harmonic of a DCR Nd:YAG laser, operated at 10 Hz. Coumarin 420 was used to excite
biacetyl, while benzene excitation was provided by frequency doubling the output of the
dye laser, operating with coumarin 500, using a WEX1 wavelength extender. Frequency
doubled rodamine 640 was used to excite naphthalene, while anthracene excitation was
provided by frequency dodubling the output of the dye laser operating with LDS-698,
using WEX-1 wavelength extender. The laser light was passed through a series of baffles
into a vacuum chamber, and intersected the molecular beam about 35 mm from the valve
orifice, where the number of molecular collisions is very small.

Excitation spectra were recorded with a Hamamatsu R1104 photomultiplier tube, using appropriate filters to exclude scattered laser light. A Bausch and Lomb 0.25 meter

monochromator was used to record dispersed emission spectra. Signals were averaged over a number of laser shots with a PAR model 162 boxcar, or with a Tektroniks 2400 digital oscilloscope interfaced to an IBM personal computer.

3. Results and Discussion

Fig. 2 shows the excitation spectrum of jet expanded anthracene–naphthalene mixture. This spectrum is very similar to that obtained by Zewail and coworkers [20] for pure anthracene and is due to the $A_{1g} \rightarrow B_{2u}$ transition of anthracene polarized along the short molecular axis of anthracene. In our study we have concentrated on sections I and II of this spectrum. As we increase the anthracene pressure new spectral features, denoted as D bands, emerge in section I of the spectrum (Fig. 3). These new bands can be attributed to the formation of anthracene dimers as is evidenced from Fig. 4 where we plot the ratio of the intensity of the dimers lines Y_{dim} to that of the $11(a_{1g})_0^1$ line of monomer anthracene Y_{mon}. This intensity ratio should follow the relation

$$Y_{dim}/Y_{mon} = K_d P \tag{9}$$

where P is the anthracene pressure and K_d is the dimerization equilibrium constant uncorrected for the difference in absorption and emission crossections between the monomer and the dimer.

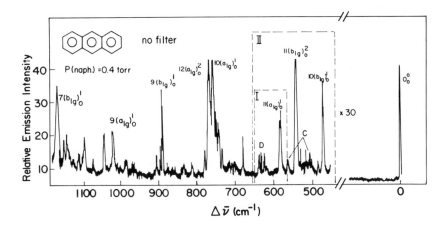

Figure 2: The excitation spectrum of jet expanded 0.8 torr anthracene and for naphthalene mixture in 3 atm of He. The O_o^o transition is measured at 30 times smaller sensitivity. Section I which shows D lines, attributed to naphthalene dimers, is shown in more detail in Fig. 3. Section II which shows C lines, attributed to the formation of anthracene-naphthalene complex is shown in more detail in Fig. 4.

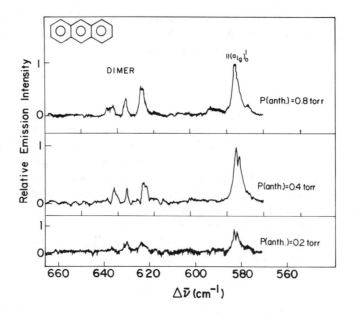

Figure 3: A blow up of section I of the excitation spectrum of anthracene showing the appeareance of dimer "Dimer" lines, as a function of anthracene pressure.

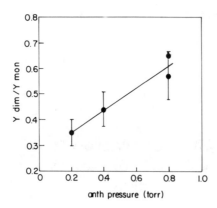

Figure 4: The ratio between anthracene dimer's emission intensity to that of anthracene monomer emission intensity as a function of anthracene pressure.

Upon adding naphthalene to anthracene new spectral features, denoted as C bands, appear in Section II of the excitation spectrum. The intensity of these bands increases with increasing naphthalene pressure with simultaneous variations in the intensities of anthracene bands. The appearance of the new C bands as well as the other spectral changes can be attributed to the formation of anthracene-naphthalene complexes in the jet expanded mixture. The C lines, denoted by an * in Fig. 5, might be attributed to a vibronic progression, of 15 cm^{-1} interval, belonging to the new vibration connected with formation of the weak anthracene-nathphalene bond, which is probably of a van der Waals nature.

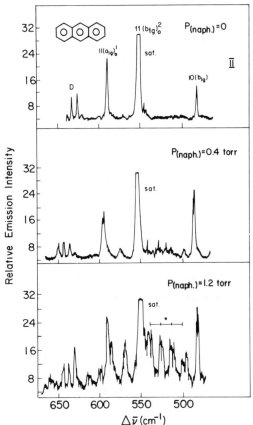

Figure 5: A blow up of section II of the excitation spectrum of anthracene-naphthalene mixture, at 1 torr anthracene as a function of added naphthalene pressure.

This picture of a bichromophoric molecular complex can be tested by examining the excitation spectrum of naphthalene and its variations upon adding anthracene to the jet expanded mixture. Fig. 6a shows the excitation spectrum of jet expanded naphthalene. This spectrum is the same as that reported previously [21]. It involves the first $A_{1g} \rightarrow B_{3u}$ forbidden-transition of naphthalene which is partially allowed due to Herzberg-Teller coupling to a_{1g} and b_{1g} vibrations, the latter gives rise to the most intense band where the transition is polarized along the short molecular axis of naphthalene. the O_0^o transition is at 32 020 cm^{-1} and the vibronic transitions of naphthalene shown in Fig. 6a have a relatively long lifetime of about 100-400 ns [43], this should be compared with the broad absorption spectrum of anthracene at this spectral region, associated with a lifetime shorter than 5 ns [20, 23, 24]. Using proper time gate we were able to distinguish between anthracene and naphthalene excitation spectra. Upon adding anthracene, the spectrum shown in Fig. 6b is obtained, while keeping the same gating conditions as those used for pure naphthalene (Fig. 6a). It exhibits the following important features:

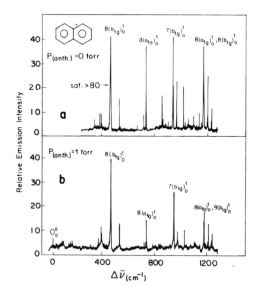

Figure 6a. Excitation spectrum of 1 torr naphthalene jet expanded with 3 atm. of He. The $8(b_{1g})_0^1$ band intensity is saturated.

Figure 6b. Same excitation spectrum upon adding 1 torr of athracene to the jet expanded mixture, showing quenching of all naphthalene bands. Both spectra are recorded during the same gated time window corresponding to naphthalene decay time. $\Delta\bar{\nu}$ measures the shift from the 0-0 transition of naphthalene.

a. By contrast to the spectrum of naphthalene-anthracene mixture monitored at the anthracene excitation range (Fig. 7), no distinct complex lines are observed.
b. Naphthalene emission is quenched by added anthracene.
c. Quenching efficiency seems to depend on the particular excitation of naphthalene.

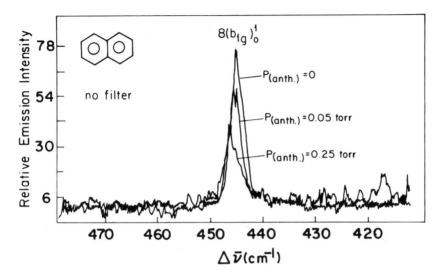

Figure 7: Fluorescence excitation spectrum of naphthalene $8(b_{1g})^1_0$ transition, observed by monitoring unfiltered fluorescence during 25 ns delayed 1 μs time window, as function of added anthracene pressure.

Unlike the benzene-biacetyl case the lifetimes of the donor and the acceptor are very different for the naphthalene-anthracene pair. Thus, in order to establish the nature of the process we have examined the time resolved excitation spectra monitored at different emission spectral regions. These emissions spectral regions correspond to either that belonging to naphthalene fluorescence [20] that belonging to anthracene fluorescence [22] Although these spectra overlap to some extent they can be time separated due to the large difference in the fluorescence lifetimes of the two molecules.

These observations are consistent with the following picture: In the jet expanded naphthalene-anthracene mixture (Fig. 6b) several different bichromophoric molecular complexes are formed. During the 25 ns delayed 1 μs time window we follow only those complexes in which a relatively inefficient intra–EET process took place, thus allowing us to monitor intra–EET during the long lifetime of excited naphthalene. In other complexes in which molecular conformation promotes strong intramolecular interaction one would expect fast intra–EET process which cannot be observed after 25 ns delay time.

Since laser excitation is performed at the collisionless region of the jet the observed quenching of the naphthalene excitation spectrum is attributed to collisionless intra-molecular process in the complexes formed by collisions near the jet orifice (see Experimental Section).

The different quenching efficiencies observed for different excitations indicate that quenching is not just the result of complex formation which reduces the number of free naphthalene molecules, but that in the complex an additional decay channel for the complexed naphthalene chromophore promotes quenching through an intra–EET process. This process is the result of weak intramolecular interaction which is not manifested by large spectral shifts (Fig. 6b). However, as shown below, a closer look at higher spectral resolution gives evidence for such shifts which are related to a slow intra–EET process.

Figure 7 shows the $8(b_{1g})_0^1$ excitation band of naphthalene as a function of added anthracene pressure. This band is recorded by monitoring the total unfiltered emission of the mixture during the lifetime of naphthalene with contributions of emission coming from both directly excited pure naphthalene and from complexed anthracene which was excited by intra–EET from excited naphthalene chromophore. Due to the 25 ns time delay no contribution to the emission from directly excited short-lived pure or complexed anthracene is observed in this spectrum. The role of intra–EET is demonstrated in Fig. 8 where the same band is observed, using the same delayed time window, by monitoring the emission of pure naphthalene excited at the $8(b_{1g})_0^1$ transition and comparing it with that of the naphthalene-anthracene complex monitored by emission filtered through a 400 nm cut-off filter which passes only anthracene emission. The $8(b_{1g})_0^1$ band of pure naphthalene at 3080.25 Å disappeared with the simultaneous appearance of a 3.5 cm^{-1} blue shifted band at 3079.9 Å. We tentatively assign this band as belonging to the naphthalene chromophore in the bichromophoric complex whose excitation is detected by monitoring anthracene emission at $\lambda > 400$ nm, due to intra–EET process, lasting during naphthalene lifetime. In other words Fig. 8 gives evidence to both bichromophoric complex formation and to intra–EET as a quenching mechanism for naphthalene. The blue shift of about 3.5 cm^{-1} is rather small and reflects the weak interaction between the two chromophores, this is the same type of interaction which leads to the observed inefficient intra–EET process. This is in agreement with the observation of small spectral shifts for other molecular complexes and with our observation of negligible shift for the benzene biacetyl complex [25]. Similar results were obtained for the naphthalene $8(a_1g)_0^1\ 8(b_{1g})_0^1$ combination band [26].

In the jet complex formation, which occurs close to its orifice, can be described by

$$D + A \;\overset{K}{\rightleftharpoons}\; D - A \tag{10}$$

where K is the association constant. The formed D–A complex is then excited in the collisionless region of the jet by the intersecting laser beam (see Experimental Section).

The observed intra–EET and all other excitation processes of the benzene-donor may be summarized in the following kinetic scheme:

D	$\xrightarrow{\sigma_D I_D}$	D*	unbound donor excitation	(11)
D–A	$\xrightarrow{\sigma_{DA} I_D}$	D*–A	donor moiety excitation in the complex	(12)
D*	$\xrightarrow{k_f^D}$	D	unbound donor fluorescence	(13)
D*–A	$\xrightarrow{k_f^{D^*A}}$	D–A	complex donor's moiety fluorescence	(14)
D*–A	$\xrightarrow{k_{ET}^{intra}}$	D–A*	intramolecular EET	(15)
D–A*	$\xrightarrow{k_f^{DA^*}}$	D–A	complex acceptor's moiety fluorescence	(16)

where σ_D and σ_{DA} are the crossections for donor absorption for the free donor and for its complexed form respectively, I_D is the excitation intensity at the donor absorption frequency and the k_f's are the corresponding fluorescence rate constants.

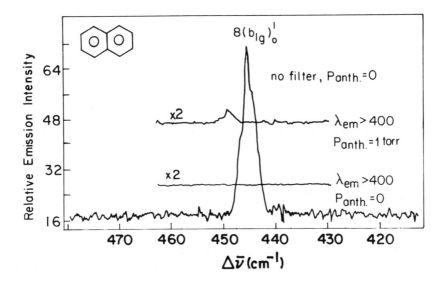

Figure 8: Comparison between the excitation spectrum of the $8(b_{1g})_0^1$ transition of pure naphthalene to that monitored by collecting unfiltered emission and that of naphthalene-anthracene complex monitored by emission collected at $\lambda > 400$ nm during excited naphthalene lifetime. Lower trace shows the base line signal at $\lambda > 400$ nm for pure naphthalene.

From an analysis of the scheme [eqs. (10)–(16)] the steady state concentrations of the donor's emitting states $[D^*]$ and $[D^*A]$ are obtained. The total donor emission intensity is given by

$$Y_D = k_r^{D^*}[D^*] + k_r^{D^*A}[D^*A] \tag{17}$$

where $k_r^{D^*}$ and $k_r^{D^*A}$ are the radiative decay rate constants of D^* and D^*A respectively. From eq. (17) we obtain the expression for the dependence of the donor's fluorescence quenching on the acceptor concentration $[A]$,

$$Y_D^0 \Big/ Y_D = 1 + \frac{\beta K[A]}{1 + \left(k_{ET}^{intra} \Big/ k_f^{D^*A}\right) + \gamma_D K[A]} \tag{18}$$

where

$$\beta = \left(k_{ET}^{intra} \Big/ k_f^{D^*A}\right) + 1 - \gamma_D \tag{19}$$

and

$$\gamma_D = \sigma_{DA}\phi_{D^*A} \Big/ \sigma_D\phi_{D^*} \tag{20}$$

where Y_D^0 is the fluorescence yield of the free D molecule, $\phi_D = k_r^{D^*} \big/ k_f^{D^*}$ and $\phi_{D^*A} = k_r^{D^*A} \big/ k_f^{D^*A}$ are the fluorescence quantum yield of the free donor and of the donor moiety in the DA complex, not including intra EET, respectively.

Three different cases may be examined:

Case a – The donor moiety of the D–A complex is not excited, i.e. $\sigma_{DA} = 0$ and, thus $\gamma_D = 0$ and eq. (18) reduces to

$$Y_D^0 / Y_D = 1 + K[A] \tag{21}$$

This equation expresses a static Stern-Volmer type quenching of the free donor molecules associated solely with D–A complex formation. it is not related to any optical molecular parameters and therefore should be the same for all naphthalene transitions. For the NAP-ANT complex $\gamma_D \neq 0$ for all of the measured lines, since excitation of the free naphthalene almost coincides with that of the NAP moiety in the complex (Fig. 6). Figure 9 shows such a Stern-Volmer plot for the two naphthalene transitions. The small deviation from linearity at high pressure for the $8(b_{1g})_0^1$ transition is probably due to formation of anthracene dimers. From this fit we obtain that $K = (14 \pm 2)$ torr^{-1}. The fit supports our assumption of 1:1 complex formation for the pressure range employed in the experiment.

Case b – Extremely efficient intra-EET manifested in $k_{ET}^{intra} \big/ k_f^{D^*A} \gg 1$ yielding

$$Y_D^0 \Big/ Y_D = 1 + K[A] \tag{22}$$

which is an apparent static quenching Stern-Volmer type relation whose origin is in complex formation followed by ultrafast intra-EET process. It predicts no dependence on the particular vibronic excitation of the donor and cannot be distinguished from a genuine static quenching process, case a. Such a fast intra-EET process is generally associated with strongly interacting donor and acceptor. Strong interaction should be manifested in marked changes in the excitation spectrum.

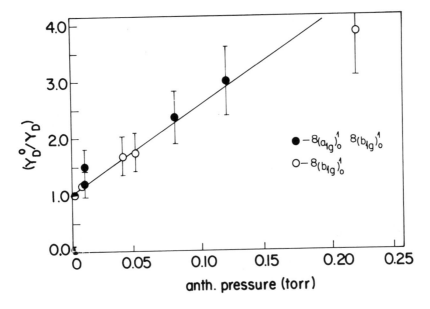

Figure 9: Ratio of pure naphthalene emission yield, without and in the presence of added anthracene, as a function of the anthracene pressure, showing the fit to eq. (21), for two different naphthalene transitions.

Case c – Weakly interacting D and A resulting in low value of k_{ET}^{intra}, which for $\gamma_D \sim 1$ yields, at low values of [A],

$$Y_D^0/Y_D = 1 + \tau_f^D k_{ET}^{intra} K[A] = 1 + \tau_f^D k_Q[A] \tag{23}$$

where $\tau_f^D = 1/k_f^{D^*}$ is the fluorescence lifetime of D^*.

Equation (23) is in a form of an <u>apparent</u> Stern-Volmer relation which usually originates from collisional quenching. However, it should be noted that here quenching is due to an inefficient intra-EET process in the collisionless region of the jet preceded by complex formation through collisions near the orifice of the jet, and eq. (23) is obtained only as the low pressure limit of eq. (18) for case c. Such a behavior was observed in the benzene-biacetyl bichromophoric molecular cluster [25].

A similar analysis can be carried out for the <u>unfiltered</u> total emission from excited complex resulting from excited donor and acceptor moieties. Its intensity Y_{DA} should be proportional to the number of D^*A species given by their partial pressure P_{D^*A} where contribution from both processes (14) and (16) should be taken into account, thus

$$Y_{DA} = P_{D^*A} \left(\phi_f^{D^*A} + \phi_{EET}\, \phi_f^{DA^*} \right) \tag{24}$$

where $\phi_f^{D^*A}$ is the quantum yield for donor's moiety fluorescence, ϕ_{EET} is the intra–EET quantum yield and $\phi_f^{DA^*}$ is the quantum yield for acceptor's moiety fluorescence which is excited by intra–EET. The partial pressure of the excited complex is given by

$$P_{D^*A} = P_{DA} \int \sigma_{DA} I dt \tag{25}$$

where the integration is done over the exciting laser pulse duration and P_{DA} is given by

$$P_{DA} = K P_D P_A \tag{26}$$

Substituting eqs. (25) and (26) in eq. (24) gives

$$Y_{DA} = \sigma_{DA} K P_D P_A \left(\phi_f^{D^*A} + \phi_{EET} \phi_f^{DA^*} \right) \int I dt \tag{27}$$

In order to obtain an expression which is independent of P_D we compare Y_{DA} to Y_D

$$Y_D = \sigma_D P_D \phi_f^D \int I dt \tag{28}$$

to obtain the ratio

$$Y_{DA} / Y_D = \frac{K \sigma_{DA}}{\sigma_D \phi_f^D} \left(\phi_f^{D^*A} + \phi_{EET} \phi_f^{DA^*} \right) P_A \tag{29}$$

where ϕ_f^D is the free donor fluorescence quantum yield

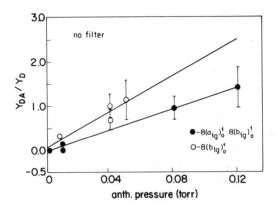

Figure 10: Ratio of intensities for emission from naphthalene-anthracene complex to that belonging to free naphthalene, for two naphthalene transitions, as a function of anthracene pressure. The fit is to eq. (29).

Figure 10 shows the fit to eq. (29) of the two naphthalene transitions that were studied. The good agreement supports the suggested model. Moreover, the dependence on the particular vibronic excitation of the donor, manifested in different slopes, is related to the ratio σ_{DA}/σ_D and to ϕ_{EET}. For the $8(b_{1g})_0^1$ excitation of naphthalene we can estimate σ_{DA}/σ_D using literature value of 800 ns for its lifetime and the value of $\phi_D = 0.5$ [22] and $\phi_f^{DA^*} = 0.15$ [23], and our measured values of $k_f^{D^*A}$ and k_{ET}^{intra} to obtain values for $\phi_f^{D^*A}$ and ϕ_{EET}. In addition, we note that K describes all of the possible complexes formed in the jet and thus the value pertaining to those sampled in the intra–EET experiment is lower. Thus using eq. (29) and the slope of Fig. 10 we obtain a lower limit of $\sigma_{DA}/\sigma_D \sim 2$ for the $8(b_{1g})_0^1$ naphthalene band.

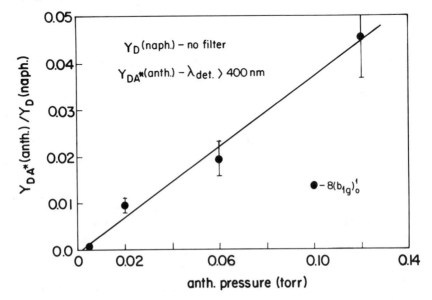

Figure 11: Ratio between the emission intensity from the anthracene moiety, in the naphthalene excited naphthalene-anthracene complex, due to intra–EET from $8(b_{1g})_0^1$ naphthalene, to that obtained from uncomplexed naphthalene, as a function of anthracene pressure. The fit is to eq. (30).

The kinetic analysis can be used also to gain additional information by looking at the emission of anthracene, monitored through the 400 nm cut off filter. From eq. (29) we obtain the acceptor's to free donor's emission intensity ratio

$$Y_{DA^*}/Y_D = \frac{0.2K\sigma_{DA}\phi_f^{DA^*}\phi_{EET}}{\sigma_D\phi_f^D}P_A \qquad (30)$$

where the factor 0.2 takes into account the actual portion of the total anthracene fluorescence sampled through the filter. The fit to eq. (30) of the sampled anthracene moiety fluorescence, resulting from intra–EET process as a function of anthracene pressure is shown in Fig. 11. Again a good fit is obtained supporting the suggested mechanism. From the slope of this plot, the estimated value of σ_{DA}/σ_D and the relation

$$\phi_{EET} = \frac{k_{ET}^{intra}}{k_f^{D^*A} + k_{ET}^{intra}} \tag{31}$$

the value of $k_{ET}^{intra} = 0.3\,\mu\,\mathrm{s}^{-1}$ is obtained.

4. Summary and Conclusions

We have presented evidence for the formation of naphthalene-anthracene (D–A) bichromophoric complexes in a supersonic jet expansion of naphthalene (D) anthracene (A) mixture. Complex formation is manifested in spectral changes in the excitation spectra of the species involved. The spectroscopic results are interpreted in terms of a bichromophoric molecular complex where at the anthracene excitation spectral region (DA^* formation) we observe the appearance of a vibronic progression associated with the interchromophore weak van der Waals bond accompanied by slight shifts and intensity variations of the various anthracene bands. While at the anthracene part of the excitation spectrum we observe all possible complexes, at the naphthalene side of the spectrum we observe only those D^*A complexes whose excited states were not totally quenched by an efficient $D^*A \rightarrow DA^*$ intra–EET process. For the complex associated with a slow intra–EET process we have observed small spectral shifts involving the same type of interaction responsible for the appearance of intra–EET. At the same time at the naphthalene part of the spectrum we can observe only small shifts and fluorescence quenching resulting from inefficient EET process. Strongly interacting excited D^*A complexes are totally quenched by fast intra–EET process and thus are not observed.

Pressure dependence of the intra–EET efficiency gives an estimated value of $0.3\,\mu s^{-1}$ for the intra–EET rate constant.

In conclusion we note that we have provided, for the first time, evidence for the existence of slow intra–EET process in bichromophoric molecular complexes formed in a supersonic jet. Previous studies of EET in supersonic jet expansion provided only spectral evidence for the existence of the process, however, without any quantitative evaluation of the transfer mechanism [27–29].

Our results point out that the mechanism of intra–EET in molecular complexes should be treated within the general framework of radiationless transitions theories, modified to include transfer from single vibronic state. It may well be that variations in the transfer efficiency upon excitation to different vibronic states (Fig. 10) testify to such an effect [30]. This point should be clarified in additional experimental and theoretical studies.

5. Acknowledgement

This study was partially supported by the Fund for the Promotion of Research at the Technion.

6. References

[1] F.C. De Schryver, N. Boens and T. Put, Adv. Photochem. **10**, 359 (1977) .

[2] B. Valeur, in *Fluorescent Biomolecules*, D.M. Jameson and G.D. Reinhardt, Edts., (Plenum Press, N.Y. 1969) p. 296.

[3] a. H. Oevring, J.W. Verhoeven, M.N. Paddon-Row, and E. Cotsaris, Chem. Phys. Lett. **143**, 488 (1988).
b. J. Kroon, A.M. Oliver, M.N. Paddon-Row, and J.W. Verhoeven, J. Am. Chem. Soc. **112**, 4868 (1990).

[4] O. Schnepp and M. Levy, J. Am. Chem. Soc. **84**, 172 (1962).

[5] S. Speiser, "Novel aspects of intermolecular electronic energy transfer in solution," J. Photochem. **22**, 195 (1983).

[6] D. Getz, A. Ron, M.B. Rubin and S. Speiser, "Dual fluorescence and intramolecular energy transfer in a bichromophoric molecule," J. Phys. Chem. **84**, 768 (1980) .

[7] S. Speiser and J. Katriel, "Intramolecular electronic energy transfer via exchange interaction in bichromophoric molecule," Chem. Phys. Lett. **102**, 88 (1983) .

[8] S. Hassoon, H. Lustig, M.B. Rubin, and S. Speiser, "The mechanism of short range intramolecular electronic energy transfer in bichromophoric molecules," J. Phys. Chem. **88**, 6367 (1984) .

[9] S. Speiser, S. Hassoon and M.B. Rubin, "The mechanism of short range intramolecular electronic transfer in bichromophoric molecules. II. Triplet-triplet transfer," J. Phys. Chem. **90**, 5085 (1986).

[10] S.-T. Levy and S. Speiser, "Calculation of the exchange integral for short range intramolecular electronic energy transfer in bichromophoric molecules," J. Chem. Phys. **96**, 3585 (1992).

[11] M.B. Rubin, S. Migdal, S. Speiser and M. Kaftory, "Synthesis and structures of substrates for investigation of intramolecular electronic energy transfer," Isr. J. Chem. **25**, 66 (1985).

[12] S.-T. Levy, M.B. Rubin and S. Speiser, "Photophysics of cyclic α−diketone-aromatic bichromophoric molecules. Structures, spectra and intra−EET," J. Am. Chem. Soc. **114**, 10747 (1992).

[13] S.-T. Levy, M.B. Rubin and S. Speiser, "Orientational effects in intra−EET in bichromophoric molecules. II. Triplet−triplet transfer," T. Photochem. Photobiol. A Chem. **69**, 287 (1993).

[14] N. Mataga and T. Kubota, Molecular Interactions and Electronic Spectra, (Dekker, New York, 1970).

[15] Th. Förster, Ann. Phys. (Leipz) **2**, 55 (1948); Th. Förster, Disc. Faraday Soc. **27**, 7 (1959) ; in *Modern Quantum Chemistry* Istanbul lectures, O. Sinanoglu Ed. (Academic Press, N.Y. 1965).

[16] D.L. Dexter, "A theory of sensitized luminescence in solids," J. Chem. Phys. **21**, 836 (1953).

[17] M. Chattoraj, B. Bal, G.L. Closs and D.H. Levy, "Conformation dependent intramolecular electronic energy transfer in a molecular beam," J. Phys. Chem. **95**, 9666 (1992) .

[18] I.B. Berlman, *Energy Transfer Parameters of Aromatic Compounds* (Academic Press, 1973).

[19] L. Li and D.M. Lubman, Rev. Sci. Instrum. **60**, 499 (1989).

[20] W.R. Lambert, P.M. Felker, J.A. Syage, and A.H. Zewail, "Jet spectroscopy of anthracene and deuterated anthracenes," J. Chem. Phys. **81**, 2195 (1984).

[21] a. S.M. Beck, D.E. Powers, J.B. Hopkins, and R.E. Smalley, "Jet-cooled naphthalene. I. Absorption spectra and line profiles," J. Chem. Phys. **73**, 2019 (1980);
b. S.M. Beck, J.B. Hopkins, D.E. Powers, and R.E. Smalley, "Jet-cooled naphthalene. II. Single vibronic level fluorescence spectra," J. Chem. Phys. **74**, 43 (1981).

[22] F.M. Behlen, D.B. McDonald, V. Sethuraman, and S.A. Rice, "Fluorescence spectroscopy of cold and warm naphthalene molecules: Some new vibrational assignments," J. Chem. Phys. **75**, 5685 (1981) .

[23] A. Amirav and J. Jortner, Chem. Phys. Lett. **94**, 737 (1990).

[24] M.M. Doxtader, E.A. Elisa, A.K. Bhattacharya, S.M. Cohen, and M.R. Topp, Chem. Phys. **101**, 413 (1986).

[25] J. Bigman, Y. Karni and S. Speiser (in preparation).

[26] Y. Karni and S. Speiser (in preparation).

[27] D.E. Poeltl and J.K. McVey, "Excited-state dynamics of hydrogen-bonded dimers of benzoic acid," J. Chem. Phys. **80**, 1801 (1984).

[28] Y. Tomioka, H. Abe, M. Ito, and N. Mikami, "Electronic relaxation in a mixed dimer of benzoid acid and *p*-toluic acid," J. Phys. Chem. **88**, 5186 (1984) .

[29] F. Lahmani, C. Lardeux-Dedonder and A. Zehnacker-Rentien, "Electronic excitation transfer in a mixed dimer of *p*-xylene and *p*-difluorobenzene," J. Chem. Phys. **92**, 4159 (1990).

[30] S.H. Lin, W.Z. Xiao, J. Bigman, Y. Karni and S. Speiser (in preparation).

ELECTRON SOLVATION: QUANTUM AND CLASSICAL ASPECTS

Alexander Mosyak and Abraham Nitzan
School of Chemistry, The Sackler Faculty of Science,
Tel Aviv University, Tel Aviv, 69978, ISRAEL

I. Introduction

Interest in the solvated electron has been revived recently due to advances in experimental techniques which have made it possible to monitor the solvation dynamics of electron on the femtosecond timescale,[1] as well as progress in numerical methods, particularly quantum dynamical simulations, which have led to numerical studies of such solvation processes.[2] Quantum simulations were used recently to analyze other phenomena associated with the solvated electron, in particular its spectroscopy, in bulk solvents and in clusters.[3]

With the currently available computer resources, an all-quantum calculation of such systems is impossible, and different versions of mixed quantum classical simulations have been used by different groups. In these approaches the electron is treated quantum mechanically, using path integral methods for ground state or thermal equilibrium problems and numerical grid techniques for solving the time dependent Schrödinger equation. The solvent is treated classically and the electron solvent interaction is included in the spirit of the Ehrenfest theorem, whereupon the electron sees the instantaneous potential induced by the solvent while the latter responds to the *expectation value* of the same potential evaluated with the instantaneous electronic wavefunction. A particularly useful version of this approach is the adiabatic simulation method (ASM),[4] appropriate for processes in which the electron remains in a single electronic state during the time evolution of the system. In this type of simulation the (classical) nuclear evolution is carried out under the expectation value of the electron-nuclei potential, and the electron is restricted to the given quantum state for the instantaneous nuclear configuration using a suitable relaxation method.

Despite some marked successes, the ASM has some severe limitations originated from two main sources. The first, obviously, occurs in situations where the adiabatic approximation breaks down. Early adiabatic simulations of electron hydration dynamics (following photoejection of electrons into water)[2] have proven inadequate by the experimental observation[1] that the process is non-adiabatic. This has led to more recent non-adiabatic simulations of these

557

J. Jortner et al. (eds.), Reaction Dynamics in Clusters and Condensed Phases, 557–568.
© 1994 *Kluwer Academic Publishers. Printed in the Netherlands.*

processes.[5] The second is associated with the failure of the mixed quantum-classical scheme described above to account for processes where the electronic wavefunction becomes strongly non-localized in configuration space with relatively slow transition between different parts of the wavefunction.[6] Mathematically these two sources of failure are related: the assumption that the nuclei respond to the *average* field of the electron breaks down under the circumstances described.

The use of classical dynamics for the nuclear motion in many chemical processes of interest has a long tradition both in fundamental discussions and in usage. In contrast, the need to use quantum mechanics to describe electrons is almost taken for granted in studies of chemical systems, although classical dynamics has been very useful in electron transport problems in solid state physics.

The present contribution focuses on the question of applicability of the classical electron concept in processes related to electron-solvent dynamics. In simulations carried out by our groups and others,[2] it was observed that *adiabatic solvation dynamics* of electron in water, as obtained from computer simulations using the methods described above is qualitatively very similar to the classical dynamics of solvation of an ion of radius similar to that of the fully solvated electron. (It should be emphasized again that the actual process of electron hydration is non-adiabatic and cannot be described by adiabatic simulations). The reason, as observed in the quantum simulation,[2a] is that the shrinking of the mean radius of the electron distribution occurs on a very fast timescale (< 20fs), and the subsequent dynamics is dominated by solvent relaxation about the newly created charge distribution of radius ~ 4au. In another simulation, Motakabbir et al[7] have compared quantum analysis of likely localization sites for electron in pure liquid water to an earlier analysis by Schnitker et al of the same problem using classical mechanics,[8] demonstrating the merit of the classical approach as a qualitative tool for the analysis of localized electronic states supported by preexisting configurational order in a liquid. Here we examine the use of classical mechanics for several other processes:

(1) The dynamics of *adiabatic* electron solvation in water is reexamined in order to clarify the importance of the electron zero point energy in this process.

(2) A classical analysis of the absorption lineshape of the solvated electron is made and compared to the quantum mechanically calculated spectrum in the same electron-water model.[3]

(3) The penetration of the electron through a static film of water molecules is calculated classically and the results compared to recent quantum calculations in the same model.[9] The interest in this problem stems from recent experiments of photoemission from ice covered metal surfaces.[10,11] The photoemission current drops exponentially as the thickness of the water film increases, however the kinetic energy distribution of the emitted electrons is almost thickness independent.[11] This indicates that the effect of the film is dominated by elastic scattering processes. Indeed, quantum dynamical simulations have shown that the average transmission probability was not affected by freezing all motions of the water atoms. Below we study the same problem using classical equations of motion for the electron, and compare the

results to quantum mechanical simulations.

(4) Thermalization of subexcitation electrons in water. Electrons which are photoinjected into water with energy less then the lowest electronic excitation energy of this solvent, loose their excess kinetic energy by transferring it to the water nuclear degrees of freedom. This process has long been studied theoretically[12] and experimentally,[13] however no firm numbers are available for the timescale of the thermalization process and for the spatial distribution of the solvated electrons which are obtained at the end of this process. Experimentally this information can be inferred indirectly from the time evolution of the subsequent geminate recombination. Theoretical estimates are obtained from stochastic theories using the inelastic crossection of electron scattering from a single water molecule as input,[12a] or from dielectric loss theories.[12b] The numerical simulation methods described above fail for this problemm because due to vigorous multiple scattering, the electron wavefunction becomes strongly fragmented, and electron density appears, after 1-2 fs, in many disconnected islands (Fig. 1). Assuming that most of this density appears in classically allowed regions, classical mechanics should be able to describe the following energy loss. In fact, given the quantum nature of vibrational energy transfer into the high frequency OH modes, it is possible that our main source of error in such a classical calculation is the imposition of classical mechanics on these modes.

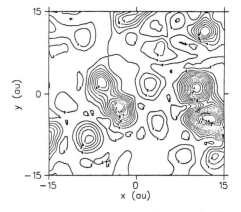

Fig. 1. A Contour plot of the electron density in the xy plane, obtained from the time dependent solution of the Schrödinger equation for an electron injected with translational energy 4.72ev into a static film of 4 layers of water molecules. The density is computed in the middle of the film, at t=2fs. See Ref. 9 for details.

Why should classical mechanics provide even qualitative information on such essentially quantum processes? We have already rationalized above the success of classical mechanics in describing the adiabatic hydration dynamics of the electron when compared to quantum simulations. The actual breakdown of the classical description of that process stems from its non-adiabatic nature.

Other sources of failure of classical descriptions are usually associated with either zero point energy, tunneling or diffraction and interference effects. We shall see that a major part of the effect of the electron localization (zero point) energy is related to the finite spreading of its wavefunction, and can be taken into account by replacing its interaction field by an average over a spherical electron density whose radius is determined by physical considerations. Tunneling does not seem to dominate the processes of interest. The role of diffraction and interference effects in the processes under consideration is unclear, but from general considerations it appears that it will not be very important for the processes under study. Consider for example electron penetration through a static amorphous water film. For a ~ 4ev electron the de Broglie wavelength is ~ 2au, of the same order as intermolecular distances. In a sense this is a generalization of the classic two slit interference experiment where the two possible pathways are replaced by many random ones. The emerging wavefunction is a linear combination of many waves with random phases. This random distribution of phases will wipe out the interference pattern, and the signal $|\psi|^2$ will be dominated by diagonal contributions, making diffraction and interference unimportant in this process.

In what follows we first describe some technical details associated with our simulations, then describe the classical simulation results, comparing them to their quantum counterparts when available.

II. Technical details

The systems studied were described before.[14] They include water molecules, using the RWKM-2 flexible water-water potential,[15] and an electron with the electron water pseudo-potential taken from Barnett et al.[14] This potential has been shown to account well, within 10 percent, for the peak position and width of the absorption spectrum of the hydrated electron. In order to save computing effort we have chosen to modify to the electron-water potential such has to bring it to the form $V(r_e,\{R_i\}) = \sum_i A_i V_i(|r_e-R_i|)$ where i=0,...,3. (R_i, i=0,1,2 are the positions of the oxygen and two hydrogens respectively, and $R_3=R_0+(R_1-R_2-2R_0)\delta$ with $\delta=0.2218$ is the position of the negative charge). This is achieved by replacing $\rho^{2/3}$ in the exclusion part of the interaction by

$$4e^{-2|r_e-R_0|/a_0} + \sum_{j=1}^{2} e^{-2|r_e-R_j|/a_0}$$ and $\rho^{1/3}$ in the exchange part by

$$4e^{-|r_e-R_0|/a_0} + \sum_{j=1}^{2} e^{-|r_e-R_j|/a_0}$$ where a_0 is the Bohr radius. (The form of the

charge density used in Ref 14 is $\rho = 4e^{-3|r_e-R_0|/a_0} + \sum_{j=1}^{2} e^{-3|r_e-R_j|/a_0}$). This modification makes it possible to tabulate the functions $V_i(R)$ and thus to avoid their repeated evaluation. This save about an order of magnitude in computer time when the "diffuse electron" model (see below) is used, while having only a small effect (about 10 percent) on the results.

Another way to reduce computational effort is to recognize that this system is characterized by vastly different timescales resulting from the large difference between the electron and the water atoms masses. We have used the procedure of Tuckerman et al[16] to handle this situation, resulting in another substantial saving in computation time.

We have used two different representations for the electron in different parts of our work. In the first, the "point electron" (PE) model, the electron is a regular point particle interacting with the water molecules according to the pseudopotential of Barnett et al.[14] In the second, the "diffuse electron" (DE) model, the finite spread of the electron was introduced by representing its density as a Gaussian centered at the electron position r_e and characterized by a given gyration radius $R_g = [<r^2> - <r>^2]^{1/2}$. In this case the electron water interaction was averaged over this distribution.

The "absorption lineshape" of the solvated electron presented below was calculated as the Fourier transform of the electron velocity autocorrelation function computed from the classical trajectory. Recalling that the ground state hydrated electron is characterized by a radius of $\sim$ 4au, it makes sense to use the DE trajectory in this calculation.

The data presented below is obtained from the average of 10-50 trajectories, each starting from a different local water configuration. The water "bulk" contains 256 water molecules at 300K and density 1gr/cm^3 with periodic boundary conditions. The water films were prepared as described in Ref. 9, by depositing water on Pt (111) surface, using the water-platinum potential of Spohr and Heinzinger.[17] Following the previous quantum simulation work,[9] once the film was prepared the penetration simulations were done with the water molecules static, and the platinum substrate was disregarded.

Finally, in some of our studies we were interested in energy transfer into the water nuclear degrees of freedom. The instantaneous water translational (center of mass motion) and rotational kinetic energies were computed directly from the center of mass velocity and from the angular momenta and moments of inertia, respectively, of the water molecules. Together they constitute the intermolecular kinetic energy of the solvent. The intramolecular, vibrational energy of the water molecules is computed as the difference between the total solvent kinetic energy (obtained from the individual atomic velocities) and the intermolecular part.

III. Results and discussion

In what follows we use the term "electron" to describe the classical entity whose mass, charge and interaction with the water environment are identical to that taken for the quantum mechanical entity that was used as a model for an electron in previous work.[9,14] By "point electron" (PE) we refer to this classical particle as is, while for the "diffuse electron" (DE) we replace this classical particle by a Gaussian distribution of a given width, so that the particle moves as a rigid body however all interactions are calculated as averages over this distribution. Unless otherwise stated, the width of this distribution was chosen so that $R_g = [<r^2> - <r>^2]^{1/2}$ is 4au, the same as the average value obtained for the ground state of the hydrated electron in the quantum simulations.[3]

Figure 2 shows the time evolution of the electron-water interaction following electron insertion to a pre-existing minimum of this potential in neutral water at 300K. The results shown are averages over 10 trajectories, each starting with zero kinetic energy from a different local configuration. The Full line corresponds to the DE and the dotted line - to the PE. While the two lines show qualitatively similar behavior, it is the DE result which is quantitatively very similar to results obtained from quantum mechanical simulations of adiabatic solvation, while the PE behavior strongly overestimates the electron-water interaction and shows considerably faster solvation. For example, the average potential energy of the equilibrated DE is -5.0ev, less then 10 percent lower in magnitude than that calculated in the quantum case, while the result obtained for the PE is three times larger.

These observations emphasize the importance of taking into account the finite extent of the electron density in calculating its interaction with its classical environment. As mentioned before, the timescale on which this finite size relaxes to what is practically its final value is short relative to that of the subsequent solvent motion. Once this size is reached the rest of the process is dominated by the classical solvent motion.

Consider now the absorption lineshape of the solvated electron. For a bound classical charged particle this lineshape is proportional to the Fourier transform of the velocity time correlation function, $<v(t) \cdot v(0)>$. For a one dimensional oscillator the period of this function provides a good estimate of the fundamental quantum absorption frequency. The absorption lineshape of the solvated electron is dominated by the broad distribution of solvent configurations, namely by the distribution of shapes of the potential well which localizes the electron. We may infer that using the classical velocity autocorrelation function, averaged over solvent configurations, in order to estimate the absorption lineshape of the solvated electron may not be as ridiculous it first sounds.

Figure 3 shows the results of this calculation. The full and dotted lines are the Fourier transforms of $<v(t) \cdot v(0)>$ for the DE and the PE, respectively. The arrow points to the position of the peak of the lineshape obtained from the quantum mechanical simulations[3] (at ~ 2.1ev, about 15% higher than the experimental result, and with width ~ 1ev). The classical simulations grossly overestimate both the peak position and the width of the absorption lineshape. It is seen that the DE lineshape is much closer to the quantum one than the PE result, however it appears that imposing the rigid Gaussian shape on the electron density without letting it adjust itself to the cavity shape fluctuations is a restriction too strong for this application, leading to overestimated width and peak position.

Results for electron transmission through water films are displayed in figures 4 and 5. The films (1-4 monolayers in the xy plane) were prepared as described above and in Ref. 9. The electron starts at time t=0 and position z=0 which is the edge of the film (the position of the metal plane on which the film was originally prepared) with velocity in the z direction, towards the film,

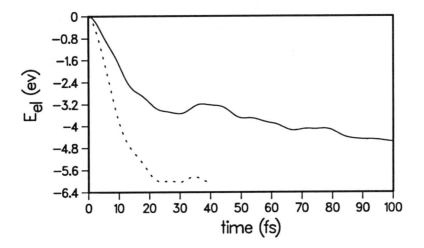

Fig. 2. The time evolution of the electron-water interaction (relative to its value at t=0) following the insertion of the electron at t=0 into a pre-existing minimum of this interaction with zero kinetic energy. Full line: the DE model. Dotted line: the PE model.

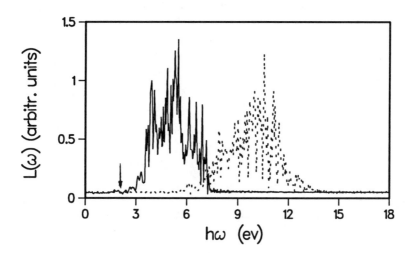

Fig. 3. The Fourier transform of the velocity correlation function obtained from an equilibrium trajectory of the solvated electron. Full line: Diffuse electron. Dotted line: Point electron.

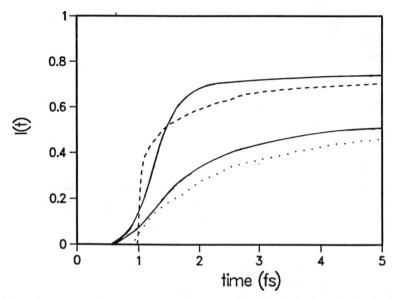

Fig. 4. The integrated current as a function of time in the simulation of the penetration experiment discussed in the text. The upper and lower full lines are results of the quantum simulations of Reference 9 for 1 and 3 monolayer films respectively. The dashed and dotted lines are the corresponding classical (PE) results.

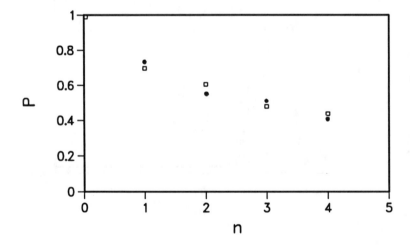

Fig. 5. The electron transmission probability at t=5fs as a function of film thickness (number of monolayers), in the penetration experiment discussed in the text. Full circles: the quantum simulation results from Ref. 9. Empty squares: the classical (PE) results. The electron kinetic energy is 4.72ev.

corresponding to momentum p_z=0.589au (E_k=4.72ev). The number of electrons that make it to the "detector", at position z=28au on the other side of the film, is counted as a function of time. Figure 1 shows an example of the distribution of electron density, $|\psi(t)|^2$, obtained from solving the time dependent Schrödinger equation for an electron entering a 4-monolayers film (see Ref. 9), as a contour plot in the xy plane in the middle of the film at t=2fs. Obviously the DE Gaussian is not relevant here. The electron wavefunction is strongly fragmented and its distribution is made of many little pieces. On the relevant timescale of ~ 1fs (see below) it is not clear that using a diffuse electron model for averaging the electron-water interaction makes sense. We have therefore used the PE in these simulations.

The results for the integrated current (the accumulated number of electrons that reach the detector) are shown as functions of time, for films of thickness 1 and 3 monolayers, in Fig. 4. The dashed line represents the results of the classical simulations for a 1-monolayer film and the dotted line - the corresponding results for a 3-monolayer film. The upper and lower full lines are the quantum results from Ref. 9 for 1 and 3 monolayers respectively. The integrated current at t=5fs (normalized by its value in the absence of film) is displayed as a function of film thickness in Fig. 5. This is in fact the transmission probability. Here empty squares are results of the classical simulations (averaged over 1500 trajectories with different starting positions in the xy plane) and full circles are the quantum results from Ref. 9.

Such close agreement as seen between the quantum and classical results in Fig. 5 should be regarded as accidental (recall that the electron-water interaction is not exactly the same in the two simulations). However the similarity between the results show that important quantum mechanical effects are indeed washed away by the multiple scattering that accompanies the transmission.

It should come as no surprise that this agreement between classical and quantum result breaks down at lower electron energies. Indeed, we have found that at the considerably lower electron energy of 0.9ev (the other energy considered in Ref.9) classical mechanics strongly overestimated the transmission probability relative to the quantum result.

Next, consider electron thermalization. Available quantum simulation methods cannot be used for this process for reason outlined above. On the other hand, we can hardly expect that classical mechanics will describe correctly processes of such highly quantum nature as energy transfer between a 0.1-5ev electron and high frequency OH vibrations. It is still of interest to check how well (or badly) will classical mechanics do in describing such average observables as the distance that the electron covers before stopping (this will be its average distance from the parent ion in a photoionization experiment), or the average time until it is captured by the evolving solvent cavity. Classical mechanics may provide at least qualitative information on other average quantities of interest, such as the efficiency of different solvent modes in accepting the electron energy and the related isotope effect (e.g. the relative efficiency of H_2O and D_2O in stopping the electron).

The question of using a PE or a DE model for these simulations cannot be answered definitively. On the very short (~1fs) timescale the PE model seems to be a better representation, while at longer times the DE model becomes relevant. In most of the following simulations we have used the PE model. Because this implies stronger electron-solvent interaction, the results should provide an upper bound estimate for the stopping efficiency of the water solvent. Figure 6 shows for comparison a result obtained for the DE case.

Fig. 6 shows the energy of the classical electron after entering a bulk of neutral water (or D_2O) with 3ev of translational kinetic energy. This and the following results are averages over 45 trajectories starting at different water configurations. It is seen that H_2O is more efficient than D_2O in stopping the electron. The same is seen in Fig. 7, where the average distance traversed by the electron is shown as a function of time. The fact that electron goes further in D_2O than in H_2O before it stops has indeed been inferred from the experimentally observed kinetics of the subsequent geminate recombination.[13] Both the distance and the timescale shown in Figs 6 and 7 are within accepted estimates for the relaxation of subexcitation electrons in water.

Finally consider the energy transfer to water modes in this classical process. Figure 8 shows the time evolution of the kinetic energy in the intermolecular translational and rotational-librational degrees of freedom, as well as in the intramolecular degrees of freedom. Obviously energy transfer between these modes make a clean assignment of energy accepting efficiencies of different types of modes impossible, however from Fig. 8 it is quite clear that (as expected) center of mass translational motion of the water is practically irrelevant for this process, which is dominated by intermolecular rotations/librations with important role played also by the intramolecular vibrations. The isotope effect observed above is easily understood in this light since the higher frequency modes of the hydrogenic molecule will be more efficient acceptors of electron energies. There is no obvious reason why this conclusion will not stand also in the corresponding quantum system.

IV. Conclusions

The processes discussed in this paper involve an electron, a clearly quantum entity. However these processes are not "strongly" quantum in the sense that they are not dominated by tunneling and/or non-adiabatic processes, while zero point motion, when important, is taken at least partly into account. In addition, the quantum mechanical phase is destroyed by multiple scattering events. Finally, the observables discussed are of coarse grained nature, in the sense that they are averages over many quantum mechanical channels. These are probably the reasons why classical mechanics appears to work reasonably well for some processes involving electron solvation and the solvated electron. A critical source of uncertainty arizes from the unknown "electron size" which determines the width of the DE distribution. For this reason classical simulations of electronic processes of the kind discussed in this work should be regarded only as tools for developing qualitative understanding

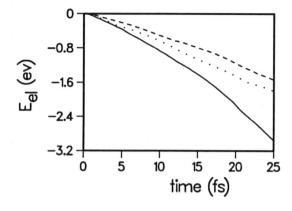

Fig. 6. The time evolution of the relative total energy of an electron injected at t=0 into neutral water. Full line: PE/H$_2$O. Dashed line: PE/D$_2$O. Dotted line: DE/H$_2$O.

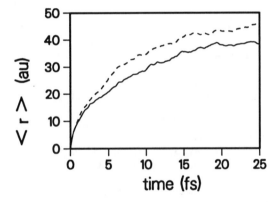

Fig. 7. The evolution of the average distance traversed by the electron following its injection at t=0 into neutral water with translational kinetic energy 3ev. Full line: H$_2$O. Dashed line: D$_2$O.

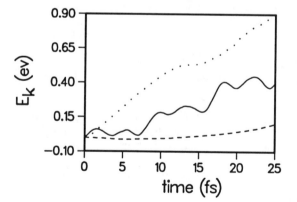

Fig. 8. The intramolecular vibrational kinetic energy (full line), the rotational librational kinetic energy (dotted line) and the center-of-mass translational kinetic energy (dashed line), as functions of time, following the injection of a 3ev (translational energy) electron into neutral water.

Acknowledgements. This research was supported in part by the Israel National Science Foundation and by the United States-Israel binational science foundation. We thank Professors R.D. Levine and R.B. Gerber for helpful discussions.

References

1. A. Migus, Y. Gauduel, J.L. Martin and A. Autonetti, Phys.Rev.Let. **58**,1559(1987); F.H. Long, H. Lu and K.B. Eisenthal, Phys. Rev. Letters **64** 1469 (1990).
2. (a) R.N. Barnett, U. Landman and A. Nitzan, J.Chem.Phys. **90**,4413(1989) and references therein; (b) P.J. Rossky and J. Schnitker, J.Phys.Chem. **92**,4277(1988) and references therein.
3. J. Schnitker, K.A. Motakabbir, P.J. Rossky and R. Friesner, Phys.Rev.Lett. **60**,456(1988); R.N. Barnett, U. Landman G. Makov and A. Nitzan, J.Chem.Phys. **93**,6226(1990).
4. See, e.g. R.N. Barnett, U. Landman and A. Nitzan, J.Chem.Phys. **89**,2242(1988).
5. F.J. Webster, J. Schnitker, M.S. Friedrich, R.A. Friesner and P.J. Rossky, Phys. Rev.Lett. **66**,3172(1991); F.J. webster, P.J. Rossky and R.A. Friesner, Comp.Phys. Comm. **63**,494(1991).; E.Neria, A.Nitzan, R.B.Barnett and U.Landman, Phys. Rev. Lett. **67**,1011(1991); E. Neria and A. Nitzan, J.Chem.Phys, in press.
6. N. Makri and W.H. Miller, J.Chem.Phys. **87**,5781(1987); Z. Kotler, E. Neria and A. Nitzan, Comp.Phys.Comm. **63**,243(1991).
7. K.A. Motakabbir, J. Schnitker and P.J. Rossky, J.Phys.Chem. **97**,2055(1992).
8. J. Schnitker, P.J. Rossky and G.A. Kenny-Wallace, J.Chem.Phys. **85**,2986(1986).
9. R.N. Barnett, U. Landman and A. Nitzan, J.Chem.Phys. **93**,6535(1990).
10. T.L. Gilton, C.P. Dehnbostal and J.P. Cowin, J.Chem.Phys. **91**,1937(1989).
11. S.K. Jo and J.M White, J.Chem.Phys. **94**,5761(1991).
12. (a). M. Michaud and L. Sanche, Phys.Rev. **A30**,6067(1984); T. Goulet and J.-P. Jay-Gerin, J.Phys.Chem. **92**,6871(1988).
    (b). I. Rips and R.J. Silbey, J.Chem.Phys. 4495 (1991) and references therein.
13. A.C. Chernovitz and C.D. Jonah, J.Phys.Chem. **92**,5946(1988).
14. R.N. Barnett, U.Landman, C.L. Cleveland and J.Jortner, J.Chem.Phys. **88**,4421(1988).
15. J.R. Reimers and R.O. Watts, Chem.Phys. **85**,83(1984).
16. M.E. Tuckerman, B.J. Berne and A. Rossi, J.Chem.Phys. **94**,1465(1991). Tuckerman et al (J.Chem.Phys. **97**,1990(1992)) have recently provided a better algorithm based on the Trotter factorization of the Liuville propagator.
17. E. Spohr and K. Heinzinger, Ber.Bunsenges.Phys.Chem. **92**,1358(1988).